NETWORK TUTORIAL

A Complete Introduction to Networks

BY STEVE STEINKE
AND THE EDITORS OF *NETWORK MAGAZINE*

CMP**Books**
San Francisco

Published by CMP Books
An imprint of CMP Media LLC
Main office: CMP Books, 600 Harrison St., San Francisco, CA 94107 USA
Phone: 415-947-6615; Fax: 415-947-6015
www.cmpbooks.com
Email: books@cmp.com

ISBN: 1-57820-302-3

For individual orders, and for information on special discounts for quantity orders, please contact:
CMP Books Distribution Center, 6600 Silacci Way, Gilroy, CA 95020
Tel: 1-800-500-6875 or 408-848-3854; Fax: 408-848-5784
Email: cmp@rushorder.com; Web: www.cmpbooks.com

Distributed to the book trade in the U.S. by:
Publishers Group West, 1700 Fourth Street, Berkeley, California 94710

Distributed in Canada by:
Jaguar Book Group, 100 Armstrong Avenue, Georgetown, Ontario M6K 3E7 Canada

Printed in the United States of America

03 04 05 06 07 5 4 3 2 1

TABLE OF CONTENTS

INTRODUCTION TO THE 5TH EDITION

The latest-written tutorial in the 4th Edition of the *Network Tutorial* ran in March 2000. Since that date we have published several tutorials on optical networking, voice-over-IP protocols, and local loop technologies—technologies that have been closely associated with carriers and telecom service providers, who in turn were betting on rapidly growing demand generated by the Internet and so-called e-commerce. These pieces have become perhaps less immediately crucial to our readers as the telecom bubble bursting followed the dot-com bubble bursting and the market for these technologies shriveled.

On the other hand, many of the tutorials (along with *Network Magazine* as a whole) over the last three years have focused on network security, a trend that began well before the attacks of September 11th, 2001, got everyone thinking about security. Which is not to say that there are technology solutions to the threat posed by terrorists like Al Qaida, or that national security has anything but a tangential relation to network security despite all the dubious stories about terrorist hackers from China, Iraq, or evildoer nations elsewhere. Major security threats to the networks of 2003 still seem to arise primarily from disgruntled employees, thrill seeking vandals with time on their hands, and criminals who want to take economic advantage of the anonymity and remote-control characteristics of The New Public Network.

The other area of growth over the past three years has been wireless networking. Wireless LANs in particular have continued to spread rapidly in recent years when most other networking markets have stagnated or declined. This edition of the *Network Tutorial* can now claim to provide a reasonable foundation for readers who need to explore network security and wireless networks.

The authors of tutorials between April 2000 and October 2002 include: Doug Allen, Jonathan Angel, Jim Carr, Elizabeth Clark, Andrew Conry-Murray, Andy Dornan, David Greenfield, Rob Kirby, and Steve Steinke.

I have eliminated a handful of elderly tutorials, most of which I almost removed from the 4th edition. Anyone who wants to pursue these relics can find them on the *Network Magazine* web site, www.networkmagazine.com. I have also eliminated a number of errors and other infelicities, though no doubt there are more of these. I encourage you to let me know if you encounter mistakes.

Steve Steinke
ssteinke@cmp.com
October 2002

INTRODUCTION TO NETWORK TUTORIAL, 4TH EDITION

The previous edition of this book included the first 88 of these two-magazine-page overviews of technical networking topics, running monthly from August 1988 until December 1995. I remember thinking at the time I compiled them just how fresh many of them had managed to remain after as much as seven years.

After five more years, years that encompassed a great many major changes in the networking field, many of the early tutorials no longer seem fresh at all. In fact, reading some of them again was like opening a time capsule. A few (Twisted Pair FDDI) I've jettisoned altogether, although they're still on the Web site for serious nostalgia junkies. A few (Which Fast LAN?) I've left because they're historically interesting, even if they're barely relevant to a network manager nowadays. And a few others I've slapped a fresh coat of paint over, taking out the most obvious bad market predictions, updating references to their 21st century equivalents, and correcting mistakes I failed to catch the last time around.

Incidentally, the 4th Edition of the Network Tutorial follows the 3rd Edition of the LAN Tutorial because we changed the name of the magazine in 1997. For better or worse, no one would admit to actually having a LAN in the late 1990s—if you didn't have an Intranet you might as well throw in the towel. Now "Intranet" is beginning to have a moldy ring to it, though I must admit I never warmed up to the term and have tried gamely but unsuccessfully to keep the magazine's editors and writers from using it. In any event, the magazine, which was never reluctant, dating all the way back to 1986, to write about wide area network technology and other topics not strictly bounded by the local area, had no strong attachment to the name "LAN," and was happy to take the title "Network Magazine," which somehow no one had locked up in the whole speckled history of computer networks.

The third edition of the LAN tutorial included each installment of an ongoing series of articles that originally appeared in *LAN Magazine* through December 1995. These tutorials were written by Rebecca J. Campbell, Lee Chae, Dave Fogle, Alan Frank, Melanie McMullen, Steve Steinke, Aaron Brenner, Jim Carr, Ken Mackin, Thomas Peltier, Patricia Schnaidt, and Bonny

Hinners. Melanie McMullen and Steve Steinke expanded and upgraded the glossary to provide an even more comprehensive quick reference for unfamiliar terms.

The third edition was edited by Steve Steinke, building on the editorial framework created by Patricia Schnaidt from the first and second editions.

The tutorials new to the 4th edition ran between January 1996 and March 2000. The authors who wrote tutorials in that time span were Lee Chae, Alan Frank, David Greenfield, Anita Karvé, Steve Steinke, and Alan Zeichick.

I have one request to make of the reader: Please grant the authors a fair degree of slack with respect to the URLs cited here. Internet links get changed all the time, for reasons trivial and profound. We tend to print URLs with as much specificity as possible in order to unambiguously identify sources. However, these detailed locators are consequently fragile things. Aside from these considerations, URLs printed in a book will lack a certain degree of freshness from the mere facts of printing technology delays and shelf lives. A bit of ingenuity with a general search engine, or with the search features at one of the domains we cite, will go a long ways toward tracking down documents that are no longer in the place they were when we found them. The good news is that as far as I can tell, hardly anything that has been posted to the Web ever goes away completely.

Steve Steinke
January 2000

SECTION I

Understanding Networks—Layers and Protocols

Networks Protocols Part One

A network protocol is a set of rules for communicating between computers. Protocols govern format, timing, sequencing, and error control. Without these rules, the computer cannot make sense of the stream of incoming bits.

But there is more than just basic communication. Suppose you plan to send a file from one computer to another. You could simply send it all in one single string of data. Unfortunately, that would stop others from using the network for the entire time it takes to send the message. This would not be appreciated by the other users. Additionally, if an error occurred during the transmission, the entire file would have to be sent again. To resolve both of these problems, the file is broken into small pieces called packets and the packets are grouped in a certain fashion. This means that information must be added to tell the receiver where each group belongs in relation to others, but this is a minor issue. To further improve transmission reliability, timing information and error correcting information are added.

Because of this complexity, computer communication is broken down into steps. Each step has its own rules of operation and, consequently, its own protocol. These steps must be executed in a certain order, from the top down on transmission and from the bottom up on reception. Because of this hierarchical arrangement, the term protocol stack is often used to describe these steps. A protocol stack, therefore, is a set of rules for communication, and each step in the sequence has its own subset of rules.

What is a protocol, really? It is software that resides either in a computer's memory or in the memory of a transmission device, like a network interface card. When data is ready for transmission, this software is executed. The software prepares data for transmission and sets the transmission in motion. At the receiving end, the software takes the data off the wire and prepares it for the computer by taking off all the information added by the transmitting end.

There are a lot of protocols, and this often leads to confusion. A Novell network can communicate through its own set of rules (its own protocol called IPX/SPX), Microsoft does it another way (NetBEUI), DEC once did it a third way (DECnet), and IBM does it yet a fourth (NetBIOS). Since the transmitter and the receiver have to "speak" the same protocol, these four systems cannot talk directly to each other. And even if they could directly communicate, there is no guarantee the data would be usable once it was communicated.

Anyone who's ever wanted to transfer data from an IBM-compatible personal computer to an Apple Macintosh computer realizes that what should be a simple procedure is sometimes anything but. These two popular computers use widely differing—and incompatible—file systems. That makes exchanging information between them impossible, unless you have translation software or a LAN. Even with a network, file transfer between these two types of computers isn't always transparent. [Editor's note: Even in the Internet age, Mac/Windows/Unix file exchange is often less than perfectly transparent.]

If two types of personal computers can't communicate easily, imagine the problems occurring between PCs and mainframe computers, which operate in vastly different environments and usually under their own proprietary operating software and protocols. For example, the original IBM PC's peripheral interface—known as a bus—transmitted data eight bits at a time. The newer 386, 486, and Pentium PCs have 32-bit buses, and mainframes have even wider buses. This means that peripherals designed to operate with one bus are incompatible with another bus, and this includes network interface cards (NICs). Similar incompatibilities also exist with software. For instance, Unix-based applications (and often the data generated with them) cannot be used directly on PCs operating under Windows or MS-DOS. Resolving some of these incompatibilities is where protocol standards fit in.

A protocol standard is a set of rules for computer communication that has been widely agreed upon and implemented by many vendors, users, and standards bodies. Ideally, a protocol standard should allow computers to talk to each other, even if they are from different vendors. Computers don't have to use an industry-standard protocol to communicate, but if they use a proprietary protocol then they can only communicate with equipment of their own kind.

There are many standard protocols, none of which could be called universal, but the successful ones can be characterized with something called the OSI model. The standards and protocols associated with the OSI reference model are a cornerstone of the open systems concept for linking the literally dozens of dissimilar computers found in offices throughout the world.

THE OSI MODEL

The Open System Interconnection (OSI) model includes a set of protocols that attempt to define and standardize the data communications process. The OSI protocols were defined by the International Organization for Standardization (ISO). The OSI protocols have received the support of most major computer and network vendors, many large customers, and most governments, including the United States.

The OSI model is a concept that describes how data communications should take place. It divides the process into seven groups, called layers. Into these layers are fitted the protocol standards developed by the ISO and other standards bodies, including the Institute of Electrical and Electronic Engineers (IEEE), American National Standards Institute (ANSI), and the International Telecommunications Union (ITU), formerly known as the CCITT (Comité Consultatif International Téléphonique et Télégraphique).

The OSI model is not a single definition of how data communications actually takes place in the real world. Numerous protocols may exist at each layer. The OSI model states how the process should be divided and what protocols should be used at each layer. If a network vendor implements one of the protocols at each layer, its network components should work with other vendors' offerings.

The OSI model is modular. Each successive layer of the OSI model works with the one above and below it. At least in theory, you may substitute one protocol for another at the same layer without affecting the operation of layers above or below. For example, Token Ring or Ethernet hardware should operate with multiple upper-layer services, including the transport protocols, network operating system, internetwork protocols, and applications interfaces. However, for this interoperability to work, vendors must create products that meet the OSI model's specifications.

Although each layer of the OSI model provides its own set of functions, it is possible to group the layers into two distinct categories. The first four layers— physical, data link, network, and transport—provide the end-to-end services necessary for the transfer of data between two systems. These layers provide the protocols associated with the communications network used to link two computers together.

The top three layers—the application, presentation, and session layers—provide the application services required for the exchange of information. That is, they allow two applications, each running on a different node of the network, to interact with each other through the services provided by their respective operating systems.

A graphical illustration of the OSI model is shown above. The following is a description of just what each layer does.

1. The Physical layer provides the electrical and mechanical interface to the network medium (the cable). This layer gives the data-link layer (layer 2) its ability to transport a stream of serial data bits between two communicating systems; it conveys the bits that move along the cable. It is responsible for making sure that the raw bits get from one place to another, no matter what shape they are in, and deals with the mechanical and electrical characteristics of the cable.

■ The OSI model is not a single definition of how data communications takes place. It states how the processes should be divided and offers several options. In addition to the OSI protocols, as defined by ISO, networks can use the TCP/IP protocol suite, the IBM Systems Network Architecture (SNA) suite, and others. TCP/IP and SNA roughly follow the OSI structure.

MYRIAD PROTOCOL STACKS

Layer	ISO	TCP/IP	IBM
7. Application	FTAM X.400 JTAM X.500 VT CASE	SMTP FTP NFS Telnet SNMP	
6. Presentation	8923		
5. Session	8327		NetBIOS APPC
4. Transport	8073 (TPO) 8602 (CONS)	UDP TCP	NetBEUI APPC
3. Network	8208 (X.25) 8473 (CLNS) 9542 (ES-IS) 8348 (CONS)	IP	APPC
2. Data-Link	8802.2 LLC 8802.3/4/5	LLC Ethernet	LLC HDLC SDLC MAC
1. Physical	8802.3 Ethernet 8802.4 Token Bus 8802.5 Token Ring	Ethernet FDDI Token Ring	Token Ring Ethernet FDDI

2. The Data-Link layer handles the physical transfer, framing (the assembly of data into a single unit or block), flow control and error-control functions over a single transmission link; it is responsible for getting the data packaged for the Physical layer. The data link layer provides the network layer (layer 3) reliable information-transfer capabilities. The data-link layer is often subdivided into two parts—Logical Link Control (LLC) and Medium Access Control (MAC)—depending on the implementation.

3. The Network layer establishes, maintains, and terminates logical and physical connections among multiple interconnected networks. The network layer is responsible for translating logical addresses, or names, into physical (or data-link) addresses. It provides network routing and flow-control functions across the computer-network interface.

4. The Transport layer ensures data is successfully sent and received between two end nodes. If data is sent incorrectly, this layer has the responsibility to ask for retransmission of the data. Specifically, it provides a reliable, network-independent message-interchange service to the top three application-oriented layers. This layer acts as an interface

between the bottom and top three layers. By providing the session layer (layer 5) with a reliable message transfer service, it hides the detailed operation of the underlying network from the session layer.

5. The Session layer decides when to turn communication on and off between two computers—it provides the mechanisms that control the data-exchange process and coordinates the interaction between them. It sets up and clears communication channels between two communicating components. Unlike the network layer (layer 3), it deals with the programs running in each machine to establish conversations between them. Some of the most commonly encountered protocol stacks, including TCP/IP, don't implement a session layer.

6. The Presentation layer performs code conversion and data reformatting (syntax translation). It is the translator of the network, making sure the data is in the correct form for the receiving application. Of course, both the sending and receiving applications must be able to use data subscribing to one of the available abstract data syntax forms. Most commonly, applications handle these sorts of data translations themselves rather than handing them off to a Presentation layer.

7. The Application layer provides the interface between the software running in a computer and the network. It provides functions to the user's software, including file transfer access and management (FTAM) and electronic mail service.

Unfortunately, protocols in the real world do not conform precisely to these neat definitions. Some network products and architectures combine layers. Others leave layers out. Still others break the layers apart. But no matter how they do it, all working network products achieve the same result—getting data from here to there. The question is, do they do it in a way that is compatible with networks in the rest of the world?

WHAT OSI IS AND IS NOT

While discussing the OSI reference model it is important to understand what the model does not specify as well as what it actually spells out. The ISO created the OSI reference model solely to describe the external behavior of electronics systems, not their internal functions.

The reference model does not determine programming or operating system functions, nor does it specify an application programming interface (API). Neither does it dictate the end-user interface—that is, the command-line and/or icon-based prompts a user uses to interact with a computer system.

The OSI standards merely describe what is placed on a network cable and when and how it will be placed there. It does not state how vendors must build their computers, only the

■ ISO has specified many different protocols at each layer of the OSI model. Some of the options are shown here.

THE OSI PROTOCOLS

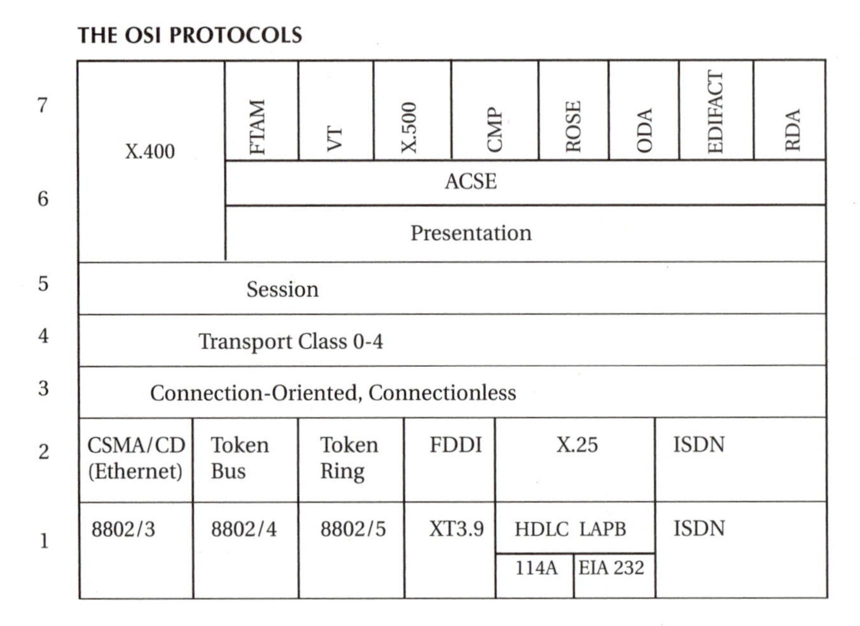

kinds of behavior these systems may exhibit while performing certain communications operations.

The OSI standards are distinct from the OSI suite of protocols. This concept permits a vendor to develop network elements that are more or less ignorant of the other components on the network. They are said to be ignorant in that they may need to know that other network components exist, but not the specific details about their operating systems or interface buses. One of the primary benefits of this concept is that vendors can change the internal design of their network components without affecting their network functionality, as long as they maintain the OSI-prescribed external attributes. The figure below shows the protocols in the OSI suite.

CONNECTION TYPES

The OSI protocol suite is inherently connection-oriented, but the services each OSI layer provides can either be connection-oriented, or connectionless. In the three-step connection-oriented mode operation (the steps are connection establishment, data transfer, and connection release), an explicit binding between two systems takes place.

In connectionless operation, no such explicit link occurs; data transfer takes place with no specified connection and disconnection function occurring between the two communicating systems. Connectionless communication is also known as datagram communication.

AT THE PHYSICAL LAYER

Let's compare some real protocols to the OSI model. The best known physical layer standards of the OSI model are those from the IEEE. That is, the ISO adopted some of the IEEE's physical network standards as part of its OSI model, including IEEE 802.3 or Ethernet, IEEE 802.4 or token-passing bus, and IEEE 802.5 or Token Ring. ISO has changed the numbering scheme, however, so 802.3 networks are referred to as ISO 8802-3, 802.4 networks are ISO 8802-4, and 802.5 networks are ISO 8802-5.

Each physical layer standard defines the network's physical characteristics and how to get raw data from one place to another. They also define how multiple computers can simultaneously use the network without interfering with each other. (Technically, this last part is a job for the data-link layer, but we'll deal with that later.)

IEEE 802.3 defines a network that can transmit data at 10Mbps and uses a logical bus (or a straight line) layout. (Physically, the network can be configured as a bus or a star.) Data is simultaneously visible to all machines on the network and is nondirectional on the cable. All machines receive every frame, but only those meant to receive the data will process the frame and pass it to the next layer of the stack. Network access is determined by a protocol called Carrier Sense Multiple Access/Collision Detection (CSMA/CD). CSMA/CD lets any computer send data whenever the cable is free of traffic. If the data collides with another data packet, both computers "back off," or wait a random time, then try again to send the data until access is permitted. Thus, once there is a high level of traffic, the more users there are, the more crowded and slower the network will become. Ethernet has found wide acceptance in office automation networks.

IEEE 802.4 defines a physical network that has a bus layout. Like 802.3, Token Bus is a shared medium network. All machines receive all data but do not respond unless data is addressed to them. But unlike 802.3, network access is determined by a token that moves around the network. The token is visible to every device but only the device that is next in line for the token gets it. Once a device has the token it may transmit data. The Manufacturing Automation Protocol (MAP) and Technical Office Protocol (TOP) standards use an 802.4 physical layer. Token Bus has had little success outside of factory automation networks.

IEEE 802.5 defines a network that transmits data at 4Mbps or 16Mbps and uses a logical ring layout, but is physically configured as a star. Data moves around the ring from station to station, and each station regenerates the signal. It does not support simultaneous multiple access as Ethernet does. The network access protocol is token-passing. The token and data move about in a ring, rather than over a bus as they do in Token Bus. Token Ring has found moderate acceptance in office automation networks and a greater degree of support in IBM-centric environments.

There are other physical and data-link layer standards, some that conform to the OSI model and others that don't. ARCnet is a well known one that only became standardized in 1998, long after the time when it had any commercial significance. It uses a token-passing bus access method, but not the same as does IEEE 802.4. LocalTalk is Apple's proprietary network that transmits data at 230.4Kbps and uses CSMA/CA (Collision Avoidance). Fiber Distributed Data Interface (FDDI) is an ANSI and OSI standard for a fiber-optic LAN that uses a token-passing protocol to transmit data at 100Mbps on a ring.

WHEN IT BEGAN

The International Standards Organization, based in Geneva, Switzerland, is a multinational body of representatives from the standards-setting agencies of about 90 countries. These agencies include the American National Standards Institute (ANSI) and British Standards Institute (BSI).

Because of the multinational nature of Europe, and its critical need for intersystem communication, the market for OSI-based products is particularly strong there. As a result, the European Computer Manufacturers' Association (ECMA) has played a major role in developing the OSI standards. In fact, before the Internet's Transmission Control Protocol/Internet Protocol (TCP/IP) began to dominate international networks, European networking vendors and users were generally further advanced in network standards, based on OSI implementations, than were their American counterparts, who relied principally on proprietary solutions such as IBM's Systems Network Architecture (SNA) or TCP/IP.

Creating the OSI standards was a long, drawn-out process: The ISO began work on OSI protocols in the late 1970s, finally releasing its seven-layer architecture in 1984. It wasn't until 1988 that the five-step standards-setting process finally resulted in stabilized protocols for the upper layers of the OSI reference model. [Editor's note: From the perspective of 2000, the primary worldwide significance of the OSI protocols was in the use of the seven layer stack model as a way of learning about networks. While there remain many implementations of OSI protocols, particularly in Europe where they were in some cases legally imposed, it's clear that worldwide, the lion's share of new development and investment is devoted to TCP/IP and will be for the foreseeable future.]

This tutorial, number 2, was originally published in the September 1988 issue of LAN Magazine/Network Magazine.

Network Protocols Part Two

The Data-Link layer (the second OSI layer) is often divided into two sublayers; the Logical Link Control (LLC) and the Medium Access Control (MAC). The IEEE also defines standards at the data-link layer. The ISO standards for the MAC layer, or lower half of the data-link layer, were taken directly from the IEEE 802.x standards.

Medium Access Control, as its name suggests, is the protocol that determines which computer gets to use the cable (the transmission medium) when several computers are trying. For example, 802.3 allows packets to collide with each other, forcing the computers to retry a transmission until it is sent successfully. 802.4 and 802.5 limit conversation to the computer with the token. Remember, this is done in fractions of a second, so even when the network is busy, users don't wait very long for access on any of these three network types.

The upper half of the data-link layer, the LLC, provides reliable data transfer over the physical link. In essence, it manages the physical link.

The IEEE splits the data-link layer in half because the layer has two jobs to do. The first is to coordinate the physical transfer of data. The second is to manage access to the physical medium. Dividing the layer allows for more modularity and therefore more flexibility. The type of medium access control has more to do with the physical requirements of the network than the actual management of data transfer. In other words, the MAC layer is closer to the physical layer than the LLC layer. By dividing the layer, a number of MAC layers can be created, each corresponding to a different physical layer, but just one LLC layer can handle them all. This increases flexibility and gives the LLC an important role in providing an interface between the various MAC layers and the higher-layer protocols. The role of the data-link's upper layer is so crucial, the IEEE gave it a standard of its own: 802.2 LLC.

Besides 802.2, other protocols can perform the LLC functions. High-level Data-Link Control (HDLC) is a protocol from ISO, which also conforms to the OSI model. IBM's Synchronous Data-Link Control (SDLC) does not conform to the OSI model but performs functions similar to the data-link layer. Digital Equipment's DDCMP or Digital Data Communications Protocol provides similar functions.

THREE TRANSPORT PROTOCOLS

The ISO has established protocol standards for the middle layers of the OSI model. The transport layer, at layer four, ensures that data is reliably transferred among transport services and users. Layer five, the session layer, is responsible for process-to-process communication. The line between the session and transport layers is often blurred.

As of yet, no ISO transport or session layer has been implemented on a widespread basis, nor has the complete OSI protocol stack been established. To make matters more confusing, most middle-layer protocols on the market today do not fit neatly into the OSI model's transport and session layers, since many were created before the ISO began work on the OSI model.

The good news is many existing protocols are being incorporated into the OSI model. Where existing protocols are not incorporated, interfaces to the OSI model are being implemented. This is the case for TCP/IP, and IPX, which are the major middle-layer protocols available today.

In the PC LAN environment, NetBIOS has been an important protocol. IBM developed NetBIOS (or Network Basic Input/Output System) as an input/output system for networks. NetBIOS can be considered a session-layer protocol that acts as an application interface to the network. It provides the tools for a program to establish a session with another program over the network. Many programs have been written to this interface.

NetBIOS does not obey the rules of the OSI model in that it does not talk only to the layers above and below it. Programs can talk directly to NetBIOS, skipping the application and presentation layers. This doesn't keep NetBIOS from doing its job; it just makes it incompatible with the OSI model. The main drawback of NetBIOS is that it is limited to working on a single network.

TCP/IP or Transmission Control Protocol/Internet Protocol is actually several protocols. TCP is a transport protocol. IP operates on the network layer. TCP/IP traditionally enjoyed enormous support in government, scientific, and academic internetworks and in recent years has dominated the commercial networking environment, too. Part of the explanation is that corporate networks began to approach the size of networks found in the government and in universities, which drove corporations to look for internetworking protocol standards. They found TCP/IP to be progressively more useful as it became more widespread. Many people once viewed TCP/IP as an interim solution until OSI could be deployed, but no one seriously believes that the OSI protocols will ever have more than a niche role in the future.

Often when TCP/IP is discussed, the subjects of SMTP, FTP, Telnet, and SNMP are also raised. These are application protocols developed specifically for TCP/IP. SMTP or the Simple Mail Transfer Protocol is the electronic mail relay standard. FTP stands for File Transfer Protocol and is used to exchange files among computers running TCP/IP. Telnet is remote log-in and terminal emulation software. SNMP or the Simple Network Management Protocol is the most widely implemented network management protocol. The figure shows the protocols of TCP/IP.

Novell traditionally used IPX/SPX as its native transport protocols, though the company introduced a "native" implementation of TCP/IP in place of IPX/SPX. Internetwork Packet Exchange (IPX) and Sequenced Packet Exchange (SPX) are both variants of Xerox's XNS

THE TCP/IP PROTOCOL STACK

Layer					
5-7	File Transfer Protocol (FTP)	Trivial File Transfer Protocol (TFTP)	Simple Mail Transfer Protocol (SMTP)	Telnet	Simple Network Management Protocol (SNMP)
4	Transmission Control Protocol (TCP)			User Datagram Protocol (UDP)	
3	Internet Protocol (IP)				
2	Logical Link Control (LLC)				
	Medium Access Control (MAC)				
1	Ethernet	Token Ring	FDDI	X.25	

■ The TCP/IP stack includes protocols that provide services equivalent to the OSI stack.

protocol. IPX provides network layer services, while SPX is somewhat rarely employed by applications that need transport layer services. Because IPX implementations prior to the introduction of NetWare Link Services Protocol (NLSP) in NetWare 4 caused a great deal of broadcast traffic and required frequent transmission acknowledgements, which can cause problems in a WAN, Novell also supported TCP/IP with gateways prior to its native TCP/IP implementation.

Other transport layer protocols include XNS and NetBEUI. XNS or Xerox Network System was one of the first local area network protocols used on a wide basis, mainly for Ethernet networks. 3Com's 3+ used a version of it. NetBEUI is IBM's transport protocol for its PC networking products. (The legacy of IBM's long-deceased partnership with Microsoft lives on in Microsoft's default implementations of NetBEUI in Windows for Workgroups, Windows 95/98, and Windows NT.)

PROTOCOL BABEL

If the number of available protocols seems like senseless confusion, it is and it isn't. Certain protocols have different advantages in specific environments. No single protocol stack will work better than every other in every setting. NetBIOS works well in small PC networks but is practically useless for communicating with WANs; APPC works well in peer-to-peer mainframe environments; TCP/IP excels in internetworks and heterogeneous environments.

On the other hand, much more is made about the differences in protocols than is warranted. Proprietary protocols can be perfect solutions in many cases. Besides, if proprietary protocols are sufficiently widespread, they become de facto standards, and gateways to other protocols are built. These include DEC's protocol suite, Sun Microsystems' Network Filing System and other protocols, and Apple's AppleTalk protocols. While these enjoy widespread use, that use is based on the computers these companies sell and not the proliferation of the protocols throughout the networking industry.

Whether it's a proprietary or standard protocol, users are faced with difficult choices. These choices are made slightly easier by the shakeout and standardization that has occurred at the physical and data-link layers. There are three choices: Token Ring, Ethernet, or FDDI. At the transport layers, IPX/SPX and TCP/IP emerged as the dominant protocols. [Editor's note: As of 2000, practically every new local network installation uses some form of Ethernet and TCP/IP, while the installed base remnants of Token Ring, FDDI, and IPX are diminishing irretrievably.]

This tutorial, number 3, by Aaron Brenner, was originally published in the October 1988 issue of LAN Magazine/Network Magazine.

The Networking Reference Library

Paper is not an obsolete medium for information on networking technology, in spite of the theoretical pronouncements of Internet ideologues, but you can't always find the information you need in a vendor's technical manual, on the Internet, or even in the pages of *Network Magazine*. Sometimes your best alternative is a reference book.

Over the years I've come across a handful of indispensable works. One interesting feature of all the following books is that they are valuable to readers with almost any level of expertise, even though they generally provide deep, advanced discussions of their subjects. These books are the places I go, or send my colleagues to, when I need a deeper background or richer understanding of a subject.

***Computer Networks (fourth edition),* by Andrew S. Tanenbaum** is the fundamental resource for explanations of general networking topics. You can learn the basic physics of optical fiber and microwave transmission and get basic explanations of the common modulation schemes, such as T1 and SONET. A full treatment of the different types of switching adds a dimension to the discussion of ATM. The book's basic strategy is to navigate through the seven layers of the OSI stack, providing deep explanations of each one. All the common technologies of local- and wide-area networks, the telephone system, and wireless systems are dealt with.

The book isn't loaded down with page after page of differential equations, though Tanenbaum is not reluctant to use appropriate math to explain such things as Shannon's law or an error correction algorithm. Some of the explanations of protocols include pseudocode examples and state diagrams. Overall, the book makes effective use of hundreds of diagrams, and a simulator for example protocols is available on the Web.

While Tanenbaum's book contains an abundance of serious information, it is a joy to read. He puts technology in its historical context, and you're never very far from a pungent anecdote or joke. People who are serious about understanding networking and protocols can work out the problems at the end of each chapter, just like the engineering students who use the book as their text. This is the book to have if you're ever stranded on a desert island with a network.

***Data Communications Networking Devices: Operation, Utilization and LAN and WAN Internetworking (fourth edition),* by Gilbert Held.** While Tanenbaum is terrific at explaining basic principles and technologies, Held is terrific at explaining the specific operations of equipment and software. Particularly for the Physical and Data-link layers, as well as for wide area networking components, this book is packed with valuable explanations and examples. Much of this kind of information is locked up in telephone company manuals and esoteric textbooks; Held makes it possible for the rest of us to understand WAN devices and operations.

You can often find examples of Held's work in the pages of *Network Magazine*.

***Interconnections, Second Edition: Bridges, Routers, Switches, and Internetworking Protocols,* by Radia Perlman.** Radia Perlman is something of a legend in the networking industry. She was the original developer of the Spanning Tree protocol for Digital Equipment; she developed DECnet Phase 5 and the OSI IS-IS routing protocol, and she worked on Novell's NetWare Link Services Protocol, which is another link-state routing protocol. In her time at Novell, she worked on security issues that affect routing. She is now a Sun Microsystems Distinguished Engineer. Along with all the remarkable engineering credentials, Perlman writes informative, easy-to-digest, amusing books. It's hard to imagine topics with more boredom potential than routing protocols and encryption algorithms, but just as she and her co-authors did with Network Security: Private Communication in a Public World, Perlman manages to hold your interest.

This book examines the OSI and TCP/IP routing protocols from the point of view of a protocol designer or perhaps a software developer, and not so much from a network administrator's perspective. It's a great place to start if you want to understand how these protocols behave and why.

***Network Security: Private Communication in a Public World,* (second edition), by Charlie Kaufman, Radia Perlman, and Mike Speciner.** *Network Security* isn't a comprehensive look at every subject affecting security (firewalls and viruses aren't discussed much, for example). But for easy-to-understand, yet thorough and rigorous, presentations of the principles of cryptography and authentication (which relies on cryptography), this book is hard to beat. It doesn't shrink from discussing the underlying math (number theory), but the heavy sledding is pulled out into its own chapter, and you can learn plenty without it. The book is lively and opinionated, despite its usefulness as a textbook, with problems at the end of each chapter. This is the place to get a firm grounding in public key and secret key cryptography, hashing, message digests, digital signatures, and the other core concepts of security systems.

Charlie Kaufman works for IBM, where he is chief security architect for Lotus Notes and Domino. Mike Speciner is a senior consulting engineer for ThinkEngine Networks.

***Hacking Exposed: Network Security Secrets and Solutions,* (third edition), by Stuart McClure, Joel Scambray, and George Kurtz.** *Hacking Exposed* provides a blow-by-blow listing of the security vulnerabilities of computers and networks. Each exploit or vulnerability is rated by popularity, simplicity, and impact, which helps provide a context for these issues. Such information is very difficult to come by elsewhere. Anyone with responsibility for network security, or anyone with a degree of curiosity about just how systems can be attacked, will find a great deal of useful information here.

***TCP/IP Illustrated, Volume 1: The Protocols,* by W. Richard Stevens.** The TCP/IP Illustrated series is probably the best reference for network administrators and programmers who want to understand the details and implementations of the world's most famous networking protocols. Volume 1 provides an overview of all the usual suspects, including IP itself, Internet Control Message Protocol (ICMP), Address Resolution Protocol (ARP), TCP, UDP, SNMP, FTP, TFTP, BOOTP; routing information protocols such as RIP and OSPF; and DNS. The protocols and the basic programs, including ping , traceroute, and telnet, are covered in detail. And if you can get a hold of the common Unix utility named tcpdump (which prints out the headers of selected packets on a network), you can dynamically recreate the conditions illustrated in the book—a useful way to enhance your learning, even if you don't have a protocol analyzer at hand.

***SNMP, SNMPv2, SNMPv3, and RMON 1 and 2 (third edition),* by William Stallings.** As is the case with other TCP/IP protocols, it's possible to go to one of the many request-for-proposal repositories and look up the details of the SNMP specification. But this process is not efficient, because later versions of the specification supersede early ones; the Management Information Base (MIB) specifications are spelled out in a number of separate documents; and you can't always find a full or clear explanation of why these protocols work as they do. Stallings has saved those who need to program with or perform their management tasks with SNMP- or RMON-based systems a great deal of trouble by collecting the relevant information in one place and wrapping it up with clear explanations of how the technology functions.

The first edition of this book covered CMIP rather than RMON; that the onetime heir apparent to the management crown is no longer mentioned in the book is an indicator of how soundly TCP/IP and SNMP whipped the OSI protocols and CMIP in the marketplace. Meanwhile, SNMPv2 has gone down for the third time, perhaps to be resuscitated as SNMPv3, and RMON-2 has come to hold many of the keys to the future of management. This is the best place to get a deep understanding of the major management protocols and related issues.

***The Essential Guide to Wireless Communications Applications (2nd edition)*, by Andy Dornan.** Andy Dornan covers a wide variety of wireless topics in his role as senior editor for *Network Magazine*. The book's title is overly modest—in fact, it comprehensively presents the technical underpinnings, the advantages and disadvantages, and the likely futures of practically every variety of wireless technology: voice as well as data; fixed wireless as well as mobile; packet-based as well as TDM-based. While the discussion of wireless applications is thorough and insightful, there is a great deal of valuable reference and tutorial material on wireless communications in general.

Microsoft Windows 98 Resource Kit, Microsoft Windows 2000 Professional Resource Kit, and Microsoft Windows 2000 Server Resource Kit. Most people install Windows 95/98 or NT/2000 without reading any manuals, or they get the operating system with their hardware. Both of these big, bumptious OSs are deep and complicated, but fortunately Microsoft makes available better than 1,700 pages of in-depth explanations for each one. (The Windows 2000 Server Resource Kit is a whole library, with 7296 pages total.) Do you need to know how to work with Banyan VINES or DEC Pathworks? Do you want to administer aspects of the operating system centrally, or perform automated installs over the network? Do you want to understand why Winsock, or the Registry, works the way it does? Do you simply need a list of the files that ought to be in a particular system directory? These books provide answers to all these questions, and few other sources do.

These aren't dumbbell books that use up a lot of trees by saying, "Now left-click once on the File Menu (see Figure such and such). Then left-click on the Save item (see Figure blah blah)." The Microsoft teams that assembled these books respect your expertise and your time, yet these books are consistently understandable as well as comprehensive. Each book comes with supplemental executable files on floppy disks or CD, which focus on providing policy-based administration for large numbers of clients.

***Peter Norton's New Inside the PC,* by Peter Norton and Scott H. A. Clark.** Peter Norton is probably responsible for more people successfully understanding PCs and their operating systems and peripherals than anyone in the world. Between his *Inside the PC* books and the various versions of Norton Utilities, enlightenment is within anyone's reach. He is a graceful, efficient writer with something illuminating to say about a vast number of technical subjects.

If you're unfamiliar with the memory model of Intel-based PCs, or you want to get a better grip on interrupt-level conflicts, you will find thorough explanations here. All the components inside the PC are discussed in detail, but Norton also explains operating systems and common PC peripherals. I can't think of a better book to give an intelligent beginner an understanding of how PCs work, though, like the other books on this list, this one has reference depth for those in the know.

***Newton's Telecom Dictionary (18th edition),* by Harry Newton with Ray Horak.** If Harry Newton's dictionary doesn't list the weird acronym or obfuscating term you've come across, there's a high probability that the term does not exist. The legacy telephone companies use an incredible volume of jargon and counterintuitive terminology, and Newton decodes it for you. The mainframe and minicomputer worlds have their own languages, and Newton translates. The Internet crowd, the PC contingent, the computer telephony circle—you will be better able to understand any of these constituencies with *Newton's Telecom Dictionary*. Newton doesn't stop with raw definitions or mere spell outs of acronyms, either. He puts you on the trail of a full understanding. Some technical dictionaries are apparently done by people who see engineering or programming as their main calling in life. Newton is highly literate and extremely accurate—qualities I greatly value in dictionaries.

Newton updates this dictionary a couple times a year. Obviously, new terms must be defined daily in our volatile world. You'll come to rely on Newton for comprehensive and correct definitions of high-tech terms.

In the interest of complete disclosure, I should mention that Harry Newton was the founder of *LAN Magazine*, which became *Network Magazine* in 1997. Furthermore, Miller Freeman/CMP bought Flatiron Publishing as well as a number of magazines and trade shows from Newton and his company in 1997.

13 CRUCIAL REFERENCE BOOKS

Computer Networks (fourth edition), by Andrew S. Tanenbaum (Prentice Hall, 2002, ISBN: 0130661023)

Data Communications Networking Devices (fourth edition), by Gilbert Held (Wiley, 1998, ISBN: 047197515X)

Interconnections, Second Edition : Bridges, Routers, Switches, and Internetworking Protocols, by Radia Perlman (Addison Wesley, 1999, ISBN: 0201634481)

Network Security: Private Communication in a Public World, (second edition), by Charlie Kaufman, Radia Perlman, and Mike Speciner (Prentice Hall, 2002, ISBN: 0130460192)

Hacking Exposed: Network Security Secrets and Solutions, (third edition), by Stuart McClure, Joel Scambray, and George Kurtz (McGraw-Hill Osborne Media, 2001 ISBN: 0072193816).

TCP/IP Illustrated, Volume 1: The Protocols, by W. Richard Stevens (Addison Wesley, 1994, ISBN: 0201633469)

SNMP, SNMPv2, SNMPv3, and RMON 1 and 2 (third edition), by William Stallings (Addison Wesley, 1999, ISBN: 0201485346)

The Essential Guide to Wireless Communications Applications (2nd edition), by Andy Dornan (Prentice Hall, 2002; ISBN: 0130097187)

Microsoft Windows 98 Resource Kit (Microsoft Press, 1998, ISBN: 1572316446)

Microsoft Windows 2000 Professional Resource Kit (Microsoft Press, 2000, ISBN: 1572318082)

Microsoft Windows 2000 Server Resource Kit (Microsoft Press, 2000, ISBN: 1572318058)

Peter Norton's New Inside the PC, by Peter Norton and Scott H. A. Clark (Sams Publishing, 2002, ISBN: 0672322897)

Newton's Telecom Dictionary (18th edition), by Harry Newton (CMP Books, 2002, ISBN: 1578201047)

This tutorial, number 116, by Steve Steinke, was originally published in the March 1998 issue of Network Magazine.

SECTION II

Physical Layer Protocols

Encoding, Modulation, and the Physical Layer

The Physical layer doesn't get any respect. This negligence results partly from the fact that there are few end-user interventions once a network's cabling or wireless infrastructure is successfully put into place. The actual behavior of electromagnetic signals and raw digital sequences is generally built into hardware interfaces and is beyond any ordinary fiddling or tuning. Nevertheless, a bit of education about the Physical layer can help prepare you for a deeper understanding of new wireless, optical, and local-loop technologies that will have important future roles in every network.

The Physical layer is responsible for turning some medium into a bit pipe. Copper cabling is the most common data-networking medium of LANs and in the local loop, while fiber optic cabling is prevalent in most wide area networks—and is perhaps the local-loop and the LAN medium of the future. Radio frequency bands also play a role in some data networks, including signals that are relayed via satellites. Some local networks and point-to-point links employ infrared signals propagated through space, and visible light lasers are sometimes used to transmit data without the aid of a fiber optic cable.

Each medium has its strengths, vulnerabilities, and quirks. Cost is obviously an important factor, which explains the prevalence of copper cable in today's infrastructures. Performance, in the form of reliable throughput, is closely related to cost. The rapidly increasing price/performance of fiber optic cabling accounts for the growing hegemony of that medium as it migrates from interstate backbones to metropolitan areas, neighborhoods, and campuses. Although wireless systems have the powerful cost advantages of no cables, trenches, or poles, their relatively low reliability, low privacy, and low throughput offset these advantages.

The main task of the Physical layer is to make the best use of the medium at hand. The laws of physics constrain a medium's potential for reliable bit throughput, but so do regulations devised by the people who need to share limited resources. For example, T1 lines interfere with analog voice, ISDN, and Asymmetric Digital Subscriber Lines (ADSLs) so much that aggregated cables or binder groups can't include T1s.

ANALOG AND DIGITAL DATA

Data that needs to be communicated may be in analog or digital form. Analog data is continuous, taking on innumerable values within a range. Voices, images, and temperature readings from a sensor are all examples of analog data. Digital data takes on a limited number of discrete values. In the limiting, and most common case, digital data takes one of two values: zero or one. Logical values such as true or false, integers, and text are commonly encountered examples of digital data.

In order to manipulate or communicate data, it must be encoded as some kind of signal, usually an electrical or electromagnetic signal. Analog data can be encoded as an analog signal. Perhaps the most common example is a plain old telephone in the local loop, though a cassette tape player, the video and audio components of a TV program, and many other household media use analog signals to represent analog data.

Analog data is also commonly encoded with digital signals. If a phone call travels beyond the local exchange carrier's central office into the long distance network, it will be digitized. If you scan an image or capture a sound on the computer, you're converting analog data to digital signals. This analog-to-digital conversion is usually accomplished with a special device or process referred to as a codec, which is short for coder-decoder.

Digital data is routinely converted to analog signals. The most common example is when you make use of the omnipresent voice infrastructure for computer connectivity and employ a modem to represent your bits in the form of audible tones. (Modem is short for modulator-demodulator, which performs the inverse of what a codec does—though in most cases, of course, both a codec and a modem perform both analog-to-digital and digital-to-analog conversions.) Modulation can be considered to be a special case of encoding, though the terms tend to overlap in ordinary usage. Technically speaking, modulation involves combining two signals, either of which can be analog or digital, to produce a resultant signal, which can be analog or digital. Encoding, then, is the representation of data by a signal using any method.

Finally, digital data is also regularly represented by digital signals. Any time you send e-mail, load a file, or download Web pages, you're encoding digital data with digital signals.

SIGNAL OBSTACLES

The enemies of both analog and digital signals include attenuation, noise, and crosstalk. Attenuation is the tendency of a signal to get weaker with distance. Analog signals must be amplified before they become too diminished to be detectable. Unfortunately, analog signals accumulate noise with repeated amplification. Digital signals, while they are degraded by attenuation, can be detected and repeated indefinitely with no loss of data. This property is one of the principal reasons digital communication became increasingly important in the last years of the 20th century.

Noise is the backdrop of the universe. Atoms and molecules in motion create random electromagnetic signals that prevent any communication channel from being perfectly clear. Of course, all sorts of events, from elevator motors and electric mixers to lightning and solar flares, also contribute noise to our communications environment. Some encoding techniques are less susceptible to particular kinds of noise than others. Crosstalk is a special form of noise that is induced by other signals on a common medium.

Digital signal transmission is used in LANs, where cable lengths are relatively short and thus not subject to severe attenuation. (Attenuation increases with increasing frequency, and digital transmissions have high frequency components, which means that channels with constrained bandwidth aren't suitable for high throughput digital transmissions.) The best known examples of digital transmission on telephone facilities are T1 lines and ISDN.

The simplest representation of digital signals is a line code known as Non Return to Zero Level, or NRZ-L (see Figure 1a). This code is the archetype of what a digital signal looks like, although there are innumerable variations in how best to transmit a digital signal across various media. NRZ-L has severe limitations in practice, but it has real-life applications in RS-232 links and in data storage on hard disk drives.

Two of the biggest shortcomings of NRZ-L are its DC component and its inability to carry synchronization information along with the data. If an NRZ-L signal has a sequence of ones, the signal can't pass through such electrical components as transformers and capacitors, which only conduct when the signal is changing. As for synchronization, correct timing is

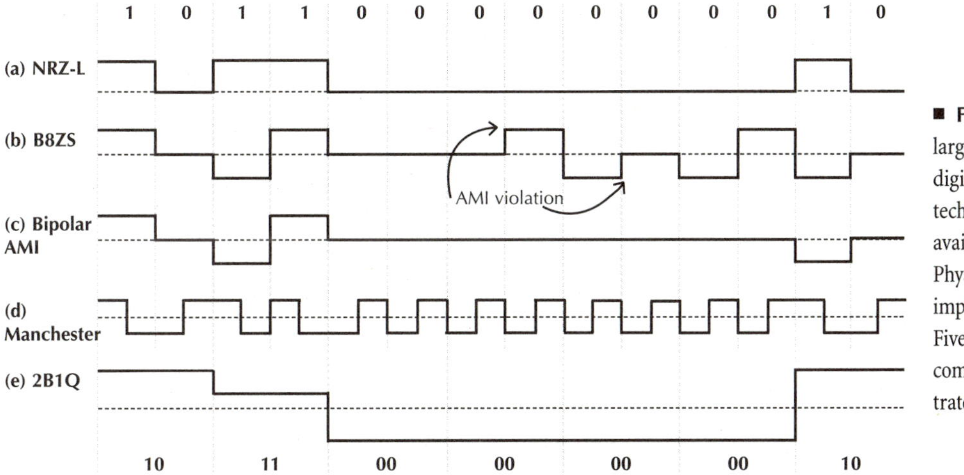

■ **Figure 1:** A large number of digital encoding techniques are available to design Physical-Layer implementations. Five of the most common are illustrated here.

essential for a receiver to identify the discrete states of the digital signal. If a series of ones appears in an NRZ-L transmission, the receiver will require an additional synchronization signal to be aware of how many there are.

T1 lines often use a line code called Bipolar with 8 Zero Substitution (B8ZS, see Figure 1b). B8ZS is a variant of Bipolar Alternate Mark Inversions (AMI, see Figure 1c). (Marks and spaces are just a terminological variation on zeros and ones.) Bipolar AMI solves the DC component problem by alternating the polarity of ones—zeros are represented by no signal, the first one is a positive signal, the second one is a negative signal, and the signal values of subsequent ones alternate. However, with a long string of zeros, Bipolar AMI signals can lose self-synchronization. The 8 Zero Substitution trick takes care of the problem by breaking the alternation rule when it comes across a sequence of eight consecutive zeros. By sticking ones in the places of the fourth and fifth zeros, and in the places of the seventh and eighth zeros, with the first substitute one incorrectly having the polarity of the previous one, and the third substitute one incorrectly having the polarity of the second substitute one, the receiver recognizes an intentional violation and concludes that there is in fact a sequence of eight zeros. This coded violation ensures that there will never be a sequence of more than seven successive no-signal bit times. The rules of mark inversion also add a degree of Physical-layer error detection to this encoding method; noncoded violations will indicate spoiled bits.

Ethernet uses a type of digital signal known as Manchester encoding (see Figure 1d). A one is indicated by a high/low transition in the middle of a bit, while a zero is indicated by a low/high transition in the middle of a bit. Based on the previous discussion, you can see that Manchester encoding has no DC component and is fully self-synchronizing. If there is no transition in a bit time, you have a Physical-layer error indication. The drawback to applying this line code more widely is that its bandwidth requirement is twice the baud rate; in other words, there are significant spectral components as high as 20MHz, which are no problem on coaxial cable or on short distances of twisted-pair cabling, but not suitable for long distances.

ISDN lines make use of a line code known as 2 Binary 1 Quaternary (2B1Q). Symmetric Digital Subscriber Line (SDSL) and High-Bit Rate Digital Subscriber Line (HDSL) also employ this encoding method. A line with 2B1Q encoding uses four distinct signaling levels, with data represented in 2-bit units (see Figure 1e). By encoding two bits with each signal transition, 2B1Q represents the distinction between bits per second and baud rate. The baud rate of a signal is the number of signal transitions per second, and it can't be higher than the bandwidth of the channel. The number of bits per signal element, represented by L, is given by log2L. Thus a code with eight signal elements could encode 3 bits per baud, a code with 16 signal elements could encode 4 bits per baud, and a code with 256 signal elements could encode 8 bits per baud. In the case of ISDN, the bandwidth the signal occupies is 80KHz, the baud rate is 80Kbaud, and the raw data rate is 160Kbits/sec.

DIGITAL SIGNALS, ANALOG TRANSMISSION

A vast infrastructure exists for analog signaling, and much of it can readily transport digital signals as well. The telephony local loop, the cable TV infrastructure, and practically every form of wireless communication are inherently analog transmission media that have been adapted for digital signals.

The earliest modems used a technique known as Frequency-Shift Keying (FSK, see Figure 2a) to represent digital data. FSK devices, such as the Bell 103 modem, used one tone (1070Hz) for zeros and another tone (1270Hz) for 1s. Amplitude-Shift Keying (ASK, see Figure 2b), is one way of describing the modulation of digital data over fiber optic cable. In this case, no light represents a zero while the presence of light above a threshold level represents a one. The third attribute of a sinusoidal signal is its phase, and Phase-Shift Keying (PSK, see Figure 2c) is widely used in modems. Nowadays, modems commonly use a combination of phase and amplitude modulation to encode multiple bits in a single signaling event or symbol.

Cable modems and ADSL make use of a signaling technique called Quadrature Amplitude Modulation (QAM). With QAM, the carrier signal is split into two signals, shifted in

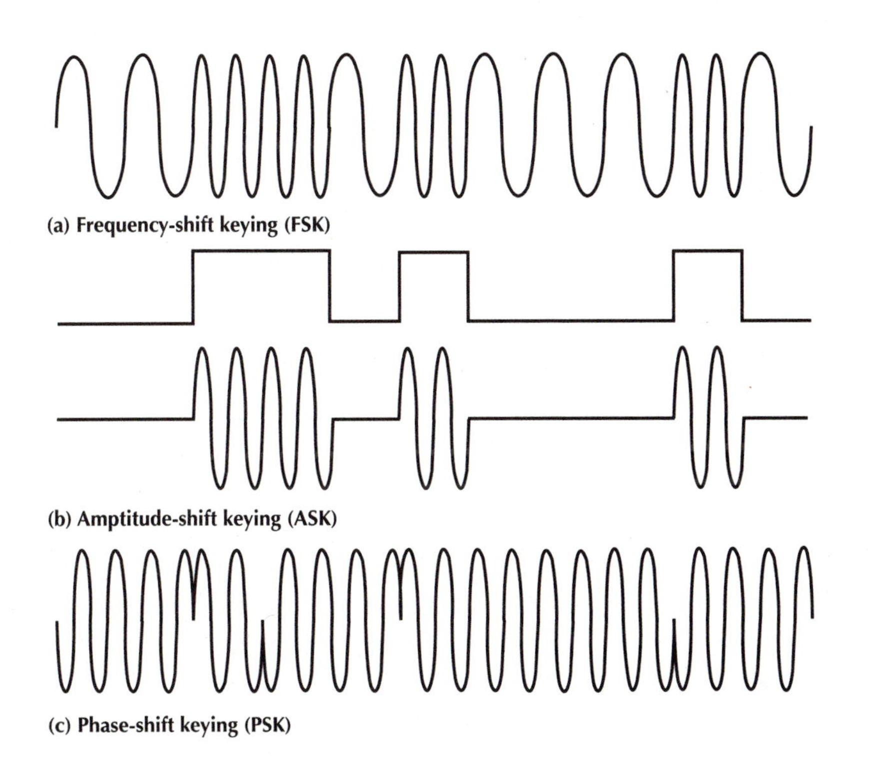

■ **Figure 2:** There are three possible basic techniques for modulating an analog signal with digital data. In practice, combinations of these methods are often employed.

phase by 90 degrees. Each component is modulated with ASK, using as many as 16 amplitude levels to represent as many as 256 different states. One ADSL modulation technique, Discrete Multitone (DMT), divides the available twisted-pair spectrum above 25KHz into 256 downstream subchannels of 4KHz each. QAM is then applied to each subchannel according to its individual performance. At best, a single subchannel may be able to carry as many as 60Kbits/sec. Theoretically then, ADSL could provide throughput rates as high as 15.36 Mbits/sec.

This tutorial, number 128, by Steve Steinke, was originally published in the March 1999 issue of Network Magazine.

Network Delay and Signal Propagation

There is a widespread impression throughout much of the networking industry that optical networks are inherently faster and more responsive than those based on copper wire, and that wholly photonic transmissions are invariably faster and more responsive than transmissions consisting of signals based on electrons. In order to explore the senses in which this claim is accurate, we must look at some of the fundamentals of physics and communications theory.

The first fundamental fact is that electromagnetic waves propagate at a rate of approximately 300,000 kilometers (km) per second (s), or 186,000 miles per s in a vacuum. (This universal upper limit to the velocity that matter or information can achieve is usually abbreviated by "c.") The velocity of electromagnetic radiation in a vacuum is independent of frequency, so gamma waves, ultraviolet radiation, visible light, infrared (which is the range where practically all optical fiber traffic fits), and radio waves travel through space at the same rate.

The propagation velocity of waves can be dramatically affected by the medium they traverse, however. Recent experiments have actually demonstrated an ability to essentially stop and restart the propagation of light. In more ordinary media, such as certain commercial single-mode optical fiber products, the propagation velocity of a signal is 68 percent of c or 205,000km/s (see table).

In comparison, electric waves or signals in commonly used copper wire travel at speeds between 55 percent and 80 percent of c. Note that this is the propagation velocity of a signal or a wave, not the velocity of electrons in a wire. Electrons, in fact, move at speeds of about a fraction of a millimeter per s. Compare a wire to a garden hose filled with water; if you turn on the water at one end of the hose, water will flow out the other end almost immediately—in fact, as rapidly as a pressure wave propagates through water. Imagining that the speed of an electric wave in a cable would be constrained by the speed of individual electrons would be similar to counting the time it takes to fill an empty hose with water before it flows out the other end.

LATENCY

The concept of latency is important for network and computer operations generally, but it isn't always consistently and tightly defined. In general, latency refers to the delay between the occurrence of two events. One important latency metric of a hard disk is the data access time: the time it takes to return a random block of data to the processor after a request is submitted, typically 8 milliseconds (ms) to 12ms in modern drives. (Drive makers sometimes restrict the term "latency" to a period called "rotational latency," the average time required to read data from a track once the head is positioned over it. Rotational latency typically contributes an average of 4ms or 5ms to a random disk access operation.)

Modems have two prominent latency producers: the time between the initial call request and the time the call is completed with the modems mating successfully (often as much as 30s), and the time between the point where the application sends data to the modem and the point where analog information begins to flow on the circuit (typically some 100ms).

Some of the most widely used latency measures for networks are end-to-end trip time, round-trip time, keystroke response time, and transaction completion time. End-to-end trip time is the time it takes a packet or other unit of data to travel from source to destination. Round-trip time adds the time for a return response or acknowledgment to the end-to-end latency. In the PSTN, round-trip time is almost exactly double the end-to-end time, but in connectionless networks the return path may be substantially different from the original transmission's. Keystroke response times measure from the time that a user presses an "enter" key or issues some other execute command until the time that an entire screen update is completed. This kind of latency is the best indicator of the responsiveness of an

CALCULATIONS

1.) $\frac{800\text{km}}{205{,}000\text{ km/sec}} = 7.8\text{ms each way}$

2.) $\frac{4{,}320\text{km}}{205{,}000\text{km/sec}} = 21\text{ms each way}$

3.) $\frac{36{,}000\text{km}}{300{,}000\text{ km/sec}} = 120\text{ms each way}$

4.) $\frac{100\text{Kbyte} \times 8\text{bits/byte}}{53\text{Kbits/sec}} = 15\text{s}$

5.) $\frac{100\text{Kbytes} \times 8\text{bits}}{64\text{Kbits/sec}} = 12.5\text{s}$

6.) $20 \times \left(\frac{1\text{Kbyte} \times 8\text{bits/byte}}{53\text{Kbits/sec}} + 100\text{msec}\right) = 5.0\text{s}$

7.) $20 \times \left(\frac{1\text{Kbyte} \times 8\text{bits/byte}}{64\text{Kbits/sec}} + 10\text{msec}\right) = 2.7\text{s}$

8.) Maximum carrying capacity= $W \bullet \log_2 (1+s/n)$ (Shannon's Law) where W is bandwidth (Hz) and s/n is the signal-to-noise ratio. A typical s/n for optical fiber of 20dB is equal to 2.0.
$12.5\text{ THz} \bullet \log_2(3) = 12.5 \times 1.6 = 20\text{Tbits/sec}$

interactive application. Transaction completion time can be measured for automatic or "headless" applications, as well as for interactive ones.

For communications networks, propagation time is a significant part of end-to-end latency. The circuit-switched Time Division Multiplexing (TDM)-based PSTN is specifically designed so that once a call is established, there is very little added latency beyond propagation latency. Most telephones, as well as Class 4 switches, convert two-wire full-duplex circuits to four-wire simplex circuits using circuits called hybrids. Imperfect impedance matching in these hybrids creates echo signals, which reflect a speaker's voice back in a delayed form. People experience acoustic-psychological problems with delayed echoes, showing greater sensitivity as the delay increases. Delays of 10ms to 20ms are generally undetectable, but greater delays are more troublesome. U.S. phone companies have traditionally installed echo-cancellation circuitry every 500 miles (800km.) An 800km circuit running over optical fiber would introduce round-trip delays of about 15.6ms.

A commonly cited rule of thumb by voice-over-IP (VoIP) vendors is that round-trip delay times for high-quality voice should be less than 150ms. Round-trip transit time for North American transcontinental calls is about 42ms (San Francisco to Boston over optical fiber). Intercontinental calls may have three or four times as much propagation delay as continental landlines. Geosynchronous satellite communications, where the forward and return paths both involve up and down legs of at least the height of the satellite above the equator—36,000km—have minimum round-trip times of 480ms or so. (Satellite signals do have the advantage of propagating at the full speed of light.)

Connectionless networks, such as the Internet and other IP-based networks, use throughput capacity efficiently, rather than minimizing latency. If I run a traceroute utility from my house in the Bay Area to a server in Boston, I typically see round-trip times of 110ms, almost three times the propagation delay for the round-trip path. You'd better believe that the difference between 42ms and 110ms is accounted for by the 26 intermediate (one-way) nodes between the two hosts. With no congestion at the intermediate points, an IP network's round-trip time could approach the propagation delay time, but an IP network with no congestion or queuing anywhere wouldn't have the cost advantages of traditional TDM phone networks for data transmission, which led to the growth of IP in recent years.

THROUGHPUT

Throughput and latency are completely independent issues. A channel with high throughput can move large quantities of data rapidly, but the first bit of data can never arrive faster than the latency permits. A file transfer operation, for example, may only have to suffer the "latency penalty" once at the beginning of the process, whereupon the full throughput capacity will be available to finish the job. Adding more throughput capacity will speed up the task. A financial transaction, on the other hand, might have numerous back-and-forth data flows, each of which must cope with the channel's latency. Throwing additional throughput at this problem may contribute little or nothing to overall performance. Improving the performance will require either reducing the latency of the individual flows or reengineering the application to reduce "ping-pong" operations.

For example, consider two channels where one has analog 56K modems at each end and the other has ISDN terminal adapters at each end. The best-case throughput of the analog modem channel will be perhaps 53Kbits/sec, while the throughput of an ISDN B channel is 64Kbits/sec. ISDN Terminal Adapters (TAs) have latencies of about 10ms, while analog modems have latencies of 100ms, as I mentioned earlier. So you can expect a 100Kbyte file transfer to take about 15s on the modem channel and 12.5s on the ISDN channel, assuming the calls are already established. The latency values are rounding errors in this scenario.

Now consider a series of 10 round-trip transaction flows of 1Kbyte, where each of the 20 flows can only occur after the previous one concludes. The analog link will take more than 5s for such a transaction, while the ISDN link will take 2.7s. No matter how many additional modem links are added to increase throughput, the transaction can never be completed in less than 2s, thanks to the cumulative latency of the channel.

With this discussion of latency and throughput in mind, let's return to the initial discussion of optical fiber and copper media. The throughput capacity of optical fiber is, of course, staggering. The theoretical carrying capacity of a single-mode fiber using the 1,550nm "window" is 20Tbits/sec, or 20,000Gbits/sec (assuming 12.5THz bandwidth and a signal-to-noise value of 20dB). Twisted-pair copper wiring is hard-pressed to carry 1Gbit/sec of data

SIGNAL VELOCITY IN VARIOUS MEDIA

Material	Propagation velocity (fraction of speed of light in a vacuum)	Index of refraction	Velocity of signal (km/s)
Optical fiber	.68	1.46	205,000
Flint glass	.58	1.71	175,000
Water	.75	1.33	226,000
Diamond	.41	2.45	122,000
Air	.99971	1.00029	299,890
Copper wire (category 5 cable)	.77	N/A	231,000

Mixed Media. Light travels at a surprising variety of speeds in different transparent media, but electrical signals travel at comparable speeds in copper wiring.

farther than 100 meters. Coaxial cable can support somewhat higher throughput than twisted-pair, but the throughput advantages of optical fiber are so great that there has been little development of coax for data-carrying applications. Note that the cable TV operators were the first to deploy optical fiber to residential networks, and they chose to supplement their existing coaxial facilities with fiber.

Optical fiber has few, if any, advantages over copper wires where latency is concerned, however. Propagation latency is more or less the same with either medium. (Free-space optics and radio channels actually propagate 40 percent to 50 percent faster than optical fiber or copper wire.) Queuing, routing, and packetization latencies are also much the same—after all, these operations are performed electronically before they are converted to optical data flows. Any place that optical fiber is installed instead of copper wire, it will be because of its high throughput capacity or some other feature, not because it has superior latency characteristics.

RESOURCES

There is a good set of slides covering this topic by Dr. Pisit Charnkeitkong of Ransit University at http://vishnu.rsu.ac.th/instructor/pisit/NetHtmlSlide/03-Media.pdf.

Cisco Systems has a valuable Introduction to Voice and Telephone Technology at www.cisco.com/networkers/nw99_pres/401.pdf.

The October 1999 issue of *Communication Systems Design* features an article entitled "Echo-cancellation for Voice over IP" by John C. Gammel, which provides a more detailed account of latency, its effects on echo creation, and efforts to minimize the problem of echoes in voice-over-IP (VoIP) systems. See www.csdmag.com/main/1999/10/9910feat4.htm.

This tutorial, number 154, by Steve Steinke, was originally published in the May 2001 issue of Network Magazine.

Basic Phone Services and Circuits

Here's how to go the last mile with the phone company.

Telephone company offerings are complex. Unless we have corporate responsibility for provisioning voice or data telecommunications services from the telephone company, most of us rarely have any involvement with telephone lines aside from the analog local loop that provides the voice services that allow telemarketers to call us during dinner.

While telephone company offerings can be really hard to understand and to distinguish between, even that is easy in comparison to understanding their costs. This tutorial can't begin to tackle the cost issue, but it will try to make the relevant distinctions among some of the more frequently encountered products and services available from telephone service providers.

ANALOG LINES, DIGITAL LINES

The first important distinction is between analog and digital lines. Analog signals vary continuously, and they represent particular values, such as the volume and pitch of a voice or the color and brightness of a section of an image. Digital signals have meaning only at discrete levels—in the most common case, the signal is either on or off, present or absent, 1 or 0.

Analog telephone lines are the legacy systems of the telephone universe. The great preponderance of residential telephone lines is analog. Fifty-year-old telephones will probably work on your local loop—the connection between your home telephone jack and the telephone company's central office. (Your central office is probably not a gigantic building downtown—the average local loop is about 2.5 miles long [four kilometers], so the "central office" is most often an inconspicuous building in or near your neighborhood.)

When you talk on the telephone, the microphone in the "receiver" (while you're talking, it's actually a transmitter) produces an analog signal that travels to the central office and is switched either to another local destination or to other switching offices that connect it to a remote destination. Dialing the telephone produces the in-band signals that tell the switching system where to route the call. The telephone companies have learned a great deal about the electrical characteristics of human voice signals over the years, and they have determined that we will be reasonably satisfied with voice signals that do not transmit frequencies below 300Hz or above 3,100Hz. Note that high fidelity is usually considered to be a system that can reproduce frequencies between 20Hz and 20KHz without distortion. While voices are recognizable with the standard telephone frequency range, that range of frequencies is likely to be inadequate for other types of sounds—for instance, music sounds lousy over the telephone. To allow for gradual rolloffs of high and low frequencies, the telephone companies allow an analog telephone channel a bandwidth of 4,000Hz to work with.

At the central office, the odds are that the analog signal will be digitized in order to be switched across the telephone network. Aside from Giblet County, AK, and Rat Fork, WY, the U.S. telephone network that interconnects central offices uses digital signaling. Although many urban business telephone users have digital services direct to their PBXs or data communication devices, and ISDN lines are digital, the local loop is sometimes referred to as "the last mile," because residences, generally saddled with analog-only transmission facilities, were traditionally limited to the 4,000Hz bandwidth that dial-up modems can use.

WHAT CAN 4KHZ DO?

Modems convert digital signals from a computer into analog signals in the telephone frequency range. There is a hard upper limit to the capacity of a channel with a given bandwidth. A channel's throughput in bits per second depends on the bandwidth and the achievable signal-to-noise ratio. The current top throughput rate for modems of 33.6Kbits/sec is quite close to the limit. As users of 28.8Kbit/sec modems know, the actual throughput achievable on normally noisy analog lines is rarely the full-rated value, and it may be much lower. Compression and caching and other tricks can mask the limit to an extent, but we'll see perpetual motion machines before we see a classically defined modem with, say, 50Kbits/sec or even 40Kbits/sec throughput on ordinary analog telephone lines. The near-56Kbit/sec V.90 modem doesn't actually send modulated analog signals like other modems do. Instead, it transmits a "raw" digital signal in the downstream direction, avoiding the quantization noise that would be created by a digital-to-analog conversion.

When the telephone company reverses the process and digitizes an analog voice signal, it uses a 64Kbit/sec channel. (This conversion is a worldwide standard.) One of these channels, called a DS0 (digital signal, level zero), is the basic building block for telephone processes. You can agglomerate (the precise term is multiplex) 24 DS0s into a DS1. If you lease a T1 line, you get a DS1 channel. With synchronization bits after each 192 bits (that is, 8,000 times a second), the DS1 capacity is 1.544Mbits/sec (the product of 24 and 64,000, added to 8,000).

DEDICATED LINES, SWITCHED LINES

The second important distinction to make about telephone lines is whether they are dedicated circuits you lease or switched services you buy. If you order a T1 line or a low data-rate leased line such as dataphone digital service (DDS), you are renting a point-to-point facility from the telephone company. You have dedicated use of such a circuit-with 1.544Mbits/sec (T1) or 56Kbits/sec (low data rate) of capacity, respectively.

While frame relay services are switched through a cloud frame by frame, they are almost invariably sold as permanent virtual circuits, where the customer specifies the end points. For purposes of designing a network layout, frame relay links have more in common with dedicated lines than with switched lines, but the cost can be substantially lower for an equivalent capacity. Switched services, such as residential analog telephone service, are services purchased from the telephone company. You can select any destination on the telephone network and connect to it through the network of public switches. You generally pay for connect time or actual traffic volume, so unlike a dedicated line, the bill will be low if usage is low. Switched digital services include X.25, Switched 56, ISDN Basic Rate Interface (BRI), ISDN Primary Rate Interface (PRI), Switched Multimegabit Data Service (SMDS), and ATM. It's also possible to set up private networks that supply these services using your own switching equipment and leased lines or even privately owned lines—for example, if you are a university, a railroad, or a municipal utility.

If the circuit provided by the phone company is already a digital circuit, there is no need for a modem to provide digital-to-analog conversion services between the terminal equipment (phone company talk for such equipment as computers, fax machines, videophones, and digital telephone instruments) and the telephone system. Nonetheless, customer premises equipment still needs to behave like a good citizen of the telephony network. In particular, it must present the correct electrical termination to the local loop, transmit traffic properly, and support phone company diagnostic procedures.

A line that supports ISDN BRI service must be connected to a device called an NT1 (network termination 1). In addition to the line termination and diagnostic functions, the NT1 interface converts the two-wire local loop to the four-wire system used by digital terminal equipment. For digital leased lines—T1 and DDS—and for the digital services, the digital subscriber line from the phone company needs to be terminated by a channel service unit or CSU. The CSU terminates and conditions the line and responds to diagnostic commands. Customer terminal equipment is designed to interface with a data service unit (DSU), which hands over properly formatted digital signals to the CSU. CSUs and DSUs are often combined into a single unit—a CSU/DSU, of all things. The DSU may be built into a router or multiplexer. So even though end-to-end digital services don't require modems, a piece or two of interfacing hardware is always required for connectivity.

MEDIA FOR PHONE SERVICES

While 33.6Kbits/sec is a stretch for most local loops configured for analog service, the same twisted-pair wiring running between your house and a central office is very likely capable of supplying ISDN BRI service, with 128Kbits/sec of data throughput capacity and another 16Kbits/sec of control and setup capacity. How is this possible? Analog telephone circuits are heavily filtered to keep the signals attenuated outside their 4KHz bandwidth because 4KHz is the highest frequency a DS0 can handle. Digital circuits don't need to be filtered the same way, so the twisted pair cable can support a much greater bandwidth, which allows greater throughput.

Leased 56Kbit/sec and 64Kbit/sec lines and services that run on these lines, such as frame relay and Switched 56, may be delivered on a two-wire digital line or on a four-wire

TABLE 1—TELEPHONE SERVICE TYPES

Circuit type	Service	Switched service	Local loop transmission medium
Analog	POTS	Circuit switched	Two-wire twisted pair
DS0 (64Kbits/sec)	DDS (leased line)	Dedicated line	Two-wire or four-wire twisted pair
	Frame relay	Packet-switched PVC	Two-wire or four-wire twisted pair
	X.25	Packet switched	Two-wire or four-wire twisted pair
	Switched 56	Circuit switched	Two-wire or four-wire twisted pair
	Switched 64	Circuit switched	Two-wire or four-wire twisted pair
	ISDN BRI	Circuit switched	Two-wire twisted pair
Multiple DS0 (between 64Kbits/sec and 1.536Mbits/sec in 64Kbits/sec increments)	Fractional T1	Dedicated line	Two-wire or four-wire twisted pair
	Frame relay	Packet-switched PVC	Two-wire or four-wire twisted pair
DS1 (1.544Mbits/sec) (24 DS0s)	T1 leased line	Dedicated line	Four-wire twisted pair or fiber
	Frame relay	Packet-switched PVC	Four-wire twisted pair or fiber
	SMDS	Packet switched	Four-wire twisted pair or fiber
	ISDN PRI	Circuit switched	Four-wire twisted pair or fiber
DS3 (44.736Mbits/sec) (28 DS1s, 672 DS0s)	T3 leased line	Dedicated line	Coax or fiber
	ATM	Cell switched	Coax or fiber
	SMDS	Packet switched	Coax or fiber

digital line (which has separate wire pairs for transmitting and receiving). T1 lines as well as ISDN PRI and frame relay are often delivered on four-wire digital lines or perhaps on optical fiber. T3 lines are sometimes coaxial cable, but most high-capacity traffic is carried on optical fiber. While ISDN is getting a lot of attention as a high-capacity, wide-area connection, it is not the last word on throughput for the "last mile." PairGain (Tustin, CA) and AT&T Paradyne (Largo, FL) market products using Bellcore's (Piscataway, NJ) high bit-rate digital subscriber loop (HDSL) technology. These products serve to equalize local loops dynamically, making it possible to support DS1 throughput—1.544Mbits/sec—over most existing twisted pair local loops, provided that HDSL devices are installed at both ends. (With standard 24-gauge wire, HDSL can be used successfully on local loops up to 2.3 miles long, with no repeaters. Ordinary T1 circuits require repeaters at least every 3,000 feet to 5,000 feet.) If you want to transport DS1 levels of traffic over the last mile, the alternatives to HDSL are to install "fiber to the curb" at great expense or to install several repeaters on each line, which is not as expensive as all-new fiber but is still costly and imposes a large maintenance cost on the telephone company (and, ultimately, on the customer).

HDSL is not even the final word on improving the throughput of the last mile. Asymmetrical digital subscriber line technology (ADSL) supports throughput as high as 8Mbits/sec in a single direction, with a much lower throughput—perhaps as low as 64Kbits/sec—in the other direction. In a perfect—or at least competitive—world, where customers pay for a telephone service based on the actual cost of delivering that service, a high percentage of analog telephone customers could receive ISDN PRI (or another T1 service) at a price comparable to today's cost for ISDN BRI.

But, perhaps today's ISDN promoters don't have much to worry about. In most cases, telephone companies can install these special line equalizers and keep the savings to themselves. As is so often the case with tariffed services, there's no requirement that the rates be rational.

This tutorial, number 89, by Steve Steinke, was first published in the January 1996 issue of LAN Magazine/Network Magazine.

Getting Data over the Telephone Line: CSUs and DSUs

When you lease a T1 line for frame relay or Switched 56 service from either the telephone company or alternative service provider, you receive access to one of the company's connectors, to which your network attaches. (In telephone company jargon, everything beyond the connector is called customer premises equipment, or CPE.)

However, you can't attach your router to the telco's connector by simply plugging a cable with the right pins into each device. To ensure that a network interfaces properly with a telephone line, several functions must be performed. CSUs and DSUs were created to handle those functions.

THE ROLE OF THE CSU

A channel service unit, or CSU, is the first device the external telephone line encounters on the customer premises. As recently as the early 1980s, CSUs were always owned by the telephone company, which leased the devices to customers. But during the course of telecommunications deregulation, users were allowed to buy and install their own CSUs.

One of the principal functions of a CSU is to protect the carrier and its customers from any weird events your network might introduce onto the carrier's system.

A CSU provides proper electrical termination for the telephone line and performs line conditioning and equalization. It also supports "loopback tests" for the carrier, meaning the CSU can reflect a diagnostic signal to the telephone company without sending it through any CPE, so the carrier can determine if a problem is one it needs to correct itself. CSUs often have indicator lights or LEDs that identify lost local lines, lost telco connections, and loopback operation.

When CSUs were provided only by the telephone company, they were generally powered by the telephone line itself. Nowadays not all telephone lines supply power. So, a CSU may need to have its own power supply, and perhaps a backup power supply, at the user site.

THE DSU'S DUTY

Data service units—sometimes referred to as digital service units—or DSUs, sit between a CSU and customer equipment such as routers, multiplexers, and terminal servers. DSUs are commonly equipped with RS-232 or V.35 interfaces. Their main function is to adapt the digital data stream produced by the customer equipment to the signaling standards of the telephone carrier equipment, and vice versa.

To get a better grasp of a DSU's function, we need to delve a bit into the mysteries of the Physical layer. For those of us who aren't electrical engineers, it's easy to fail to give the appropriate respect to layer 1 and take the attitude "It's just electrons on the wire."

In fact, a large part of a telephone company's investment is devoted to delivering a Physical layer that's compatible with many years of legacy hardware, but also able to interoperate with state-of-the-art components such as fiber optic cable, Switched Multimegabit Data Service, and ATM.

The digital streams produced by many customer devices—especially those with throughput less than 56Kbits/sec—are asynchronous, which means that each byte is distinguished by start and stop bits and that the time interval between bytes is arbitrary. However, the preponderance of customer devices in the public telecommunication infrastructure uses synchronous signaling, in which senders and receivers coordinate local clocks with each other in order to identify the boundaries between units of data.

In this case, the DSU may be called upon to parcel out incoming asynchronous data at the stable rate the carrier line expects, and to wrap start and stop bits around incoming synchronous data before passing it along to the user network.

The signaling techniques of the telephone network are quite different from those used by many customer premises devices. It's more or less natural to think of digital signals as positive voltage for a *1* and a zero voltage for a *0*. This type of signaling is known as *unipolar nonreturn to zero* (see Figure 1A).

There are several objections to a unipolar nonreturn to zero approach from the point of view of the telephone system, even though it works for RS-232 equipment. Unipolar nonreturn to zero signaling tends to build up a direct current (DC) signal component on a line over time. This signal component can be blocked by some types of electrical components—for example, transformers. Furthermore, a more or less random level of DC on the telephone wire interferes with the task of providing power to devices, such as repeaters and CSUs, that derive operating power from the line.

Using a signaling method where each *1* or *0* returns to zero makes it easier to detect the correct digit, but does nothing to compensate for the DC (see Figure 1B). Using a polar signaling method where *1*s are positive and *0*s are negative can diminish the DC buildup (see Figure 1C). But lengthy strings of *1*s or *0*s will still have that result. Like unipolar signaling, returning each bit to zero doesn't solve the DC problem (see Figure 1D).

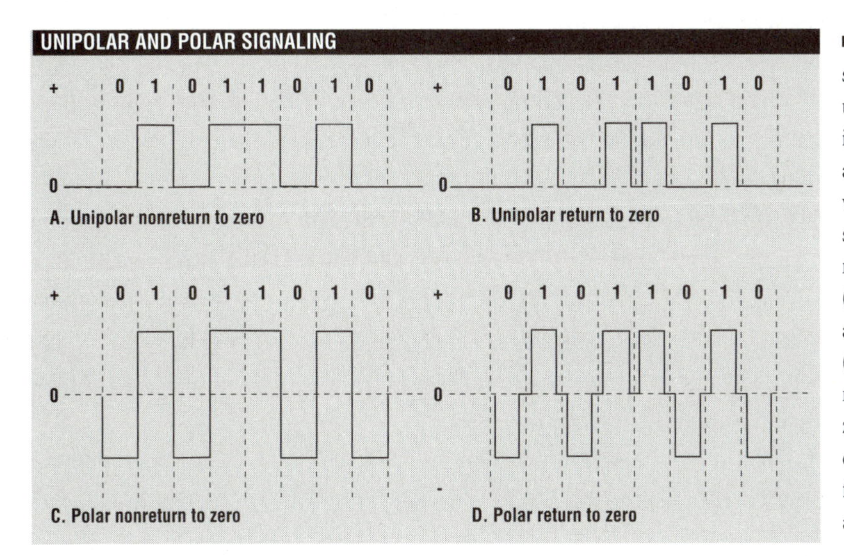

■ **Figure 1:** With unipolar signaling, the mark signal, usually representing a 1, is indicated by a positive voltage or current, while zero voltage or current represents a space or 0. Polar signaling represents marks (or1s) with a positive voltage or current, while spaces (or 0s) are represented by a negative value. Return to zero signaling requires that each mark signal start and finish at zero within the alloted time for a bit.

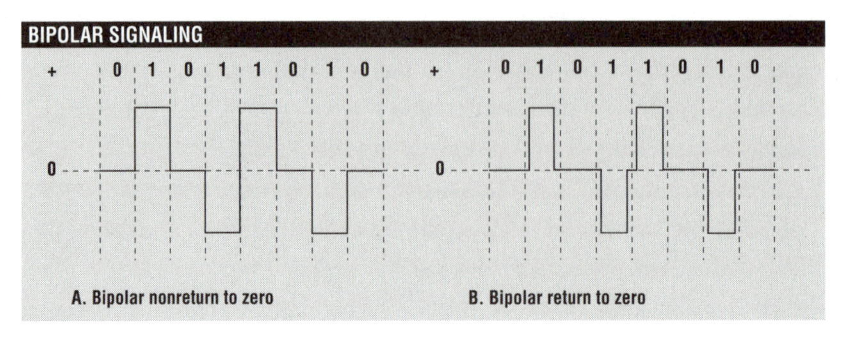

■ **Figure 2:** Bipolar signaling indicates a mark or 1 with alternating positive and negative valves—alternate mark inversion (AMI)—and spaces or 0s are indicated by zero values. Two successive marks with the same polarity constitute a bipolar violation, or BPV.

The solution is to indicate a *0* with no voltage and indicate a *1* with alternating positive and negative voltages (see Figure 2A). No DC voltage can build because any residual charge from a positive-going *1* will be canceled by the succeeding negative-going *1*. Failure of a *1* pulse to have the opposite sign of the preceding *1* can be readily detected by simple circuitry and is known as a bipolar violation (BPV).

By making each bit return to zero, digits are easier to detect (see Figure 2B). In addition, this method allows wider separation of repeaters than other signaling methods. Bipolar signaling is also referred to as alternate mark inversion or AMI. *Return to zero* is often abbreviated RZ or RTZ, while *nonreturn to zero* is NRZ.

KEEPING IT TOGETHER

Another problem with the synchronous telephone network is the need to ensure that synchronization is maintained across any possible circuit. Telephone company devices, including the repeaters that rejuvenate the signal along the way, receive their clock, or synchronization, cues from the stream of bits.

One general rule for interconnecting older devices is that no more than eight bit times should pass without a signal. But it's not too rare for a stream of data to have eight or more successive *0*s and thus provide no heartbeat for these devices. Without synchronization, transmission is impossible.

One solution has been to dedicate one bit out of every eight to a control role. If all the other bits in an eight-bit string are *0*s, then the control bit will be set to *1* and synchronization will be maintained. The need for this control bit is the reason that dataphone digital service (DDS) lines and Switched 56 service, which after all run over 64Kbits/sec DS0 circuits, provide only 56Kbits/sec service to users. The eighth bit keeps the equipment running, so only seven out of eight bits are available to users.

An alternate solution to the synchronization problem, which has been widely adopted since the mid-1980s, involves forcing a deliberate bipolar violation to maintain the clock function while indicating that the BPV pulses are not to be interpreted as data. This technique, *binary eight zero substitution*, is called B8ZS. DS0 circuits that support B8ZS from end to end can carry 64Kbits/sec of user data.

Digital signals sent over telephone lines also need to be framed. Framing makes it possible to maintain the separation of multiple data streams that have been merged into a single circuit, with each stream getting its own time slot. Special overhead bits serve to demarcate the beginning of a frame. For example, on a DS1 (or T1) circuit, every 193rd bit is a framing bit, followed by 24 DS0 segments of eight bits each.

In light of these functions—AMI signaling, zero substitution, and framing requirements—we can return to the DSU. The role of the DSU is to convert the unipolar digital signal from the user device to one with the properties demanded by the telephone circuitry.

MIXING THE TWO

DSUs are often built into other devices, such as multiplexers or channel banks, and are often combined with a CSU to form a single unit referred to as a CSU/DSU or a DSU/CSU. CSU/DSUs may have built-in compression, and they may include analog or ISDN dial-up ports for backup.

Roughly speaking, a CSU has a similar role to that of an NT1 on ISDN lines, while a DSU is comparable to an ISDN terminal adapter. Sometimes CSU/DSUs (and ISDN terminal adapters) are described as "digital modems." This terminology is misleading and crude because CSU/DSUs don't modulate or demodulate anything.

Modulation is the process of using one signal (either analog or digital) to modify another analog signal, known as the carrier. The three fundamental ways a carrier can be modulated are known as amplitude modulation, frequency modulation, and phase modulation. For AM (amplitude modulation) broadcast radio, an analog audio signal modulates a carrier in the range between 540KHz and 1640KHz. For FM (frequency modulation) broadcast radio, an analog audio signal modulates a carrier in the range of 88MHz to 108MHz. Broadcast TV modulates a carrier using AM for the luminance or black and white video, using FM for the sound, and using a form of phase modulation for chrominance or color information.

Modems modulate an analog audio carrier that can pass through the POTS analog telephone line with digital signals. The modulation technique used by modern modems is a sophisticated combination of phase and amplitude modulation.

When CSU/DSUs are used on switched lines, they need to signal the destination of the call to the telephone company switch. In these cases, it sometimes makes sense for the DSU to use the AT command set originally employed by Hayes modems because software developers are familiar with the commands. Other than this minor overlap, and the fact that modems sit between a serial interface and the telephone network, the functions of modems and CSU/DSUs are totally different.

A modem is designed to transform digital signals into analog audio signals that fit into a circuit designed for voice transmission. CSU/DSUs adapt one kind of digital signaling to another type that is capable of fitting into the digital telephony system.

This tutorial, number 92, by Steve Steinke, was originally published in the April 1996 issue of LAN Magazine/Network Magazine.

The ISDN Connection

To really understand the Integrated Services Digital Network (ISDN), you need to put it into context. You need to see what's come before it to more fully comprehend what the new technology brings. This tutorial briefly reviews the existing conventional telephone system before discussing how ISDN departs from the norm.

The conventional telephone system—often referred to by the acronym POTS (meaning plain old telephone system)—is an analog system. The pattern of voltage variations on the local telephone line is a direct analog of the acoustical pressure variations that produced them.

The telephone line in your home is typically a single unshielded twisted pair (UTP) of copper wires that runs to the telephone company's central office (CO). This line is usually referred to as the subscriber loop or local loop.

The central office is a key concentration point for telecommunications. Inside the central office is a telephone switch that connects your line with that of the person you want to call, establishing the circuit.

ISDN SPELLED OUT

What's different about ISDN compared to POTS? First, it's digital instead of analog. Second, the "integrated services" portion of the name is a reference to one of the basic goals of ISDN—to provide integrated, multipurpose voice, data, and video communications, all over a single system. Contrast that with today's norm, where you typically need three services to provide the same communications ability: dial-up analog phone lines to carry voice, digital leased lines for data, and coaxial-cable networks (cable TV) for video. The idea that one network can handle all communications types is obviously appealing.

There are two levels of ISDN service. The Basic Rate Interface (BRI) is lower in cost than the other option, and it can run over typical residential subscriber loops without rewiring. BRI consists of three separate channels—two bearer channels (B channels) and one data (or

D) channel. Each B channel has a data rate of 64Kbits/sec, while the D channel is 16Kbits/sec. The D channel is used for signaling, such as call setup and tear-down. The B channels carry user data, which might be digitized voice, binary data, or digitized video.

The Primary Rate Interface (PRI) might be used at a business headquarters site. It consists of 23 64Kbits/sec B channels, plus one 64Kbit/sec D channel. Again, the B channels carry the payload (voice, data, or other user information), while the D channel carries telco signaling. To get PRI service, you'll need a T1 line from your site to your local exchange carrier's CO.

In Europe, the commonly used digital data line that's roughly equivalent to the T1 line used in North America is known as E1, and it has a higher capacity than the T1—2.048Mbits/sec vs. 1.544Mbits/sec. European ISDN PRIs take advantage of the increased capacity and carry 30 64Kbit/sec B channels, plus one 64Kbit/sec D channel.

With POTS, it's a given that one phone line lets you make only one call at a time. If you're on the line when other people try to call you, they'll get a busy signal and will have to wait until you've finished your call before they can get through. With ISDN, it's a little different. With an ISDN BRI, you get two phone numbers—one for each B channel. Consequently, you could make two separate phone calls to different destinations, even though you've got only a single phone line.

PRI B channels are the same as BRI B channels; you can place a call from a BRI to a PRI, or vice versa, linking one of the BRI's B channels to one of the PRI's B channels. With 23 B channels, you could simultaneously have 23 separate ISDN calls going out over a single Primary Rate Interface to the telco, even though each of these calls has a separate destination.

ISDN is a digital, synchronous, full-duplex link. When voice is carried over ISDN, it's digitized at the telco standard of 8,000 eight-bit samples per second. This analog-to-digital conversion is performed by a compression/decompression (codec) device in the ISDN phone. If a conventional analog phone is used, it must be adapted to the network by an ISDN terminal adapter (TA). In this case, the codec is part of the TA.

The sampling rate and sampling precision is the reason 64Kbits/sec was chosen for bearer channel data rates. Eight bits sent 8,000 times per second equals 64,000 bits per second.

ISDN'S BIG PICTURE

The block diagram in the figure shows how user equipment interfaces with the telco-provided ISDN BRI connection. A key feature of the BRI is that it can run over most existing subscriber loops, which are two-wire (single-pair) UTP. On the left side of the diagram (everything left of the letter U) is the customer premises, while the right side is the telco side, including the central office ISDN switch.

The telcos don't want customer equipment to connect directly to their networks, for fear of faulty or poorly designed equipment disrupting operation of the network. So the telcos interpose a device known as a network terminator (NT1) between the subscriber loop and customer equipment.

The customer's equipment is referred to as terminal equipment (TE), and there are two classifications of these: TE1 and TE2. TE1 equipment is digital equipment that's ISDN-capable. An ISDN phone is one example of a TE1 device. A TE1 can plug directly into the NT1.

A conventional analog phone or an ASCII data terminal (or the RS-232 serial port on a computer) represents equipment that's not ISDN-ready. Such terminal equipment is classified as TE2. These devices must have an ISDN terminal adapter interposed between them and the NT1. The terminal adapter makes the TE2 equipment ISDN-capable.

If a customer premises switch, also called a PBX, is used to connect an ISDN line to the terminal equipment or terminal adapter, it will reside between that equipment and the NT1. PBXs are classified as NT2 devices.

The figure shows an example of a TE1 and TE2 connected, through a PBX, or NT2, to the NT1 and the subscriber loop. For the TE2 connection, the setup calls for a terminal adapter. At the other end of the subscriber loop is the telco's central office switch. Various reference points have been defined by the standards, and these are shown in the figure as the letters R, S, T, and U.

In the United States, the demarcation point between the telco's network and the customer premises is located at the customer end of the subscriber loop—the point marked U in the figure. Martha Haywood, director of engineering at Telebit (Sunnyvale, CA), a manufacturer of both ISDN- and analog-based remote LAN access equipment, reports that in Europe, the NT1 is traditionally owned by the telecommunications provider, even though it's installed on the customer's premises. Therefore, it is part of the telecommunications network. Because of this, the International Telecommunications Union (ITU) standards body recognizes the T reference point as being the demarcation between the customer's and the telecommunication provider's respective portions of the network.

■ An ISDN network terminator (NT1) must be placed between the telco's network and customer equipment. Customer equipment (also called terminal equipment or TE) is shown on the left side of the figure. TE1 represents ISDN-capable terminal equipment, while TE2 is non-ISDN equipment, for which an ISDN terminal adapter (TA) is required. Some sites will have a PBX switch on the premises, as shown in NT2.

CONNECTING TERMINAL EQUIPMENT TO ISDN

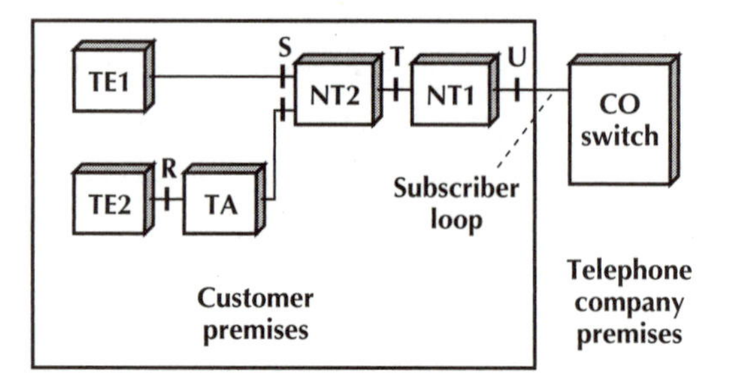

A customer premises switch (NT2) will most likely be used with PRIs, rather than BRIs, but it's shown in the figure to indicate its place in the overall scheme of things. On most BRIs, TE1s and TAs will be plugged directly into the NT1. The S and T reference points are thus essentially interchangeable; the switch merely serves as a line selector.

Some manufacturers of ISDN terminal equipment and terminal adapters incorporate an NT1 right into the product. In this case, the device is said to have a U interface, since it can plug directly into the subscriber loop. Having a built-in NT1 means there's one less thing to buy, but Telebit's Haywood points out that there can be drawbacks, too. For example, the ISDN standards permit up to eight TEs to be connected to a single NT1. (This doesn't mean that all eight can be active at the same time; only one TE at a time can use a B channel. Thus, a BRI can support only two calls at a time.)

In the case of the BRI, which has two B channels, a terminal adapter and an ISDN phone could both be online at the same time, with each device using one of the B channels. But if the terminal adapter had a built-in NT1, you couldn't plug in another device—unless the designers of the terminal adapter had the foresight to put in a port for a phone—because the subscriber loop itself cannot be shared. In other words, there can be only one NT1 per BRI.

Haywood says that initially, there may be strong interest for products with built-in NT1s, but as customers gain experience with ISDN, they'll begin to see the advantage of keeping the NT1 as a separate device.

RESOURCES

There's a wealth of information available on ISDN. If you're interested in delving further into the subject, refer to the following sources:

Bell Communications Research (Bellcore), the research arm of the Regional Bell Operating Companies (RBOCs), maintains a World Wide Web site that contains a great deal of information compiled by the National ISDN Users Forum (NIUF). This information is in encapsulated PostScript (EPS) format. To access it, set your Web browser to: http://info.bellcore.com.

Pacific Bell's Market Applications Lab (510-823-1663) can answer questions about the suitability of ISDN for your application, and what types of equipment you should consider.

Pacific Bell Applications BBS (510-277-1037) has an ISDN tutorial that can be downloaded.

For more information, try the following books on ISDN:

Digital Telephony and Network Integration, Second Edition, by Bernhard E. Keiser and Eugene Strange, Van Nostrand Reinhold, New York, ISBN 0-442-00901-1.

ISDN, Second Edition, by Gary C. Kessler, McGraw-Hill, New York, ISBN 0-07-034247-4.

ISDN and Broadband ISDN, Second Edition, by William Stallings, Macmillan, Englewood Cliffs, NJ, ISBN 0-02-415475-X.

Telecommunications, Second Edition, by Warren Hioki, Prentice Hall, Englewood Cliffs, NJ, ISBN 0-13-123878-7.

This tutorial, number 82, by Alan Frank, was originally published in the June 1995 issue of LAN Magazine/Network Magazine.

Inverse Multiplexing

IT shops are scrambling to accommodate the convergence of data and the WAN. LANs pump more and more bandwidth to the WAN's edge, but few network engineers have optimized their systems to a WAN environment, where bandwidth is limited and you pay for every transmission.

The traditional WAN solution was leased from one of the phone companies. Leased services, although they provide dedicated bandwidth, are expensive. Depending on the distance between the two points, T1 lines can run you a five-figure bill per month.

In response to the constricted economy and the advent of LANs, WAN-service providers developed options more attractive to data users both in price and technology. The Regional Bell Operating Companies, interexchange carriers, and value-added network providers are heavily promoting these services, including frame relay, ISDN, and SMDS.

But you don't have to venture into a new and unproven technology to get better service; you can recycle what you already have. The public-switched network offers time-proven service, and the prices are low. A 56Kbit/sec data call now costs about the same as a voice call.

Bandwidth on demand dictates that users can access dial-up lines to call up additional WAN bandwidth. So when a LAN application bursts and more data needs to be sent than there is pipe to send it through, another switched-service line can automatically be called into action.

From bandwidth on demand comes inverse multiplexing.

MORE ELEGANT THAN ITS NAME

What inverse multiplexing accomplishes is more elegant than its moniker suggests. Ordinary multiplexing takes multiple low-speed lines and puts them onto one high-speed line. Inverse multiplexing spreads a high-speed output over multiple low-speed (and presumably lower-cost) lines, while maintaining the appearance of a high-speed transmission. Hence the name "inverse."

Inverse multiplexing works as follows. A T1 line is made up of 24 channels, each running at 64Kbits/sec, for a total throughput of 1.544Mbits/sec. Each channel only has 56Kbits/sec of throughput, because 8Kbits/sec is necessary for in-band signaling. The key is that a T1 line consists of 24 separate channels.

For example, you can inverse multiplex two 56Kbit/sec lines into what appears to the application to be one 112Kbit/sec pipe. (The number and speed of lines is indicated using Nx, where the "N" denotes the number of lines and the speed follows the "x." So two 56Kbit/sec lines are referred to as 2x56.) Inverse multiplexing typically uses multiple 56Kbit/sec leased or switched lines, although 64Kbit/sec lines can be used internationally. You can also use ISDN Basic Rate Interface, ISDN Primary Rate Interface, ISDN HO Switched 384, or full T1 (see Figure 1).

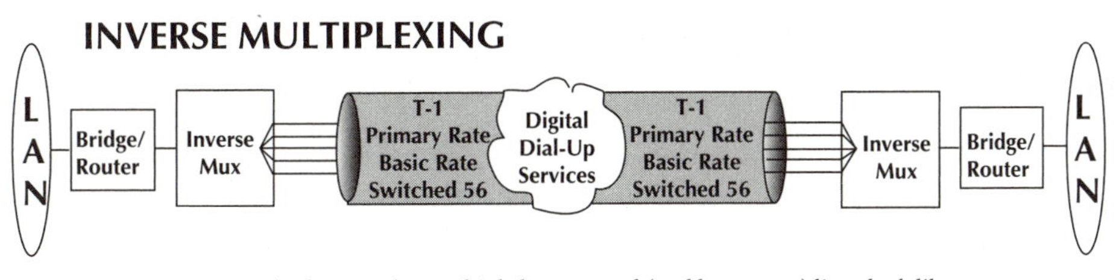

■ **Figure 1:** An inverse multiplexer makes multiple lower-speed (and lower-cost) lines look like a single higher-speed (but still lower-cost) line to the applications at either end.

The benefit is that you are buying low-cost, low-speed lines and ratcheting up the speed yourself. With inverse multiplexing, you can get the bandwidth equivalent of a T1 but at a much lower price. You can buy fractional T1 today, but the applications view the channels as separate.

Inverse multiplexing will be most beneficial if you are trying to internetwork several LANs, but it is also applicable for videoconferences, traffic overflow, and disaster recovery.

INSERT THE INVERSE MUX

Figure 2 shows how an inverse multiplexer works. Multiple high speed data streams are input into the inverse mux. These data streams can come from the different ports on the same device, perhaps a router, or they may come from different sources, perhaps a router, a terminal server, and a video codec. The inverse mux uses time-division multiplexing to split the data stream into multiple 56Kbit/sec channels.

The separate 56Kbit/sec channels take diverse paths through the public switched network—unbeknown to the application. For example, Figure 2 shows that the "abcdefghijklmnop" message is broken into four streams: "aeim," "bfjn," "cgko," and "dhlp." The "aeim" stream goes from the switch in San Francisco to Chicago and New York before it reaches Washington, DC. The "cgko" packet goes to Los Angeles and Atlanta on its way to the

INSIDE INVERSE MULTIPLEXING

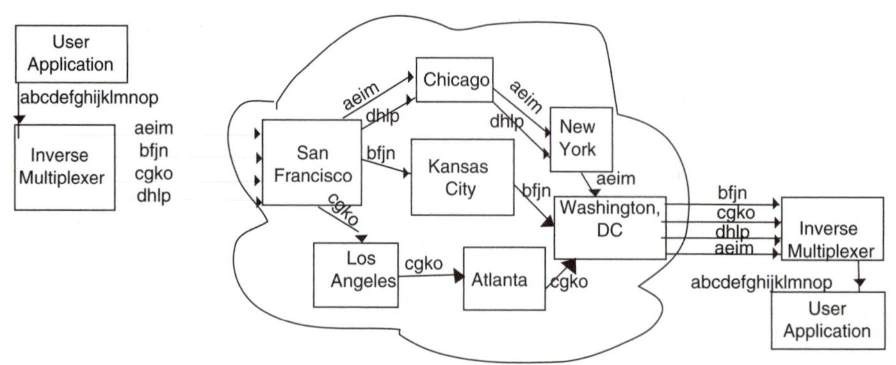

■ **Figure 2:** An inverse multiplexer aggregates different data streams over higher-speed lines. The seperate channels take diverse paths through the network and arrive at their destination but not necessarily at the same time or in the right order. The inverse mux puts the packets back in the proper order and adjusts for any delay.

District. The four data streams arrive at the same destination but not necessarily at the same time or in the right order. Then, like a router, which segments and reassembles packets, the inverse mux buffers the arriving packets and puts them in the proper order. It then passes the intact information to the router, then the user's application.

A STANDARD METHOD

Inverse multiplexing improves on a feature some routers offer—load balancing or load sharing. With load sharing, the router tries to balance the traffic across multiple outgoing ports; however, load sharing adds about 30 percent overhead to the transmission. Also, each vendor's implementation is unique, so one vendor's implementation will not work with another's. You're locked into a proprietary solution.

Inverse multiplexing accomplishes the same function but uses a standard called Bonding. This specification was written by the Bandwidth on Demand Interoperability Group, a consortium of 40 manufacturers, and has been passed to the American National Standards Institute TR41.4 group.

Bonding supports four modes. Mode 0 is a special mode for dual 56Kbit/sec calls, which offers special economies when both sides are using 56Kbit/sec lines. Mode 1 works with any number of lines (called Nx) and has no subchannel. The subchannel can be used for control signaling and other types of information. Mode 2 is also Nx but has an in-band subchannel overhead of 1.6 percent. Finally, Mode 3 is also Nx with an out-of-band subchannel.

Ascend, a leading manufacturer of inverse muxes, has its own protocol— Ascend Inverse

Multiplexing (AIM)—which it recommends for higher performance. AIM supports four modes: Static, which has no subchannel; Manual, which has an in-band subchannel overhead of 0.2 percent; Delta, which is an out-of-band subchannel; and Dynamic, for dial-up bandwidth.

The ISDN H.221 specification can also be used for inverse multiplexing, but it is primarily used for video, not LAN, data.

IMMEDIATE GRATIFICATION

Inverse multiplexing offers a way to "roll your own" dial-up bandwidth service immediately. You don't have to wait for the phone companies to tariff new service offerings.

Later this year, the WAN-service providers will step up their offerings of commercial services that will deliver the equivalent of inverse multiplexing. For example, AT&T currently offers Switched 384, which provides the same bandwidth as six 64Kbit/sec lines and an inverse mux. Yet 384Kbits/sec is a rather unwieldy chunk of bandwidth and so less attractive to LAN users.

The carriers may make inverse multiplexing less attractive in another way. When you dial up additional lines using an inverse multiplexer, you are charged for each call setup. So if you dial up six 64Kbit/sec lines, you are charged for six separate calls. If you buy Switched 384 or similar service, you are charged for only one call.

This tutorial, number 55, by Patricia Schnaidt, was originally published in the April 1993 issue of LAN Magazine/Network Magazine.

Always On/Dynamic ISDN (AO/DI)

When you need a wide area link that's faster than an analog telephone line, what comes to mind? Many think of ISDN in its Basic Rate Interface (BRI) incarnation as the next step up from the Plain Old Telephone System (POTS). Unlike leased lines, the endpoints of which are fixed, ISDN (like POTS) is a switched-circuit technology, so you can place a call to anyone who has an ISDN telephone number. (As a matter of fact, you can call POTS numbers from an ISDN number, and vice versa, but this is a least common denominator situation, so you have to settle for analog line performance.)

The classic ISDN BRI is divided through Time Division Multiplexing (TDM) into three independent channels: one 16Kbit/sec Data (D) channel, and two 64Kbit/sec Bearer (B) channels. The D channel is used by the telco for signaling, and the B channels carry user data. When you dial a number on an ISDN line, the D channel handles the linking of one, or both, of the B channels to the desired destination telephone number; this is known as call setup. Similarly, when you hang up and terminate the call, the D channel is used for call tear-down.

WHAT'S NEW

A new technology twist that ISDN equipment vendors and service providers are now unveiling allows for some user data to be carried over the D channel. Unlike the B channels, which are operational only during the course of a call, the D channel is always active, and always connected to the switch at the telephone company's Central Office (CO). Always On/Dynamic ISDN (AO/DI) takes advantage of this feature. Under AO/DI, a portion of the D channel is operated as a link in an X.25 packet-switched network. The Point-to-Point Protocol (PPP) is used over this link. Of the D channel's 16Kbit/sec total capacity, 9.6Kbits/sec is devoted to AO/DI user data traffic; the balance of the link continues to be used for telco signaling (call setup and tear-down, for example).

At the telco's CO, user data packets sent across the D channel link are stripped off by packet handlers placed in front of the ISDN switch; the data packets are then sent across an X.25 network to their final destination.

One frequently cited application for AO/DI is Internet access. Packets ride the D channel to the telco's CO, and are then put onto the X.25 packet network to be forwarded to the ISP. The ISP, in turn, routes the packets to the Internet.

EXPANDING BANDWIDTH-ON-DEMAND

Under AO/DI, the D channel plays a role in some of the bandwidth-on-demand features that ISDN devices are noted for. If there is a minimal amount of traffic (that is, if there are a small number of packets to be sent), the B channels will be inactive, but the D channel will still be carrying packets. If the amount of data waiting to be sent across the 9.6Kbit/sec D channel exceeds 7500 bytes, a circuit-switched B channel link will be established, providing a throughput of 64Kbits/sec. At this point, the D channel will be idled for user data purposes; it will continue to be used for telco signaling. If utilization of the B channel exceeds a preconfigured threshold (typically 70 percent utilization), a second B channel will be established, providing a 128Kbit/sec data transfer rate.

If utilization of the 128Kbit/sec link should fall below some preset percentage, one of the B channels will be dropped, so the link throughput will fall back to 64Kbits/sec. If traffic falls further, the remaining B channel will be dropped, and the ISDN device will revert to using the 9.6Kbit/sec D channel.

KEEP IN TOUCH

Those of us who have a T1, or even a mere 56Kbit/sec leased-line, connection to the Internet know how useful it is to be constantly connected, and how frustrating it can be to have to wait for a modem to dial and establish a connection. The connection time for an analog modem operating on a POTS line ranges from about 10 seconds to about 30 seconds. ISDN cuts the connection time to a second or two, but ISDN with AO/DI eliminates it entirely.

At a time when users consider a 14.4Kbit/sec modem to be obsolete, a 9.6Kbit/sec communications channel may seem so slow as to be not worth the bother, but this is not necessarily the case. A low data-rate connection that's always available can enable some applications that just aren't feasible (or as useful) with occasionally-connected links. As one example, burglar and fire alarm companies could use the 9.6Kbit/sec X.25 link to poll alarm sensors at the customer's site—without the expense of placing a B channel call. It may require only one packet to poll a sensor, and one packet for the sensor's response, so this can be a very low data-rate application. Stock quotes and news headlines are also examples of information that can be pushed to user's computers at a data rate that's a mere trickle.

It's also possible to move data in the other direction: Utility companies are able to eliminate the expense of sending meter readers to each home and business if the meters are outfitted with a digital interface to the ISDN D channel, and retailers can use the D channel for credit card authorization.

Small business users and telecommuters that need electronic mail often depend on their ISP to run the mail server. If these customers are connected with analog lines, or even with non-AO/DI ISDN lines, they will have to periodically call up to see if they have new messages. With AO/DI, users could be notified automatically when they get a new mail message. The notification would come in the form of a packet (or several) that is sent across the X.25 D channel link. If the message in question is a short text file, it may be possible to download it without crossing the threshold set for the 9.6Kbit/sec D channel.

Development of AO/DI technology has been shepherded by the Vendors ISDN Association (VIA), a San Ramon, CA-based trade association. VIA members consist primarily of ISDN equipment providers. The organization has been putting together the proposed specification for AO/DI.

Clearly, for AO/DI to work, ISDN equipment vendors and service providers must support it. Bell Atlantic, BellSouth, and SBC Communications (the parent company of Southwestern Bell and Pacific Bell) have announced plans to support AO/DI in the first quarter of 1998. Equipment vendors who have announced their intention to support it include 3Com, Adtran, Arescom, Ascend Communications, BinTec Communications, Cisco Systems, Digi International, ECI Telecom, Eicon Technology, ITK Telecommunications, Jetstream Communications, Shiva, and Virtual Access. Release dates for products range from the first quarter of 1998 through the third quarter of 1998, depending on the vendor and product.

Eicon Technology is already shipping a product. Eicon's Diva T/A ISDN terminal adapter is an external terminal adapter that supports AO/DI. It also has Auto-SPID and SPID-Wizard features, which automate the task of getting an ISDN device connected and talking with the telco's CO switch.

If you're planning to take advantage of AO/DI for your Internet connections, your telecommunications carrier and your ISP must both support AO/DI. The telco will need to strip packets off the D channel and put them onto the X.25 network. Whichever ISP you use will also have to have a connection to your telco's X.25 network to receive and send your packets.

ISPs use high-density ISDN routers that let them terminate a large number of B channels on a single chassis. Ascend Communications has a big chunk of the service provider market with its Max product. Mike Baccala, senior technical manager for product engineering at Ascend, says the company will offer support for AO/DI in the "first part of the second quarter of 1998." It will be a free firmware upgrade for existing equipment, he says.

Telco pricing for X.25 packet data over the D channel is still in flux, but figures around $5 per month over the price of regular ISDN service are being discussed; this could make it quite attractive for users.

AO/DI's attraction for telcos is that it could dramatically lower B channel usage. Telcos base their projections of how many switch ports will be needed for each local exchange upon the number of users in that exchange, as well as upon the assumption that each line will be used only occasionally. This is a reasonable assumption for voice calls, but Web surfers can be connected for long periods of time.

Many telcos price ISDN using a pricing model that's similar to analog lines: local calls are not charged over and above the monthly rate for service. This has led most ISPs to place a POP in each local exchange in which they have customers. Subscribers to the ISP's service can place local calls and remain connected to their ISP's POP for long periods of time, without incurring significant per-minute charges, so there's little incentive for users to keep their connect-times short. AO/DI can give users constant connectivity, while simultaneously freeing up B channels for data transfers that truly require B channels' higher throughput.

Because the D channel X.25 packets are handled at the CO by the X.25 packet handler, it's possible to route these packets without first crossing the circuit-switched fabric of the CO's switch. This reduces the impact on the telephony network because, from the switch's viewpoint, there's no activity until a B channel is raised.

READY WHEN?

When will AO/DI be available to you? It's somewhat difficult to predict whether the ramp-up for AO/DI will match ISDN's initial slow, painful birth and development, or whether it will all happen quickly and smoothly. From the equipment providers' standpoint, support for AO/DI appears to be coming quickly. It's also possible that much existing ISDN equipment can be upgraded by loading new software—another spur to rapid deployment. We're not talking about AO/DI competing with ISDN; it's simply an extension to ISDN's existing feature set. As such, AO/DI can leverage ISDN's hard-won gains in market development and telco deployment.

Interoperability between different vendors' equipment will also be a key issue. VIA has already sponsored an early interoperability test, but deployments of production units into the field will be the litmus test.

The critical path in AO/DI deployment and usage is likely to be telco and ISP support for it. Although you may have customer premises equipment that's AO/DI-enabled, your local telephone company will need to offer support for AO/DI before you can use it. According to Ascend's Baccala, the telco's CO switch must be AO/DI-enabled. However, he says, "all the switches are already X.25-capable, so this is not a big deal."

In addition, your ISP will need ISDN routers or concentrators that support AO/DI, as well as have the necessary connection to your telco's X.25 network.

RESOURCES

If you'd like to explore Always On/Dynamic ISDN (AO/DI) in more depth, check the Vendors ISDN Association (VIA, San Ramon, CA) Web site at www.via-isdn.org. VIA maintains an FTP site that contains, among other things, the proposed RFC for AO/DI. Look for the document named aodirfc.doc.

For more ISDN information, you can also surf to these Web sites:

European ISDN Users Forum www2.echo.lu/eiuf/en/eiuf.html

North American ISDN Users Forum www.niuf.nist.gov

Dan Kegel's ISDN Web Page www.alumni.caltech.edu/~dank/isdn

This tutorial, number 117, by Alan Frank, was originally published in the April 1998 issue of Network Magazine.

ADSL and SDSL

The best technologies solve problems with existing equipment. Residences shackled with 56.6Kbit/sec dial-up lines want access to broadband applications, and businesses with expensive T1/E1 connections to the Internet want some budget relief. Where available, enter Digital Subscriber Line (DSL).

DSL connects the customer premises to a service provider's Central Office (CO) using existing copper telephone lines. When complications don't render lines ineligible, DSL modems can boost a poky 56.6Kbit/sec dial-up connection to speeds that meet and surpass 1.54Mbits/sec. However, DSL's major drawback is that availability is contingent upon distance from the service provider's CO.

DSL is not a one-size-fits-all technology—there are many, many sizes, though they won't all be in stock in any given area (see "Key Players in the Extended Family" at the end of this tutorial). Varieties of DSL generally fall along one of two main distributions, though all differ in specifics. The two primary models—Asymmetric DSL (ADSL) and Symmetric DSL (SDSL)—stem from an early fork in the road: Asymmetry favors downstream data flow, while symmetry maintains equal upstream and downstream speeds.

ADSL has found a home in residential circles, and SDSL has made friends among the suit-and-tie crowd. Both systems possess strengths and limitations that directly trace back to their fundamental differences in symmetry.

THE SHARING SIDE OF ASYMMETRY

ADSL has made considerable inroads to the residential market, where it competes with cable modems for customers starved for high-bandwidth connections. Remarkably tailored to the appetite of the Web-surfing home user, ADSL delivers between 384Kbits/sec and 7.1Mbits/sec of data in the downstream direction. Upstream throughputs hit 128Kbits/sec to 1.54Mbits/sec.

An asymmetrical model complements the residential profile of Internet use: Masses of multi-media and text course downstream, and undemanding levels of traffic make their way upstream. Costs for ADSL generally scale from $40 to $200 per month, depending on expected data rates and service-level guarantees. Cable modem services often cost less, at an average price of $40 per month, but lines are shared among customers, not dedicated like DSL.

Carrier loops accommodate ADSL alongside analog voice service by assigning digital signals to frequencies above the voice audio spectrum (see Figure 1). To do this, a splitter must be installed. A splitter uses a low-pass filter to separate telephone frequencies on the low end of the audio spectrum from the higher frequencies of ADSL signals. Bandwidth available to ADSL remains unaffected whether or not analog voice frequencies are in use.

Splitters are required at both the customer premises and the CO for full-rate ADSL. They generally do not require power, and therefore do not obstruct "lifeline" voice service in the event of a power outage.

ADSL speeds are not an exact science, although they scale downward at somewhat predictable intervals. Service providers deliver a "best effort" service whose results depend heavily on distance from the CO. "Best effort" generally means that providers only guarantee about 50 percent bandwidth. Attenuation and interference factors such as crosstalk prey on the lines after 10,000 feet, and can render them unsuitable to data transfer at distances greater than 18,000 feet.

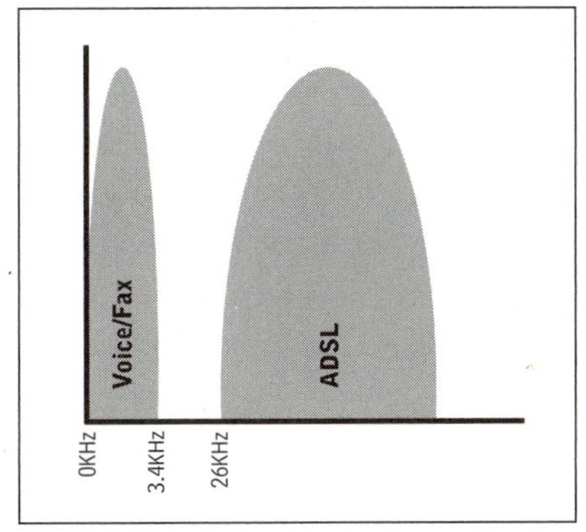

Source: ©2Wire, 1999

■ **Figure 1:** Asymmetric Digital Subscriber Line (ADSL) sends data at frequencies of 26KHz to 1100KHz, maintaining analog voice service on the same copper wire in the 0KHz to 3.4KHz range. Symmetric DSL (SDSL) uses the entire frequency of the line for data and does not coexist with analog voice service.

ADSL speeds can reach up to 7.1Mbits/sec downstream and 1.5Mbits/sec upstream at distances up to 12,000 feet from the CO. However, DSL Reports (www.dslreports.com) business editor Nick Braak says not to expect the upper limit. "It's virtually impossible to get 7.1Mbits/sec, even in a lab," says Braak. ADSL gets reined back to rates of 1.5Mbits/sec downstream and 384Kbits/sec upstream after 10,000 feet; as the line approaches 18,000 feet, speeds drop yet further—384Kbits/sec downstream and 128Kbits/sec upstream.

Service contracts for ADSL may prohibit the use of home networks or any use of Web servers. However, DSL technology does not prohibit operability with household LANs. For example, even if the provider restricts the customer to a single IP address, Network Address Translation (NAT) enables multiple users to share that one IP address.

One DSL connection is sufficient for a home full of computers. Some DSL modems will include an Ethernet hub, and dedicated devices known as "residential gateways" act as bridges between the Internet and home networks.

ADSL is available in two modulation schemes: Discrete Multitone (DMT) and Carrierless Amplitude and Phase (CAP).

DMT slices available frequencies into 256 channels of 4.3125KHz each, within the range of 26KHz to 1100KHz (as noted in Figure 1).

THE COPPER BONE CONNECTS TO THE ATU-R BONE

So there's the CO, the twisted-pair copper, and the remote location. What connects to what?

At the customer site is a device known as the ADSL Transmission Unit-Remote (ATU-R). Though originally specified for ADSL, ATU-R now refers to the remote unit for any DSL service. In addition to providing DSL-modem functionality, some ATU-Rs can perform bridging, routing, and Time Division Multiplexing (TDM). An ADSL Transmission Unit-Central Office (ATU-C), located at the CO, terminates the other side of the copper line and coordinates the link on the CO side.

The DSL provider feeds multiple DSL loops into a high-speed backbone network by means of a DSL Access Multiplexer (DSLAM). At the CO, the DSLAM aggregates data traffic from several DSL loops and feeds that traffic to the service provider backbone, which sends the traffic to destinations elsewhere on the network. Typically, the DSLAM connects to an ATM network with PVCs established to ISPs and other networks.

G.LITE: ADSL, HOLD THE SPLITTER

A modified form of ADSL, known as G.lite, eliminates the necessity of a splitter at the customer premises.

G.lite throughputs fall short of full-rate ADSL speeds, though they still put 56.6Kbit/sec dial-up modems to shame. Bandwidth capabilities shrink as a result of potentially increased noise interference, and remote management introduces higher levels of interference. (Types of interference that besiege data transmission are discussed in the section "Cut Out That Racket.")

Using DMT, the same modulation technique used in ADSL, G.lite offers maximum downstream full-rate speeds of 1.5Mbits/sec and upstream rates of 384Kbits/sec.

The ITU's G.992.1 recommendation, also known as G.dmt, was published in 1999, along with G.992.2, or G.lite. G.lite became available in 1999 and costs less than ADSL in large part because it doesn't require a "truck roll," the industry term for a service technician visit to a customer site for installation or troubleshooting. Service providers find it difficult to justify hundreds of dollars in labor for a single $49-per-month residential connection, and any modification that reduces costs receives their undivided attention.

BUSINESS-GRADE DSL

An enterprise has distinct needs from a home user, and SDSL, operating on a symmetric model, lends itself naturally to office applications.

Corporate upstream capacity can fill up quickly due to heavy Web-server traffic and the resource demands of employees who send large volumes of PDFs, PowerPoint presentations, and other documents. Outgoing enterprise traffic can equal, if not exceed, incoming traffic. Furnishing downstream and upstream data rates of 1.5Mbits/sec in North America and 2.048Mbits/sec in Europe, SDSL resembles the T1/E1 connection that pervades enterprise network architectures worldwide.

Where ADSL utilizes unoccupied frequencies and averts conflict with analog voice frequencies, SDSL takes over the whole line. SDSL eliminates analog voice capabilities in favor of full-duplex data transmission. No splitter, no analog voice—nothing but data.

A viable alternative to T1/E1 pipes, SDSL has gotten a fair amount of attention from Competitive Local Exchange Carriers (CLECs) seeking to woo a business market ripe for value-added services. SDSL service as such is almost always marketed by CLECs. However, HDSL is commonly employed by ILEC to implement T1 service.

Under optimal conditions, SDSL rivals T1/E1 data rates and triples ISDN speeds (128Kbits/sec) at its furthest reaches. Figure 2 depicts the distance/speed relationship of SDSL. Data rates taper off in direct proportion to distance from the CO; they also vary depending on the vendor.

SDSL distance from the Central Office (CO)	Maximum data rate
Up to 10,000 ft.	1.5Mbits/sec
11,000 ft. to 12,000 ft.	1Mbit/sec
13,000 ft. to 15,000 ft.	784Kbits/sec
16,000 ft. to 18,000 ft.	416Kbits/sec

Source: DSL.net

■ **Figure 2:** Symmetric Digital Subscriber Line (SDSL) speeds depend on the distance from the Central Office (CO); they also vary depending on the vendor. SDSL lines are provisioned at a fixed data rate, while ADSL can adapt dynamically to line conditions.

SDSL utilizes the 2 Binary, 1 Quaternary (2B1Q) modulation scheme adapted from ISDN BRI. Each pair of binary digits represents one quaternary symbol. Two bits are sent per hertz (Hz).

SDSL may be better adapted to its business niche than ADSL in the residential domain. Whereas cable modems entice home users with lower prices than ADSL, SDSL keeps pace with T1/E1 data speeds at substantially reduced costs. Typical T1 prices range between $500 and $1,500, depending on distance—the equivalent cost for SDSL ranges from $170 to $450. In the SDSL pricing scheme, the lower end of the spectrum denotes the cost for lower guaranteed data rates.

CUT OUT THAT RACKET

The factors that deteriorate signal quality are numerous and varied, though many are not exclusive to DSL. However, some devices that made life easier in the purely public-switched past have come to haunt present-day DSL deployment.

Crosstalk. Electrical energy radiating from bundles of wire converging at a service provider's CO produces an inconvenient disturbance known as Near-End Crosstalk (NEXT). When signals wander between channels of different cables, line capacity takes a dive. "Near end" specifies that the interference derives from an adjacent pair of cables at the same location.

Segregating DSL lines from T1/E1 lines significantly reduces the negative effects of crosstalk, but there is no guarantee that a service provider will apply that design principle.

NEXT has a related counterpart, Far-End Crosstalk (FEXT), an interference phenomenon whose source originates from another pair of cables at the far end of the connection. With regard to DSL, the degree of FEXT interference is significantly lower than NEXT.

Attenuation. Signal intensities fade as they traverse copper cable, especially signals at high bit rates and higher frequencies. This limitation places a cap on the distances DSL is capable of.

Lower-resistance wire can minimize attenuation, but any given service provider may not deem the expense justifiable. Thick wires have less resistance than thin wires, but they also carry a higher price tag. The most popular cable gauges are 24 gauge (approximately 0.5 millimeter) and 26 gauge (approximately 0.4 millimeter). Due to reduced attenuation, 24-gauge copper cable permits conduction over greater distances.

Load coils. When the PSTN was populated only by voice calls, load coils squeezed more distance out of phone lines—a wholly commendable pursuit. The problem nowadays is that they interfere with DSL.

Sacrificing frequencies above 3.4KHz for improved transmission of frequencies in the voice range, load coils are mutually exclusive to DSL. Prospective DSL subscribers cannot get DSL service until load coils no longer reside on a copper run.

Bridged taps. When the phone company doesn't feel like totally disconnecting an unused section of cabling, it takes a shortcut by installing a bridged tap. These practices never seemed to bother anyone until the rabid demand for DSL emerged. Bridged taps cripple DSL's ability to make use of the line, however, and they must often be removed to qualify the line for DSL use.

Echo suppressors. An echo suppressor permits transmission in only one direction at a time. The devices inhibit potential echoes but also render full-duplex communication impossible. Modems can send a 2.1KHz answer tone at the beginning of a connection to deactivate an echo suppressor.

Fiber optic cable. Distance limitations and noise interference are not the only potential DSL pitfalls. If fiber, in the form of a digital loop carrier, is employed in the local loop, the route is ineligible for DSL. Fiber supports digital transmission, but DSL was developed only for analog copper line. The local loop of the future will be built on a hybrid fiber/twisted pair foundation, with short copper runs to neighborhood fiber nodes.

DANCE ON COMMAND

To initiate services, contact one of the following sources: your local telephone company (also known as an Incumbent Local Exchange Carrier, or ILEC), a CLEC, or a local ISP.

Monopolies die hard. An ILEC can supply a splitter and provision ADSL over an existing line, but until recently, customers who brought their business to a CLEC were forced to pay for an additional phone line. Apparently this was the penalty levied for not going to the ILEC. Extra charges sting on a monthly basis, but more frustratingly, the extra hoop can necessitate a pointless wait for installation of an additional circuit.

However, as of June 6, 2000, by FCC ruling, the additional phone line is no longer required, enabling CLECs to reduce prices, if deemed desirable, by approximately $20 per month.

VOICEOVERS

Everyone wants to extricate themselves from the high costs of local (and indirectly, long distance) voice charges with Voice over DSL (VoDSL). ADSL already supports analog voice frequencies by allocating digital data communications to higher frequencies, but VoDSL takes an alternate tack. VoDSL converts voice from analog to digital format, sending voice as just a fraction of its digital load.

ADSL and SDSL both support VoDSL, but G.lite is generally considered unsuitable for the task.

LIFE WITH DSL

DSL does not guarantee that as soon as you reach for the Internet, your bidding will be done instantly. DSL is merely a last mile solution, an attractive option for high-speed Internet access. Once off the DSL circuit at the CO, you're typically on a third-party network, and thus are vulnerable to all the bottlenecks and traffic problems inherent to that service provider.

The wise user will opt for backup with automatic switchover, even when he or she finally receives DSL service, choosing regular dial-up V.90 technology or ISDN, if available. DSL can go down periodically.

A choice based solely on price may eventually provoke regret. The lower the monthly fee, the less responsive you should expect your service to be.

Security is also a concern with DSL, just as with every other data pipe. Unlike cable modems, DSL users enjoy dedicated connections that remain immune to the activities of other users. Neighborhoods do not share lines, as with cable modems, a point which DSL provisioners cite as a security advantage. Both technologies can be at risk from intrusion and Denial of Service (DoS) attacks, however, due to always-on connections and fixed IP addresses.

If data transmission systems ever evolve into living organisms, twisted-pair copper may lead the pack in adaptive survivalism. From within the communications gene pool, ordinary copper stepped up its act and colonized previously underserved local loop niches. The last mile is a large and eager market, especially receptive to affordable technologies that go heavy on the bandwidth.

Free, unlimited broadband access for all may not materialize in this lifetime, but if you can get DSL, you're moving in the right direction.

KEY PLAYERS IN THE EXTENDED FAMILY

Other strains of Digital Subscriber Line (DSL) include High Bit Rate DSL (HDSL), ISDN DSL (IDSL), and Very High Bit-Rate DSL (VDSL).

HDSL. The first symmetric form of DSL, HDSL emerged in the early 1990s as an alternative to T1/E1 lines. Its primary benefit is that it requires no signal repeaters. Like SDSL, HDSL achieves maximum rates of 1.5Mbits/sec, but it does not extend beyond distances of 15,000 feet and requires two cable pairs to SDSL's one. As with SDSL, HDSL does not permit line sharing with analog phones.

IDSL. A hybrid of DSL and ISDN technologies, IDSL bypasses the PSTN and travels on the data network instead. IDSL employs the data-encoding technique of ISDN and delivers up to 144Kbits/sec of bandwidth, 16Kbits/sec more than ISDN alone.

VDSL. An asymmetrical technology like ADSL, VDSL operates at rates of 13Mbits/sec to 52Mbits/sec downstream and 1.5Mbits/sec to 2.3Mbits/sec upstream. Currently in the experimental phase, the catch with VDSL is that a recipient must be located within 1,000 feet to 4,500 feet of the Central Office (CO) or other access point.

RESOURCES

An excellent all-around resource for Digital Subscriber Line (DSL) information and peer reviews of DSL service providers can be found at www.dslreports.com. A section devoted entirely to business use of DSL is expected to launch in August 2000.

The xDSL Resource Web site has a comprehensive FAQ listing on each of the varieties of DSL, their modulation techniques, and the interference minefield on the local loop. Go to www.xdslresource.com/xDSLFAQ.shtml.

A special report on DSL appeared in the April 2000 issue of *Network Magazine*. It covers implementation hurdles and Voice over DSL (VoDSL). It also provides a comparative analysis of DSL costs and deployment from U.S. sites. This special report and other DSL articles from back issues of *Network Magazine* are available at www.networkmagazine.com.

This tutorial, number 145, by Rob Kirby, was originally published in the August 2000 issue of Network Magazine.

Fundamentals of Optical Networking

The Bell system and other nations' telephony providers began deploying optical fiber for long distance voice trunks as early as 1977. Those first commercial deployments carried DS3 traffic (45Mbits/sec) a few dozen miles before requiring a repeater to regenerate the signal. Today's fiber optic systems support throughputs as high as 1Tbit/sec, with unassisted propagation distances of 3,000 miles and farther.

The story of the huge increases in capacity that optical networking enables is the story of a handful of technical breakthroughs, beginning with the laser in the late 1950s, followed by huge advances in the clarity of glass fibers, then by native optical amplification technology, and finally by Wavelength Division Multiplexing (WDM).

Any sort of light will travel along an optical fiber. The boundary between two light-conducting media with differing indices of refraction serves as an efficient mirror to light that hits the boundary at a sufficiently low angle—the same phenomenon accounts for the glare of the sun across a body of water in the late afternoon or early morning.

Optical fiber cabling encases a narrow thread of glass inside a layer, or "cladding," made of material with an index of refraction different from that of the fiber. Thus, properly introduced light will travel great distances, so long as the fiber itself doesn't absorb the light.

However, ordinary white light sources or even the single-wavelength light emitted by LEDs can only carry limited quantities of data. White light's many wavelengths each travel through the fiber at different speeds, limiting the resolution of a series of digital pulses. While the wavelengths of LED-generated light are bunched together around a single value, they are still not as tightly bunched as those produced by a laser.

The coherent light produced by lasers has a uniform phase, as well as a uniform wavelength. The upper throughput limit of LED-generated signals is about 300Mbits/sec, whereas lasers haven't shown any sign of pooping out at 10Gbit/sec (OC-192) and higher data rates. Lasers have also traditionally been capable of creating higher levels of power than LEDs.

A technology on the verge of real deployment is the tunable laser. Historic lasers, of course, produce a very pure, single bandwidth. In the short term, a laser capable of generating multiple wavelengths would cut the operating and inventory costs of carriers that operate networks using many wavelengths. More significantly for the long term, tunable lasers hold the promise of all-optical, high-speed switching and routing.

The second key to the development of optical networking was the development of low-loss cable. In the 1970s, fiber losses were on the order of 20 decibels (dB) per kilometer. State-of-the-art fiber optic cable nowadays has losses in the range of .2dB to .3dB per kilometer.

It turns out that silica or glass aren't the limiting factors, but that trace impurities, especially such fairly common elements as iron, copper, nickel, and chromium, are the biggest problem. In fact, one part per billion of these elements is the threshold of tolerability.

Furthermore, hydroxyl ions (OH), present in every water molecule, will absorb important wavelengths unacceptably if they are present in one part per 100 million. The development of fabrication techniques that could achieve these exceedingly high levels of purity were key to increasing the performance of optical fiber.

ELECTRONICS PLUS OPTICS

The typical applications of the 1980s used fiber optics purely for transport. Inputs to cable were always electronic, and at the destination the data was converted back to an electronic signal. Furthermore, whenever the signal weakened along the way, an electronic repeater regenerated it. There was no purely optical regeneration capability until the development of the Erbium Doped Fiber Amplifier (EDFA) in the late 1980s.

An EDFA is a purely optical device in the sense that there is no conversion of optical signals to electronic signals, and all the relevant steps of its operation are the actions of photons—no electron flows are required.

Erbium is special because it, like several other rare earth elements, has a complex system of electron shells or bands, including a metastable band whose energy level difference from the base configuration of erbium ions is close to the energy of a photon in the 1,550 nanometer (nm) range.

The 1,550nm area is central to most modern optical developments because optical signal losses are at their lowest values in this range. (The longest visible wavelength, where red turns into infrared, is approximately 770nm. The wavelengths that carry signals in optical fiber are not visible to the naked eye.)

A section of erbium-doped fiber is exposed to intense levels of light—980nm and 1,480nm wavelengths are workable choices—and the erbium ions respond by absorbing photons, whereupon some electrons jump into higher-energy metastable orbits.

The arrival of a signal photon causes an electron in a metastable energy band to jump back to the base band and give up a photon with precisely the same wavelength and phase as the incoming signal photon. Thus a cascade of photons with the same direction, wavelength, and phase can result from a single photon, resulting in the purely optical amplification of the input signal.

Purely optical operations aren't desirable just for their technical elegance. For one thing, optical operations can take place faster than electronic ones. As the commonly encountered throughput rate of a backbone link reaches 10Gbits/sec and higher, there is some question whether electronic devices can keep up, especially when cost is taken into account.

EDFAs amplify rather than repeat, as traditional electronic regeneration devices do. Both amplification of the incident signal and digital repeating get the job done, but amplification has a strong flexibility advantage. For example, if you decide to change the line code of the signal, perhaps encoding two or three bits per symbol instead of one, an amplified link requires no change in configuration.

A repeater, on the other hand, has to be replaced or reconfigured to accommodate line code changes. Furthermore, as WDM signals become widespread, a link with EDFAs needs no upgrade—it simply amplifies wavelengths as they arrive, cascading out appropriate flows of photons for each wavelength.

If a link with repeaters is converted to WDM, repeaters for each wavelength have to be installed at each regeneration point, along with splitting and recombining hardware to get all the signals off to the correct electronic processor. In general, an EDFA is likely to cost substantially less than an electronic repeater solution.

PASSIVE OPTICAL NETWORKS

EDFAs make it possible to get optical signals to just about any destination without electronic processing. However, there are few purely optical solutions for routing or switching optical signals, whether the solution resembles circuit switching or packet switching, and whether the links are connection-oriented or connectionless.

One purely optical technique employs some of the characteristics of shared-medium LANs or cable-TV systems, however.

You can use optical couplers and splitters to distribute every wavelength to every node of the network, then you can filter all the inappropriate wavelengths at the end point. (This filtering step is analogous to shared Ethernet, where every node on the network has electronic access to each frame, but nodes process only the frames intended for them.) Thus, every end point has access to one or more wavelengths over the common medium.

On a hybrid fiber/coax cable-TV system, the final multiplexing step is performed electronically, but the network between the head end and the optical end nodes is a Passive Optical Network (PON). Access and metro-area networks that don't have to cover long link distances may be able to avoid amplification, and thus have more choices for usable wavelengths. PONs often employ wavelengths in the 1,310nm range.

WDM and Dense WDM (DWDM) have been the big news in optical networking for the last 10 years or so. The number of wavelengths that can be accommodated on a single fiber has recently doubled every couple of years. The first-to-be-standardized 1,530nm to 1,560nm basic range supports only 40 channels with a channel spacing of 100GHz, with another 40 channels defined in the 1,560nm to 1,600nm "extended" range that can be amplified with enhanced EDFAs. The number of channels doubles as the channel spacing is halved, but there is an upper limit eventually determined by various noise, distortion, and crosstalk factors.

AT&T Bell Labs demonstrated 1,022 wavelengths in a recent test, but not with an OC-192 throughput rate on each wavelength. Because EDFAs operate across a fairly narrow window in this wavelength range, using other wavelengths would necessitate either an alternative optical amplification technology or electronic regeneration, with the increased costs and complexity mentioned earlier.

Wavelength "routing," which might be more accurately described as wavelength circuit switching, is a crucial part of a DWDM network (see figure 1). Today's commercial products

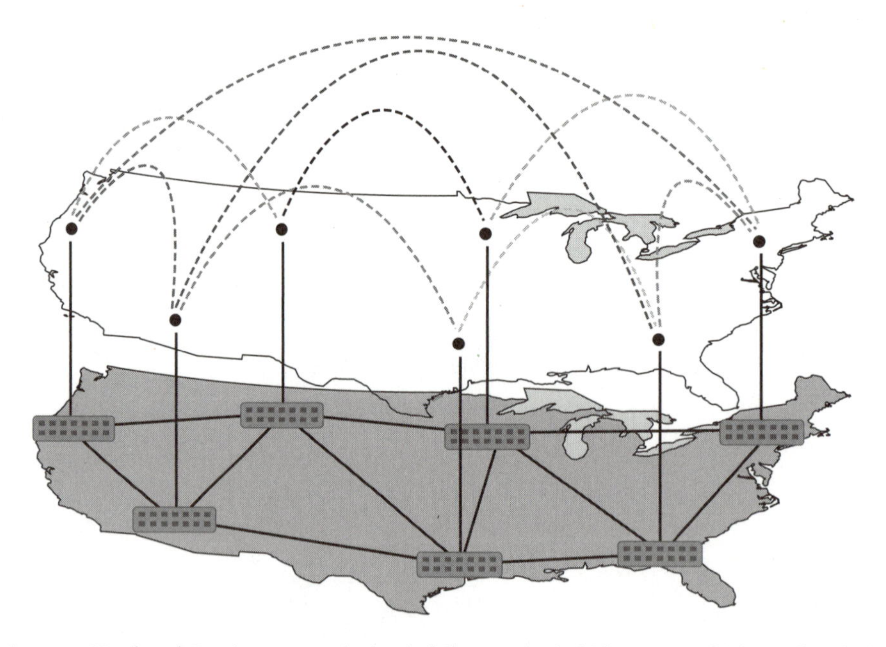

■ **Figure 1:** Wavelength Routing. A network of optical fibers, each of which carries multiple wavelengths (or lambdas), can be used much more effectively if data paths can be switched from one lambda to another at each node. Today's wavelength routers require an optical-to-electronic-to-optical conversion in order to switch to a different node.

require an electronic conversion to move traffic off one wavelength and onto another. Given the latency and relative slowness of the electronic conversion, switching a wavelength using this system takes a millisecond or longer.

Several solutions that more closely approach the pure-optical ideal have been demonstrated, however. One involves some form of tiny mirrors that are reconfigured to reflect demultiplexed wavelengths along their intended paths. Another system, recently demonstrated by Agilent Technologies, involves injecting inkjet-like bubbles into a pattern of junctions to steer optical traffic.

While both of these approaches avoid the dreaded optical-to-electronic conversion, they are still rather mechanical and slow. The estimated time for their operation also takes milliseconds, rather than the microseconds or nanoseconds required to create ATM- or frame-relay-like virtual circuits.

These technologies are suitable for manual provisioning and for the sort of protection switching that SONET provides, but the pure optically routed network is still years away.

RESOURCES

A truly magnificent book that covers all the contributory technologies affecting optical networking is *Understanding Optical Communications*, by Harry J. R. Dutton, Prentice Hall PTR, 1998, ISBN 0130201413.

Cisco Systems acquires optical networking companies for breakfast. The white papers at the URL below are concerned with wavelength routing. Go to www.cisco.com/warp/public/cc/ cisco/mkt/optical/wave/prodlit/index.shtml.

Alcatel has a collection of white papers on various optical networking topics at www.usa.alcatel.com/telecom/transpt/optical/techpaps/techpaps.htm.

Nortel has multiple classes of optical offerings, including end-to-end optical Ethernet. You can find several white papers on this topic at www.nortelnetworks.com/products/01/endtoendethernet/doclib.html.

This tutorial, number 142, by Steve Steinke, was originally published in the May 2000 issue of Network Magazine.

Fiber and Optical Networking

The foundation of any optical system is the fiber optic cabling, where many challenges of reaching higher transmission speeds are played out. Understanding the types of fibers available, how they work, and their limitations is crucial to mastering next-generation networks.

Basic fiber cables consist of five components. At the center of the cable is the silica core and cladding used for carrying the optic signal. A coating, strength members, and a plastic jacket enclose the fiber, providing the necessary tinsel and scratch resistance to protect the fibers. Attached to both ends of the core are transceivers for emitting and receiving the light pulses that form the information bits in the optical network. The ability of clear glass to contain light is the key behind optical transmissions and is based around the principle of total internal reflection.

Here's how it works: By injecting light at a specific angle, the glass cladding acts as a supermirror, reflecting light within the silica. The trick here has to do with the Refractive Index (RI), or the change in the speed of light in a substance (in this case, silica) relative to the speed of light in a vacuum.

Light normally travels at about 300,000 kilometers (km) per second in a vacuum. However, when light moves from a substance of lower density to one of higher density (say, between air and water), the light changes and is refracted. This is why a stick appears to bend when one half is placed in water.

How much the beam will bend depends on two things: the angle at which light strikes the water and the RI. At some point an angle is reached so that the light reflects off the water like a mirror. This is the critical angle, and the reflection of all the light is the total internal reflection.

This same principle determines how light propagates down a clear fiber. Fibers are manufactured so that the core contains a slightly higher RI than the surrounding cladding. If light travels through the core and hits the cladding at a particular angle, it will stay in the fiber. The exact size of the angle depends on the difference in RIs, but if a typical RI difference of 1 percent is assumed, all light striking the cladding at eight degrees or under will continue on in the fiber (see Figure 1).

LIGHT TRAVEL

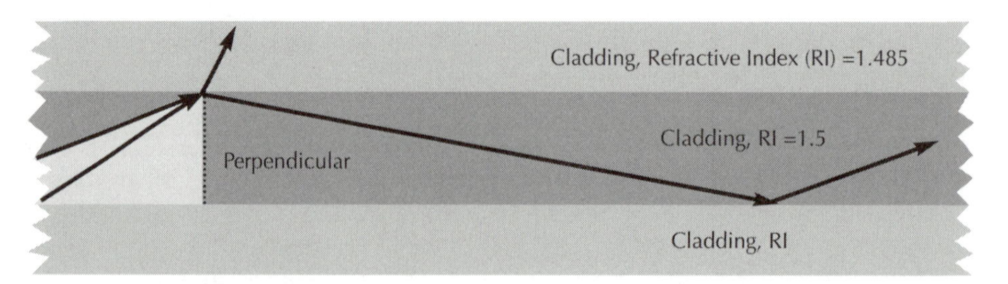

■ **Figure 1:** All waves that strike the cladding beyond 82 degrees from the perpendicular propagate down the fiber. Waves that fall within 82 degrees go into the cladding.

TRANSMISSION WINDOWS

Of course, the light used in fiber optic transmission isn't the same as the light in a flashlight. The light sources in an optical network are more precise. The ITU has specified six transmission bands for fiber optic transmissions. These bands are expressed in terms of wavelength sizes measured in nanometers (nm), one billionth of a meter, or microns (µm), one thousandth of a meter. The six bands are the O-Band (1,260nm to 1,310nm), the E-Band (1,360nm to 1,460nm), the S-Band (1,460nm to 1,530nm), the C-Band (1,530nm to 1,565nm), the L-Band (1,565nm to 1,625nm), and the U-Band (1,625nm to 1,675). A seventh band, not defined by the ITU, but used in private networks, runs around 850nm. To put those measurements into perspective, the human hair is about 100µm wide. Typically, the higher the transmission window, the lower the attenuation or signal degradation (see Figure 2), but also the more expensive the electronics.

The earliest fibers operated in the first transmission window of 850nm. These standard multimode fibers, or step-index multimode fibers, contained a relatively large core of 50.5µm or 62.5µm, depending on the cable. They're called multimode because the waves that make up the light take multiple paths or modes through the fiber. How many modes requires some complex calculations, but those numbers easily reach into the hundreds.

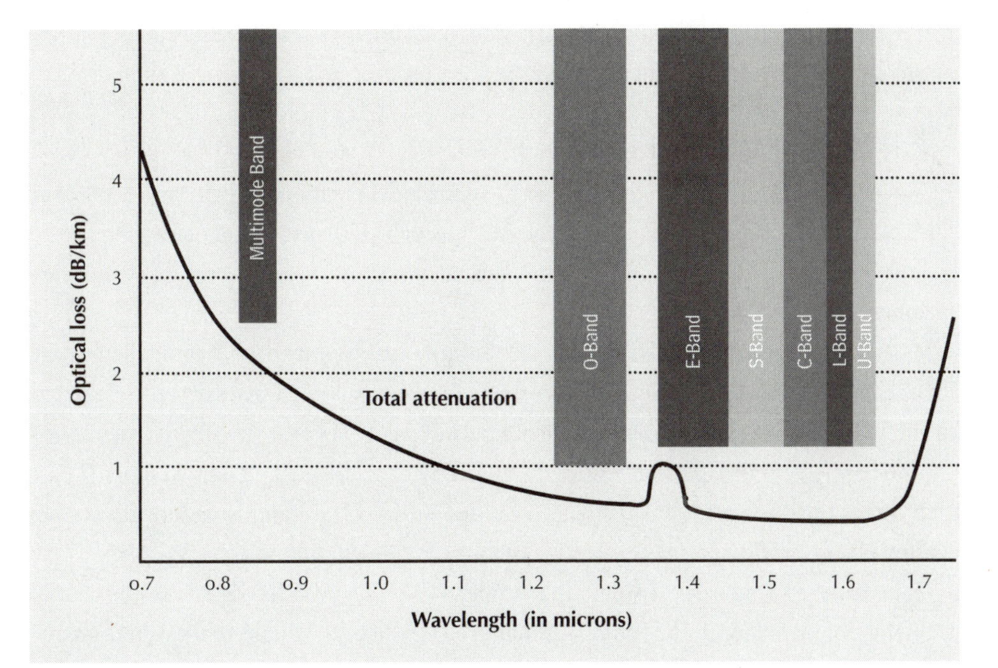

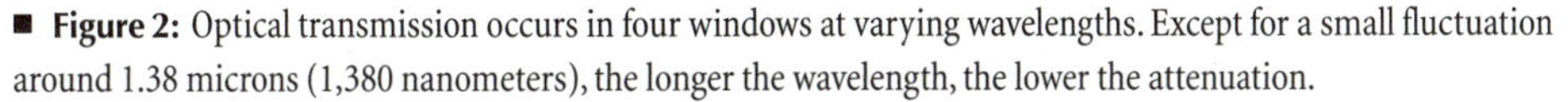
■ **Figure 2:** Optical transmission occurs in four windows at varying wavelengths. Except for a small fluctuation around 1.38 microns (1,380 nanometers), the longer the wavelength, the lower the attenuation.

These modes cause many problems for step-index fiber, resulting in limited distances. One problem is modal dispersion. Since modes extend through the fiber at different angles, their lengths are slightly different. The result is that light takes less time to travel down some modes (the shorter ones) than others, dispersing or spreading out the light pulse. With short fiber spans there is little spreading out of the pulse, but on longer distances (over a kilometer) the light pulse spreads out so much that it's unreadable.

To increase the range, manufacturers developed Graded Index (GI) fiber, an improved multimode fiber that operates in the second transmission window at around 1,300nm. GI fiber virtually eliminates modal dispersion by gradually decreasing the RI out toward the cladding where the modes are longest. Then waves on the longer modes travel faster than on the shorter modes, so the entire pulse arrives at the receiver at about the same time.

Today, the vast majority of fiber installed in local premises is GI fiber, and with good reason: With GI fiber, manufacturers can extract 200MHz of bandwidth over 2km. "A stepped index fiber would be a tenth of that or worse," says Eric Anderson, product line manager in the technology development department at Microtest (www.microtest.com), a manufacturer of cable test equipment.

But on distances over 2km, GI fibers need high-powered lasers. This combination introduces the problem of modal noise. With modal noise, the fiber and connectors interact so that there are power fluctuations at the receivers. This often greatly increases the signal-to-noise ratio in a link, limiting the length of the fiber.

SINGLE-MODE FIBER

By using a much narrower core of 8µm to 10µm, only one mode of light travels over the fiber. Actually, there are two modes, but they are normally independent of one another. With a single mode, many multimode problems, such as modal noise and modal dispersion, are no longer an issue.

Cost, however, is an issue. The minute size of these cores demands that components have much tighter tolerance, which increases costs. These costs, however, are easily outweighed by the increased bandwidth and distances of 80km and longer. There are two main types of single-mode fiber: Non Dispersion Shifted Fiber (NDSF) and Dispersion Shifted Fiber (DSF). DSF comes in two types: Zero Dispersion Shifted Fiber (ZDSF) and Non Zero Dispersion Shifted Fiber (NZDF).

Like GI fibers, the earliest single-mode fiber, NDSF, also carries signals in the second transmission window at 1,310nm. Although NDSF, like all single-mode fiber, greatly improves on GI fiber's range, there are other problems, namely chromatic dispersion. Any light pulse, no matter how precise the laser, contains a range of waves at different frequencies. Since the impact of the RI varies with the wave's frequency, the waves end up propagating down the fiber at different velocities, and the pulse disperses until the signal becomes unintelligible.

At the same time, "waveguide" dispersion also affects wavelength velocity. As waves move down the wire, parts of the electric and magnetic fields extend into the cladding, where the RI is lower. The longer the wavelength, the more energy carried in the cladding—and the faster the wave travels. At the second transmission window (around 1,310nm), chromatic dispersion and waveguide dispersion cancel each other. Outside of that window, dispersion increases, limiting the length of the fiber.

Here lies the problem. The third and fourth transmission windows are better suited for longer distances than the second transmission window. They have lower attenuation and can work with the best optical amplifiers, Erbium-Doped Fiber Amplifiers (EDFAs).

Dispersion Shifted Fibers (DSFs) move optimal dispersion points to higher frequencies by altering the core-cladding interface. ZDSF moves the zero dispersion frequency from 1,310nm by increasing the waveguide dispersion until it cancels out chromatic dispersion at 1,550nm. The problem? DWDM gear and EDFAs operate in this window. Signals traveling over ZDSF can combine to create additional signals that may be amplified by EDFAs and superimposed onto Dense Wavelength Division Multiplexing (DWDM) channels, causing noise in a problem known as "four-wave mixing."

NZDF avoids four-wave mixing by moving the zero dispersion point above the range of EDFAs. The signal continues to operate in the third or fourth transmission windows with only a moderate amount of dispersion, which actually helps—as it provides a minimal level of interference needed to separate DWDM channels from one another. Therefore, NZDF is the preferred cable for new optic installations.

Today, manufacturers have reduced the attenuation and the transmission window restriction on NZDF cables. Lucent Technologies' All Wave, for example, works at a frequency range of 1,310nm to 1,550nm without the attenuation peak of around 1,380nm. Similarly, the company optimized its Metrowave cable for 1,310nm and 1,550nm, offering providers an easy migration path to higher spectrums.

RESOURCES

There's a huge amount of material about optical transmissions both in print and on the Web. Two books to check out are *Understanding Optical Communications* by Harry Dutton and *Understanding Fiber Optics* by Jeff Hecht. Hecht's book is more readable, but Dutton's covers everything about optical transmissions, with all of the gory engineering details.

Plenty can be found on the Web, as well. A great primer on physics can be found at www.colorado.edu/physics/2000/index.pl?Type=TOC/. A good optical primer can be found at www.vislab.usyd.edu.au/photonics/fibres/index.html.

This tutorial, number 156, by David Greenfield, was originally published in the July 2001 issue of Network Magazine.

Passive Optical Networks

When it comes to last-mile access, network operators must choose between their favorites—will it be low cost, long range, high performance, or rapid deployment? Choose DSL or cable-modem access and they gain the benefits of installed infrastructure, but sacrifice range and multimegabit performance. Go with fixed wireless and they sacrifice performance or distance.

Passive Optical Networks (PONs) offer an alternative. By eliminating regenerators and active equipment normally used in fiber runs, PONs reduce the deployment and installation costs of fiber. Those costs still require laying fiber—making PONs far more expensive than DSL—but the potential performance gain and long-term prospects make PONs well-suited for new neighborhoods or installations.

PONs may never knock down consumer doors, but that doesn't mean the technology won't appear in existing neighborhoods. PONs are a strong complement to enabling Very High Data Digital Subscriber Line (VDSL). Today, VDSL modems can reach 52Mbit/sec over 300 meters. PONs enable carriers to extend that distance by only running VDSL to the customer premises, and then going over fiber back to the Central Office (CO). What's more, a PON's use of ATM—the protocol favored by DSL—simplifies the interconnection of the two technologies. Finally, new QoS and fault tolerant features add critical technologies needed to deliver real-time video streaming, an often talked about application for VDSL.

A PON, or an Optical Access Network (OAN), is a tree-like structure consisting of several branches, called Optical Distribution Networks (ODNs). The ODNs run from the CO to the customer premises using a mix of passive branching components, passive optical attenuators, and splices.

Three active devices can be used in a PON. At the CO, carriers install a special switch, called an Optical Line Terminal (OLT), which either generates light signals on its own or takes in SONET signals (such as OC-12) from a colocated SONET cross-connect. The OLT then broadcasts this traffic through one or more outgoing subscriber ports. Depending on where the fiber terminates, either an Optical Network Unit (ONU) or an Optical Network Termination (ONT) receives the optical signal and converts it into an electrical signal for use in the customer premises. (ONTs are used when the fiber extends into the customer premise, whereas ONUs are used when fiber is terminated outside of the home.) DSL then brings the signal to the customer premises (see Figure 1).

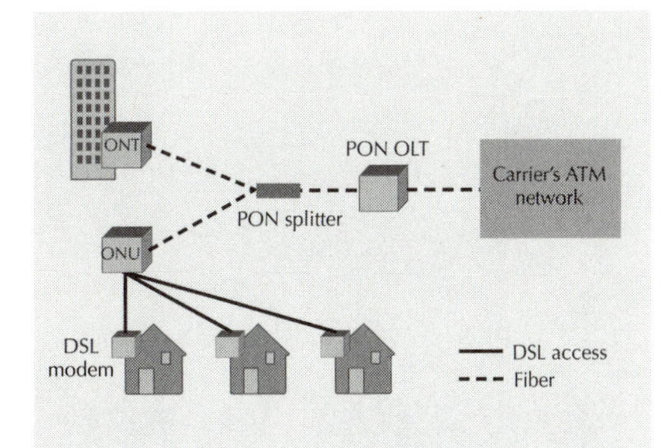

■ **Figure 1:** PONs consist of four key elements: An OLT sits in the central office controlling the PON network. Splitters sit at various fiber junctions dividing the signal into multiple channels. ONU or ONTs terminate the signals at or just before the customer premises.

WDM, TDM, AND ATM

PONs use a combination of an access protocol, either ATM or increasingly Gigabit Ethernet, Wavelength Division Multiplexing (WDM), and Time Division Multiplexing (TDM). ATM-based PONs (APONs) work like big ATM networks. Customers establish Virtual Circuits (VCs) across the APON to a destination, such as another office or the ISP's premises. These VCs are bundled into what's known as Virtual Paths (VPaths) for faster switching in the carrier's network. The IEEE is just now standardizing Ethernet-based PON (EPON).

The speed of operation depends on whether the APON is symmetrical or asymmetrical. Symmetrical APONs operate at OC-3 speeds (155.52Mbit/sec). For Asymmetrical APONs, the downstream transmission from the OLT in the CO to the customer premises ranges from 155.52 to 622.08Mbit/sec. Upstream transmission from the customer premises to the CO occurs at 155.52Mbit/sec. All network topologies support both configurations, unless the fiber extends to the home, in which case asymmetrical operation isn't supported.

The upstream and downstream transmissions occur over two channels, which can be different wavelengths on the same cable or two different cables. The original APON specification called for downstream transmission on a single fiber occuring between 1,480 and 1,580 nanometers (nm); on dual fibers, between 1,260nm and 1,360nm. Upstream transmissions always occur between 1,260nm and 1,360nm.

A new ITU standard adds Wide WDM (WWDM) capabilities to APONs. With the specification, downstream transmission is split into two bands: regular transmissions range from 1,480nm to 1,500nm, and a new enhanced band for future applications (such as video services) occurs at 1,539nm and 1,565nm. APONs can now support 16 wavelengths (with 200GHz spacing between channels) or 32 wavelengths (with 100GHz spacing between channels). However, since customers need one wavelength for transmission and another for receiving, these numbers are effectively cut in half.

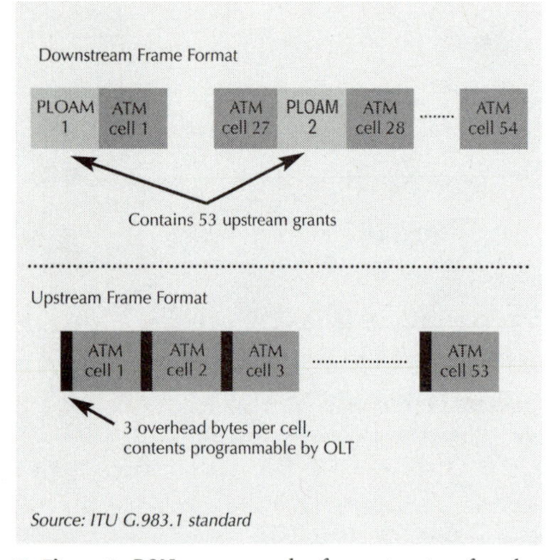

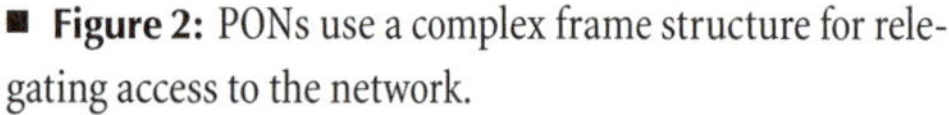

■ **Figure 2:** PONs use a complex frame structure for relegating access to the network.

If there's bad news, it's that implementing these new channels will require expensive equipment. Since the new spectrum is narrower, there's less allowable drift in the laser. Therefore, PON equipment delivering WWDM will likely need more expensive lasers. Across the main band from 1,480nm to 1,500nm, PONs run a TDM architecture across the send and receive channels. Within these time slots, the network behaves like a LAN, implementing Ethernet or ATM.

The OLT regulates network access with a complex frame structure. The upstream and downstream channels are divided into frames and time slots. The exact size of each depends on the configuration. Within a symmetrical network, for example, downstream frames consist of 56 time slots—54 for data and two for management information. Each time slot is equal to an ATM cell, or 53 bytes. Upstream frames are 53 time slots long, but run 56 bytes each. The additional three bytes are overhead which the OLT can program for various purposes, such as requesting an ONU to transmit a Physical Layer Operations and Maintenance (PLOAM) cell or a minislot to gather management information (see Figure 2). Asymmetrical APONs use frame time slot configurations that are four times greater on the downstream to accommodate the 622Mbit/sec speeds.

The APON specification uses two types of cells. Data cells carry information including user data, signaling information, and Operations and Management (OAM) information. The second, PLOAM cells, pass physical infrastructure information around the network. They also carry grants, codes supplied by the OLT that allow access to the network. OLTs can also divide up the time slot into minislots to gather information about the ONUs' traffic queues in order to implement QoS or Dynamic Bandwidth Allocation (DBA).

PON WORKINGS

One benefit of APONs is that the system automatically detects new devices. Periodically, the OLT creates a gap in the upstream time slots of a few hundred microseconds and sends a special permission, a ranging grant, which anyone can answer. A new ONT will see the ranging grant and respond with its unique serial number. The OLT receives this number and configures the ONT with the information needed to join the APON. This information includes an APON ID and two kinds of grants—a PLOAM grant and a data grant, which is used to access the network.

Then ATM kicks in. When the OLT sends an ATM cell down the APON, each ONT compares the cell's VPath identifier against its own. If there's a match, the ONT copies the cell, removes it from the network, and sends it to the customer premises. Each customer premises then compares the cell's VC identifier against its own, and if there's a match, the node copies the data and removes the cell.

When an ONT needs to send information, it waits for the OLT to send a PLOAM cell. Each PLOAM cell has 26 or 27 grants that anyone can read. The ONT checks the data grant number in the PLOAM cell, and when it matches its own, the ONT uses the grants to send the data. The cell is then transmitted upstream. The OLT receiver receives the bits and, using the preamble to recover the clock, reads out the cells and passes them to the ATM switch for delivery onto the provider's metro core network.

SECURITY AND QOS

Security and QoS have been hot areas in APONs. APON security is tricky because any ONT can read any cell. Like the voice network, there's no inherent security mechanism to prevent intrusion.

However, the current APON specification does provide some rudimentary form of security, called scrambling. Each cell's payload runs through encryption algorithms that mix up (or "scrambles") the data using a 24-bit encryption key. The shortness of the encryption key makes it very weak (even 40-bit keys are considered weak), but changing the encryption key's key on the fly improves the security somewhat.

As for QoS, while ATM enables different types of circuits, the current APON specification doesn't. Each traffic flow is treated equally. A new ITU standard will add DBA to APONs. DBA will allow the OLT to determine who has the data, and adjust the number of grants sent accordingly.

OLT will accomplish this by sending a minislot grant. The ONT will reply with the queue information, indicating the number of cells to be transmitted. The OLT will then measure the amount of data to be sent against the type of service the customer purchased, existing traffic conditions, and so on. Based on this information, the OLT will then issue a grant, and the ONT will respond in the normal manner.

MANAGEMENT AND SURVIVABILITY

Public networks aim for reliability above all. Forget the increased performance, enhanced security, and improved QoS that PONs can offer. Without adequate uptime, no service will be in business for long, so providing adequate management and fault tolerance is crucial.

With APONs, the PLOAM cells supply the management. They allow ONTs to report back to the OLT about themselves. The same cells also provide certain bit patterns that can align lasers and measure power output.

For fault tolerance, an emerging APON specification will enable redundancy in the network. The APON will work similar to SONET's dual-ring configuration: Each ONT will connect to two OLTs. Two protection scenarios are then possible. One-to-one protection lets the transmitter choose which line to use, leaving the other untouched. Providers could use this configuration to put Unspecified Bit Rate (UBR) traffic on the backup line, with the understanding that a failure could cause the traffic to be bumped from the network.

The second approach is called one-plus-one protection, in which transmitters and receivers on both sides of the network pick the cleaner line to communicate over. Therefore, in the event of a switch-over, no traffic is lost from the backup line.

RESOURCES

To catch up on the details of the Passive Optical Network (PON) spec, visit www.itu.org and download the PON standard, g.983.1.

You can find the Full Service Access Network (FSAN) organization at www.fsanet.net.

There are plenty of articles on PONS in *Network Magazine*, including "PON Progress Report: Operating on RBOC Time" (June 2001, page 68, www.networkmagazine.com/article/NMG20010518S0008) and "Passive Optical Networks in Action" (November 2000, page 124, www.networkmagazine.com/article/NMG20001103S0008).

Additional detail on the PON specification and other optical network technologies can be found in *The Essential Guide to Optical Networking* written by this author.

This tutorial, number 161, by David Greenfield, was originally published in the December 2001 issue of Network Magazine.

Synchronous Optical Network

A standard for high-speed communications over fiber optic cable brings relief to telcos and customers.

The phenomenal growth of the Internet as a place of commerce and information exchange has led to more and more traffic traveling over wide area links. If you talk to some of the larger service providers like MCI and UUNET, you will hear tales of how network size is doubling every few months due to the vast amount of electronic traffic coming and going from customers' sites.

Although most of the traffic clogging the Internet is still data, we're starting to see a mixture of voice, video, and other multimedia applications vying for precious bandwidth. For some time, most companies were content with leased 56Kbit/sec or T1 lines, and in rare instances a company required T-3 (45Mbits/sec) pipes. Although T1 still fulfills most companies' needs, service providers that have to aggregate multiple T1s and provide connections across the country or around the world have needed more transport capacity.

That much-needed bandwidth has come in the form of Synchronous Optical Network (SONET), a broadband transmission standard. It provides an optical signal format for the high-speed transfer of data, video, and other types of information across great distances without regard to the specific services and applications it supports.

HIGH-SPEED NETWORK

SONET was first conceived of in the mid-1980s, when the telecommunications industry and various standards bodies—including CCITT, ANSI, IEEE, and EIA—determined the need for a fiber optic-based standard that could handle much more than just voice at gigabit-per-second rates. Before then, fiber optic products from different manufacturers were not compatible with each other, forcing carriers to do business with only one fiber supplier. In addition, the connections between various networks could be tricky if products from more than one vendor were being used.

Before SONET standards were developed and started becoming widely implemented, an older infrastructure, known as the North American Digital Hierarchy (NADH), was the primary Physical layer used by T1 lines (see Table 1).

TABLE 1—NORTH AMERICAN DIGITAL HIERARCHY RATES

Digital signal	Number of 64Kbit/sec channels	Line rate
DS-0	1	64Kbits/sec
DS-1	24	1.54Mbits/sec
DS-2	96	6.31Mbits/sec
DS-3	672	44.74Mbits/sec
DS-4	2,016	139.26Mbits/sec

Using NADH, telephone company central offices were able to send DS-1 signals, which break down into 24 64Kbit/sec segments (or DS-0 signals), over copper T1 lines. But, thousands or even tens of thousands of voice and data transmissions can occur simultaneously. When this happens, the only viable option is fiber optic cabling. The NADH infrastructure still exists and continues to be widely used, even with a lack of standards and inherent bandwidth limitations, but eventually SONET will become the primary transport vehicle.

In general, copper lines are fine for transmitting voice and data. However, as the number of such transmissions over a single piece of copper cabling increases, bandwidth becomes limited. Fiber optic cabling brings much more bandwidth to the picture. An analog line on the local loop provides just 4KHz of bandwidth for voice, while a single optical fiber can carry 3Gbits/sec. ([Editor's note: The capacity of a single optical fiber has increased into the terabit range in 2002.] While you cannot directly compare analog bandwidths with digital bit rates, the best current analog modem technologies can deliver only 56Kbits/sec over an analog phone line.) Fiber also comes in two varieties, multimode and single-mode, giving customers the option of using this medium over the LAN or over a distance of several kilometers.

In addition to the advantages of using fiber optic cabling, going with a synchronous transmission method such as SONET offers several advantages. Synchronous transmissions support bandwidth (or circuit) provisioning, which allows providers to have control over individual DS-0, DS-1, and DS-3 channels. This ability allows the provider to add more channels to meet traffic demand and then remove them when they are no longer needed.

Synchronous transmission environments also allow customers to do real-time routing around nodes that are experiencing a lot of traffic. Customers can reconfigure routes without having to end voice and other sessions. Another benefit of synchronous transmission includes the ability for service providers to conduct automatic testing and maintenance of network performance. This is accomplished through overhead channels that are tacked onto the transmission itself. These channels can perform automated functions such as maintenance, testing, and issuance of reports.

BUILDING BLOCKS

The basic foundation of SONET consists of groups of DS-0 signals (64Kbits/sec) that are multiplexed to create a 51.84Mbit/sec signal, which is also known as STS-1 (Synchronous Transport Signal). STS-1 is an electrical signal rate that corresponds to the Optical Carrier line rate of OC-1, SONET's building block (see Table 2).

Subsequent SONET rates are created by interleaving (at the byte level) STS-1 signals to create a concatenated, or linked, signal. For example, three STS-1 frames can form an STS-3 frame (155Mbits/sec). Rates above STS-3 can be created by either directly multiplexing STS-1 signals or by byte-interleaving STS-3 signals.

This ability to use direct multiplexing is an improvement over the NADH scheme, which doesn't always allow it because the signals that need to be multiplexed to get a higher signal rate are asynchronous to each other and could be operating at slightly different frequencies.

TABLE 2—SONET LINE RATES

Electrical signal	Optical Carrier line	Line rate
STS-1	OC-1	51.84 Mbits/sec
STS-3	OC-3	155.52 Mbits/sec
STS-12	OC-12	622.08 Mbits/sec
STS-24	OC-24	1.24 Gbits/sec
STS-48	OC-48	2.48 Gbits/sec
STS-192	OC-192	9.95 Gbits/sec

With NADH, signals can't just leap from DS-1 to DS-3; instead, a two-step process must be taken, which involves adding overhead to the transmission. This often leads to loss of integrity, making it difficult to create higher signal rates.

If the OC-3 and OC-12 rates look suspiciously like the rate for two particular flavors of ATM, it's no coincidence. ATM at the 155Mbit/sec and 622Mbit/sec rates was designed specifically to use SONET as the transport mechanism, and in fact, you cannot run higher bit rate ATM over anything except SONET. Because OC-3 has the same line rate as 155Mbits/sec and OC-12 has the same rate as 622Mbits/sec, many people erroneously equate them.

A designation of OC-3 simply means the SONET pipe itself is operating at a data rate of 155Mbits/sec. The pipe could be transporting 155Mbit/sec ATM, but because SONET is independent of the service it carries, the service could be FDDI, ISDN, or SMDS. The same applies to OC-12, but when you see the designation OC-12c, it means the bandwidth capacity is actually a concatenation of several smaller pipes instead of a single 622Mbit/sec ATM connection. For example, bandwidth capacity might consist of four OC-3 signals. (For more on high-speed ATM, see "ATM in the Fast Lane," July 1997, page 48.) Initially, SONET was defined only up to the OC-48 level, but now we're hearing talk of OC-192 (almost 10Gbits/sec).

Just because the basic level of SONET starts at 51Mbits/sec doesn't mean lower bit rate signals are ignored. The basic STS-1 frame contains 810 DS-0s, 783 of which are used for sending data (including slower asynchronous signals) and 27 of which are overhead. The overhead in this case is information concerning framing, errors, operations, and format identification.

Signals with speeds below STS-1, such as DS-1 and the European E1 (2.048Mbits/sec) can be accommodated by dividing the STS-1 payload into smaller segments that are known as Virtual Tributaries (VTs). The lower data rate signals are combined with overhead information, which leads to the creation of Synchronous Payload Envelopes (SPEs). SPEs allow these signals to be transported at high speeds without compromising integrity. Each VT on an STS-1 signal includes its own overhead information and exists as a distinct segment within the signal.

For example, VT-1.5, which is commonly used in North America, supports a line rate of 1.728Mbits/sec. Because this rate is slightly greater than the 1.54Mbit/sec rate of a DS-1 circuit, VT-1.5 is specially designed to carry a DS-1 along with the overhead information discussed above.

TRANSPORT AND TOPOLOGY

The SONET standard includes a definition of a transmission protocol stack which solves the operation and maintenance problems often found when dealing with networks that have component streams lacking a common clock.

The photonic layer is the electrical and optical interface for transporting information over fiber optic cabling. It converts STS electrical signals into optical light pulses (and vice versa, at the receiving end). The section layer transports STS frames over optical cabling. This layer is commonly compared with the Data-Link layer of the OSI model, which also handles framing and physical transfer.

The line layer takes care of a number of functions, including synchronization and multiplexing for the path layer above it. It also provides automatic protection switching, which uses provisioned spare capacity in the event of a failure on the primary circuit.

The highest level, the path layer, takes services such as DS-3, T1, or ISDN and maps them into the SONET format. This layer, which can be accessed only by equipment like an add/drop multiplexer (a device that breaks down a SONET line into its component parts), takes care of all end-to-end communications, maintenance, and control.

SONET supports several topologies, including point to point, a hub and spoke star configuration, and the ring topology. The ring topology, which is by far the most popular, has been used for years by such network technologies as FDDI and Token Ring and has proven quite robust and fault-tolerant. A SONET ring can contain two pairs of transmit and receive fibers. One pair can be designated as active with the other one functioning as a secondary in case of failure. SONET rings have a "self-healing" feature that makes them even more appealing for long distance connections from one end of the country to another.

PRACTICAL PURPOSES

The telecommunications industry was the driving force behind defining a fiber optic system standard, and so far, SONET has remained the domain of carriers such as MCI, AT&T, Sprint, and WorldCom, all of which continue to send large amounts of voice traffic, but have also experienced a dramatic rise in Internet traffic. In the last year or two, large

ISPs, such as UUNET, have installed SONET rings to bring more bandwidth and reliability to congested networks.

For much of the corporate world, SONET may carry too hefty a price tag to be practical. Companies can call their local carrier and ask about SONET equipment and service, but with an investment of tens or even hundreds of thousands of dollars, even traditional early adopters such as the financial world may pass for now. But, because SONET leverages existing fiber cabling already in use by many companies and brings stability and reliability to wide area networks, the move may not be so painful once providers become more aggressive about pricing.

This tutorial, number 108, by Anita Karvé, was originally published in the August 1997 issue of Network Magazine.

Wave Division Multiplexing

The amount of traffic—data, voice, and multimedia—traveling over the Internet and other networks grows at rates that are hard to quantify, but everyone agrees the increase is beyond anyone's wildest dreams.

Once all this traffic leaves a local network (or home or small business), it goes into the hands of a carrier or service provider, and typically is sent over a fiber optic cabling infrastructure. Even service providers too small to have their own fiber infrastructure use fiber built and run by someone else.

Fiber optic cabling moves lots of traffic quickly, but even these fat pipes feel the bandwidth pinch. So rather than pay exorbitant amounts to lay new fiber cabling, providers rely on technologies that increase the amount of data a single piece of fiber can handle.

EARLY SOLUTIONS

You can multiplex multiple signals on a single medium by assigning different frequencies to each signal (Frequency Division Multiplexing or FDM), or by assigning different time slots to each signal (Time Division Multiplexing or TDM). Early telephony multiplexing systems used FDM (which is also the standard multiplexing technique used for cable television), but TDM has become the dominant method used by carriers for accommodating multiple data streams on a single cable.

In a TDM-based network, the channels are scanned in sequence, and each channel is given access to the data link at a particular time slot. If a channel has no data to send when its turn comes up, that slot goes unused and the next channel gets its turn.

The TDM network inserts a frame slot at the beginning and end of a group of channels. One pass across all channels results in a frame. Any empty channels get filled with bits to keep frames a consistent size.

Because frames arrive at predictable intervals, TDM is used extensively in the circuit-switched world to send voice traffic. Some examples of TDM are T1 (1.5Mbit/sec) and T3 (45Mbit/sec) circuits, which are used for voice traffic as well as for data.

Another technology that relies heavily on TDM is SONET, a digital fiber optic transport standard used worldwide. It's known internationally as Synchronous Digital Hierarchy (SDH). SONET/SDH is prized because it scales to larger and larger levels of bandwidth. It is an effective Physical-layer technology for such data types as ATM and, more recently, TCP/IP traffic.

The basic building blocks of SONET are groups of DS-0 (64Kbit/sec) circuits that are multiplexed to create a faster Optical Carrier (OC) line rate of 51Mbits/sec. This corresponds to an OC-1 circuit.

From there, OC-1s can be multiplexed to create even faster connections. For example, an OC-3 circuit can carry 155Mbits/sec. An OC-12 circuit (622Mbits/sec) is made up of three OC-3s, and an OC-48 can carry four OC-12s.

The trouble with SONET, and with TDM in general, is that as you multiplex more and more signals on fiber and increase its capacity, you need to add and upgrade the equipment that switches and routes the signals. Add/drop multiplexers and other high-speed transmission equipment are needed to manage the overhead created by the complex system.

Also, as you move to higher and higher speeds, the combination of lower-speed trunks requires SONET add/drop multiplexers to deal with all the data on the new, combined pipe, because none of the multiplexers can deal directly with their own data traffic. All of this makes adopting ever-higher-speed versions of SONET a costly and complicated way to maximize pipes.

TDM devices are required to convert data from light waves to electronic signals and back again—presently there are no commercially available methods for directly switching beams of light. This conversion process adds to the system's complexity, but does not do anything to alleviate bottlenecks.

Another related concern with TDM-based systems is scalability. Currently, the common top speed for TDM is 10Gbits/sec. If you look at the SONET hierarchy, the next logical step for TDM is all the way to 40Gbits/sec (equivalent to OC-768). However, there are doubts as to whether current TDM equipment can handle such a leap.

MAKING WAVES

While TDM-based networks thrive, there is another method for expanding fiber capacity beyond TDM without the bottlenecks and complexity. Wave Division Multiplexing (WDM) divides the light traveling through fiber into wavelengths, also known as lambdas. Lambda is the Greek letter used to represent wavelength. (Because wavelength is inversely proportional to frequency, WDM is logically equivalent to FDM.) Each wavelength can support the high speeds that once required entire optical fibers—even as high as 10Gbits/sec each. In other words, on WDM systems, multiple channels can be transmitted over a single fiber because they are sent at different wavelengths.

WDM wavelengths can each carry independent signals—OC-3 voice on one wavelength, analog video on another wavelength, and OC-12 ATM on yet another one.

Currently, WDM systems can carry as many as two dozen channels, but in the future, capacity should increase to 128 channels or more on a single fiber. The potential bandwidth on WDM systems is mind boggling. For example, a system with 24 channels, each running at OC-48, would have a total capacity of 60Gbits/sec, and a system with 40 such channels would carry 100Gbits/sec.

Today, Dense Wave Division Multiplexing (DWDM) has subsumed WDM. Technically they are similar, but as the name implies, DWDM systems contain many more channels, and therefore much more bandwidth capacity. But most discussions of WDM and DWDM don't strictly define how many channels constitute one or the other, so the terms have become interchangeable (although DWDM is more commonly used).

While DWDM is far from a new technology, major telecommunications vendors and carriers have only recently employed it to boost fiber capacity. So far, DWDM has been used mostly in point-to-point links over great distances. The cost benefits in this scenario are significant because DWDM doesn't require nearly as much equipment as traditional TDM-based systems. Although DWDM has found a strong position in the long distance carrier space, the cost savings are not nearly as compelling for shorter links. Instead, options such as single-fiber SONET are still the best deal for short hauls.

DWDM is starting to attract interest in the medium-haul market, which could include metropolitan area networks. But because of the large amounts of TDM equipment already out there, you'll likely see hybrid deployments of TDM and DWDM technologies.

Because DWDM is economical and a relatively simple method of transporting network data, this technology will probably be used as the basic rails that other types of traffic ride on.

WORKING TOGETHER

In the future, fiber networks will probably have a combination of SONET/SDH that runs over a DWDM infrastructure. For example, data traffic would be multiplexed by SONET add/drop multiplexers before going over a DWDM system. The advantage is that SONET/SDH is easier to manage than raw DWDM. SONET allows for network provisioning and repair, which occurs by means of communications among various points on a SONET

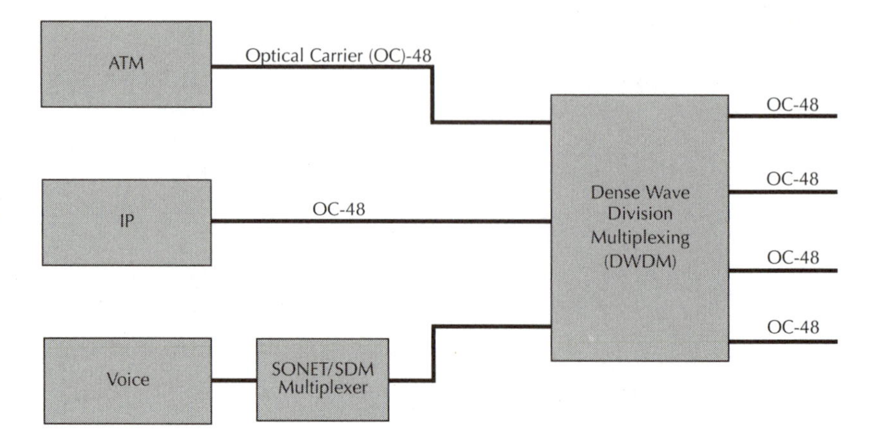

■ **Direct to the Mainline** Connecting the ATM and IP traffic directly to an optical-based network such as Dense Wave Division Multiplexing (DWDM) can reduce costs and work around potential WAN traffic bottle-necks.

network. Also, no current standards define how to manage the different wavelengths in a DWDM system, while SONET/SDH is much more established. It is therefore easier to manage bandwidth using SONET/SDH.

Today, most wide area traffic, including IP, is converted to ATM cells and then transported over SONET. This usually occurs because IP traffic needs to be multiplexed with ATM traffic or other TDM traffic to make it cost-effective to send. However, this method does include a degree of overhead—the infamous cell tax. (For more on IP over ATM over SONET and other wide area options, see "ATM and Alternatives in the Wide Area Backbone," July 1999.)

Dense Wave Division Multiplexing steps in nicely here by lowering the cost of necessary equipment and by solving the bandwidth bottleneck that could begin plaguing the higher levels of TDM (see Figure). In this scenario, high-speed interfaces (such as OC-48 or OC-192) could let routers and switches connect directly into a DWDM system without going through SONET TDM based on add/drop multiplexers. So you'll see systems in which data and voice traffic are multiplexed by SONET before going over DWDM, and you'll see systems in which IP and ATM data connect directly to DWDM without any interference from SONET.

DOWN THE ROAD

An infrastructure based on DWDM should let service providers increase capacity within specific places on their networks. Providers can offer different levels of bandwidth to customers by making individual wavelengths available, which should be a more cost-effective solution.

While many businesses, especially very small ones, are doing just fine with a T1 line, many large corporations aren't satisfied with a T3. For the latter group, the option to lease wavelengths might enable new applications that require mass quantities of bandwidth.

Over the next several years, as more and more traffic goes onto existing wide area infrastructures, the need to optimize the fiber already in place will grow. TDM has provided a way to increase fiber capacity, and when combined with the more efficient and cost-effective DWDM, the move toward a system that supports all types of data and voice traffic without running out of bandwidth comes closer to reality.

RESOURCES

For tutorials on SONET, Synchronous Digital Hierarchy (SDH), Dense Wave Division Multiplexing (DWDM), and optical networking, visit www.sonet.com/edu/edu.htm.

A number of white papers on optical networking can be found at www.usa.alcatel.com/telecom/transpt/optical/techpaps/techpaps.htm.

This tutorial, number 133, by Anita Karvé, was originally published in the August 1999 issue of Network Magazine.

GSM and TDMA Cellular Networks

Cellular networks have traditionally been the preserve of voice, a tradition set to change over the next few years. Most cell phones worldwide already have the ability to send and receive short text messages, and an increasing number now incorporate more advanced Internet capabilities. However, how far can existing infrastructure take the wireless Internet? To understand this, it's necessary to explore how today's mobile phone systems work.

Despite the marketing terms used by different operators, all mobile systems except point-to-point radio are cellular: they divide their coverage area into cells, at the center of which sits a Base Transceiver Station (BTS).

Each operator is licensed by government to use only a small slice of the available radio spectrum (typically around 20MHz), but no single BTS can use more than one third of this slice. This is because adjacent cells would interfere with each other if they used the same frequencies, so at least three different radio channels are needed. The most efficient cell arrangement uses a honeycomb pattern, as shown in Figure 1.

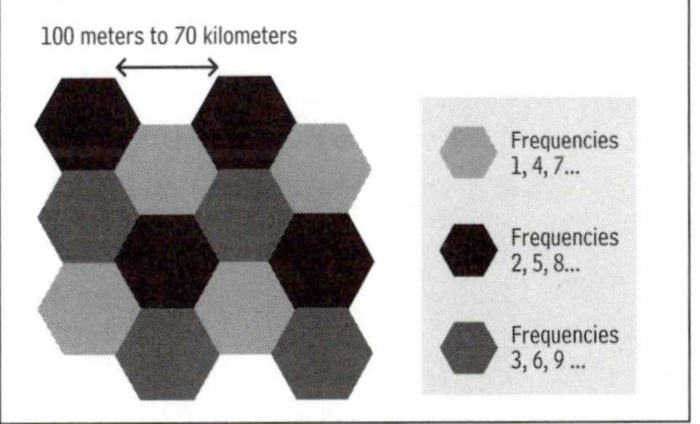

■ **Figure 1:** In an ideal cellular network, hexagonal cells are arranged in a honeycomb structure, each using a third of the available spectrum to prevent interference.

The simplest cellular systems use Frequency Division Multiplexing (FDM), which gives each conversation its own waveband, just like a radio station. For example, the analog American Mobile Phone System (AMPS), still the most widely used wireless standard in the United States, gives each user 30KHz. This limits the number of calls to the number of 30KHz frequencies available: If an operator has a license for 10MHz of spectrum, it can support at most 111 calls per cell.

It's possible to increase capacity by making cells smaller, but this is expensive and perhaps dangerous. Current technology also limits cell diameters to at least 100 meters (330 feet) because the internal circuitry of a BTS generates so much radiation that it would interfere with very low power transmissions. The only way to carry more traffic without building more cells is to pack more calls into a single frequency. This means that the voice must be converted into data.

TIME OFF

There are at least three ways to share a frequency, the simplest of which is further FDM. The trouble with this, as radio listeners know, is that nearby frequencies interfere with each other.

A second (and more sophisticated) approach is Code Division Multiple Access (CDMA), recently developed by Qualcomm. CDMA encodes each transmission to provide, in theory, unlimited capacity—it will form the basis of future broadband systems.

Finally, most existing mobile data networks use Time Division Multiple Access (TDMA), a method by which each conversation only uses a frequency for part of the time.

To most U.S. consumers, TDMA means a particular mobile system, the one technically known as the Digital-American Mobile Phone System (D-AMPS) or IS-136. However, to engineers, this is just one among many TDMA technologies. The most important is Global System for Mobile Communications (GSM), the cell phone standard developed by European governments in the 1980s. GSM now accounts for more than half the world's mobile market.

TDMA also forms the basis of both Personal Digital Cellular (PDC), the second-most-popular mobile standard (even though it's only used in Japan), and Motorola's proprietary Integrated Digital Enhanced Network (IDEN).

All TDMA systems work by dividing a frequency into a number of time slots, each of which corresponds to one communication channel. A cell phone transmits and receives in only one slot, remaining silent until its turn comes around again. This means that a D-AMPS phone only transceives for one-third of the time, and a GSM for one-eighth (see Figure 2). The number of slots, the cycle length, and the frequency width all depend on the particular technology: As shown in Table 1, GSM uses wider frequencies than the others, with much shorter time slots.

GSM's wide frequencies give it scalability advantages, but the short time slots cause problems in keeping phones synchronized with each other. Radio signals take just over 0.003 microsecond to travel a kilometer (km), which adds up to a round-trip delay of around 0.4 microsecond for a phone only 60 km (40 miles) from the base station. The time slot only lasts 0.577 microsecond, so this delay is enough to make the phone miss its slot entirely, even though it would be unnoticeable to a human listener. In practice, GSM phones cannot be used more than 35 km (22 miles) from a BTS, no matter how strong the signal.

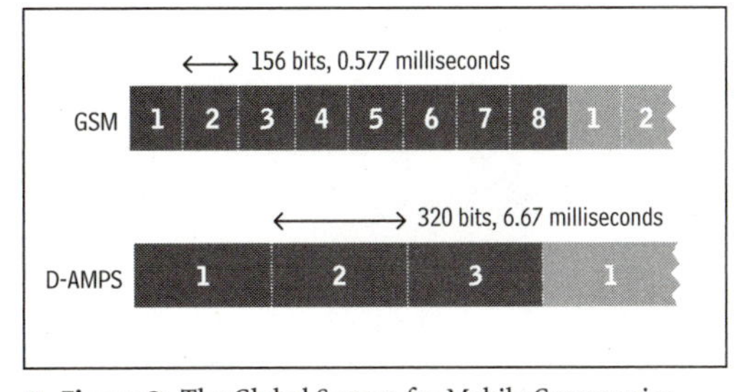

■ **Figure 2:** The Global System for Mobile Communications (GSM) multiplexing scheme uses eight time slots, compared to Digital-American Mobile Phone System's (D-AMPS) three.

Each 0.577-microsecond GSM time slot carries 156 bits, which in theory amounts to a capacity of 33.8Kbits/sec for each of the eight channels. Unfortunately, most of this is not available for data applications. The GSM protocol is very wasteful in terms of its overhead, requiring a lot of bandwidth for such functions as signaling and cryptography. To prevent eavesdroppers, every transmission is encrypted using a secret algorithm known as A5, which is thought to use keys of up to 56 bits.

The raw bandwidth of a D-AMPS channel is only 16Kbits/sec, of which much less is used for protocol overhead. This is partly because it doesn't include strong encryption, thanks to U.S. export regulations. Another reason is that call setup and location monitoring are handled by a separate analog channel, for compatibility with the older AMPS networks.

TDMA TECHNOLOGIES

Time Division Multiple Access (TDMA) technology	Frequency width	Number of slots	Slot length	Bits per slot	Cycles per second	Coverage
Global System for Mobile Communications (GSM)	200KHz	8	0.577 microsecond	156	217	Worldwide
Digital-American Mobile Phone System (D-AMPS)	30KHz	3	6.67 microseconds	320	50	America
Personal Digital Cellular (PDC)	25KHz	3	6.67 microseconds	430	50	Japan
Integrated Digital Enhanced Network (IDEN)	25KHz	6	7.25 microseconds	464	23	America

■ **Table 1:** Time Division Multiple Access (TDMA) technologies used by current mobile networks.

DATA DELAYS

On the wired telephone network, toll-quality voice requires a full DS0 channel, operating at 64Kbits/sec. Cellular systems don't have this kind of capacity, so the bitstream has to be compressed. Each technology has its own codec (coder/decoder), and as with all compression, there is a tradeoff between quality and compactness.

GSM's codec gives the best results, producing a 260-bit packet every 20 milliseconds (ms) for a data rate of 13Kbits/sec. The D-AMPS codec requires only 8Kbits/sec, which is lower quality but still intelligible. Vendors have recently developed "half-rate" codecs for both systems, which aim to double the voice capacity of a network, though so far these are rarely used.

However, a cell phone can't just transmit the raw output of its codec. Wireless networks suffer from interference caused by everything from cosmic rays to microwave ovens, and they need some way to ensure that data arrives intact. Both GSM and D-AMPS use Forward Error Correction (FEC), which enables the bits judged most important to be recovered without retransmission if they are lost. GSM's algorithm is again the more sophisticated and bandwidth-hungry: including FEC, its speech rate climbs to 22.4Kbits/sec, compared to only 13Kbits/sec for D-AMPS.

The earliest schemes for sending data over mobile networks had to route it through the voice codec, reducing capacity to around 5Kbits/sec. However, recent upgrades to both GMC and D-AMPS enable data to bypass the codec entirely, as well as some of the error correction, pushing the rate to 14.4Kbits/sec over GSM and 9.6Kbits/sec over D-AMPS (see Table 2). Data traffic is actually more tolerant of errors than highly compressed voice because it doesn't matter in which order packets arrive.

BANDWIDTH ALLOCATION

	Digital-American Mobile Phone System (D-AMPS)		Global System for Mobile Communications (GSM)	
	Voice	**Data**	**Voice**	**Data**
Information	8Kbits/sec	9.6Kbits/sec	13Kbits/sec	14.4Kbits/sec
Error correction	5Kbits/sec	3.4Kbits/sec	9.4Kbits/sec	8Kbits/sec
Protocol overhead	3Kbits/sec		11.4Kbits/sec	
Total	**16Kbits/sec**		**33.8Kbits/sec**	

■ **Table 2:** Where does the bandwidth go?

To increase data capacity further, carriers are turning to two new technologies: High Speed Circuit Switched Data (HSCSD) and General Packet Radio Service (GPRS). They were intended for GSM but can also be applied in a more limited way to D-AMPS and PDC systems.

Both technologies grant each channel access to more than one time slot in the multiplex. GPRS also adds packet-switching to use bandwidth more efficiently and permit "always-on" connections.

Regular GSM calls use only one slot out of eight, so in theory GPRS could provide an eightfold acceleration to 115Kbits/sec. However, the first GPRS services, expected in Europe by early 2001, will only offer a fraction of this rate. The main reason is that a full-speed device would generate too much microwave radiation, overheating the phone and possibly threatening the user's health.

The problems for GPRS over D-AMPS and PDC are even more serious. These only use three slots, compared to GSM's eight, limiting the maximum bandwidth to only 28.8Kbits/sec. This means that while Europe's existing cellular system will eventually be upgraded to ISDN speeds, American mobile operators who want landline performance will have to roll out entirely new networks.

RESOURCES

The Global System for Mobile Communications (GSM) Association, a Dublin, Ireland-based alliance of GSM operators and vendors, is located at www.gsmworld.com.

Intel runs two Web sites designed to aid anyone wanting to send data over digital cellular networks. The sites contain many articles on existing and future wireless systems. Intel has an obvious interest in promoting data applications but is not tied to any particular phone technology or vendor. Go to www.gsmdata.com and www.pcsdata.com.

This tutorial, number 141, by Andy Dornan, was originally published in the April 2000 issue of Network Magazine.

CDMA and 3G Cellular Networks

For all the hype about the wireless Internet, today's mobile data networks are rather primitive. Usually bolted onto cell phone systems built for voice, they offer low bit rates and poor interoperability. All this will change over the next three years, as operators construct third-generation (3G; see "Mobile Generations" below) mobile networks. These aim to provide packet-switched data to a handheld terminal with throughputs measured in hundreds of Kbits/sec.

MOBILE GENERATIONS

1G: Analog voice
2G: Digital voice and messaging
3G: Broadband data and Voice over IP (VoIP)

3G has been in gestation since 1992, when the International Telecommunications Union (ITU) began work on a standard called IMT-2000. IMT stands for International Mobile Telecommunications; the number 2000 initially had three meanings: the year that services should become available, the frequency range in MHz that would be used, and the data rate in Kbits/sec. The name has stuck, though all these criteria were eventually abandoned. The year slipped back to 2002, North America is already using the recommended frequencies for other services, and high speeds will only be available through special indoor "pico-cells."

The ITU envisaged IMT-2000 as a single global standard, but the world's regulators, vendors, and carriers were unable to reach a unanimous agreement. The path to 3G will be gradual, and everyone wants to ensure compatibility with their existing systems. The FCC's failure to license any new spectrum further complicated things: American operators must upgrade their existing networks rather than build new ones.

In October 1999, representatives from different countries finally agreed to disagree. The result is a "federal standard," or, more accurately, a fudge. IMT-2000 will have at least three modes of operation, with no guarantee that phones for one mode will work with the others.

ACCESS CODES

Two of the three IMT-2000 modes are based on Code Division Multiple Access (CDMA), a system that enables many users to share the same frequency band at the same time (see Table 1). Each signal is encoded differently so that it can be understood by a receiver with the same code. CDMA actually predates both computers and mobile phones, though it was considered too complicated for use in cellular phone networks until the late '90s. Many of the innovations in cell networks were due to the company Qualcomm (www.qualcomm.com), whose growth in the cellular technology market has echoed that of Microsoft and Cisco Systems.

CDMA is often compared to an airport transit lounge, where many people are speaking in different languages. Each listener only understands one language, and therefore concentrates on his or her own conversation, ignoring the rest. The analogy isn't exact, because a roomful of people all talking at once soon becomes very loud. Everyone ends up trying to shout above the background noise, which just makes the problem worse.

To prevent such a vicious circle, CDMA codes are chosen so that they cancel each other out. For exact cancellation, signals must be perfectly timed; base stations need to make very

CDMA FLAVORS

Code Division Multiple Access (CDMA) system	Channel bandwidth	Chip rate	Maximum capacity	Real capacity
cdmaOne IS-95b	1.25MHz	1.2288MHz	115Kbits/sec	64Kbits/sec
cdma2000 1XMC	1.25MHz	1.2288MHz	384Kbits/sec	144Kbits/sec
cdma2000 1Xtreme	1.25MHz	1.2288MHz	5.2Mbits/sec	1.2Mbits/sec
cdma2000 HDR	1.25MHz	1.2288MHz	2.4Mbits/sec	621Kbits/sec
cdma2000 3XMC	3.75MHz	3.6864MHz	4Mbits/sec	1.117Mbits/sec
Wideband CDMA (W-CDMA)	5MHz	4.096MHz	4Mbits/sec	1.126Mbits/sec

Source: Motorola

■ **Table 1: Prick Up Your Ears.** Code Division Multiple Access (CDMA) systems enable many users to share the same frequency band at the same time. Currently, cdmaOne is the only such system in use.

precise measurements of their time and location. They do this by using signals from Global Positioning System (GPS) satellites, which can pinpoint anywhere on Earth to within four meters and measure time more accurately than the Earth's own rotation.

The only CDMA system in use so far is cdmaOne, developed by Qualcomm but now supervised by an independent organization called the CDMA Development Group (CDG). It has been standardized by the Telecommunications Industry Association (TIA) as IS-95a, and is very popular among cellular operators in America and Asia. Because it already uses CDMA, it is easier to upgrade to 3G compared to rival systems based on Time Division Multiple Access (TDMA).

cdmaOne spreads every signal over a 1.25MHz channel, transmitting on the entire bandwidth at once. It uses a set of 64 codes, known as Walsh sequences, so in theory up to 64 phones could use the channel at once. In practice, that number depends on the data throughput. The basic system offers voice and 14.4Kbit/sec data rates, which facilitates between 15 and 20 users. An upgrade called IS-95b offers data rates of up to 115Kbits/sec, which would mean only two users per channel.

OPEN WIDE

To reach the IMT-2000 target of 2Mbits/sec, CDMA systems need to use more codes, a different modulation scheme, and wider bandwidths. The official upgrade, developed by Qualcomm and ratified by the ITU, is known as cdma2000 3XMC. The 3 comes from its 3.75MHz bandwidth, the result of three cdmaOne 1.25MHz channels joined together.

As an intermediary step, most cdmaOne operators are deploying a technology called cdma2000 1XMC, which uses the same 1.25MHz channels. It doubles the number of codes to

128, thus doubling either the throughput per user or the number of users in a cell. Qualcomm and Motorola are also pushing rival schemes that enhance 1XMC, known respectively as High Data Rate (HDR) and 1Xtreme.

Both 1Xtreme and HDR work by altering the modulation scheme, or the way data is actually represented in radio waves. Most cell phones use a system called Phase Shift Keying (PSK), which interrupts a wave and moves it to a different point in its cycle. The bit rate depends on the frequency of these interruptions, known as symbols, and on the number of shapes that each symbol can take.

The symbols in quadrature PSK, the system used by cdmaOne, can take four different shapes. This means that each shape can represent two bits, since two bits can take four combinations. The 8-PSK variation could represent three bits per symbol, increasing the data rate by half. HDR and 1Xtreme automatically increase the number of shapes to the highest number supported, depending on their connection quality: Poor connections cannot cope with a system using many shapes per symbol, as they become hard to distinguish.

To make things more confusing, many parts of the world are rolling out an entirely new system called Wideband CDMA (W-CDMA), or Universal Mobile Telecommunications System (UMTS). This requires the new spectrum assigned by the ITU, and thus can't be used in the United States. It is technically very similar to cdma2000 3XMC but uses a slightly wider bandwidth, hence the name. The wider bands are necessary so that the system can interoperate with Global System for Mobile Communications (GSM), the popular second-generation (2G) standard.

Though W-CDMA is not a direct upgrade from GSM, it was designed for use in areas that are already well-covered by GSM networks. The theory is that operators can build W-CDMA "hot spots" in major cities, and rely on GSM in the beginning for the rest of their coverage. All W-CDMA phones sold in Europe will work with GSM; they'll even be able to hand over calls so that users can move between cells based on the different systems without any interruption.

Even in countries where GSM is not prevalent, W-CDMA is likely to prove popular because it offers instant coverage nearly worldwide. GSM-free Japan is rolling out the world's first W-CDMA networks because its existing Personal Digital Cellular (PDC) system has run out of capacity. Mobile operators NTT DoCoMo and J-Phone both began trials in early 2000 and hope to have commercial services in operation during 2001.

TIME AGAIN

Not all 3G systems use CDMA. A third variant of IMT-2000, called EDGE, applies the same modulation tricks as HDR to the TDMA-based General Packet Radio Service (GPRS, described in Network Magazine's April 2000 Tutorial). Operators of GSM and Digital Advanced Mobile

Phone System (D-AMPS) who don't have a license for a new spectrum need GPRS; this includes all mobile providers in the United States, as the FCC hasn't issued any licenses.

EDGE requires a channel bandwidth of 200KHz, the same as GSM. This makes it more difficult for D-AMPS operators, who are used to 30KHz channels, but still more attractive than the alternative: A CDMA system requires a minimum of 1.25GHz, which a busy network may not be able to spare from its existing service. Dual-mode handsets are also easier to make if they support two TDMA systems—rather than both CDMA and TDMA—though this problem must be overcome by European manufacturers if UMTS is to be compatible with GSM.

From layer 2 upward, EDGE inherits virtually every property of GSM and GPRS. At the radio level, it works by replacing the Gaussian Minimum Shift Keying (GMSK) modulation of GSM with 8-PSK. GMSK is a binary system, permitting only one bit per symbol compared to 8-PSK's three. The raw data rate is thus tripled from GPRS's 171Kbits/sec to 513Kbits/sec. Some of this must be allocated toward error correction, with nine different throughput speeds defined depending on the quality of the connection.

There are also 29 different terminal classes; each class grants users access to the channel at different times while transmitting, receiving, or both. The result: Users have access to a wide range of speeds, from 8.8Kbits/sec to 474Kbits/sec. This doesn't quite meet the original IMT-2000 wish list, but the ITU has ratified it anyway, on the understanding that the 2Mbit/sec requirement can be satisfied by combining it with a wireless LAN technology.

RESOURCES

The official International Mobile Telecommunications (IMT)-2000 site at www.itu.int/imt2000/ has plenty of background information on the standardization process.

The Third Generation Partnership Project is an alliance of vendors and regulators that tries to stay neutral in the standards war. Its Web site is located at www.3gpp.org.

The Code Division Multiple Access (CDMA) Development Group's Web site (www.cdg.org) promotes all forms of CDMA and provides detailed mathematical information about how they work.

Detailed information on the European Wideband CDMA (W-CDMA) system can be found at www.umts-forum.org.

The Universal Wireless Communications Consortium explains how EDGE works and promotes Time Division Multiple Access (TDMA) as a third-generation (3G) technology at www.uwcc.org.

This tutorial, number 146, by Andy Dornan, was originally published in the September 2000 issue of Network Magazine.

Ultra Wideband Wireless Networks

A handful of press accounts describing a novel kind of wireless technology began appearing in the latter part of 2000. The most extreme claims about Ultra Wideband (UWB) technology were: that it could deliver hundreds of Mbits/sec of throughput; that its power requirements to link to destinations hundreds of feet away were as little as 1/1,000th that of competing technologies such as Bluetooth or 802.11b; that transceivers could be small enough to tag grocery items and small packages; and that traffic interception or even detecting operation of the devices would be practically impossible.

A slightly different way to look at the difficulty of detection and interception would be to claim that UWB devices wouldn't interfere with other electromagnetic spectrum users.

While we're certainly years away from any significant deployment of UWB devices, each of the stupendous claims made for the technology has at least a modicum of supporting evidence.

UWB devices operate by modulating extremely short-duration pulses—pulses on the order of 0.5 nanoseconds. Though a system might employ millions of pulses each second, the short duration keeps the duty cycle low—perhaps 0.5 percent—compared to the near-100-percent duty cycle of spread spectrum devices. The low duty cycle of UWB devices is the key to their low power consumption. Intel Architecture Labs has calculated the comparative spatial throughput capacity of various technologies (see Figure 1); UWB clearly has by far the highest potential for this particular metric.

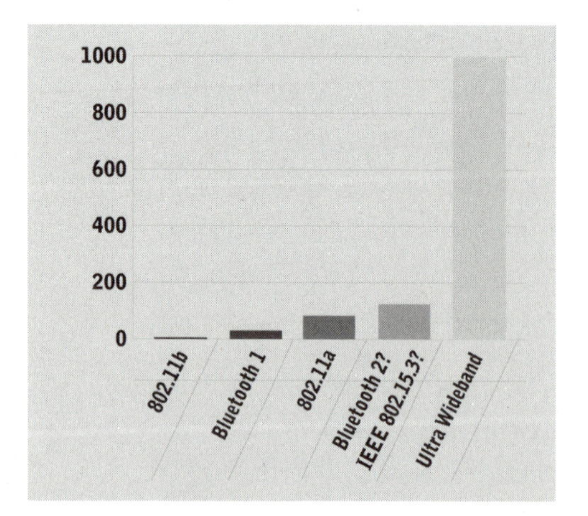

■ **Figure 1:** Intel has calculated the spatial capacity (or throughput over an area) for several potentially popular wireless technologies. Ultra Wideband systems display a tremendous relative advantage based on this metric.

In principle, pulse-based transmission is much like the original spark-gap radio that Marconi demonstrated transatlantically in 1901. Unlike most modern radio equipment, pulse-based signals don't modulate a fixed-frequency carrier. If you examine a carrier-based signal with a spectrum analyzer, you'll generally see a large component at the carrier frequency and smaller components at frequencies above and below the carrier frequency based on the modulation scheme (see Figure 2, page 32). Pulse-based systems show more or less evenly distributed energy across a broad range of frequencies—perhaps a range 2GHz or 3GHz wide for existing UWB gear. With low levels of energy across a broad fre-

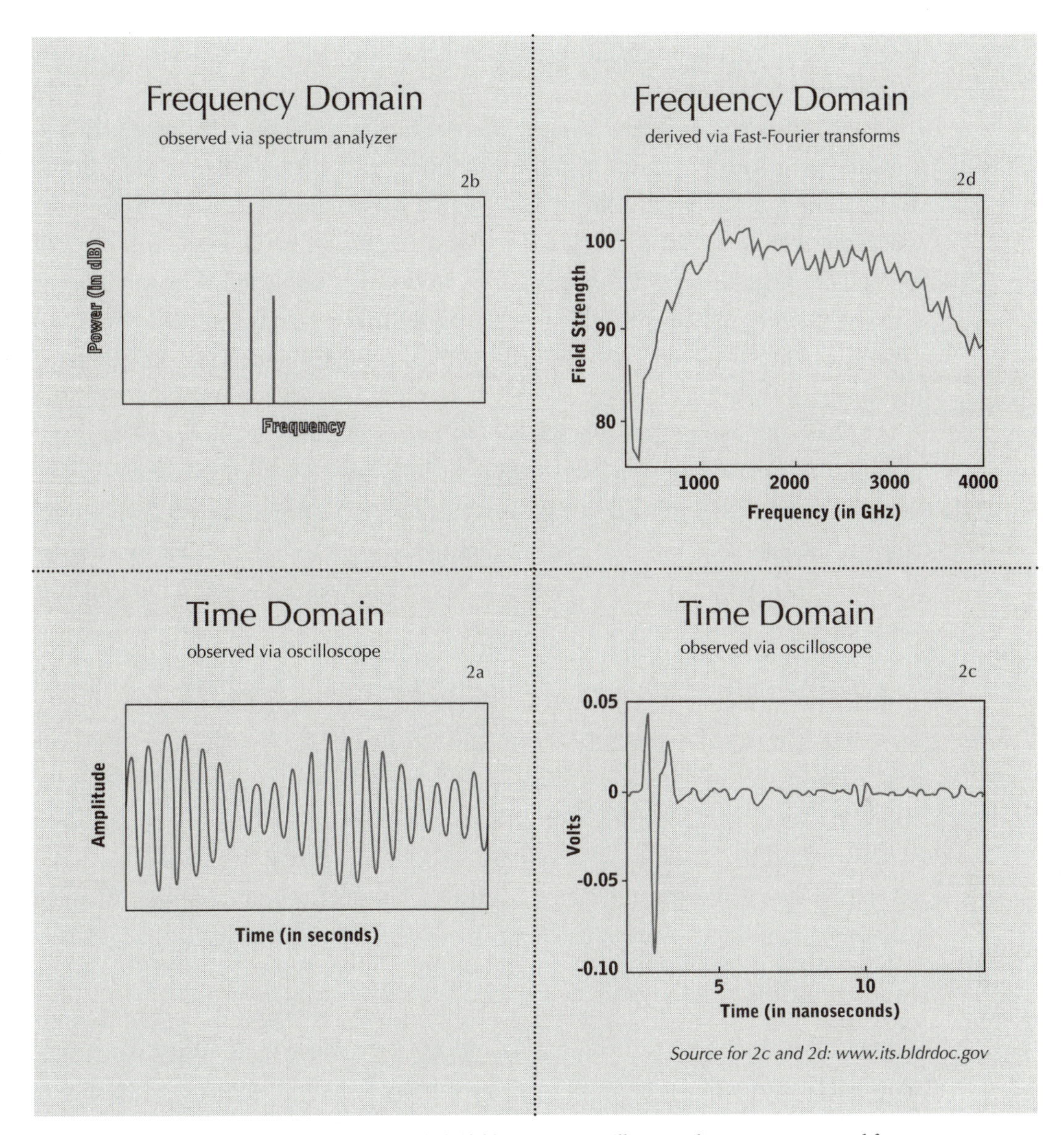

■ **Figure 2:** An amplitude-modulated signal might look like 2a on an oscilloscope, the most common tool for observing the time-domain elements of a signal. The same signal might look like 2b on a spectrum analyzer, the most common tool for observing the frequency-domain characteristics of a signal. The mathematical functions that describe time-domain signals can be converted to their frequency-domain functions by means of Fourier transforms. The time-domain function in 2c was generated by an Ultra Wideband device, while the frequency-domain graph in 2d was produced by Fast-Fourier transforms operating on 2c. The spread-out frequencies in 2d resemble ordinary noise, unlike the narrowband signals in 2b.

quency range, UWB signals are extremely difficult to distinguish from noise, particularly for ordinary narrowband receivers.

One significant additional advantage of short-duration pulses is that multipath distortion can be nearly eliminated. Multipath effects result from reflected signals that arrive at the receiver slightly out of phase with a direct signal, canceling or otherwise interfering with clean reception. (If you try to receive broadcast TV where there are tall buildings or hills for signals to bounce from, you've likely seen "ghost" images on your screen—the video version of multipath distortion.) The extremely short pulses of UWB systems can be filtered or ignored—they can readily be distinguished from unwanted multipath reflections.

Alternatively, detecting reflections of short pulses can serve as the foundation of a high-precision radar system. In fact, UWB technology has been deployed for 20 years or more in classified military and "spook" applications. The duration of a 0.5 nano-second pulse corresponds to a resolution of 15 centimeters, or about 6 inches; UWB-based radar has been used to detect collisions, "image" targets on the other side of walls, and search for land mines.

ULTRA WIDEBAND VS. SPREAD SPECTRUM

Spread spectrum technologies do artificially what UWB does naturally: array signals across a wide spectrum so that the power concentrated in any particular band is below the threshold where it would interfere with other users of that band, or even be detectable by a narrowband receiver. Spread spectrum signals begin life as ordinary narrowband modulated waveforms. Then a spreading function—direct sequence and frequency hopping are the two most common methods—rapidly cycles the original signal through multiple individual narrowband slots.

Because spread spectrum signals are "artificially" spread, their duty cycles are close to 100 percent. UWB technology has it all over spread spectrum where transmission power consumption is concerned. Furthermore, spread spectrum devices require more complex electronics than UWB, first because UWB circuits can be fundamentally simpler than narrowband circuits, but also because spread spectrum requires additional components and processing operations with pseudo-random noise generators and the associated task of synchronization. This added complexity has power-usage, device-size, and cost repercussions, too.

GOTCHAS

As with any other technology, UWB technology's strong points determine its shortcomings. UWB emissions can potentially interfere with many other consumers of the electromagnetic spectrum. Users of Global Positioning Systems (GPSs), particularly those in the aviation industry who use GPS data for navigation and landing, have serious reservations about widespread mobile devices that introduce even very low levels of interference into the 1.2GHz and 1.5GHz bands. Sprint and Qualcomm, whose Code Division Multiple Access (CDMA) technology underlies Sprint's PCS systems, conducted studies that showed degradations of cell phone service in the presence of UWB devices. The National Association of Broadcasters opposed FCC approval for UWB out of concern for spectrum used by remote camera crews and for satellite-based content distribution.

While it's hardly surprising that current owners would resist even minuscule incursions into their spectrum, UWB technology doesn't provide a free-lunch windfall of hitherto unused spectrum. Its ready observability in the time domain doesn't carry over to the frequency domain, but that doesn't mean there's no impact on existing services whose frequency domain signatures are easier to observe. It's also possible for UWB vendors to filter specific frequencies that are important for public safety and other overriding concerns, though if the owners of all the relevant spectrum had to be filtered, there wouldn't be any left for UWB to operate in.

The FCC has issued a Notice of Proposed Rule Making with respect to UWB technology, with a decision expected before the end of 2001. UWB proponents have requested that their devices be governed by Part 15.209 rules, which set emission limits for such things as hair dryers and laptop computers. FCC approval would only be the beginning of a process of regulatory activity, however. Once UWB is on track with regulatory approval, it seems likely that there would also be a process of standardization aimed at minimizing interference with other technologies, among other concerns.

In many respects, the excitement over UWB technology turns out not to be unrealistic at all. It seems unlikely that unreasonable regulatory obstacles will impede further development, though it's dangerous to underestimate the amount of delay a standards committee can add to a technology introduction. And no matter what, Bluetooth and 802.11b devices will never be able to find studs in your walls, detect your cat's presence several rooms away, or prevent your SUV from colliding with a cement truck.

RESOURCES

Probably the best-detailed introduction to Ultra Wideband (UWB) technology is an article in Intel's Developer Journal for the second quarter of 2001. Go to http://developer.intel.com/technology/itj/q22001/articles/art_4.htm.

Several other vendors active in UWB technology offer various white papers, links to patents, product information, and historical information:

Aether Wire and Location www.aetherwire.com/Aether_Wire/aether.html

MultiSpectral Solutions www.multispectral.com

Pulse-Link www.pulselink.net

Time Domain www.timedomain.com

The somewhat desultory Web site for the Ultra Wideband Working Group can be found at www.uwb.org.

This tutorial, number 160, by Steve Steinke, was originally published in the November 2001 issue of Network Magazine.

SECTION III

Data-Link Protocols

Topologies

Understanding the topology of LAN technologies can tell you a lot about your alternatives when installing or expanding a LAN. At its basic level, the topology of a network refers to the way in which all of its pieces have been connected. That is, it refers to the layout of the computers, printers, and other equipment hooked to the network in your building.

Because cable connects these scattered computing resources together into a network, your network's topology is also a function of the way in which the cabling is organized, whether it is arrayed in a bus, ring, or star, which are the three basic physical topologies available to LAN designers (see Figure 1). Although recent technological advances have blurred the distinctions between the physical and logical arrangements, the topology you select (or are forced to select) may also dictate the media-access control method (that is, Ethernet or Token Ring) under which your network will operate.

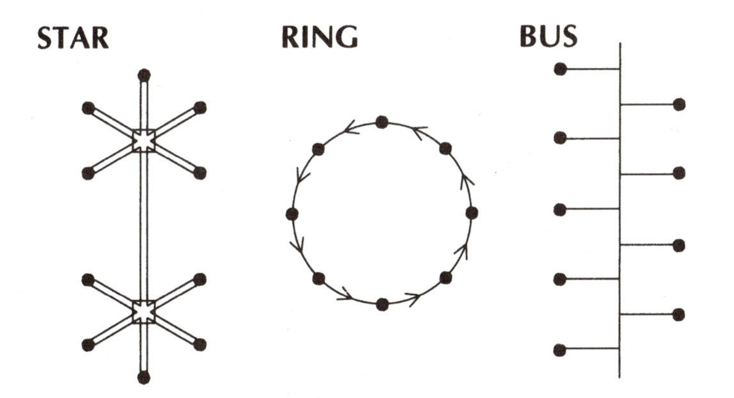

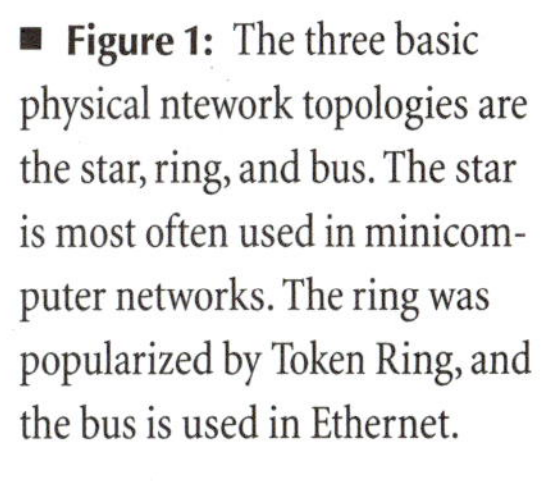

■ **Figure 1:** The three basic physical ntework topologies are the star, ring, and bus. The star is most often used in minicomputer networks. The ring was popularized by Token Ring, and the bus is used in Ethernet.

A network's logical layout may differ from its physical layout. The logical topology defines the electrical path; the physical path defines how the cables, concentrators, and nodes are arranged. For example, Ethernet must be a logical bus network; however, it can be physically configured as a bus or star. Token Ring is a logical ring, but is physically configured as a star. FDDI, a logical ring, is physically configured either as a ring or a star.

THE STAR ROUTE

Until recently, the star topology has been found mostly in minicomputer and mainframe environments. These typically consist of a system of terminals or PCs, each wired to a central processor. It is also used by AT&T in both its StarLAN network and its Private Branch Exchange (PBX) based network. The star topology is ideal for wide area network (WAN) applications in which outlying offices must communicate with a central office.

A principle advantage of the star topology is that it not only allows centralizing key networking resources—concentrators or line conditioning equipment—but also gives the network administrator a focal point for network management. When something goes wrong with the network, the administrator can troubleshoot it from one place, usually a wiring closet, but possibly from a remote management console.

The star-based network requires a substantial investment in cable, however. Each workstation is connected to the central concentrator by its own dedicated line. In some star-based network technologies (ARCnet, for example) this line is coax cable that runs from an active hub to a workstation. (ARCnet can also operate as a bus.)

The 10Base-T Ethernet standard permits operating traditionally bus-based Ethernet in a star-wired configuration using unshielded twisted-pair (or high grade telephone) wiring.

CASCADED STARS

The use of a modular multiport repeater (also known as a hub or concentrator) with Ethernet allows creating large networks made of what can be called cascaded stars. In this arrangement, one centralized multiport repeater serves as the focal point for many other multiport repeaters, in effect creating a series of star-based Ethernets (see Figure 2).

Using modular multiport repeaters also permits mixing star- and bus-based Ethernet workgroups into a single large network. In this instance, the modular repeater must only be able to accept modules that support the many Ethernet-compatible cable types.

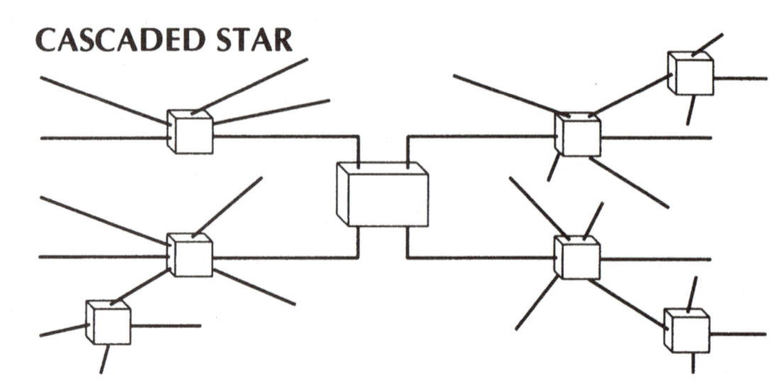

■ **Figure 2:** Twisted-pair Ethernets are often composed of cascaded stars, in which multiport repeaters (represented by boxes) are connected to one another and to a central repeater.

THE RING

IBM popularized the ring topology with its Token Ring technology. Like the bus, a Token Ring network uses a single cable. Unlike the bus, the cable's ends are looped to form a complete logical circle or ring. Physically, Token Ring is a star-wired network. Each workstation is connected directly to a central device called a Media Access Unit (MAU). Logically (or electronically), however, the Token Ring remains a true ring (see Figure 3).

Unlike the bus, Token Ring uses a deterministic, rather than a contention-based, access method. In the Token Ring access method, an electronic signal called a token is passed from station to station on the ring, with each station regenerating the token as it passes by.

TOKEN RING

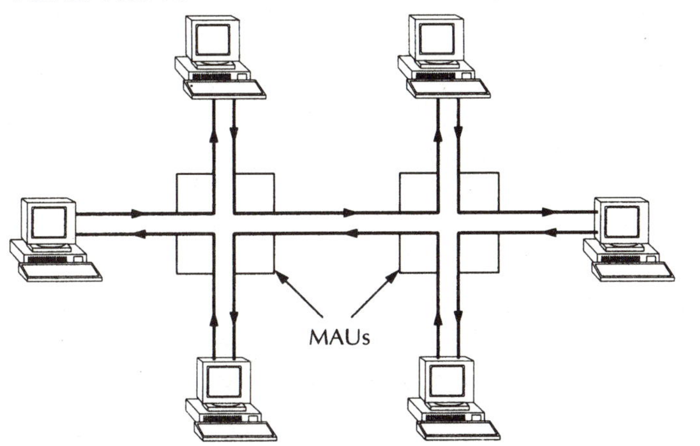

■ **Figure 3:** Although a physical star, IBM's Token Ring Implementation is logically or electronically a ring. The arrows represent a data packet as it moves from one station to another around the ring.

When a station wishes to transmit data over the network, it must wait until the token is passed to it by its neighboring station. It takes control of the station and then places a data packet on the network. Only after the data packet has made a full circuit of the ring, returning to its originator, does the station release the token for the next workstation.

Token Ring can also be expanded by linking multiple rings together, just like Ethernet. In these arrangements, one Token Ring, usually a 16Mbits/sec ring, is dedicated as an internetwork loop, with work group or departmental rings connecting to the company-wide ring via a PC running bridging software.

TAKING THE BUS

In a bus topology, all workstations on the network are attached to a single cable. Ethernet, AppleTalk, and IBM's PC Network are examples of bus-based networks. This sharing of the transmission media (or cable) has several important ramifications. Most importantly, it means that the cable can carry only one message at a time, and each workstation on the network must be capable of knowing when it can and cannot transmit using this shared medium.

Ethernet employs what is called a Carrier Sense Multiple Access/Collision Detection (CSMA/CD) access method to arbitrate use of the cable and to maximize its throughput. In this method, each station on the bus is always listening on the cable for transmission for other stations. It only transmits when the cable is not busy with another transmission. It is able to sense the collision that occurs when it and another station on the bus transmit at the same moment. Having sensed that a collision has occurred—and that the transmission has miscarried—each workstation waits a random time period (usually several microseconds) before retransmitting.

Naturally, frequent retransmission can slow down an Ethernet LAN; this limits the number of workstations that can be placed on any network segment. Fortunately, network managers have access to numerous devices, such as switches, bridges, and routers, that divide Ethernets in small segments to mitigate this problem.

One common arrangement is to run numerous secondary segments off a backbone bus. In a typical installation, Fast or Gigabit Ethernet segments running between a building's floors would serve as the backbone. PCs on each floor would be connected to each other and the backbone via twisted-pair cable.

These secondary or horizontal segments can be linked to the primary bus via a repeater, a bridge, or a router. Each of these devices has its own benefits and disadvantages, principally in the amount of traffic control they provide and the amount of administration they require.

Two major shortcomings of the traditional coaxial cable bus topology are that it requires lots of cable and troubleshooting the length of several thousand feet of cable can be time-consuming and frustrating. The bus topology is, however, highly expandable.

REALITY INJECTION

Unfortunately, reality often dictates the choice of network topology and access method. For example, organizations with a large installed base of IBM equipment once generally opted for an IBM Token Ring network because IBM heavily endorsed Token Ring. Or it may be physically or financially impossible to install a particular cable type. For example, cable raceways may be jammed full with cable, dictating that you use a cable with a narrow diameter, such as fiber, or even a wireless, radio-based transmission method.

This tutorial, number 19, was originally published in the February 1990 issue of LAN Magazine/Network Magazine.

Access Methods

Access method is the term given to the set of rules by which networks arbitrate the use of a common medium. It is the way the LAN keeps different streams of data from crashing into each other as they share the network.

Networks need access methods for the same reason streets need traffic lights—to keep people from hitting each other. Think of the access method as traffic law. The network cable is the street. Traffic law (or the access method) regulates the use of the street (or cable), determining who can drive (or send data) where and at what time. On a network, if two or more people try to send data at exactly the same time, their signals will interfere with each other, ruining the data being transmitted. The access method prevents this.

The access method works at the data-link layer (layer 2) because it is concerned with the use of the medium that connects users. The access method doesn't care what is being sent over the network, just like the traffic law doesn't stipulate what you can carry. It just says you have to drive on the right side of the road and obey the traffic lights and signs.

Three traditional access methods are used today, although others exist and may become increasingly important. They are Ethernet, Token Ring, and ARCnet. Actually, these technologies encompass wider-ranging standards than their access methods. They also define other features of network transmission, such as the electrical characteristics of signals, and the size of data packets sent. Nevertheless, these standards are best known by the access methods they employ.

ETHERNET

Ethernet is the most common network access method. It was developed by Xerox in the mid-1970s. It describes data transmission at 10Mbits/sec, 100Mbits/sec, 1Gbits/sec, and other throughput rates using the CSMA/CD protocol. Ethernet gained its popularity in engineering, scientific, and university environments. The shipments of Ethernet interface adapters grew substantially faster than those of Token Ring through the 1990s. ARCnet unit shipments have declined, and little new development of ARCnet-based solutions should be expected. Flurries of interest in FDDI and ATM for high speed backbones occurred in the late 1990s, but the multiple flavors of Ethernet dominate the great majority of LAN environments at the turn of the millennium.

The Ethernet access method is Carrier Sense Multiple Access/Collision Detection (CSMA/CD). This is a broadcast access method, which means every computer "hears" every transmission, but not every computer "listens" to every transmission.

Here's how CSMA/CD works. When a computer wants to send a message it does, as long as the cable isn't in use by another transmitting node. (This is the carrier sense part.) The signal it sends moves up and down the cable in every direction, passing every computer on the network segment. (This is multiple access.) Every computer can hear the message, but unless the message is addressed to it, the computer ignores it. Only the computer to which the message is addressed receives the message. The message is recognized because it contains the address of the destination computer.

A "collision" occurs if two computers send at the same time (because there is a narrow window of time in which the second computer may have begun transmitting but the "busy signal" has not yet reached the first computer, which blithely begins to transmit). A collision doesn't make any noise, but the signals become garbled and the messages can't be understood. In fact, nodes that detect a collision automatically transmit a special "jam" signal, which unambiguously destroys the colliding transmissions. When this happens, each of the colliding computers "backs off" or waits for a random amount of time, then tries to retransmit. This wait/retransmission sequence can repeat until both messages are transmitted successfully. The whole process takes a small fraction of a second.

Ethernet's detractors characterize it as an inefficient access method because frames are prone to collisions. But while collisions occur, they don't consume very much throughput capacity in most cases. Since the whole process of transmitting, colliding, and retransmitting takes place so quickly, the delay a collision causes is normally minuscule. Of course, as the traffic on a network approaches the total throughput capacity, the number of collisions will mount and the network will slow considerably. This happens with some large-scale imaging or engineering applications, or network segments with too many nodes. As long as an Ethernet network has a low traffic load, traditionally the most common environment, delay caused by collisions is seldom noticeable. In a switched environment, especially on Ethernet backbones, most links will have only two nodes, and the incidence of collisions on these segments will be minuscule.

TOKEN RING

When Token Ring was introduced in 1984, it was not the first token-passing, ring network, but because it was endorsed by IBM, it has had a tremendous impact on the network industry. Token Ring became part of IBM's connectivity solution for all its computers—personal, midrange, and mainframe. IBM's specifications match those of the IEEE 802.5 standard.

Token Ring unit shipments are still increasing in 1995, though this growth is unlikely to continue for long. IBM has had a stranglehold on the Token Ring market, though it no longer supplies the 90 percent share of Token Ring network interface cards that characterized this market in the 1980s.

The original Token Ring system transmitted data at 4Mbits/sec; the newer specification calls for 16Mbit/sec transmission. In Token Ring, the computers are arranged in a logical ring, but all data passing between work stations is routed through a hub. A multi-station access unit (MSAU or MAU) acts as the hub, and each work station is connected to it. Token Ring uses a token-passing access method to prevent data collisions—a token being a series of data bits created by one of the computers. The token moves around the ring, giving successive computers the right to transmit. When a computer receives the token, it may transmit a message of any length as long as the time to send does not exceed the token-holding timer (this combination of token and data is called a frame). As this message (frame) moves around the network, each computer regenerates the signal. Only the receiving computer copies the message into its memory, then marks the message as received. The sending computer removes the message from the token and recirculates it.

Token Ring's advantages include reliability and ease of maintenance. It uses a star-wired ring topology in which all computers are directly wired to a MAU. The MAU allows malfunctioning computers to be disconnected from the network. This overcomes one disadvantage of token-passing, which is that one malfunctioning computer can bring down the network, since all computers are actively passing signals around the ring.

ARCNET

ARCnet was developed by Datapoint in the early 1970s. It was especially popular in very small networks, since it was inexpensive and easy to maintain. ARCnet uses a token-passing access method that works on a star-bus topology. Data is transmitted at 2.5Mbits/sec. The network cable is laid out as a series of stars. Each computer is attached to a hub at the center of a star, and the hubs are connected in a bus or line. ARCnetPlus was designed as a backbone technology and can transmit data at 20Mbits/sec.

When a computer wants to send data on an ARCnet network, it must have the token. The token moves around the network in a given pattern, which in ARCnet's case is a logical ring. All computers on the network are numbered with an address from 0 to 255. (The maximum number of computers on each ARCnet segment is thus 256.) The token moves from computer to computer in numerical order, even if adjacent numbers are at physically opposite ends of the network. When the token reaches the highest number on the network it moves to the lowest, thus creating a logical ring.

Once a computer has the token, it can send one 512-byte packet. A packet is composed of the destination address, its own address, up to 508 bytes of data, and other information. The packet moves from node to node in sequential order until it reaches the destination node. At the destination, the data is removed and the token released to the next node.

The advantage of token passing is predictability. Because the token moves through the network in a determined path, it is possible to calculate the best and worst cases for data transmission. This makes network performance predictable. It also means introduction of new network nodes will have a predictable effect. This differs from Ethernet, where the addition of new nodes may or may not seriously affect performance. However, claims for a "predictable" network can be misleading—for example, lost tokens will affect worst-case delivery times.

A disadvantage of the token-passing access method is the fact that each node acts as a repeater, accepting and regenerating the token as it passes around the network in a specific pattern. If there is a malfunctioning node, the token may be destroyed or simply lost, bringing down the whole network. The token must then be regenerated.

This tutorial, number 4, by Aaron Brenner, was originally published in the November 1988 issue of LAN Magazine/Network Magazine.

Ethernet Frame Types

GETTING THE PICTURE ON FRAMES

For most network administrators, Ethernet frame types rate little or no attention. If they think about them at all, it's usually as a possible source of workstation problems—for example, a user's net.cfg or Windows network settings configured with the wrong frame type can prevent the workstation from connecting to a server. Recently, however, Novell pushed the topic into the foreground. By changing the default Ethernet frame type from 802.3 to 802.2 for its NetWare operating systems 3.12 and 4.x, the company forced administrators of existing networks to at least consider migrating their users to the newer frame type. The reasons behind Novell's change center around a need to ensure compatibility of Novell's IPX/SPX with future demands, such as increased security and reliability. Also, the increasing diversity of many corporate networks—which often have devices from several vendors connected to the same network, using several different protocols—prompted Novell to encourage its users to move toward a more standardized frame type, one that readily coexists with other frame types.

IF IT AIN'T BROKE, WHY FIX IT?

The 802.3 frame type (and the way Novell uses it) isn't really such a bad thing. In fact, it may offer slight (though probably negligible) advantages on NetWare-only networks due to its slightly lower overhead per packet. Furthermore, because the 802.2 standard deals only with the Data-link layer, the newer frame type uses an 802.3 packet as its Physical-layer skeleton (or perhaps we should call it an exoskeleton, since the Physical-layer header is added to the outside of the 802.2 frame). The problem is that Novell used the Physical-layer 802.3 frame type without a standard Data-link layer header when it created its proprietary IPX/SPX packet structure—hence the term 802.3 Raw.

In a single-vendor environment, this omission doesn't cause any problems. But in a corporate network, where several packet types may coexist, the lack of a clearly defined field stating the frame's type can make the jobs of routers and bridges more difficult and may result in lost packets. Keep in mind that Ethernet packets don't arrive at a workstation with neat labels above each field describing the contents; a router or NIC can only examine the contents of predefined locations to determine which protocol or frame type a packet uses. To decode and route a packet that doesn't conform to standard protocols, a router must be designed to specifically search for the defining characteristics of the nonconforming packet.

Adopting a more standard Data-link layer protocol format gave NetWare more flexibility in coexisting with other network operating systems. But Novell's change of default frame type wasn't a big surprise. NetWare has fully supported other Ethernet frame types—Ethernet_802.2, Ethernet_II, and Ethernet_SNAP—since version 3.x. NetWare 2.x was the last version to require a complete relinking of the operating system to enable support for other frame types.

For better insight into Novell's reasoning, let's examine the structure of an 802.3 packet and see how it differs from the other frame types used on Ethernet networks.

ETHERNET II

The precursor to the IEEE 802.3 frame type was the Ethernet II packet. This frame type, originally developed by Digital, Intel, and Xerox, relies on Ethernet's Carrier Sense Multiple Access with Collision Detection (CSMA/CD) access scheme (the same as the 802.3 type). To allow all workstations on the network to synchronize their receiving clocks and to help the transmitting station detect transmissions from other stations, a seven-byte preamble and a one-byte Start of Frame Delimiter precede each frame. The preamble's seven-byte length ensures that the transmitting stations can detect another transmission (or the resulting jam signal, indicating a collision), no matter how far away the competing station resides on the network segment. (The calculations required for setting the preamble length depend on the propagation speed of the signal on the wire and the minimum amount of time required to transmit a signal that covers the entire length of the network segment.) The preamble can usually be considered part of the hardware mechanics used to send a packet of data across the wire, rather than part of the packet itself.

Immediately following the preamble of the Ethernet II frame format is a six-byte field that contains the address of the destination station, a six-byte Source Address field, and a two-byte

Frame Type field. Together, these three fields compose the Ethernet II header (see Figure 1). When transmitting each byte of the address fields, the least significant (rightmost) bits are transmitted first. For the destination address, the first bit transmitted (bit 0 of byte 0) indicates whether the address represents a single station or a multicast address. Thus, if the first byte of the destination is odd, the packet is destined for a group of workstations rather than one unique physical address.

A special kind of multicast is the broadcast, in which all bits of the address field are set to 1. For individual addresses, the first three bytes identify the manufacturer of the network interface card, and the last three bytes are a unique number assigned to the individual card. For example, addresses on 3Com cards all start with the three hexadecimal bytes 02 60 8C. This address is called the physical, or MAC (Media Access Control), address. The destination address identifies the immediate recipient on the network—not necessarily the ultimate recipient.

The Frame Type field identifies the higher-level protocol used to create the packet, such as TCP, Xerox Network System (XNS), or AppleTalk. For example, a hexadecimal value of 08 00 indicates a TCP/IP packet, 06 00 indicates an XNS packet, and 81 37 indicates a Novell NetWare packet formatted for Ethernet II.

The next field in the frame contains the actual frame data. This field can contain up to 1,500 bytes, including the headers for the higher-level protocols used to encapsulate the original data. Complete Ethernet packets range from 64 bytes to 1,518 bytes. There isn't a minimum size for the actual data, however. When transmitting smaller data packets, a Pad field must be added to bring the total size of the Ethernet packet up to at least 64 bytes.

The last field in the packet definition is the four-byte Frame Check Sequence (FCS). This value is calculated from the rest of the packet's data, using a 32-bit cyclic-redundancy check (CRC-32) algorithm. As it receives the packet data, the receiving station performs the same

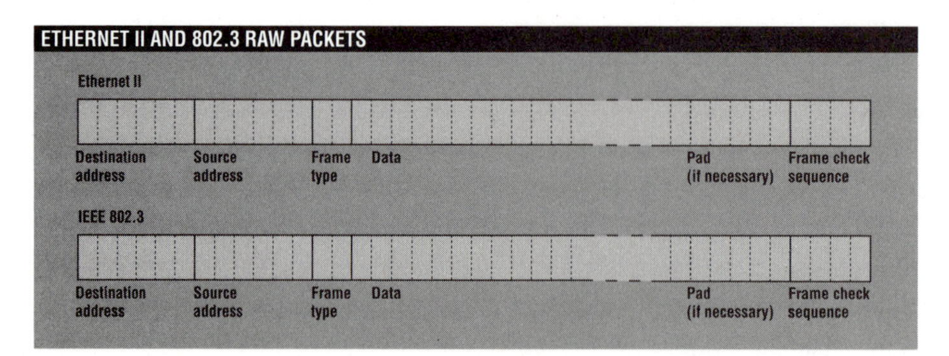

■ **Figure 1:** Ethernet II and 802.3 Raw packets have similar structures, except for Frame Type and Frame Length fields. The different fields can coexist because all assigned frame types are greater than 05FE.

calculations, then compares its results with the FCS transmitted with the packet. If the results are different, the packet is rejected.

RAW AND PHYSICAL

Closely resembling the Ethernet II format, the IEEE 802.3 specification defines the physical layout of CSMA/CD packets. For its Ethernet_802.3 packet format, Novell uses the 802.3 frame type without adding an IEEE 802.2 LLC header (in this case, NetWare adds its own proprietary higher-level information). This type of packet can be called an 802.3 Raw format. NetWare's 802.3 format is the only CSMA/CD packet type that doesn't incorporate a corresponding standard header for logical-link control or data-link control information.

The main difference between the Ethernet II and IEEE 802.3 specifications is in one two-byte field. Where Ethernet II specifies a Frame Type field, 802.3 specifies a Frame Length field. While this may seem to make Ethernet II and IEEE 802.3 packets incompatible on the same wire, they can coexist quite well. This is possible due to the 1,518-byte limit on the size of an Ethernet or 802.3 frame and the fact that all Ethernet II Frame Types (assigned and managed by Xerox) are values greater than 1,518 (decimal). Thus, if a packet has a decimal value of 1,518 or less (05 FE hexadecimal) in byte positions 13 to 14, it will be considered an 802.3 packet.

Another difference shows up in the Destination Address and Source Address fields. Whereas Ethernet II uses one bit to indicate multicast addresses, 802.3 uses two bits. The first bit is similar to the multicast bit in that it indicates whether the address is for an individual or for a group, and the second bit indicates whether the address is locally or universally assigned. The second bit is rarely used on Ethernet (CSMA/CD) networks. Novell's use of a Raw 802.3 packet format makes bridging between Ethernet and Token Ring networks difficult. Without specialized instructions for translating packets into the different frame types, a bridge cannot transfer 802.3 Raw packets between Token Ring and Ethernet (CSMA/CD). However, because Token Ring (using the IEEE 802.5 Physical-layer specification) uses the IEEE 802.2 Data-link layer format, Token Ring-to-Ethernet bridges need only convert the Physical-layer information for NetWare packets formatted with the higher-level 802.2 Data-link layer information.

In Novell's 802.3 Raw format, the Data field begins with IPX header information. The first two bytes in this header (for this format) are always hexadecimal FF FF. These two bytes help confirm that an 802.3 Raw packet contains encapsulated IPX information, but they correspond to IPX's Checksum field. Because this static information interferes with use of the IPX Checksum field, 802.3 Raw packets will not be able to use the security features, such as packet signing, planned for the IPX format. Packets incorporating 802.2 link information are free to use the IPX Checksum feature. Note that IEEE does not recognize Novell's 802.3 Raw format; it recognizes only 802.3 packets encoded with 802.2 and 802.2 SNAP headers.

A LOGICAL ADDITION

Adding IEEE 802.2 LLC information to an 802.3 physical packet format requires three additional fields at the beginning of the Data field: a one-byte Destination Service Access Point (DSAP) field, a one-byte Source Service Access Point (SSAP) field, and a one-byte Control field. IEEE assigns Service Access Point numbers (SAPs); among those currently defined are E0 for Novell, F0 for NetBIOS, 06 for TCP/IP, and AA for the Subnetwork Access Protocol (SNAP). (See Figure 2 for a description of 802.2 and 802.2 SNAP frame types.) NetWare packets using the Ethernet_802.2 format have DSAP and SSAP values of E0, and the Control field is set to 03 (denoting the 802.2 unnumbered format).

An additional frame type was developed from the 802.2 format to provide support for more than 256 protocol types. The newer type, SNAP, is identical to a standard 802.2, except that it adds a five-byte Protocol Identification field. On any SNAP packet, both the DSAP and SSAP fields are set to AA, and the Control field is set to 03.

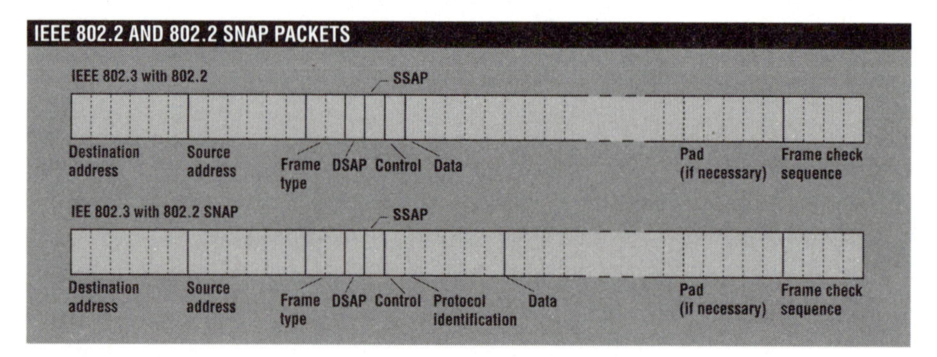

■ **Figure 2:** IEEE 802.2 and 802.2 SNAP packets start with the basic 802.3 Physical-layer frame type and add 802.2 LLC headers.

THE END PACKET

Novell's conversion of its default packet type from 802.3 Raw to the 802.2 LLC format on the 802.3 physical packet type may create more work for administrators of existing networks. However, the additional chore of converting existing workstations to the 802.2 format provides healthy returns, including better support for diverse networks, tighter network security, and greater flexibility in configuring workstations to interoperate with multiple network operating systems.

This tutorial, number 90, by Dave Fogle, was originally published in the February 1996 issue of LAN Magazine/Network Magazine.

Fast Ethernet: 100BaseT

The idea of Fast Ethernet was first proposed in 1992. In August 1993, a group of vendors came together to form the Fast Ethernet Alliance (FEA). The goal of the FEA was to speed Fast Ethernet through the Institute of Electrical and Electronic Engineers (IEEE) 802.3 body, the committee that controls the standards for Ethernet. Fast Ethernet and the FEA succeeded, and in June 1995, the technology passed a full review and was formally assigned the name 802.3u.

The IEEE's name for Fast Ethernet is 100BaseT, and the reason for this name is simple: 100BaseT is an extension of the 10BaseT standard, designed to raise the data transmission capacity of 10BaseT from 10Mbits/sec to 100Mbits/sec. An important strategy incorporated by 100BaseT is its use of the Carrier Sense Multiple Access with Collision Detection (CSMA/CD) protocol—which is the same protocol that 10BaseT uses—because of its ability to work with several different types of cable, including basic twisted-pair wiring. Both of these features play an important role in business considerations, and they make 100BaseT an attractive migration path for those networks based on 10BaseT.

The basic business argument for 100BaseT resides in the fact that Fast Ethernet is a legacy technology. Because it uses the same transmission protocol as older versions of Ethernet and is compatible with the same types of cable, less capital investment will be needed to convert an Ethernet-based network to Fast Ethernet than to other forms of high-speed networking. Also, because 100BaseT is a continuation of the old Ethernet standard, many of the same network analysis tools, procedures, and applications that run over the old Ethernet network work with 100BaseT. Consequently, managers experienced at running an Ethernet network should find the 100BaseT environment familiar, meaning less time and money must be spent by the company on training.

PROTOCOL PRESERVATION

Perhaps the shrewdest strategy taken with Fast Ethernet was the decision to leave the transmission protocol intact. The transmission protocol, in this case CSMA/CD, is the method a network uses to transmit data from one node to another, over the cable. In the OSI model, CSMA/CD is part of the Media Access Control, or MAC, layer. The MAC layer specifies how information is formatted for transmission and the way in which a network device gains access to, or control of, a network for transmission.

The name CSMA/CD can be broken down into two parts: Carrier Sense Multiple Access and Collision Detection. The first part of the name describes how a node equipped with a network adapter determines the appropriate time to send a transmission. In CSMA, a network node first "listens" to the wire to find out if any other transmission is currently being broadcast over the network. If the node receives a carrier tone, which implies that the network is busy with another transmission, it holds onto its transmission and waits until the network is clear. When it does sense a quiet network, it begins its transmission. The broadcast is actually sent to all nodes on the network segment, but only the node with the correct address accepts the transmission.

The Collision Detection feature is meant to remedy situations in which two or more nodes inadvertently broadcast a transmission at the same time. Under the CSMA structure, any node that is ready to transmit first listens to determine whether the network is free. If two nodes were to listen at the exact same moment, however, both would perceive the network to be free and both would transmit their packets simultaneously. In this situation, the transmissions would interfere with each other (network engineers call this a collision) and neither would reach its destination intact. Thanks to Collision Detection, a node will listen to the line again after it has broadcast its packet. If it detects a collision, it waits a random period of time and rebroadcasts the transmission, again listening for a collision.

THREE KINDS OF FAST

Another important consideration, along with the adoption of the CSMA/CD protocol, was the decision to design 100BaseT so it uses many basic forms of cabling—those used by older versions of Ethernet and newer forms of cabling, as well. To accommodate the different types of cables, Fast Ethernet comes in three forms: 100BaseTX, 100BaseT4, and 100BaseFX. Both 100BaseTX and 100BaseT4 work with twisted-pair cabling standards, while 100BaseFX was created to work with fiber optic cabling.

The 100BaseTX standard is compatible with two pairs of UTP or STP. One pair is designated for reception and the other for transmission. The two basic cabling standards that meet this requirement are EIA/TIA-568 Category 5 UTP and IBM's Type 1 STP. The attractiveness of 100BaseTX lies in its ability to provide full-duplex performance with network servers and the fact that it uses only two of the four pairs of wiring, leaving two pairs free for future enhancements to your network.

However, if you plan on using Category 5 cabling with 100BaseTX, be forewarned of the drawbacks related to Category 5. The cable is more expensive than other types of four-pair cabling, such as Category 3, and it requires the installation of punchdown blocks, connectors, and patch panels that are all Category 5-compliant.

The 100BaseT4 standard requires a less sophisticated cable than Category 5. The reason is that 100BaseT4 uses four pairs of wiring: one for transmission, one for reception, and two that can either transmit or receive data. Therefore, 100BaseT4 has the use of three pairs of wiring to either transmit or receive data. By dividing up the 100Mbit/sec data signal among the three pairs of wiring, 100BaseT4 reduces the average frequency of signals on the cable, allowing lower-quality cable to handle the signal successfully. Categories 3 and 4 UTP cabling, as well as Category 5 UTP and Type 1 STP, can all work in 100BaseT4 implementations.

The advantage of 100BaseT4 is its flexible cable requirements. Category 3 and 4 cabling was once more prevalent in existing networks, and if they aren't already being used in your network, they cost less than Category 5 cabling. The downside is that 100BaseT4 uses all four pairs of wiring, and it does not support full-duplex operation.

Fast Ethernet also offers a standard for operation over multimode fiber with a 62.5 micron core and 125 micron cladding. The 100BaseFX standard is designed mainly for backbone use, connecting Fast Ethernet repeaters scattered about the building. The traditional benefits of fiber optic cabling are still valid with 100BaseFX: protection from electromagnetic noise, increased security, and longer distances allowed between network devices.

THE SHORT RUN

Although Fast Ethernet is a continuation of the Ethernet standard, the migration from a 10BaseT network to a 100BaseT network isn't a straight, one-to-one conversion of hardware—some changes to the network topology may be required.

Theoretically, Fast Ethernet limits the end-to-end network diameter or the network segment diameter to 250 meters; only 10 percent of the 2,500-meter maximum theoretical size of Ethernet. Fast Ethernet's restriction is based on the speed of 100Mbit/sec transmission and the nature of the CSMA/CD protocol. Figure 1, Figure 2, and Figure 3 illustrate why a 100BaseT network can't be longer than 250 meters. In the figures, Workstations A and B represent the farthest ends of the network or network segment.

For Ethernet to work, a workstation transmitting data must listen long enough to make sure the data has reached its destination safely. In a 10Mbit/sec Ethernet network, such as 10Base5, the length of time a workstation listens for a collision is equivalent to how far a 512-bit frame (the frame size is specified in the Ethernet standard) travels before the workstation is finished processing it. In a 10Mbit/sec Ethernet network, that distance is 2,500 meters (see Figure 1). However, a 512-bit frame (the 802.3u standard specifies the same frame size, 512 bit, as the 802.3 standard) being transmitted by a workstation in the faster 100Mbit/sec Ethernet network travels only about 250 meters before the workstation is finished processing it (see Figure 2). If the receiving workstation was located farther than 250 meters from the transmitting workstation, the frame may collide with another frame down the line, and the transmitting workstation, having finished processing the transmission already, would not be listening for the collision. For this reason, the maximum network diameter for a 100BaseT network is 250 meters (see Figure 3).

To take advantage of the 250 meters, though, you will need to install two repeaters to connect all of the nodes. And, a node cannot be located farther than 100 meters away from a

MAXIMUM NETWORK DIAMETER FOR 10BASE5

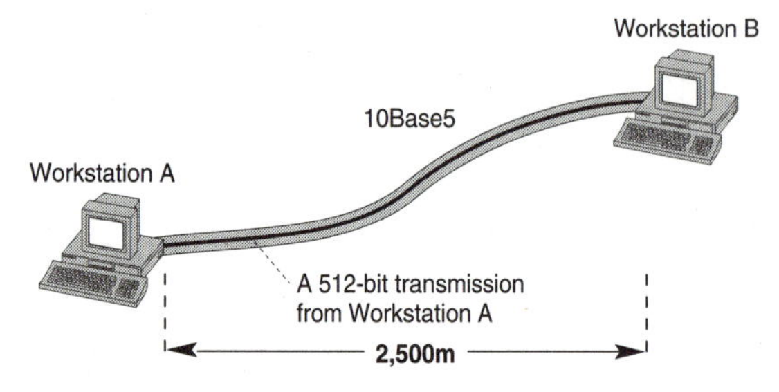

■ **Figure 1:** In traditional 10Mbps Ethernet networks, the maximum distance between two end stations is 2,500 meters.

LENGTH LIMITATIONS FOR 100BASET

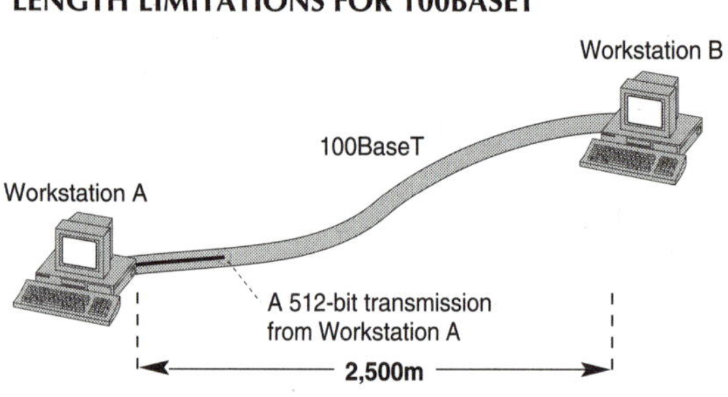

■ **Figure 2:** The maximum length limitation for 100BaseT is 250 meters, only 10 percent of the 2,500-meter maximum theoretical size of Ethernet.

MAXIMUM NETWORK DIAMETER FOR 100BASET

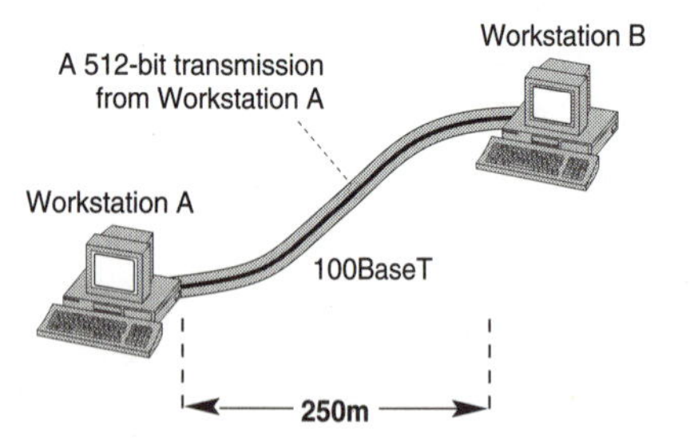

■ **Figure 3:** Because of the increased throughput capabilities of 100BaseT, workstations A and B can be no farther than 250 meters apart.

repeater—Fast Ethernet adopted the 10BaseT rule that determines 100 meters to be the farthest allowable distance a workstation can be from a hub. Due to latency introduced by connection devices such as repeaters, the actual operational distances between nodes will probably prove to be less than those stated. So, it would be prudent to measure distances on the short side.

To incorporate longer runs in your network, you'll have to invest in fiber cabling. For example, you can use 100BaseFX in half-duplex mode to connect a switch to either another switch or an end station located up to 450 meters away. A full-duplex 100BaseFX installation will allow two network devices up to two kilometers apart to communicate.

A SIMPLE SETUP

Aside from the cabling, you'll need to deal with network adapters in workstations and servers, 100BaseT hubs, and possibly some 100BaseT switches.

The adapters you'll need for a Fast Ethernet network are called 10/100Mbit/sec Ethernet adapters. These adapters take advantage of an autosensing feature—provided for in the 100BaseT standard—that allows the adapters to automatically sense 10Mbit/sec and 100Mbit/sec speed capabilities. You'll also need to install a 100BaseT hub to service the group of workstations and servers you've converted to 100BaseT.

When a PC or server equipped with a 10/100 adapter is turned on, it will emit a signal that broadcasts its 100Mbit/sec capabilities. If the receiving station, most likely a hub, is similarly suited for 100BaseT operation, it will return a signal that automatically places both the hub and PC or server in 100BaseT mode. If the hub is only 10BaseT-capable, it won't return the signal, and the PC or server will automatically go into 10BaseT mode.

For a small-scale 100BaseT setup, you can use a 10/100 bridge or switch to allow the 100BaseT part of your network to operate with the 10BaseT installed base.

DECEPTIVE SPEED

As a last piece of advice, Fast Ethernet's capabilities seem best suited for peak traffic problems. For example, if you have some users running CAD or imaging applications who need to have a higher peak throughput, Fast Ethernet can help you out. But, if your problem is caused by an overload of users, 100BaseT starts to drag at about the 50-percent utilization rate—in other words, at the same traditional threshold as 10BaseT. After all, it is an extension of 10BaseT.

This tutorial, number 88, by Lee Chae, was originally published in the December 1995 issue of LAN Magazine/Network Magazine.

Ethernet in the Local Loop

Ethernet has reigned as the market leader in the LAN for longer than most dot-commers have been high-school graduates. However, the once all-LAN technology is now encroaching on Metropolitan Area Network (MAN) turf. Ethernet's local loop appeal—whether Fast Ethernet, Gigabit Ethernet, or in-the-planning-stages 10 Gigabit Ethernet (10GbE)—comes from the economic incentive of using inexpensive Ethernet hardware and phasing out costlier WAN technologies, such as ATM and SONET, in the metropolitan area. Ethernet hardware costs a tenth of that required by the SONET and ATM competition, says Yipes (www.yipes.com), a Competitive Local Exchange Carrier (CLEC).

As compelling as the near-ubiquity and affordability of Ethernet are, limited fiber availability in the local loop has thus far kept the technology on the outskirts of the enterprise. Ethernet in the local loop is, however, a reality for a cherry-picked few.

GREAT DEAL IF YOU CAN GET IT

Optical Ethernet outside of the LAN has four primary benefits: less-expensive hardware, unification of multiple LANs in a MAN under one protocol, flexibility of bandwidth provisioning, and a familiarity that requires no new training. Trumpeting the various selling points of Ethernet, CLECs such as Yipes, Cogent Communications (www.cogentco.com), IntelliSpace (www.intellispace.net), and Telseon (www. telseon.com) offer bandwidth-hungry businesses the access they covet. These start-up providers are generous with bandwidth and conjure visions of native Gigabit Ethernet connections between LANs in a metropolitan area. However, availability is limited to just a few urban markets.

In what is referred to as a transparent LAN, an Ethernet-based MAN can be used to connect multiple LANs within a 200-kilometer diameter. With WAN lines as fast as LAN speeds, for all practical purposes, what once was a collection of disparate LANs connected over expensive leased lines becomes one big LAN. The MAN sustains the traffic in its native format—needless conversion is unwieldy and inefficient. Since Ethernet carries IP and legacy LAN traffic such as SNA, IPX, and others, all traffic stays just as it would on the LAN, but it's no longer restricted to one street address. VLAN tagging is the principle method for constructing VPNs between locations.

Businesses receive generous doles of bandwidth and experience rapid provisioning times when they want to boost bandwidth capacity. Critics of ATM and SONET contrast this flexibility with the relative difficulty of performing the same tasks with ATM switches or SONET gear. Over half the demand for Ethernet in the local loop is for high-speed Internet access.

SHOW ME THE FIBER

Early subscribers to the Ethernet local loop fit a certain profile, largely because providers have strategically dug fiber in direct proximity to the companies with the biggest wallets. According to International Data Corp. (IDC, www.idc.com), fiber availability is primarily confined to downtown areas of large cities, where approximately 8 percent to 10 percent of buildings have existing fiber connectivity. Customers of the metropolitan-area Ethernet service providers are not solely enterprises; in fact, many are themselves service providers, whether Web-hosting providers, Application Service Providers (ASPs), or others.

The metropolitan-area service provider start-ups that currently offer Ethernet in the local loop are relatively new to the game, so they generally lease the fiber from multiple sources, such as utility companies and firms like Metro Media Fiber Network (www.mmfn.com). Given the volatility of the CLEC space and their difficulty in turning any kind of a profit, ownership of one's fiber could turn out to be a huge liability for a CLEC; instead, they ink long-term leases. It's impossible to get a comprehensive footprint from just one leased fiber source, though—providers cobble together fiber lines from multiple sources to achieve adequate reach.

The enormous expense of land rights, construction labor, and equipment for digging trenches freights every inch of a proposed fiber extension with cost. Providers will break out the shovels if the expense can be cost-justified, but they focus much more on signing up customers directly on the path of existing fiber rings. Most service providers are reluctant to reveal specific fiber-dig prices (and charges vary wildly depending on logistical barriers and topography), but $90,000 per tenth of a mile seems to be a ballpark figure; $400,000 for the same distance is not unheard of.

ETHERNET OUTSIDE OF THE LAN

When Ethernet ventures outside of the LAN, it does so on fiber optic cable. Gigabit Ethernet can run over fiber in one of three ways: in its native fiber-based format, via Dense Wavelength Division Multiplexing (DWDM), or packed into SONET frames. Depending on the service provider's network, Ethernet data may utilize all three methods. Until the adoption of 10GbE, projected for IEEE ratification in March 2002, Gigabit Ethernet and Fast Ethernet will serve as the Ethernet ambassadors to the outside world. Both standards can support Carrier Sense Multiple Access with Collision Detection (CSMA/CD) in half-duplex mode, but due to distance constraints, Ethernet outside of the LAN operates in full-duplex mode. 10GbE will support only full-duplex transmission.

CSMA/CD-based Ethernet acts as if it had a strict upbringing as a child. A node listens to the line before transmitting data. In the event that two or more nodes transmit at the same time, they each back off a random amount of time before again peeking out and listening to the line, waiting their turn to be heard in an orderly fashion. On high-speed point-to-point links, however, the method is out of place.

Ethernet networks must conform to inevitable distance limitations. If the network is too large or convoluted, delays will adversely affect error detection capabilities. When collision notifications are received late, packets aren't re-sent, and data comes across as garbled. To counteract round-trip propagation delay, the IEEE increased the minimum frame size for Gigabit Ethernet from 64 bytes to 512 bytes and the maximum frame size from 1,514 bytes to 9,000 bytes. Larger frame sizes effectively buy transmitting devices more time to receive collision notifications and lessen the processing time of header information and so forth.

Regarding distance capabilities, 100Base-FX maintains data integrity up to distances of 2 kilometers on multimode fiber. Per the IEEE 802.3z specification, Gigabit Ethernet reaches up to 5 kilometers on single-mode fiber and 550 meters on multimode fiber. 10GbE should pass the 40-kilometer marker on single-mode fiber.

Switches in the MAN's backbone portion contain Gigabit Interface Converters (GBICs) that perform optical-electronic conversion and regenerate optical signals. Fiber runs between switches generally don't exceed a maximum of 70 kilometers, but some vendors such as Cisco Systems say that Gigabit Ethernet transmission over their switches can reach distances between 100 kilometers and 150 kilometers with appropriate lasers and GBICs.

Vendors that provide equipment to this metropolitan Ethernet space include Cisco Systems, Juniper Networks, Extreme Networks, Nortel Networks, Foundry Networks, and Sycamore Networks.

IMAGINE YOURSELF IN THE DRIVER'S SEAT

Let's assume you have the good fortune to work in a building right off the path of a fiber ring. No less, you also use Ethernet throughout your business. You know the technology, you trust the technology, and you already have Ethernet equipment throughout your company's LAN or LANs.

Most start-up providers that offer Ethernet in the local loop start the bidding at 1Mbit/sec and don't flinch up to 1Gbit/sec —in fact, the higher the data rate, the broader they smile. The provider brings fiber to the building and connects it directly to a 1Gbit/sec or 100Mbit/sec port of an onsite Ethernet switch. (It is also possible to locate the switch in a rack at a service provider POP or at a colocation facility.) The customer then distributes the bandwidth resources through the different areas of his or her business via the switch. For those accustomed to choosing between 1.54Mbits/sec of a monthly $1,000 T1 line or 45Mbits/sec of a $5,000-a-month T3, the name-your-fancy bandwidth increments of an Ethernet provider such as Yipes should be mighty appealing. Not to be outdone, Cogent has garnered countless headlines for its monthly flat rate of 100Mbits/sec for $1,000.

Within the fiber optic network the provider has assembled for the MAN, routers aggregate streams of Gigabit Ethernet and shuttle the data traffic throughout switches in the network. Because the routers' reach extends to the Ethernet switch at the customer premises, providers can leverage Ethernet rate-limiting capabilities that lend themselves to flexible service adjustment. Ramping up bandwidth capacity for a particular time frame of anticipated increased demand is easily accommodated—provisioning is processed in near real time.

RETAIN BABY, DRAIN BATHWATER

Don't throw out the baby with the bathwater, as the saying goes. The fact is that no provider will waste equipment that works. Ethernet in the local loop does not eliminate the need for ATM and SONET. Despite the appeal of all-Ethernet, no one's about to forklift out SONET equipment just to have one common protocol throughout the metropolitan area. Many service provider networks have embedded investment in packet-over-SONET technology. For these reasons, it is important that 10GbE incorporates OC-192 interfaces for easier integration of Ethernet with SONET.

SONET is considered inefficient because one of its two fiber rings is reserved purely for failover, but when it comes to restoration capabilities and long-haul transmission, Ethernet zealots eat their words. Likewise, ATM is adept at delivering QoS, a claim that Ethernet cannot make. It's not an all-or-none proposition: Ethernet as the sole transport method offers certain advantages, but the technology can leverage SONET and DWDM when the situation requires. Furthermore, Ethernet must concede traffic with strict QoS requirements to a separate ATM WAN connection.

The very ring-based redundancy that earns SONET criticism for inefficiency is the same feature that provides rapid traffic restoration in the event of a fiber cut. (Incidentally, some providers put Ethernet traffic over both rings, which doubles the ring's capacity but sacrifices the medium's restoration capabilities.) For the long haul, SONET encapsulates Gigabit Ethernet traffic within its payload—for example, on a crosscontinental backbone. This scenario delegates the traffic to a larger carrier, but the relationship is rarely one of peering, due to the disparity in what the two service providers can offer one another.

CRACKS IN THE ETHERNET VASE

Ethernet in the local loop can furnish high-capacity access for a low cost per bit, but the technology is not without its flaws, namely in the areas of QoS and restoration capabilities.

When a pipe is at capacity, and important traffic isn't getting through, there will be QoS issues, something Ethernet is ill-prepared to handle. Similarly, congestion in the service provider's network resulting from the aggregation of multiple gigabit streams in the core can cause problems. Providers using Ethernet will usually pile on the bandwidth to maintain acceptable QoS, but this does not scale well. Once capacity is reached, delay-sensitive applications suffer without ATM's QoS levels.

VLAN tagging based on the IEEE 802.1q and 802.1p protocols is the most promising method of assigning priority to Ethernet traffic. DiffServ marking can also be used on IP packets to support different Class of Service (CoS) levels.

In January 2001, the IEEE 802.17 Resilient Packet Ring working group met to develop an approach analogous to the ring formations in SONET. More attuned to the needs of Ethernet, the dual counter-rotating rings should mimic the quick restoration times of SONET, which can respond to a fiber cut by fully rerouting in under 50 milliseconds.

Ethernet in the local loop is gaining momentum—largely because it gives customers the high-speed access they want in the way they want it. But keep your fingers crossed regarding your proximity to existing fiber.

RESOURCES

The IEEE Web site at www.ieee.org contains up-to-date information on recent Ethernet standards, as well as information on interesting projects such as the IEEE 802.17 Resilient Packet Ring working group and the IEEE 802.3 EFMEthernet to the First Mile study group.

An online section by Sycamore Networks hosted at www.lightreading.com/opticalintellect/ has several in-depth white papers that cover optical Ethernet, including the providers and equipment manufacturers in the space.

A useful resource on Ethernet in the Metropolitan Area Network (MAN) can be found at the aptly titled Optical Ethernet Web site, located at www.optical-ethernet.com. The site contains links, news on standards activity, and various white papers.

This tutorial, number 152, by Rob Kirby, was originally published in the March 2001 issue of Network Magazine.

Frame Relay

Frame relay is a byproduct of the ISDN standards work. Frame relay was initially expected to be offered on Broadband ISDN networks, but it is currently marketed as an upgrade to X.25 packet-switched networks.

Frame relay makes sense if your enterprise network interconnects more than four sites and network traffic is fairly heavy. If your network traffic is primarily from terminals, then X.25 is a better service option; however, if your network traffic is from workstations and PCs, then frame relay is probably your best bet.

BACKTRACK TO X.25

Packet-switched networks are suitable for connecting geographically dispersed locations, because instead of directly connecting each remote site to the headquarters or connecting each site to every other one, packet switching provides a "cloud" network into which sites can connect. Once attached to the cloud, any site can communicate with any other. Each site has a local connection, which is then connected into a larger packet-switched network.

Packet-switched technologies maintain virtual connections between two users. It appears to the end users that they are directly connected, when, in reality, the connection may go through intermediate points. Virtual connections can either be permanent or temporary. A permanent virtual circuit is guaranteed to be available at all times. With a temporary, or switched virtual circuit, the connection is set up for the duration of the transmission, then broken down.

X.25 is a packet-switched technology that has been used since the early 1970s. It was designed to transport character data from host terminals, and for such low-bandwidth applications, X.25 delivers adequate performance. But stuffing Ethernet or Token Ring traffic through a 56/64Kbit/sec X.25 virtual circuit is much more difficult.

X.25 was developed when the telephone system was dominated by copper lines, and it necessarily included a large number of error-detecting and -correcting routines to handle the effects of noise, dropped calls, and other problems. However, since the vast majority of the

U.S. phone network is fiber-optic, the error-detecting and -correcting code in X.25 can be unnecessary overhead.

And so frame relay was developed—originally to run in the ISDN "D" channel. Frame relay retains many of X.25's good traits, including packet switching, while dropping other traits, such as the overhead. Frame relay was developed under the aegis of the CCITT I.122 and the corresponding American National Standards Institute committee 1.606.

JUMP TO FRAME RELAY

Frame relay's benefits include low overhead, high capacity with low delay, and reliable data transfer over existing public networks. It is largely designed to be a public service for interconnecting private local area networks, although private networks can be built.

Frame relay is generally considered to be able to provide service at speeds up to 2.048Mbits/sec, although some service providers claim that they will be able to squeeze out 45Mbits/sec. Public frame relay carriers have rolled out frame relay services in the increments of 56Kbits/sec, fractional T1, and full T1. Frame relay services will come from the RBOCs, interexchange carriers, and value-added networks (VANs), including British Telecom, Pacific Bell, CompuServe, and Wiltel.

Frame relay operates on OSI Layers 1 and 2, whereas X.25 operates on Layers 1, 2, and 3. Because frame relay operates under the network layer, it is independent of the upper-layer protocols, such as TCP/IP or IPX. This arrangement delivers greater flexibility.

Frame relay achieves high throughput with low delay by eliminating the overhead of error detection and correction. Data integrity is insured only through a cyclic redundancy check, and any corrupted packet is discarded. When necessary, the network-layer device—usually a router—handles retransmissions. That device must detect the transmission failure and request a retransmission. Although errors will occur less frequently because of the more reliable phone lines, the responsibility of handling them has simply moved to another device that must be capable of handling the added work. Frame relay will also drop packets when the network is too congested.

Frame relay offers some pricing advantages over other WAN technologies, which is priced on a distance- and capacity basis. T1, for example, is priced according to the distance between two sites; the greater the distance, the more expensive the T1 line. X.25 is priced based on usage; the higher the volume of traffic, the more you pay.

X.25 is used rather infrequently in the United States because leased lines are plentiful and relatively inexpensive; however, X.25 is widely used internationally. X.25 is often the only reliable WAN method, as many countries' phone systems are outdated and leased lines are not readily available. As a follow-on to X.25, frame relay will be popular in Europe, as well as for connecting U.S. sites to those abroad.

FRAME RELAY AND YOU

Like LANs, frame relay uses a variable length packet, with sizes ranging from a few bytes to more than 4,000 bytes. A variable packet length enables it to accommodate the LAN's bursts in traffic; however, because the delay is unpredictable, frame relay is not especially suitable for voice or video traffic—only data.

Each frame includes an 11-bit address field, the Data Link Connection Identifier (DLCI), which supplies the virtual circuit number that corresponds to a particular port on a switch. Frame relay offers independent packet addressing, which also reduces overhead. Private virtual circuits may be set up between addresses.

A variety of devices can be used to bring frame relay to your LAN. If you choose a public frame relay service, you'll need frame relay equipment on your premises, as well as a connection to the frame relay network via the local loop.

Most corporations use a router with a frame relay interface card. These routers may use specialized hardware or run on a PC.

You can also use a multiplexer, commonly referred to as a mux, that accepts data from the LAN and forwards it to the frame relay network. A mux is handy in a point-to-point configuration, but if your network uses multiple protocols or has a complex configuration, then you should use routers.

If you implement a private frame relay network because of security, reliability, or control issues, you will need to purchase frame relay switches. The added benefits may offset the costs of investing in your own switching equipment.

Frame relay equipment has the advantage of being relatively similar to X.25 equipment. In many cases, the manufacturer does little more than change the software to give an X.25 device frame relay functionality. This similarity can make upgrading from X.25 to frame relay more cost-effective than switching to another WAN technology.

FRAME RELAY FOIBLES

Frame relay comes in two versions: *Frame relay 1* sets up a permanent virtual circuit, in which the connection must be maintained at all times. One of the touted benefits is that frame relay offers "bandwidth on demand." This benefit should not only mean that you can dial up extra bits per second when the LAN traffic surges, but also that you can dial up the service itself as needed.

The next version of the standard, *frame relay 2,* will offer this switched virtual circuit service. When switched virtual circuits do become available, frame relay will become more flexible, since users can dial up the service.

Frame relay does not handle congestion with grace. When there's too much traffic on the frame relay network, the frame relay equipment drops frames. Because network traffic travels in bursts, the receiving device's buffer can easily become overloaded. The device discards the packets and waits for the end nodes to notice and retransmit. This strategy is not reliable for timing-dependent networks.

The two methods for handling congestion within frame relay are Explicit Congestion Notification (ECN) and Consolidated Link-Layer Management (CLLM). Neither method was initially standardized, and it remained the prerogative of each vendor and service provider to implement it differently or not at all. This lack of conformity makes communication among different vendors' frame relay products difficult at best. ECN offers the ability to communicate traffic overload problems downstream and upstream. ECN can be done from the source node in the direction of the data flow, which is called *Forward ECN.* Or it may be done from the data flow back toward the sending node. This process defines Backward ECN. The congestion flag is set in the DLCI address, so that all sending and receiving nodes know that the network is congested. With CLLM, one of the DLCIs is used to send link-layer control messages. These messages can be coded to include the type of congestion and the addresses of the DLCIs involved.

In addition to ECN and CLLM, the router can handle the congestion, although errors can be handled more quickly at the lower layers of the OSI stack.

Each frame relay carrier offers a different set of access rates, pricing, and price discounts. When selecting a frame relay carrier, check the link speed and the committed information rate, or the data rate between two sites, that the carrier guarantees. You should specify a committed information rate that's below the link speed you think you'll need; this way the frame relay network can accommodate bursts of LAN traffic. In some instances, you'll have to lease a line between your site and the frame relay point of presence.

WHAT TO DO, WHAT TO DO

Frame relay is a good interim media. Technically, it's better for LAN traffic than is X.25, and it has particular applicability for international networks. Yet its ultimate success will be determined by factors beyond technology. If the service providers price frame relay aggressively, then users will be encouraged to try a new service. Frame relay's success may be dampened by technologies such as ATM and SMDS that not only offer high speeds but, more importantly, are able to carry voice and video as well as data. For it is in the high-speed integration of data and voice that a true enterprise LAN/WAN can be constructed.

This tutorial, number 54, by Patricia Schnaidt, was originally published in the February 1993 issue of LAN Magazine/Network Magazine.

ATM (Asynchronous Transfer Mode)

Asynchronous Transfer Mode promises to deliver vast amounts of bandwidth to network users. While ATM was envisioned as technology for public network carriers, its application has been recast, and you can expect to see ATM deployed in private as well as public networks over the next decade.

ATM is the purported solution to the LAN/WAN integration quandary. Companies are looking for an efficient and cost-effective method of integrating their dispersed multiprotocol LANs, and frame relay, SMDS, and T3 are vying as contenders. So far, none has been wholly successful. LAN technologies, with their ability to carry large amounts of data over limited distances, are inherently unsuitable in a geographically large network. WAN services, although able to efficiently carry voice and to a lesser extent data over long distances, offer limited bandwidth. ATM, however, can effectively integrate the benefits of LAN and WAN technologies while minimizing the side effects of both.

ATM offers a high bandwidth service that is capable of carrying data, voice, and video over great distances. ATM can provide interfaces to transmission speeds ranging from 1Mbit/sec to 10Gbits/sec. It offers low latency, making it suitable for time-sensitive or isochronous services such as video and voice. Plus, it is protocol- and distance-independent.

WHAT'S WRONG WITH LANS

There's nothing wrong with LANs, as long as their users are local to their server and their applications don't require vast amounts of bandwidth. But beyond that, LANs are shared media that don't scale gracefully.

Most LANs are relatively low speed. Ethernet and Token Ring were designed when LANs were primarily occupied with file transfer. LANs stand to be the delivery mechanism of a whole host of distributed, client-server applications, but these applications must be viable over a wide area network. While 10Mbits/sec may have once seemed extravagant, MIPS and RAM are inexpensive, and workstations and PCs are pumping vast amounts of data onto the network. Five users on an Ethernet or Fast Ethernet aren't uncommon; some Fast Ethernet segments have just one user.

FDDI, designed as a 100Mbit/sec backbone technology and rescaled into a high-speed workgroup technology, can accommodate the high bandwidth needs of workstations. Adapter cards and hubs are costly to purchase, and fiber is expensive to install, although twisted-pair FDDI was once expected to bring down the pricing to be competitive with the more expensive

Token Ring products. But at 100Mbits/sec, FDDI is inherently a LAN technology and limited to transmitting information a few kilometers, not cross-country.

LANs don't scale well. When traffic becomes overwhelming, network managers segment the network with bridges and routers, thereby reducing overall traffic. Segmentation can increase delay—as anyone sitting on the other side of a slow router from the desired server can attest.

LANs don't scale well when they are based on a shared medium. All users on a shared medium LAN must share the available bandwidth, whether it is a 10Mbit/sec or 100Mbit/sec pipe. When a station transmits, it occupies the entire bandwidth, and all other stations that want to transmit must wait until the sender has finished. Even on a high-speed LAN such as Fast Ethernet or FDDI, only one station may use the bandwidth at any given time. Two stations that want to transmit 50Mbits/sec each cannot transmit in parallel.

THE KILLER APPLICATION

The next "killer" application may do more than spur product sales, as killer applications are supposed to do; it may kill the network. The multimedia applications looming on the horizon will consume every spare Kbit/sec on the LAN but will not be sated. Apple, Microsoft, IBM, and many other developers are writing applications that will integrate voice, video, and data. Image-enabled software, such as Lotus Notes, is only the beginning. Novell is integrating image capability into NetWare. E-mail will come equipped so users can make voice annotations. Imagine having workers improve their job skills or students expand their knowledge by downloading video clips from video servers that reside in different cities. Consider the possibilities of interactive video. Now ruminate on these applications' effects on the LAN and WAN.

FDDI was once supposed to be the medium for multimedia applications, but because of its hard-to-expand bandwidth and insensitivity to time delays, FDDI is suitable in only limited applications. Whereas data can tolerate some delays in transmissions, video and voice cannot. FDDI, like all legacy forms of LAN technologies, cannot guarantee that quality of service.

The proposed FDDI-II is sensitive to the needs of voice and video; however, the specification is still under development and is completely incompatible with FDDI-I (or what we think of as FDDI).

THE TECHNICAL DETAILS

ATM can be used to integrate disparate LANs across the WAN as well as on the LAN. ATM is a switch technology, not a shared media technology, which diverges from traditional LAN architecture, but is quite common in telecommunications.

Specifically, ATM is a CCITT and ANSI standard for cell switching that operates at speeds from 1.544Mbits/sec to 10Gbits/sec, with several specified interim speeds. Cells are short fixed-length packets, and in ATM, the cells consist of 48 bytes of user information plus 5 bytes for the header. Because the cell size is fixed, network delays and latencies can be predicted, making ATM suitable for carrying real-time information. LANs use variable length packets, which makes delays unpredictable and unsuitable for carrying voice and video. With the fixed cell size, large packet switches can be built rather inexpensively.

With an ATM network, any one user can directly connect to any other user by establishing a virtual circuit, which can transfer data over the link with no added overhead. Contrast this with a LAN internetwork, where each packet must find its way through every intermediate router. In other words, ATM is connection-oriented, while most LAN technologies, including IP and IPX, create connectionless networks.

ATM is scalable. With ATM, additional switches can be added to increase the network capacity. In a switch-based architecture, the aggregate capacity of the network goes up as more ports or lines are added. With a shared medium such as LANs, the aggregate capacity remains the same.

Logical connections between users are made via virtual circuits and virtual paths. Virtual circuits can be permanently established, thereby guaranteeing a level of access, or set up dynamically, allowing for the network service to adjust itself to the demand. For each call, the user application specifies the average and peak traffic rates, peak traffic duration, and burstiness of the traffic. By setting these parameters, network designers can ensure that the voice, video, and data traffic get the required quality of service. For example, the network can respond to a traffic burst by automatically allocating additional bandwidth to a particular virtual circuit. Or, certain types of traffic or calls can be prioritized according to their importance or sensitivity to time delay.

Cell relay separates the relaying of data cells from the management of logical connections. Hardware will process the cells, while software will establish the virtual circuits, manage the resources, route calls, and handle billing. This separation into layers enables you to upgrade the hardware and software separately, thereby allowing a longer life for the network infrastructure.

ATM is independent of the upper-layer network protocols, supporting IP and other Layer 3 protocols. ATM can use a variety of transmission speeds and protocols at the physical layer.

WHERE ATM FITS

ATM may make its appearance in several places on your network: hubs, routers, to the desktop, and eventually, as a publicly offered service.

ATM has made a marginal appearance on the desktop. Some workstation vendors have equipped their machines with ATM interfaces. Multimedia and other bandwidth intensive applications suitable for ATM to the desktop include engineering, financial analysis, medical imaging, and multimedia. Initially, prices of ATM adapter cards were high—in the $1,500 to $2,000 range, but prices decreased, though never to the point where a substantial market imperative developed for desktop ATM. Router vendors are incorporating ATM into their products, primarily to create campus backbones and to access wide area links.

NOT JUST ANOTHER ACRONYM

It seems like every time you turn around, there's another LAN or WAN technology claiming to be the ultimate solution. A barrage of standards and a slew of acronyms assail network planners and administrators. ISDN didn't work, then there was frame relay and SMDS. Why should ATM be any different?

ATM is designed to handle the needs of both voice and data, thereby reintegrating the communication that was disjointed by computers and telephones. ATM has strong support from manufacturers, telephone companies, and users, both domestically and internationally. Whereas the U.S. and Europe use different speed T1, ISDN, and other WAN services, ATM—at 155Mbits/sec and higher rates—provides a common ground for a single global infrastructure. Also, ATM is at the beginning of its lifecycle. ATM supports speeds up to 10Gbits/sec but also accommodates slower speeds. You can install it today, and with the increasing adoption of SONET, the infrastructure and transmission system should be viable for the next decade.

This tutorial, number 50, by Patricia Schnaidt, was originally published in the October 1992 issue of LAN Magazine/Network Magazine.

ATM Basics

Asynchronous Transfer Mode—what a mouthful of vague abstractions. If there weren't so much money and job security riding on ATM, it would be awfully tempting for many of us to ignore it. A transfer mode is a method of transmitting, multiplexing, and switching data in a communications network. (Multiplexing is combining multiple streams of data on a single circuit; if it weren't for multiplexing, the view of the sky in our cities would be blocked by telephone wires.)

It's only possible to understand "asynchronous" by getting a grasp on "synchronous." The terms refer to digital signals; more specifically, they identify two ways that units of data are framed or blocked within a stream of bits. Synchronous signals are closely tied to some sort of clock, so each unit of data begins, for example, precisely at 0.0ms, then 7.5ms, then 15.0ms, then 22.5ms, and so forth. Asynchronous signals are not bound tightly to a clock—perhaps their data units have a start and a stop bit, or some kind of unique bit pattern to identify the beginning and end of a character or a packet.

Most serial communications and practically all LAN communications are asynchronous, but most data transfers in and out of your microprocessor, the traffic on your parallel port, and the traffic on your computer's bus, are synchronous. Given a steady stream of data, synchronous transmission tends to be more efficient than asynchronous, while asynchronous transmission tends to be more flexible and resilient.

The telephone companies, who incidentally built the infrastructure for wide-area computer data communications when they built their voice networks, can be forgiven for concentrating on a data type—human speech—that is highly intolerant of timing variations. As they built equipment to handle thousands and millions of simultaneous conversations, they developed techniques for multiplexing numerous digital voice circuits on single lines—first copper wires, then fiber optic cable.

Time division multiplexing turned out to be the best way to combine many telephone circuits on a single physical cable. At each level of concentration, a 64Kbits/sec telephone circuit is tied to a specific time slot. If the timing of a telephone call is disrupted, perhaps through being routed via a satellite or as a result of some malfunction, we might hear a slightly delayed echo that makes it nearly impossible to keep talking, or the whole conversation may be unintelligible. Synchronous communication is well matched to the voice data type.

Video and much multimedia material also matched well with the characteristics of synchronous communication. Not only is there often human speech or other sound involved, the sound can't be allowed to wander away from the image; if it does, the video or multimedia session starts to feel like a badly dubbed Japanese sci-fi movie.

Unfortunately, most computer data communications do not fit well with synchronous methods. For one thing, they tend to be bursty, meaning the ratio of the peak data rate to the average data rate is high. (Synchronous data links have no time gaps—the peak data rate is the same as the average data rate.) Thus, data communicators face the quandary of either paying huge amounts for data pipes that are mostly idle except at peak usage times or suffering through long delays at peak times with a less-expensive small data pipe. The telephone network is not well suited to supply bandwidth-on-demand.

ATM is the result of a compromise among all of the data-type constituencies to find a single common denominator for all types of data. One alternative to time division multiplexing is to use packet or cell multiplexing. A stream of bits is broken up into discrete packets or cells, each of which has a header indicating its path and other worthwhile information. If the cell size is made small, and the overall throughput of the circuit is high, delay-sensitive traffic can be carried along with bursty types of data successfully, and everyone gets what they need from the data link. Voice and video work without glitches, and data customers (potentially) get bandwidth-on-demand. As a universal transport, ATM can plausibly be installed on the desktop, on departmental and campus backbones, on high-capacity wide area services, and even on a global information superhighway system.

During the development of the fundamental ATM definition, the voice interests—particularly the European telephone providers—wanted a 32-byte cell with a 4-byte header, while many North American interests preferred a more efficient 64-byte cell with a 5-byte header. The compromise of a 48-byte cell with a 5-byte header was reached, so, an ATM cell is a 53-byte entity.

ATM AND NETWORKS

Like frame relay and X.25, ATM protocols are connection oriented. ATM sessions take place over virtual circuits (virtual because they need not use particular physical paths, although once the virtual circuit is established, it stays in place for the duration of a session). (For a graphical representation, see Figure 1.) Most, if not all, of today's ATM services offer only permanent virtual circuits (PVCs); setting up and tearing down PVCs is a job for the telephone company unless the ATM network is completely private. The real promise of bandwidth-on-demand will be fulfilled when switched virtual circuits (SVCs) become available. PVCs are comparable to leased lines, while SVCs are comparable to dial-up voice service. An ATM SVC will typically take only a fraction of a second to be established, however.

With its connection orientation, ATM does not readily compare with shared medium protocols, such as Ethernet and Token Ring, or with connectionless protocols that perform routing, such as IP and IPX. With the development of LAN emulation standards, ATM services can be made available to Ethernet and Token Ring networks. Products for translating frame relay data to ATM have been announced. IP and Address Resolution Protocol over ATM are described in the Internet RFC1577. In general, ATM fits into the data link and physical layers, but because connec-

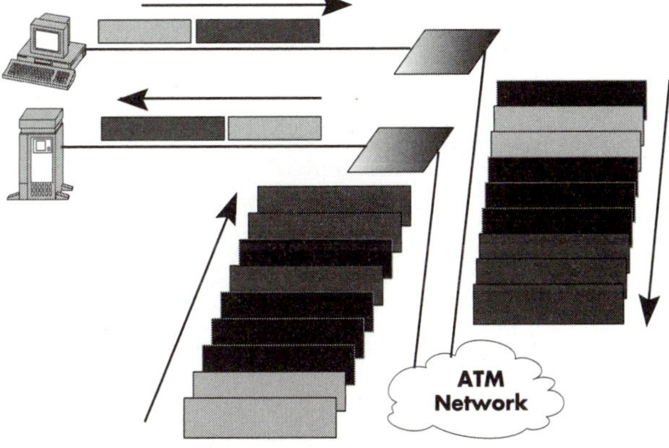

■ **Figure 1:** Computer Data: Packets of computer data are chopped up into ATM cells for transit over the ATM virtual circuit, then reassembled at the receiving end. Source: ATM Forum.

tion-oriented protocols don't require routing, it is possible for ATM to provide services to the upper layer protocols directly. This is the sense in which ATM is supposed to sound the death knell for all routers.

ATM PROTOCOLS

The top layer of the ATM protocol stack is the ATM Adaptation Layer (AAL). Different AALs correspond to the different data types ATM supports. Thus AAL1 permits the ATM device to closely resemble a constant bit-rate voice circuit; AAL3/4 and AAL5 are used for variable bit-rate data types, which are those typically found on computer networks. The AAL is also responsible for integrating the inherently connection-oriented ATM with connectionless data sources, enabling ATM clients to emulate broadcasting and multicasting.

The ATM layer is the common core of all ATM technology. There are multiple AALs and multiple physical layer options, but the protocol that describes the cell header layout and governs the actions of switches on the cells is a constant. The ATM layer is responsible for cell routing, multiplexing, and demultiplexing.

Before any user data can flow over an ATM virtual circuit, each intermediate switch must create a local routing table entry that maps the inbound virtual channel identifier to an outbound port. In order to simplify the routing overhead for intermediate links, ATM defines virtual paths (VP), which are basically virtual channels defined over two or more physical links that are treated as a unit. VPs are semi-permanent connections, and their routing tables may be set up in advance. A packet that travels along a VP will not have to be rerouted at each of the component VCs.

Based on the type of data carried in cells, the ATM layer interleaves multiple streams together based on the priority of each type. It is also responsible for identifying congestion, managing faults, and managing traffic.

At the physical layer, now customarily referred to as PHY, ATM supports (or will support) multimode optical fiber, single mode fiber, STP, coaxial cable, and UTP, at throughputs as high as 10Gbits/sec, although the speed of ATM can be extended as far as the market's ability to pay for it. ATM traffic can readily fit into SONET or SDH (Synchronous Digital Hierarchy, the international superset of SONET standards) data streams—the 155Mbits/sec single- and multimode fiber physical layer standards are based on SONET frames. A 45Mbits/sec standard for the DS3 interface, which uses coaxial cable, has also been defined. DS3 facilities are much more widely installed in North America than SONET facilities. At 100Mbits/sec, ATM can use the physical standards defined for FDDI.

The ATM Forum has chosen to adopt existing physical layer standards wherever it can. Potential future standards may include: 52Mbits/sec over Category 3 (or higher) UTP;

155Mbits/sec over Category 5 UTP; and 1.544Mbits/sec for T1 (or DS1) lines. IBM and several allied companies have also proposed a 25Mbit/sec standard for desktop connections using Category 5 UTP cabling.

This tutorial, number 81, by Steve Steinke, was originally published in the May 1995 issue of LAN Magazine/Network Magazine.

MPLS in Brief

Why Multiprotocol Label Switching (MPLS)? A better question is, why not ATM? Despite IP's early popularity, ATM's connection-oriented features have always held an allure for networkers. Early ATM switches were faster than routers and much better at handling different types of traffic. But IP bigots never liked ATM's efficiency and scalability problems in an IP environment.

Even setting up and maintaining ATM was a pain for network operators. Someone had to be responsible for managing all the individual Permanent Virtual Circuits (PVCs) used to create an ATM network. Even if operators were prepared for that complexity, clients had to be prepared to pay ATM's famed "cell tax," the 20 percent overhead incurred when ATM switches segment large IP packets into small, fixed-length ATM cells.

ATM switches also ran into scalability problems. With the Segmentation and Reassembly (SAR) overhead inherent in chopping IP packets into cells, the maximum link performance depends on high-speed electronic components. These electronic components get expensive and difficult to manufacture at speeds greater than 2.5Gbits/sec. The highest-performing routed links can handle speeds four times faster.

Routing became a huge problem in large ATM networks when ATM routers were connected through ATM virtual circuits. Normally, engineers look to maximize network performance and create a full mesh of circuits connecting the routers; each router becomes adjacent to every other router on that network, regardless of its physical location.

The problem with typical routing protocols is that adjacent routers update each other with information about network changes. Networks comprised of a few routers generate nominal amounts of routing information, but the information increases as the number of routers increases. Every n (number of new routers) entails as much as n4 new updates. Since the amount of routing information grows so quickly, large networks can reach the point where the routing traffic overwhelms the router. Work-arounds are possible, but typically not without sacrificing performance or simplicity.

THE MPLS ANSWER

MPLS aims to give routers not only greater speed than ATM switches but also the sophistication of a connection-oriented protocol. MPLS was designed to do this by enabling routers to make forwarding decisions based on short labels, thereby avoiding the complex packet-by-packet look-ups used in conventional routing.

Today, the need exists for the turbo-boost that MPLS gives routing. Advances in ASIC design have dramatically improved the speed of even conventional routers. It's clear, however, that MPLS enables carriers to better control the traffic flows in their networks. With traffic engineering, network designers can assign various parameters to links and then use that information to maximize the efficiency of their networks.

Since MPLS supports traffic engineering, carriers can eliminate multiple layers in their networks. Instead of running IP over ATM and worrying about configuring and maintaining both networks, MPLS practitioners aim to migrate many of ATM's functions to MPLS, and perhaps even eliminate the underlying ATM protocol. Finally, MPLS segregates different customers' traffic into separate VPNs. Although MPLS VPNs aren't encrypted, the very act of segregating the traffic provides a first line of defense.

UNDER THE COVERS

So just how does MPLS work this magic? Unlike typical routing, MPLS works on the idea of flows, or Forwarding Equivalence Classes (FECs) in MPLS parlance. Flows consist of packets between common endpoints identified by features such as network addresses, port numbers, or protocol types. Traditional routing reads the destination address and looks at routing tables for the appropriate route for each packet. Each router populates these routing tables by running routing protocols—such as RIP, OSPF, or Border Gateway Protocol (BGP)—to identify the appropriate route through the network.

By contrast, MPLS calculates the route once on each flow (or FEC) through a provider's network. The MPLS-compatible router embeds a label consisting of short, fixed-length values inside each frame or cell. Along the way, routers use these labels to reduce look-up time and improve scalability.

Of course, that's a gross simplification of a complex process. Let's look at the path that a flow might take between two points. When a packet leaves your PC, it makes its way across the network and ultimately hits a Label-Edge Router (LER), likely located at the entrance to the carrier's network.

The LER is the doorkeeper to the MPLS network and classifies packets as a member of an FEC. An FEC may be defined by its IP header, the interface through which a packet arrives, the packet type (multicast or unicast), or other information such as the Type of Service (ToS) field used to mark packets by DiffServ. Routing protocols modified for MPLS's unique requirements, such as OSPF with Traffic Engineering (OSPF-TE) and BGP-TE, gather the routing information needed to identify where to send the packet.

Once the LER determines the FEC's route, it inserts a label in each frame or cell. Typically, this label gets appended to the layer-2 (Ethernet, for example) header, but if there's no room, a small reference field is added that directs the router to the label's location inside the data field. If the underlying network is based on ATM, the label populates the Virtual Path Identifier/Virtual Channel Identifier (VPI/VCI) field. If the underlying network is frame-based, the label is enclosed in a shim between the data-link header and the IP header.

To ensure that capacity for the transmission is reserved end to end, the LER uses a Label Distribution Protocol (LDP)—such as Constraint-based Routing LDP (CR-LDP) or the RSVP-TE—to distribute the necessary labels that direct traffic along this route. Then the Label Switched Path (LSP) is established. Traffic sent onto this LSP traverses the desired route specified by the LER. Each Label-Switched Router (LSR) reads the specific label, finds the route where the packets should be forwarded in its table, and acts accordingly.

TRAFFIC ENGINEERING

As with any new protocol, MPLS carries its share of options and features, which some users have adopted with nearly Talibanesque fervor. An example is the specific LDPs used in MPLS networks. Everyone agrees that LDPs enable LSRs to reliably discover peers and establish communication using four different kinds of messages. Networkers can use these protocols to direct their traffic in accordance with certain predefined weights they've assigned to links in their networks.

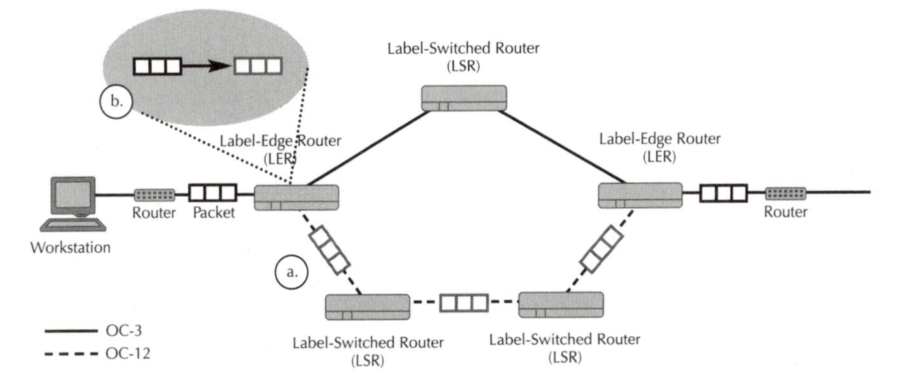

MPLS Operation. With MPLS, the LER can decide on the optimal path by accounting for considerations other than routing hops, such as line speed (a). Once the LSP is established, the packet is then properly labeled (b) and sent through the network.

Beyond that, there's been considerable debate as to the best approach. A number of vendors, most notably Cisco Systems and Juniper Networks, have advocated using RSVP-TE. Other vendors, chiefly Ericsson and Nortel Networks, have pushed CR-LDP.

The two approaches have a lot in common. Both protocols use similar Explicit Route Objects (EROs). Both protocols use ordered LSP setup procedures, and both include QoS information in their signaling messages to enable automatic resource allocation and LSP establishment.

The differences largely come down to orientation. As RSVP-TE extends the pre-existing protocol, RSVP, to support label distribution and explicit routing, there's already an installed base. CR-LDP, however, extends LDP, a comparatively new protocol, which was originally meant for hop-by-hop label distribution for explicit routing and QoS signaling. RSVP runs over IP and uses forward signaling. CR-LDP runs over TCP and uses reverse signaling.

Today, the debate has simmered down. Ericsson and Nortel support RSVP-TE as well as CR-LDP. The result is that RSVP-TE standards tend to be pushed through faster than CR-LDP. However, networkers can still expect to hear about LDP, says Paul Brittain, a senior network architect at Data Connection (www.dataconnection.com), a developer of MPLS gear. Carriers will still use LDP to distribute labels within their networks for establishing large flows between their POPs. RSVP-TE will likely be used at the customer premises to establish customer-specific flows between sites, for example.

MPLS OVER EVERYTHING

MPLS gives providers better performance and better control over their networks. At the same time, though, MPLS cannot establish or alter physical connections.

Enter Generalized MPLS (GMPLS), once called Multiprotocol Lambda Switching. GMPLS extends MPLS to control not only routers but also Dense Wavelength Division Multiplexing (DWDM) systems, Add/Drop Multiplexors (ADMs), photonic cross-connects, and so on. With GMPLS, providers can dynamically provision resources and provide the necessary redundancy for implementing various protection and restoration techniques.

MPLS deals only with what GMPLS calls Packet Switch Capable interfaces; GMPLS adds four other interface types. Layer-2 Switch Capable interfaces can forward content-based data within frames and cells. TDM Capable interfaces forward data based on the data's time slot. Lambda Switch Capable interfaces, such as a photonic cross-connect, work on individual wavelengths or wavebands. Finally, Fiber Switch Capable interfaces work on individual or multiple fibers.

These different LSPs utilize the "nesting" feature inherent to MPLS: Within MPLS, smaller flows are aggregated into larger flows. The same basic concept applies to GMPLS, but think of the LSPs as virtual representations of physical constructs. So LSPs representing lower-order SONET circuits might be nested together within a higher-order SONET circuit. Similarly, LSPs that run between Fiber Switch Capable interfaces might contain LSPs that run between Lambda Switched Capable ones, which could contain those that run over TDM, which could include Layer-2 Switch Capable LSPs, which could, finally, include Packet Switch Capable LSPs.

Otherwise, GMPLS functions much like MPLS. LSPs are established by using RSVP-TE or CR-LDP to send a path/label-request message. This message contains a generalized label request, often an ERO, and specific parameters for the particular technology. The generalized label request is a GMPLS addition that specifies the LSP encoding type and the LSP payload type. The encoding type indicates the type of technology being considered, whether it's SONET or Gigabit Ethernet, for example. The LSP payload type identifies the kind of information carried within that LSP's payload. The ERO more or less controls the path that an LSP takes through the network.

The message traverses a series of nodes, as in MPLS, to reach its destination. The destination replies with the necessary labels, which are inserted into LSR's tables along the way. Once the reply reaches the initiating LER, the LSP can be established, and traffic is sent to the destination.

RESOURCES

At the Multiprotocol Label Switching (MPLS) Resource Center (www.mplsrc.com), you'll find a great FAQ with loads of information about MPLS. The resource center's well-organized list of standards is also quite helpful.

Of course, if MPLS standards are really what you want, visit the IETF, the originator of the MPLS standard, at www.ietf.org.

Many MPLS vendors offer in-depth technical white papers. Data Connection (www.dataconnection.com), a developer of MPLS gear, offers papers that are both clear and unbiased.

This tutorial, number 163, by David Greenfield, was originally published in the February 2002 issue of Network Magazine.

PPP and PPP over Ethernet

The first protocol used for carrying IP packets over dial-up lines was Serial Line IP (SLIP). SLIP is an almost embarrassingly simple protocol. Basically, you send packets down a serial link delimited with special END characters (0xC0). Because the END character might appear in the data stream, there is an escape character to signal that a sequence of 0xDB, 0xDC should be translated back to an END character at the other end of the link. Finally, if there is an escape (0xDB) character in the input data, the special 0xDB, 0xDD sequence indicates that the output should be the escape character only. This is all there is to SLIP.

SLIP doesn't do a number of desirable things that data-link protocols can do. It's customary (and when connecting lines are noisy, essential) to perform error checking at layer 2. It's commonly desirable to authenticate users who are dialing in to an access router. Two computers connected with a serial link may want to pass traffic formed with protocols other than IP. Because it deals with these issues, the Point-to-Point Protocol (PPP) has become the predominant protocol for modem-based access to the Internet. Furthermore, a multilink version of PPP is the most common ISDN access protocol; it is also employed for inverse multiplexing analog phone lines. And even on high-speed optical lines, PPP is the protocol of choice for those who want to disintermediate ATM on point-to-point links.

PPP is one of the many variants of an early, internationally standardized data-link protocol known as the High-level Data Link Protocol (HDLC). HDLC was derived from a proprietary protocol developed by IBM called the Synchronous Data Link Protocol (SDLC).

The HDLC-derived parts of PPP are its frame definition and its ability to support asynchronous data streams (that is, those with start and stop bits) and byte-synchronous data streams, as well as the most commonly encountered bit-synchronous data. The HDLC frame is shown in Figure 1. The Frame Check Sequence (FCS) field in an HDLC frame provides the potential for layer-2 error checking, one of the significant missing capabilities in SLIP. The two-byte Protocol field permits the identification of a layer-3 protocol in the payload, thus providing part of the necessary capability for multiprotocol support.

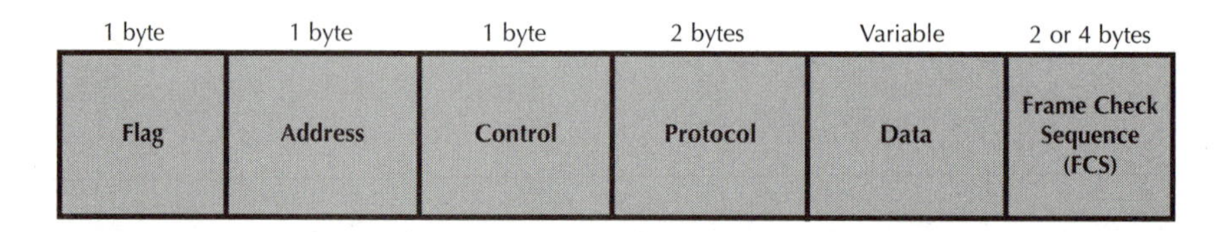

■ **Figure 1:** PPP frames are composed of six fields. The flag is a fixed pattern that delimits frames. The Address and Control fields are also fixed values. The Protocol field provides a layer of indirection so that the Data field can represent actions as well as user data. The Frame Check Sequence (FCS) field provides 16- or 32-bit error tracking.

LCP AND NC

The protocols that differentiate PPP from HDLC are the Link Control Protocol (LCP) and the Network Control Protocol (NCP). LCP establishes, configures, maintains, and terminates point-to-point links. It kicks in before any data transfer operations can begin. The protocol opens the connection, negotiates the configuration details of the link, and passes a configuration acknowledgment frame. LCP can also optionally test link quality prior to data transmission. User authentication is generally performed by LCP as soon as the link is established. Finally, LCP tears down connections when a session is finished.

LCP supports two authentication protocols, the Password Authentication Protocol (PAP) and the Challenge Handshake Authentication Protocol (CHAP). PAP sends a password across the link in the clear, so the circumstances where it can play a useful role are seriously limited. CHAP, at least in principal, has the ability to provide reliable, secure authentication.

For CHAP, the authenticator and the authenticatee share a secret, typically a password. However, unlike PAP, the password never goes over the link. Instead, the authenticator sends the authenticatee a "challenge" in the form of a random number. The authenticatee uses the challenge number and the secret to create a "hash," a message digest unique to the challenge, by means of mathematical functions that are difficult to apply inversely and thereby derive the secret.

The message digest is then transmitted to the authenticator, who applies the same algorithms to the secret and the challenge. If the results of this local operation match the value of the transmitted message digest, then the authenticator acknowledges that the authenticatee has been successfully authenticated, and the next connection stage can proceed.

Though CHAP is a much more credible authentication method than PAP, it's not perfect. Like other "shared secret" arrangements, the secret must be conveyed via a parallel communications channel. Thus, potentially onerous administrative procedures are necessary, and in many cases new authenticatees can't be set up immediately. Furthermore, Message Digest 5 (MD5), the prevalent technology for CHAP, has recently been subject to unexpectedly rapid brute force attacks.

With an established, authenticated link, NCP takes over. PPP has specific layer-3 protocol support for IPX, SNA, DECnet, Banyan VINES, and AppleTalk, as well as IP. One typical IP-specific function is the provision of dynamic IP addresses for remote dial-in users. ISPs commonly employ this PPP feature to assign temporary IP addresses to their modem-based customers.

The IP-specific NCP protocol is the IP Control Protocol (IPCP). Aside from dealing with the calling peer's IP address, this protocol can also negotiate whether or not to use header compression, providing a significant speed improvement for low-speed links. If the calling peer has an IP address, it tells the called peer what it is; if the calling peer doesn't have an IP address, the called peer can assign the caller one from a pool of addresses.

The PPP mechanism is not as complete as DHCP, which can set up a lease period and provide such things as subnet masks, default router addresses, DNS server addresses, domain names, and much more to a newly connected computer. An ISP or other provider of remote connectivity services can use DHCP as the source of IP addresses that NCP hands out.

A Multilink PPP (MP) session begins with an LCP negotiation that determines that both ends of the link have the ability to support MP. The protocol supports independent authentication of multiple links. In effect, an MP layer interspersed between the Network layer and PPP is created.

PPP OVER ETHERNET

It's common for small offices or home users to want to share an Internet or enterprise access link. The sharing urge is particularly strong with high-speed lines like Digital Subscriber Line (DSL). (I'll leave cable modems out of the discussion because they don't use point-to-point links.) DSLs don't particularly cry out for PPP because of customer need to configure IP addresses at connect time—after all, DSL is generally configured as an "always on" service. Nor is the multilink support likely to be especially valuable, as the throughput capacity is typically high enough to push the most significant bottlenecks elsewhere on the network.

A typical DSL network looks like Figure 2. On the customer premises, there is an Ethernet network, which connects to an Ethernet port on the DSL modem. The modem acts as a MAC layer bridge, relaying traffic to a DSL Access Multiplexer (DSLAM) in a phone company's central office over an ATM Permanent Virtual Circuit (PVC). From the central office to an ISP or an enterprise facility, the connection may continue to run over ATM, or it may be converted to frame relay or a leased line TDM protocol for slower-speed links.

The main problems pushing PPP over Ethernet (PPPoE) adoption include the shortage of IP addresses and the problem of connecting multiple users on a network efficiently through a shared DSL. There are multiple potential solutions, but a number of major ISPs and product manufacturers are convinced that PPPoE offers the best combination of parsimony with IP addresses, easy configuration, efficient use of the ATM network, and low-cost customer premises equipment.

ISPs typically have a limited number of IP addresses. When these scarce resources can be pooled and shared among users who dial in and stay online for an hour or two, the ISP can get by with fewer addresses than it can with fixed-address, always-on customers.

In fact, DHCP doesn't really accomplish much with DSL customers who leave their computers on all the time—DHCP leases automatically renew themselves as their expiration time approaches. So one principal thing PPPoE does is emulate the dial-up networking experience, in the sense that you can log off from Internet connectivity even though the computer remains on.

PPPoE maps individual PPP sessions onto Ethernet-connected computers. Thus, each user gets his or her own IP address, if only for the duration of a session. The alternatives to this approach have worse problems: Customer premises Network Address Translation (NAT) software is getting easier to use all the time but is nevertheless too esoteric for ordinary mortals, and static IP addresses are perhaps too precious to be tied up around the clock—unless users are willing to pay extra for them.

If multiple Ethernet clients need to access the DSL link concurrently, an ATM PVC will have to be provisioned for each user-destination pair. That is, if an Ethernet user has more than one ISP or needs to connect over a VPN to an enterprise site, then PVCs will be required for every combination. NAT gets around this problem by running all the traffic through a sin-

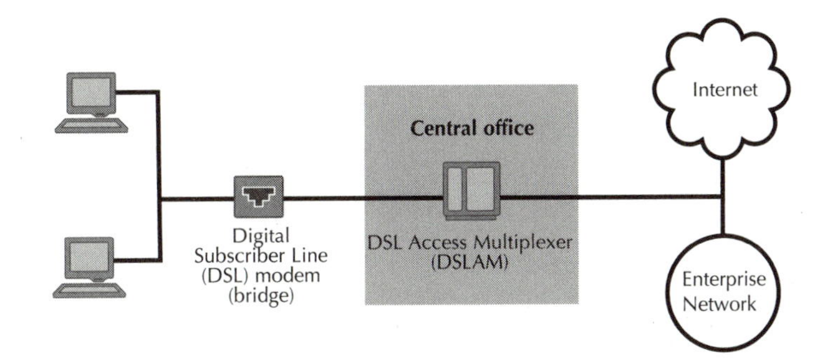

■ **Figure 2:** Practically every Digital Subscriber Line (DSL) modem uses ATM to link to a DSL Access Multiplexer (DSLAM) at the central office of a Local Exchange Carrier (LEC). In many cases, ATM Permanent Virtual Circuits (PVC) will be provisioned all the way from an ISP or an enterprise network to the DSL modem at the customer premises.

gle IP address, and thus a single PVC. Another dubious solution is to install a customer premises ATM switch between the DSL modem and the DSL line.

PPPoE requires special client software for users, and the service providers' termination equipment (DSLAMs) must also be configured to support PPPoE. This additional layer of protocol software may prove troublesome. Furthermore, as PPPoE solutions roll out, software will primarily be available for Windows and Macintosh clients, although Linux support will also be available.

Perhaps more significantly, users accustomed to static IP addresses or DHCP are likely to be unhappy with their pooled and shared addresses. Many new PPPoE users are convinced that their ISPs are the primary beneficiaries of the protocol, while the customers are rather worse off than before.

RESOURCES

The PPPoE specifications are spelled out in RFC 2516. As of this writing, this RFC is informational only and has not been put on a standards track.

RFC 1661 has the specifications for PPP; RFC 1332 covers IPCP; RFC 1334 covers PPP authentication protocols; and RFC 1990 covers Multilink PPP (MP).

There are extended accounts of HDLC and PPP in *Computer Networks, Third Edition*, by Andrew S. Tanenbaum, Upper Saddle River, NJ, Prentice-Hall PTR, 1996. HDLC and related data-link protocols are discussed in *Data & Computer Communications, Sixth Edition*, by William Stallings, Upper Saddle River, NJ, Prentice-Hall, 2000.

ACC, now a unit of Ericsson, has a good white paper on PPP here: www.acc.com/internet/about_apu/userprofiles_whitepapers/white_papers/ppp/ ppp.html.

This tutorial, number 139, by Steve Steinke, was originally published in the February 2000 issue of Network Magazine.

Cable Modem Systems

One of the most promising solutions to the demand for reasonably priced high-speed Internet access comes from the cable television world. Cable operators, who sometimes call themselves Multiple Service Operators (MSOs) when they deliver access to things other than TV programming, can devote a small fraction of their bandwidth resources to digital data and collect about as much money from subscribers as they would charge for basic cable service and a couple of premium channels.

The transformation of the cable TV network to a digital data network has been fraught with obstacles, though none so serious that money can't overcome them. First, cable TV distribution was designed to transport analog signals via Frequency Division Multiplexing (FDM). Conversion to digital format is not a problem: There are many ways to modulate a carrier to transport digital material—in fact, part of the problem has been that cable operators have been slow to adopt a standard means of digitizing content.

Second, digital streams tend to be somewhat less hardy than TV signals on long segments of coaxial cable. In most cases, MSOs that want to provide data services have found it necessary to upgrade their physical networks with fiber optic cable at the core, creating hybrid fiber-coax infrastructures.

This fiber build-out, which must be performed system-wide, regardless of the number of subscribers who want data services, has been estimated to cost at least $150 per home along the fiber route. It should be noted, however, that fiber trunks tend to increase the overall bandwidth of cable networks, permitting MSOs to provide substantially more channels than a pure coax system could (see Figure 1).

The third pothole is the problem of upstream transmission. Cable systems were designed for one-way, downstream signal distribution. Provisions for upstream communication require a separate channel—a stream of data on a cable system is not full duplex, as it is on a phone line.

The tree and branch topology of the cable system complicates the upstream transmission problem by exacerbating "ingress noise." Loose connectors, poor shielding, and similar points of high impedance on the coaxial cable (particularly on the customer premises) develop noise signals from amateur radio transmissions, electric motors, and other sources of electrical impulses.

With the cable system's branching structure, ingress noise is additive—in the upstream direction, the problem gets worse as you approach the head end of the system. On typical cable systems, the low end of the spectrum—between 5MHz and 40MHz—is not used for television signals, primarily because this band is noisy, and tracking down the sources of interference is time consuming.

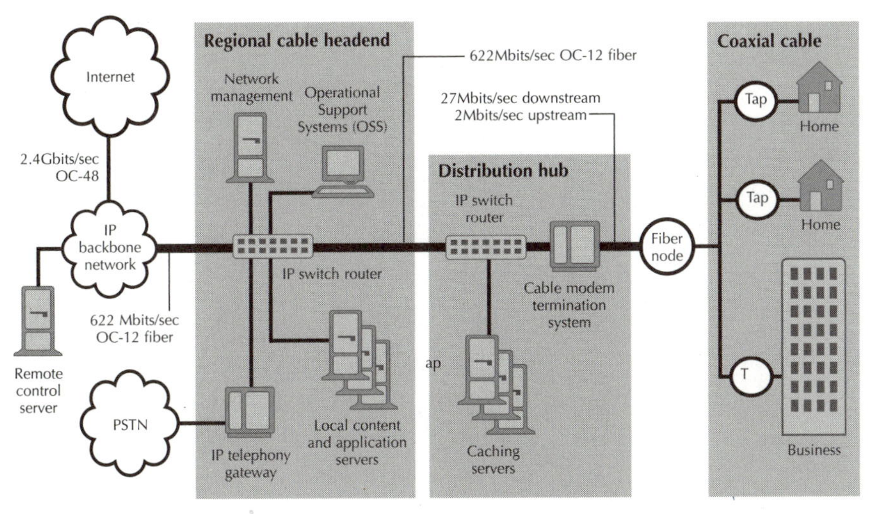

Figure 1: Cable Architecture. Large cable providers typically have termination systems at distribution hubs, with centralized management and Operational Support Systems (OSS) servers.

However, MSOs have chosen to use these frequencies as the upstream channel for data services. Several partial solutions have been deployed. One is to install a high-pass filter on customer lines. This step blocks low-frequency noise, but it requires some sort of "windowing" so that upstream data transmissions won't be filtered out along with the noise.

The principal solution to ingress noise has been the modulation technique cable modems and termination systems employ. While downstream traffic is usually modulated using 64 Quadrature Amplitude Modulation (QAM) or 256 QAM, with 6-bit or 8-bit symbols, respectively, upstream traffic uses Quadrature Phase Shift Keying (QPSK) or 16 QAM, with 2-bit or 4-bit symbols, respectively. (The efficiency of a modulation scheme is higher with greater numbers of bits per symbol, but noise interferes with the resolution of those symbols with more bits per symbol.)

Thus, while a 6MHz TV channel can sustain downstream data throughputs as high as 27Mbits/sec, upstream data throughput, limited to narrower bands between 200KHz and 3.2MHz, typically hits the 320Kbit/sec to 10.24Mbit/sec range.

An added problem for upstream flows is contention resolution. Because there is only one downstream transmitter for the whole cable data system, there is no question of downstream contention. On the upstream path, contention is inevitable. Cable data networks typically solve the problem by assigning time slots at the head end. The cable modem termination system is responsible for dividing the available upstream channels into Time Division Multiplexed (TDM) slots and assigning those slots to endpoints that need to send data.

Some MSOs take a shortcut to deal with the upstream conundrums. They use cable modems that are designed to work with an analog modem and dial in to the cable company's router over a telephone line. This option is sometimes referred to as telephone, or telco, return.

Telco return isn't unthinkable because, first of all, common Internet actions such as Web surfing and file downloading are asymmetrical in the same direction as this technique. For these activities, the ratio of downstream traffic to upstream traffic is often greater than 10 to 1. With telco return, MSOs can basically get into the high-speed data business without upgrading their networks for two-way communications.

Disadvantages for users of telco return systems include the need for the phone line to go online; the delay imposed by modem dialing and mating; and, in some instances, the inability to use the cable capacity for symmetrical sorts of activities, such as Internet gaming, IRC, or videoconferencing. For MSOs, telco return systems are complex to configure, often occupying two router ports for a single customer. The dual pathways also complicate troubleshooting and support.

IT'S THE ARCHITECTURE

The fact that the cable system medium is shared among many subscribers leads to two principal challenges for system operators and users. The first issue is declining performance as incremental users are added. The second issue is security from the activities of other users.

A typical fiber node—the point where optical transmissions are converted to electrical signals that are distributed over coaxial cable—serves between 500 and 1,000 residences. While not every residence will subscribe to high-speed data services, and not all of those who do subscribe will be online at once, cable subscribers often experience service degradations at peak hours—typically in the early evening.

Cable operators can minimize these degradations by allocating additional channels to the data service and by extending fiber nodes farther toward neighborhood end nodes. Both of these solutions are costly, however, and devoting additional bandwidth to downstream data doesn't help if the bottleneck is upstream. In fact, some cable operators, including Excite@Home, have found it necessary to cap the upstream throughput of users at 128Kbits/sec, ostensibly because of subscriber abuses such as operating Web servers in defiance of the operating agreement subscribers must sign.

The security problem may be invisible to many users, but could have an even greater impact on users than a service slowdown. With some models of cable modems, any subscriber with a protocol analyzer can see all the traffic attached to a particular fiber node. Taking financial data, capturing passwords submitted in the clear, and simply eavesdropping on the content of another user's downloads are all potential threats that cable systems must take steps to address.

Another potential vulnerability associated with a shared-medium network arises when Windows or Macintosh users set up file-sharing, peer-to-peer networks. All the subscribers on a fiber node will be able to see these shared files, directories, and folders. If they aren't password protected, they can be copied and erased at the whim of your neighbors.

Unlike Digital Subscriber Lines (DSLs), which are not physically shared among subscribers, file-sharing services need not be bound to IP to become visible to others. Thus, a NetBEUI-based peer-to-peer network that is not bound to IP is still relatively secure on a DSL system, but not on a cable modem system. If you really need a home network with cable modem service, you may want to use servers with hearty access control features, such as NetWare, Windows NT/2000, or Unix, or else install a firewall between the cable modem and your home network.

The Data Over Cable Service Interface Specification (DOCSIS), a cable industry standard developed by Cable Labs, defines a Baseline Privacy Interface (BPI) in version 1.0. DOCSIS 1.0-compliant cable modems and termination systems from multiple vendors began shipping in 1999. The BPI calls for 56-bit DES encryption of all cable modem traffic and provides for key exchange based on 768-bit RSA public key encryption. BPI solves the eavesdropping problem, though there remain IP and Address Resolution Protocol (ARP) spoofing issues that could be exploited if cable modem vendors don't protect against them.

DOCSIS also defines Physical and Data-link layer protocols, and version 1.1 supports Data-link layer prioritization, which is an important step toward providing reliable QoS for voice and video data.

The cable industry has threatened for years to compete with the Incumbent Local Exchange Carriers (ILECS) in the voice market—after all, it's the only service besides the LECs to run physical media throughout most parts of North America. Until the cable network was converted to two-way traffic with optical fiber, these threats of competition were utterly hollow. Now, however, the possibility of competition in the local loop grows ever more plausible, especially with AT&T's resources poised to convert the rest of the former TCI network to a modern hybrid fiber-coax infrastructure.

The cable modem industry is maturing rapidly. It has faced numerous obstacles and managed to work its way around most of them. Excite@Home and RoadRunner, the two largest providers of cable modem service, have both attempted to be providers of special content as well as raw high-speed access. Given both AT&T's interest in distancing itself from content and AOL's acquisition of Time Warner (which includes RoadRunner), it appears likely that cable systems will be opening up to numerous ISPs, much as DSL services have.

In the coming struggle against DSL, cable modem systems have several advantages. MSOs have deployed optical fiber closer to subscribers—especially residential subscribers—than the phone companies have. At least in theory, MSOs could have a downstream throughput advantage over DSL, as long as they move aggressively to maintain that advantage as they add new subscribers. Cable providers also have a marketing advantage:

Once a system has been upgraded, they can be confident that every subscriber's premises will support a connection.

Telcos, on the other hand, face a number of anomalies in their facilities, including distance from a Central Office (CO) and the presence of undocumented past wiring practices, which may make it impossible to connect some customers.

DSL is not without its own advantages over cable modem systems. The absence of the shared medium is foremost, though this advantage will be somewhat neutralized as DOCSIS-compliant cable modems come to dominate the installed base.

Cable modem advocates sometimes claim that DSL systems have the same shortcomings as cable because DSL service providers will be tempted to oversubscribe the link between the CO and the Internet POP. While it's true that a shortsighted service provider could hurt its customers' performance with an undersized uplink, this potential bottleneck is completely different from a medium with a fixed throughput at the core. In other words, the subscriber part of a cable network is a performance bottleneck, while the subscriber part of a DSL network is not.

DSL is more likely to be available in many business districts because cable build-outs have been primarily restricted to residential neighborhoods. Users with a choice between these two high-speed options are fortunate indeed.

This tutorial, number 140, by Steve Steinke, was published in the March 2000 issue of Network Magazine.

Signaling System 7

Next time you pick up your telephone handset, think about what happens after you dial but before you hear the telephone ringing at the other end.

If you're making a call using a regular area code, the local phone exchange can determine the call's destination simply by looking at the first six digits of the 10-digit phone number.

However, if you're calling a toll-free number, the numbers themselves mean nothing. For this type of call, your local phone service provider first needs to query a database to determine where that number is physically located. The provider then contacts phone switches along the destination to set up the call.

If you use a calling card during a business trip, the local telephone central office needs to communicate with that calling card provider's database, as well as with phone switches between you and your call's destination central office.

If you use caller ID or call forwarding, other control signals must be communicated through the Public Switched Telephone Network (PSTN) as the call is set up. Billing information also must be tracked and sent to the appropriate database after the call.

The common thread that ties the PSTN and these services together is Signaling System 7 (SS7), a communications protocol first defined by the ITU-T in 1980.

SS7 SIGNALING POINTS

SS7 is an out-of-band protocol, meaning its signals travel on their own data paths from the actual voice or data call. The protocol is used within the PSTN to communicate between three classes of devices: Service Switching Points (SSPs), Signal Transfer Points (STPs), and Service Control Points (SCPs). Collectively, these devices are referred to as signaling points, or SS7 nodes.

SS7 messages originate at an SSP, which is a telephone switch that places or receives a call. The SSP is usually found at a telco's central office, but SS7 messages may also be used by an enterprise PBX. An ISDN PRI's D channel can also send messages that are compatible with SS7. To set up the call, the SSP sends an SS7 message to a directly connected STP (see Figure 1).

Not all STPs are identical. Local STPs handle only domestic traffic within a Local Access and Transport Area (LATA), while internetwork STPs provide connectivity between LATAs. International STPs provide translation between the slightly different American version of SS7, defined by ANSI (T1.111), and the international version, defined by the ITU-T (Q.700-Q.741). Gateway STPs provide the interface between the PSTN and other services, such as cellular telephone service providers.

If the call-originating SSP knows how to route the call, it will ask the STP to create a connection to the receiving SSP. If the route is unknown, such as with toll-free numbers or local number portability (a proposed way of taking your phone number with you, even if you move across the country), the STP will contact an SCP, which has access to databases with call-routing information.

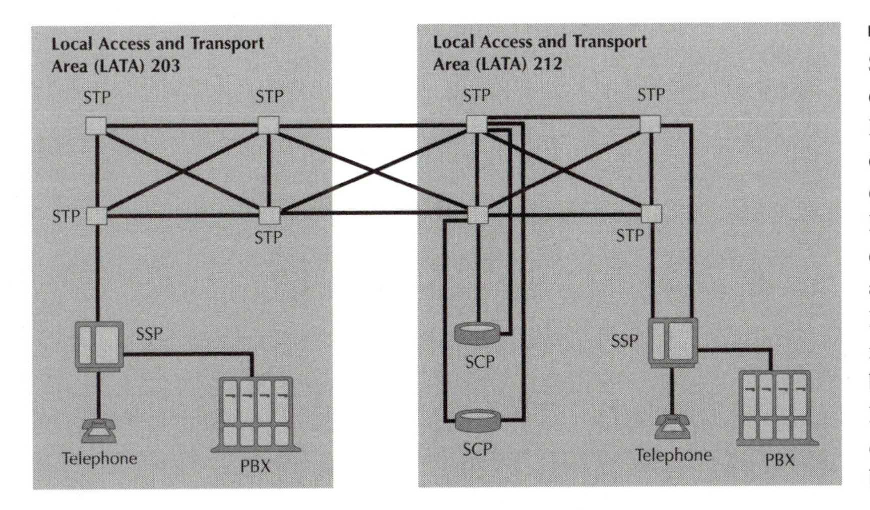

■ **Figure 1:** Signaling System 7 (SS7) consists of Service Switching Points (SSPs), which originate or terminate calls; Switching Control Points (SCPs), which offer acess to databases; and Switching Transfer Points (STPs), which route SS7 information between SSPs and SCPs. Note the many redundant links, particularly between mated STPs.

The STP uses these databases to perform a function called global title translation, which determines a call's destination SSP by translating the global title digits (the number dialed by the caller, including toll-free numbers, 900 numbers, calling-card numbers, or cellular telephone numbers) into a route to the destination SSP. In the case of cellular telephones, the global title translation process also determines the mobile identification number of the receiving phone, as cellular telephones don't really have telephone numbers. SSPs can also send billing information to the SCPs using SS7.

The robustness of the PSTN is partly due to the variety of redundant links between SS7 nodes. Nearly all STPs and SCPs are deployed in pairs, and most SSPs are linked to two or more STPs. In many cases, multiple connections between various STPs run over separate physical media. The SS7 links themselves have special names, as shown in "Linking SS7 Signaling Points (Figure 2)."

Linking SS7 Signaling Points

Link type	Description
Access (A) link	Connects a Signal Transfer Point (STP) to either a Service Switching Point (SSP) or to a Service Control Point (SCP).
Bridge (B) link	Connects one STP to another STP.
Cross (C) link	A pair of redundant bridge links between two STPs; one link is used, the other is for failover. STPs linked in this way are known as mated STPs.
Diagonal (D) link	Similar to a bridge link, but connects STPs within a Local Access and Transport Area (LATA) with an internetwork gateway STP (used to bridge LATAs).
Extended (E) link	Connects an SSP to an STP that's in a different LATA to increase failover redundancy.
Fully associated (F) link	Connects two SSPs directly, bypassing STPs, to accommodate a high volume of traffic.

■ **Figure 2**

THE SS7 PROTOCOL STACK

The SS7 protocol stack consists of four layers, or levels (see Figure 3). The lowest three levels are combined into one set of protocols, referred to as the Message Transfer Part (MTP). The MTP is split into levels 1 through 3, corresponding to the lowest three levels in the OSI seven-layer protocol stack.

MTP level 1, analogous to the OSI Physical layer, defines various physical interfaces between signaling points. Between STPs and their local SSPs and SCPs, the MTP-level-1 links are primarily DS0A (56Kbit/sec) and V.35 (64Kbit/sec) links; DS1 (1.544Mbit/sec) and faster are often used for links between STPs.

Corresponding to OSI's Data-link layer, the MTP level 2 provides node-to-node message error detection and correction (using a 16-bit Cyclic Redundancy Check, or CRC), as well as message sequencing. If an error is detected, level 2 requests a retransmission.

MTP level 3 performs similar functions to the OSI Network layer. This protocol level handles message discrimination to determine who the SS7 message is for. If the message's destination is within the local signaling point, level 3 performs message delivery; if not, it performs message routing to determine the next step toward the destination.

Level 3 is responsible for detecting whether SS7 nodes or links have failed, or whether they are unreliable, congested, or have been shut down or restarted. It determines alternate routes for messages, and sends management messages to adjacent signaling points about changes in link status.

At level 4 in the SS7 stack, two sets of protocols span OSI layers 4 to 7: ISDN User Part (ISUP), for managing phone connection-oriented communications; and Signaling Connection Control Part/Transaction Capabilities Application Part (SCCP/TCAP), to handle all other messaging.

ISUP originates, manages, and terminates ISDN and non-ISDN (also known as POTS, or Plain Old Telephone Service) connections between telephone devices. Thanks in part to ISUP, the days when a caller forgot to hang up the phone—and thus tied up the recipient's phone until the problem was solved—are over. That's because either the originating or destination SSPs can send ISUP messages to terminate the connection. ISUP also transmits caller ID information.

SCCP is used for services between STPs and databases. Corresponding to the OSI Transport layer, SCCP provides more detailed addressing information than the MTP, which only identifies a signaling point; SCCP also indicates which database within the SCP is to be queried.

The database query itself, addressed by SCCP, is sent and returned by TCAP. TCAP messages can send and receive database information, such as a credit card validation or routing information, before a call is placed using ISUP. After a call is completed, TCAP can send billing information to the appropriate accounting database. For mobile users, TCAP transmits user-authentication messages and sends messages notifying SCP databases about the location of cellular telephones. With certain new phone features, such as repeat dialing (which redials a call if the line is busy), the caller's SSP can notify the destination's SSP that this feature is requested; when the line is no longer busy, the destination's SSP notifies the caller's SSP, which redials the call.

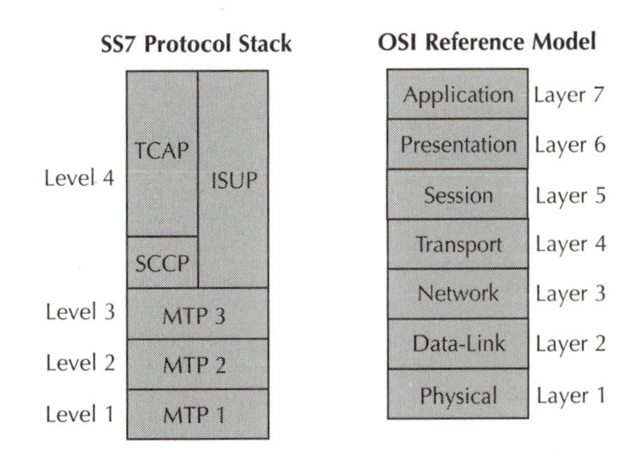

■ **Figure 3:** The SS7 protocol stacks consists of four layers, which roughly correspond to the OSI model.

INTELLIGENT NETWORK

SS7's advanced features, including TCAP messaging and more sophisticated databases connected to SCPs, are part of the phone system's evolution to a set of functions referred to by the industry as the Intelligent Network. The Intelligent Network goes beyond simple telephony, with new features such as caller ID, selective call blocking, and local number portability (all of which depend on SS7 messages). Another Intelligent Network feature enables subscribers to instantly reconfigure their own services (for example, setting up call forwarding using a Touch-Tone phone). Many more features will be phased in over the next few years, thanks to prodding from the FCC—and SS7.

RESOURCES

A good online introduction to SS7, with details about the format of the MTP, ISUP, SCCP, and TCAP protocols, is at www.microlegend.com/whatss7.htm.

Bell Atlantic offers a short self-study course on SS7, complete with quiz questions, at www.webproforum.com/bell-atlantic2/full.html.

For an overview of the telephone network, including SS7, SONET, and broadband ISDN, check out *Telecommunications Protocols,* by Travis Russell (McGraw-Hill, 1997).

For more about the specifics of SS7's components, the best in-depth reference is Travis Russell's book *Signaling System #7, 2nd Edition,* (McGraw-Hill, 1998).

This tutorial, number 125, by Alan Zeichick, was originally published in the December 1998 issue of Network Magazine.

Remote Access

Two distinct technologies for out-of-the-office LAN users.

There are two basic ways to access your network from a hotel or from your office at home: a remote control connection or a remote node connection. The first option is conceptually simple. In effect, the dial-up telephone line becomes an extension cord for the monitor, the keyboard, and the mouse, connecting them to a computer at the host site that can log on to the network. (Remote control is possible with nonnetworked PCs, too.)

In this approach, the remote PC performs no processing operations beyond executing the remote control software itself—it simply updates the display and sends keyboard and mouse input to the host. Exchanges between remote control computers and their hosts take place at the Application level, although some remote control applications can run over networks and therefore use the network protocol stack.

Over telephone lines, direct serial links, and wireless connections, remote control programs are indifferent to whether the host is on a network. Except for the ability of remote control software to perform file transfers to and from a host, remote control is a kind of terminal emulation.

Some of the most popular remote control programs are Symantec's (Cupertino, CA) pcAnywhere, Microcom's (Norwood, MA) Carbon Copy, Traveling Software's (Bothell, WA) LapLink for Windows, Stac Electronics' (San Diego) ReachOut, and Farallon's (Alameda, CA) Timbuktu.

The principal problem with remote control solutions is their expense. A host computer with a network interface, modem, and telephone line, and a remote PC and modem, must be dedicated to every remote session. If the host machines are distributed among different offices, there are potentially serious management problems—primarily the difficulty of providing adequate security and technical support. Finding someone to physically reboot a host machine after it crashes is a potentially difficult problem. Logged-in machines spread throughout the building—with no local users—are a big security risk.

A chassis- or rack-based system of host computers on plug-in boards, such as those provided by Cubix (Carson City, NV) and J&L Information Systems (Chatsworth, CA), can enable network managers to install a high density of remote-session hosts in a central, controlled location and provide some extras such as remote rebooting capabilities and remote management. Unfortunately, the initial cost per host of these systems can be higher than the cost of most standalone PCs with comparable horsepower. (Of course, reduced support, management, and administration costs will offset the initial price difference and save money in the long run.)

In light of the obstacles to providing broad-based remote control capabilities to users, remote node software offers some significant advantages. Unlike remote control, which requires a remote user to have a dedicated modem, a PC, and a network interface, remote node sessions need only a dedicated modem and a serial port. Multiple sessions can readily share a single processor and network interface.

A remote node server is essentially a router (or sometimes a bridge) that translates frames on the serial port to a frame layout that the LAN can accommodate and then passes them along. The processor is required only to route or bridge incoming and outgoing traffic, and it would take more than 150 64Kbit/sec ISDN sessions and more than 1,000 9,600bit/sec asynchronous connections to match the throughput of a 10Mbit/sec Ethernet link—supporting a few remote nodes doesn't take a state-of-the-art microprocessor.

An approach that some software vendors have taken is to combine remote control and remote node software in a single package. Users of these products gain flexibility in matching their access method to various applications and systems.

Remote node servers can be purchased with built-in modems or with serial ports that can be connected to external modems (or to other wide-area links, such as ISDN lines).

ACCESS SERVER HARDWARE

Remote node services can be provided through numerous hardware configurations (see Figure). Router or bridge processing can be accomplished on a processor in a dedicated box, on a communications server (including those that provide remote control services), or on a file server doing communication duty, as well as on a product sold as a router or bridge. (The file server must be able to accept add-on software that performs the routing or bridging jobs.) Multiple serial ports may be built in to a dedicated access box or mounted on boards that plug in to the communications server or file server. Standalone modems may be mounted in an external rack and plugged into serial ports. Internal or modular modems may be installed on plug-in boards, or mounted in PC Card (PCMCIA) slots.

Remote connection ports may also support high-speed ISDN connections. ISDN ports may be supplied on plug-in boards for PC-type servers, on modular boards for dedicated servers, or on standalone terminal adapter units (sometimes called "digital modems") that connect to a serial port. Some terminal adapter boxes (and terminal adapter boards that plug into a PC's bus) have integral NT1 functions, so they can connect directly to the ISDN telephone jack. Others require an external NT1 unit to provide an interface between the terminal adapter and the telephone line. To achieve the maximum possible throughput on low-cost dial-up lines, some ISDN access devices support inverse multiplexing of the two 64Kbit/sec B channels to provide as much as 128Kbits/sec throughput.

In many cases, dedicated remote access boxes are the easiest way to provide network access to remote users. Configuration problems are minimal; in fact, some of these products approach the plug-and-play level of installation ease. They may not be the least expensive solution, however, and they are not likely to offer the most flexibility for upgrading and expanding. Routers and remote bridges may be effective remote node solutions when an organization has standardized on a family of devices and wants to maintain consistency. Some devices that have been designed specifically for remote access are just about as easy to set up as a dedicated box—the primary difference is that modems or ISDN attachment equipment must be configured. PC servers that function as remote node servers will offer the biggest configuration challenges—software will have to be configured, as well as the ports that attach to the dial-up lines.

SOFTWARE OPTIONS

Remote access software also comes in many flavors. Dedicated access boxes, such as those made by Shiva (Burlington, MA), Telebit (Sunnyvale, CA), and 3Com (Santa Clara, CA),

REMOTE NODE SERVERS

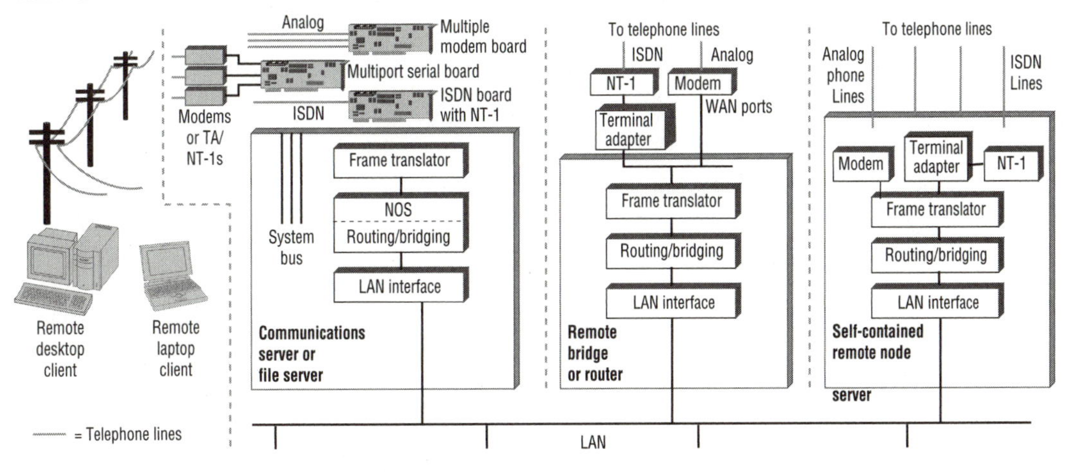

■ **Remote Node Hardware:** Self-contained remote node servers are a complete solution, with telephone and LAN interfaces built in. Remote routers and bridges often are supplied with a generic WAN port, which requires an external modem or, if ISDN lines are used, an external terminal adapter. Server-based remote nodes provide the ultimate in flexibility—even the software must be configured.

run their own routing software. The U.S. Robotics (Skokie, IL) remote node server runs Cisco's (San Jose, CA) Internetwork Operating System. IBM's LAN Distance server software runs on OS/2, and AppleTalk Remote Access (ARA) can provide remote node services to Macintosh clients.

Supporting DOS and Windows remote node connections and ARA clients, Novell's NetWare Connect also provides dial-out capabilities (for DOS, Windows, and OS/2 machines) so local network clients can share modems. Windows NT includes Remote Access Server. Attachmate's (Bellevue, WA) Remote LAN Node (RLN) software supports ARA clients as well as DOS and Windows. (There is also an RLN Turnkey Server, replete with a dedicated box.)

While most traditional remote node applications use some form of proprietary interface for remote client-to-node server sessions, PPP is beginning to dominate the field. Unlike its predecessor, SLIP, PPP supports most ordinary LAN protocols—not just TCP/IP. Defined by the Internet Engineering Task Force (IETF) request for comments (RFC) process, PPP is a variant of the High-level Data Link Control (HDLC) protocol and can be thought of as a (usually slower) alternate to Ethernet or Token Ring at the Data-link layer. PPP can work with Physical layers that consist of modems and analog telephone links as well as with ISDN telephone lines. In remote node clients and servers, PPP support holds out the promise of interoperability among products from multiple hardware and software vendors.

WHICH ONE WHEN?

Remote node access is ideal when the remote client machine has applications installed on it and the purpose of the connection is to transfer or manipulate data in relatively small doses. SQL-based database operations and desktop applications that use shared directories on a LAN for data storage lend themselves to remote node services. Because the remote pipe is a thin one—especially if modems sit between the remote client and the server—it is best to avoid 1MB- or even 100KB-sized data transfers. For example, login.exe has some 111KB. Executing this program to log in remotely can be slow if it must first be loaded over a 9,600bits/sec connection. Remote node users should keep local copies of both essential network utilities and their applications.

Traditional desktop databases, such as dBase, want to load data files into local RAM—the client itself is the database "server", and the network drive is simply a file server. Over a remote node link, access to a 1MB database will be painfully slow even if the application is installed on the remote client machine. Graphics applications also involve notoriously large files.

With large data files, and when applications are installed only on the host system, remote control provides a huge performance advantage. Because the files move across only high-speed local links and program execution is local, the thin pipe—which updates only the screen and keyboard/mouse actions—doesn't become a bottleneck. In fact, a remote control client with a pitiful 386 can take over a state-of-the-art processor at the host.

Remote e-mail is a special case. The remote clients provided by e-mail vendors such as Lotus cc:Mail (Mountain View, CA) connect via a sort of application-specific remote node connection. These connections can be assigned to a specific telephone number and the security of the connection can be managed. The remote e-mail software posts mail to the mother ship post office and downloads new messages, but will not necessarily display old messages, BBSs, and other material that would appear locally. For full-featured e-mail access to a single, complete in-box, remote control is the best choice.

With the increasing availability of ISDN connections and advances in compression technology, the demand for remote control may not grow as fast as the demand for less-costly remote node connections. Nevertheless, programs keep getting bigger, and multimedia files are generally many times the size of traditional text files, so the need for remote control won't disappear soon.

This tutorial, number 83, by Steve Steinke, was originally published in the July 1995 issue of LAN Magazine/Network Magazine.

SECTION IV

Network and Transport Layer Protocols

The TCP/IP Protocol Suite

Just about everyone in the networking industry talks about interoperability; the U.S. Department of Defense (DOD), in the guise of the ARPANET (Advanced Research Projects Agency Network) project, actually did something about it when it created the Transmission Control Protocol/Internet Protocol (TCP/IP) family of networking protocols.

TCP/IP is the DOD's answer to connecting its rapidly proliferating—and widely dissimilar—computers and networks into a loosely associated wide area network (now called the Internet). TCP/IP is the DOD's vehicle for providing distributed computing capabilities across a large area.

TCP/IP might also be called the less talented but still much in demand ugly stepsister to the International Standards Organization's (ISO) Open System Interconnection (OSI) protocols. Though the OSI protocols were designed to dominate the computer environment, TCP/IP remains the central piece in the complex interoperability puzzle.

A PLENITUDE OF PROTOCOLS

As its two-part name implies, TCP/IP encompasses more than one protocol. It includes a range of protocols that provide distinct services and capabilities necessary for communication between and control of otherwise incompatible computers and networks. In addition to the Transmission Control Protocol (TCP) and Internet Protocol (IP), these include the File Transfer Protocol (FTP), the Simple Mail Transfer Protocol (SMTP), the Internet Control Message Protocol (ICMP), and the Simple Network Management Protocol (SNMP).

Other protocols within the TCP/IP family are the Address Resolution Protocol (ARP), the Reverse Address Resolution Protocol (RARP), the Exterior Gateway Protocol (EGP), and the User Datagram Protocol (UDP). IP, TCP, FTP, SMTP, and Telnet were part of the original DOD military standard, TCP/IP protocol suite promulgated in the late 1970s. Although TCP/IP was the brainchild of and for the military, it has become the de facto protocol for general-purpose intersystem communication.

THE TCP/IP FRAMEWORK

The body of standards making up the TCP/IP suite fit within a four-layer (network access, internet, host-to-host, and process layers) communications framework, shown in Figure 1.

TCP/IP ARCHITECTURE

Process Layer
FTP
SMTP
TELNET

Host-to-Host Layer
TCP

Internet Layer
IP

Network Access Layer

■ **Figure 1:** The TCP/IP body fits within a four-layer framework.

OSI AND TCP/IP COMPARED

OSI	TCP/IP
Application	Process
Presentation	
Session	
Transport	Host-to Host
Network	Internet
Data Link	Network Access
Physical	

■ **Figure 2:** The TCP/IP network-access layer services correspond to those provided by the physical, data-link, and parts of the network layers in the OSI reference model.

Before examining these layers individually, however, it's important to first understand several other concepts.

The DOD based its model of data communication on three agents, called processes, hosts, and networks, with processes as the fundamental communications entities. Processes are executed on hosts, which are internetworked computers that can generally support multiple processes. Hosts in turn communicate with each other via a network. Successful completion of an operation on the internet requires action by all three agents.

The transfer of data from one process to another requires first getting the data to the host in which the process resides, then to the process within the host. In this model, a communications facility must be concerned only with routing data between hosts, with the hosts concerned with directing data to processes.

The network-access layer handles the exchange of data among a host, the network that host is attached to, and a host within the same network. The sending host provides the network with the network address of the receiving host to ensure that the network routes the data properly. The TCP/IP network-access layer services correspond to those provided by the physical, data-link, and parts of the network layers in the OSI reference model (see Figure 2).

The specific physical, or media-access, protocol used to put TCP/IP data on the wire is independent of TCP/IP's top three layers. This means that TCP/IP can operate over virtually any media-access protocol, including Ethernet, Token Ring, or FDDI.

The separation of the physical-layer functions from the higher layers also means that the services provided by the internet, host-to-host, and process layers are not affected by the specifics of the underlying network protocol used. The same high-level software can function properly regardless of the network type a host is connected to.

The internet layer provides services that permit data to traverse hosts residing on multiple networks. The internet routing protocol runs not only on "local" hosts, but also on gateways that connect two networks. A gateway's primary responsibility is to relay data from one network to the other, making sure it gets to the appropriate destination host.

The host-to-host layer ensures the reliability of the data and between two TCP/IP hosts. And the process layer provides protocols needed to support various end-user applications, such as file transfer or electronic mail.

THE TCP/IP PROTOCOLS

Each TCP/IP protocol provides a specific service or set of services to move data from one computer to network to computer. The services some of these provide—the File Transfer Protocol (FTP), for instance—are self-explanatory. Others aren't so obvious.

In the lexicon of the TCP/IP world, an interconnected set of networks is called an internet; the Internet Protocol (IP) is responsible for accepting segmented data (in the form of a Protocol Data Unit, or PDU) from a host computer and sending it across the Internet through the required gateways until the data reaches its destination.

The IP delivery process provides what is known as an unreliable connectionless service; proper delivery is not guaranteed by IP. Even PDUs that are delivered may arrive at the destinations out of sequence. TCP must ensure reliable delivery of PDUs. TCP provides the transport mechanism that ensure that data is delivered error-free, in the order it was sent, and without loss or duplication.

TCP's basic role is providing reliable end-to-end data transfer between two processes, called transport users (these include FTP and SMTP). In specific terms, the TCP standard describes five levels of service: multiplexing (the ability to support multiple processes), connection management, data transport, error reporting, and a variety of special capabilities.

In the basic data-transfer process, a transport user such as FTP passes data to TCP, which encapsulates the data into a segment that contains user data and control information (e.g., the destination address). TCP ensures reliable data delivery by numbering outgoing segments sequentially and then having the destination TCP module acknowledge arrival by number. If segments arrive out of order, they can be reordered via sequence numbers, and if a segment fails to arrive, the destination TCP module will not acknowledge its receipt, and the sending TCP module resends it.

TCP allows the transport user to specify the quality of transmission service it requires, permits special urgent data transmissions, and provides security classifications that can be used in routing segments to data-encryption devices. In trying to provide high-quality transmission services, TCP attempts to optimize the underlying IP and network resources. Parameters available include timeout delays and message-delivery precedence. Interrupt-driven urgent transmissions include terminal-generated break characters and alarm conditions.

The services provided by TCP and IP are defined by primitives and parameters. A primitive is a mechanism for specifying the function to be performed, while parameters are used to pass data and control information.

Only two primitives—SEND and DELIVER—are used to define the IP services. Parameters available with these primitives include source and destination host addresses, the recipient protocol (usually TCP), an identifier that distinguishes one user's data from another's, and user data.

TCP offers two primitives and associated parameters: service request and service response primitives. A TCP client sends service request primitives to TCP; TCP issues the service response primitives to the client. Many of these primitives set off an exchange of TCP segments between host processes or computers, and TCP passes the segments to IP in a SEND primitive and receives them from IP in a DELIVER primitive.

FILES AND TERMINALS

FTP exists to transfer a file or a portion of a file from one system to another under orders from an FTP user. Typically, a user executes FTP interactively through an operating system interface, which provides the input/output facilities that allow exchanging files between systems.

FTP options allow transferring ASCII and EBCDIC character sets and using transparent bit streams that permit exchanging any sort of data or text file. FTP also provides data-compression options and has password/identifier mechanisms for controlling user access.

SMTP provides the underlying capabilities for a network electronic mail facility. It does not, however, provide the user interface. Primarily, it provides mechanisms for transferring messages between separate systems. SMTP accepts e-mail messages prepared by a native mail facility (such as cc:Mail) and—making use of TCP to send and receive messages across the network— delivers them.

With SMTP, users can send mail to users anywhere in the local network as well as to those on the Internet.

TELNET outlines a network terminal-emulation standard. It allows terminals to connect to and control applications running in a remote host just as if it were a local user of the host.

In implementation, TELNET takes two forms: user and server modules. The user module interacts with the terminal I/O module, providing translation of terminal characteristics into the network-specific codes and vice versa. The server module interacts with processes and applications, serving as a terminal handler to make remote terminals look as if they are local.

SNMP AND OTHER PROTOCOLS

Among the other TCP/IP protocols, one of the most widely applied is SNMP, the Simple Network Management Protocol. SNMP supports the exchange of network management messages among hosts, including a central host that is often called a network management console.

SNMP was designed to operate over UDP, the User Datagram Protocol. UDP operates at the same level as TCP, providing a connectionless service for the exchange of messages while avoiding the overhead of TCP's reliability facilities.

ARP and RARP provide mechanisms for hosts to learn MAC and Internet addresses. The former allows a host to discover another host's MAC address, and the latter permits a host to find out its own Internet address, an important capability for diskless PCs without permanent ways to store their Internet addresses.

The Exterior Gateway Protocol allows neighboring gateways in different autonomous systems to exchange information about which networks are accessible via a particular gateway. Industry observers once predicted that most TCP/IP users would eventually migrate to OSI. The question is, when? Few commercially available products offer complete OSI functionality. Most OSI protocols remain in the standards-setting phase, and users continue to be satisfied with the level of service provided by TCP/IP.

This tutorial, number 28, by Jim Carr, was originally published in the November 1990 issue of LAN Magazine/Network Magazine.

IP Addresses and Subnet Masks

As the Internet takes on an ever larger role during our working hours—and perhaps our after-work hours as well—the more people need a better understanding of how IP addressing works and the closely related topic of subnets and how they are defined.

For example, installing IP connectivity software, which the current versions of Windows, OS/2 Warp, and the Macintosh OS all include, on a desktop system generally requires that you know your IP address and your organization's subnet mask. While most organizations administer these setup parameters centrally—typically through the IT department—it can be useful for people other than the Internet guru to understand how addresses and subnets work. While the topic may seem extraordinarily complex, a little facility with binary arithmetic is sufficient to work out how the system operates.

The Internet address is the first concept you need to grasp. The parsimonious founding fathers of the Internet allowed 32 bits to identify all Internet addresses. (2^{32} =4,294,967,296 addresses, which seems to be a plentiful supply at first glance.) IP addresses are normally written as a sequence of four decimal numbers separated by periods, with each number ranging from 0 to 255. (Each number is an 8-bit byte, or octet in Internet jargon. $2^8 = 256$, and 4 x 8 bits = 32 bits.)

So the most common representation of an Internet address looks something like this: 192.228.17.62. Sometimes the four digits of an Internet address are represented in hexadecimal (base 16) notation. As we shall see, it sometimes makes sense to represent them in binary or base 2 notation.

Internet addresses are made up of a network number and a host number. A host can be any device that runs an application. Most hosts are computers—traditionally the term "host" has been reserved for multi-user mainframes and minicomputers, but PCs and workstations are hosts in the Internet sense of the term, as are intelligent hubs, RMON monitors, and other devices.

Hosts whose addresses share a network number can send local broadcasts to one another and communicate without a router. Hosts with differing network numbers can communicate only via an IP router.

The network ID number part of the IP address can be split off from the host ID part in several ways. If the first octet of the Internet address is used for the network number, the last three octets can be used for host addresses. Such networks are called Class A networks, and they can have $2^{24} - 2 = 16,777,214$ host addresses.

If the first two octets are used for the network number, the networks are called Class B networks, and they can have $2^{16} - 2 = 65,534$ host addresses. If the first three octets are used for the network number, the networks are Class C networks, and they can have $2^8 - 2 = 254$ host addresses.

Table 1 is the binary representation of the first octet of an Internet address. The convention is that if the first (or most significant) binary digit is 0, then the address is a Class A address. Network addresses of 0 (00000000 binary) and 127 (01111111 binary) are reserved, so there are 126 potential Class A networks, recognizable by their having a first dotted decimal digit in the range of 1 to 126. These very large networks have practically all been assigned to large international organizations and government agencies.

TABLE 1. BINARY REPRESENTATION OF THE FIRST OCTET OF AN INTERNET ADDRESS

Powers of 2	2^7(128)	2^6(64)	2^5(32)	2^4(16)	2^3(8)	2^2(4)	2^1(2)	2^0(1)
Class A	0	x	x	x	x	x	x	x
Class B	1	0	x	x	x	x	x	x
Class C	1	1	0	x	x	x	x	x
Class D	1	1	1	0	x	x	x	x
Class E	1	1	1	1	x	x	x	x

Class B addresses are indicated by a 1 in the first bit and a 0 in the second bit of the first octet. Thus they can be identified by their having a first decimal digit in the range of 128 to 191 (10000000 binary to 10111111 binary). The second octet of a Class B address is also part of the network number. Thus there are potentially 16,382 Class B addresses.

Class C addresses have 1s in the first two bits and a 0 in the third bit of the first octet, and use both the second and third octets for the network number. Thus the first decimal digit of Class C addresses ranges from 192 to 223 (11000000 binary to 11011111 binary). There are 2,097,150 Class C addresses. Network numbers with the first digit higher than 223 are reserved for special purpose Classes D and E.

WHY DO WE NEED SUBNETS?

A network with 16,777,214 host addresses (Class A) , or even one with 65,534 (Class B), is likely to be unwieldy; even a Class C network with 254 addresses may well be undesirably large for many organizations. As a result of traffic patterns and congestion, upper limits on the number of allowable nodes in a network, distance limitations on LANs, and other reasons, many organizations divide their networks into subnets.

In effect, some number of the leftmost (or most significant) bits of the host addresses are expropriated and used to designate subnets. The subnet is part of the network identified by the network number, but only those hosts that are on the same subnet can communicate without a router. Members of different subnets will not see each other's local broadcasts, and they will need to go through a router to communicate, even though they may be on the same network.

The subnet mask is the method IP software uses to mark off which host bits will be transformed into subnet numbers. A subnet mask is a 32-bit number, often written in dotted decimal form like an IP address—for example, 255.255.255.192. If we write the subnet mask in binary notation (see Table 2), it's easier to see how it works. Think of the subnet mask as a strip of masking tape laid over the host part of the binary IP address wherever there are 0s in the mask.

TABLE 2. DOTTED DECIMAL AND BINARY REPRESENTATIONS OF IP ADDRESS AND SUBNET MASK

	Dotted decimal	Binary representation
IP address	192.228.17.126	11000000.11100100.00010001.01111110
Subnet mask	255.255.255.192	11111111.11111111.11111111.11000000
Bitwise AND of address and mask (resultant network number)	192.228.17.64	11000000.11100100.00010001.01000000
Subnet number	1	00000000.00000000.00000000.01
Host number	0.0.0.62	00000000.00000000.00000000.00111110

■ **Table 2:** Network numbers with the first digit higher than 223 are reserved for special purpose Classes D and E.

If the subnet mask stops at an octet boundary, it may be the same as the default mask, which is no mask at all (see Table 3). In other words, if you perform a bitwise logical "AND" with a Class B address and the Class B default mask, you get precisely the same network number you would get by simply looking at the first two octets, and there is no subnet number carved out of the Class B host octets.

It is considered good practice to use contiguous bits, starting from the left, for subnet mask values, though it is not always an absolute requirement. Following this rule, decimal representations of subnet mask values follow the sequence 128, 192, 224, 240, 248, 252, 254, 255.

■ Table 3

TABLE 3—DEFAULT SUBNET MASKS

Class A default mask	255.0.0.0	11111111.00000000.00000000.00000000
Example Class A mask	255.192.0.0	11111111.11000000.00000000.00000000
Class B default mask	255.255.0.0	11111111.11111111.00000000.00000000
Example Class B mask	255.255.248.0	11111111.11111111.11111000.00000000
Class C default mask	255.255.255.0	11111111.11111111.11111111.00000000
Example Class C mask	255.255.255.252	11111111.11111111.11111111.11111100

A SAMPLE SUBNET MASK

If you have a Class C network address and you would prefer to configure six networks of 30 hosts each, you would use the subnet mask 255.255.255.224. Subnet 0, with the binary representation 000, is reserved to refer to "this subnet," and subnet 7, with the binary representation 111, is often reserved for broadcasts to all subnets on this network. Thus subnets one through six are available. Host number 0 (00000 binary) and host number 31 (11111 binary) are reserved, so there are 30 potential host IDs, numbered 1 to 30, that could be assigned.

With this subnet mask, there are 180 total IP addresses possible on this Class C network (six subnets times 30 host numbers). With no subnetting, a Class C network has 254 total addresses. Subnets clearly can result in wasted IP addresses, although the alternative of not employing subnets at all will likely result in even worse waste—for example, multiple Class C networks being used to segment a small number of users.

In most cases, every host and router on a particular IP network should have the same subnet mask. Exceptions to this rule should be made only by those who understand fully what they want to accomplish in their network design and how using multiple subnet masks on the network achieves that goal.

It's not a good idea to experiment with subnet masks, or for that matter, with IP addresses. Each organization with an IP network connected to the Internet is obligated to request network IDs from the Network Information Center, to which the Internet Activities Board has delegated the responsibility of administering addresses.

Duplicate network numbers, duplicate host addresses, and inconsistent subnet masks can create havoc, and if you are using unregistered numbers, you are presumed to be at fault. If you are installing Internet applications on an organization's network, you must get an appropriate address and the organization's subnet mask from the person in your organization who is in charge of administering them. Aside from the hard-to-find problems that may result from ignorant experimentation with IP addresses and subnet masks, it is a lot of work to go back and reinstall IP software that has been configured improperly.

This tutorial, number 86, by Steve Steinke, was originally published in the October 1995 issue of LAN Magazine/Network Magazine.

TCP and UDP

The two Transport layer protocols in the TCP/IP family, TCP and UDP, provide network services for applications and Application layer protocols (including HTTP, SMTP, SNMP, FTP, and Telnet.) These two protocols perform those services by employing the IP to route packets to their destination networks. TCP provides connection-oriented, reliable, byte-stream packet delivery, while UDP provides connectionless, unreliable, byte-stream packet delivery. These terms need explanation.

Connection-oriented protocols establish an end-to-end link before any data moves. ATM and frame relay are connection-oriented protocols, but they operate at the Data-link layer rather than the Transport layer. Placing an ordinary voice phone call is also connection-oriented.

Reliable protocols safeguard against several forms of transmission mishaps. You can compare checksums included with a packet's data payload with a recalculation of the checksum algorithm at the destination to detect corrupted data. You must retransmit corrupted or lost data, so the protocol must provide methods for the destination to signal the source when retransmission is needed. Packetized data can arrive out of sequence, so the protocol must have a way to detect out-of-sequence packets, buffer them, and pass them to the Application layer in the correct order. It must also detect and discard duplicate transmissions. A collection of timers enables limiting the wait for various acknowledgements, so you can initiate retransmissions or link re-establishment.

Byte-stream protocols don't specifically support data units other than bytes. TCP can't structure bytes of the data payload in a packet, nor can it cope with individual bits. As far as TCP is concerned, it's responsible for transporting an unstructured string of 8-bit bytes.

A connectionless protocol doesn't establish paths across the network before data can flow. Instead, the protocol routes connectionless packets or datagrams individually at each intermediate node. Without an end-to-end link, a connectionless protocol such as UDP isn't reliable. When a UDP packet moves into the network, the sending process can't know whether the packet arrives at its destination unless the Application layer acknowledges this fact. Nor can the protocol detect duplicate or out-of-sequence packets. The standard jargon describes UDP as "unreliable," though a more descriptive term might be "nonreliable." On modern networks, UDP traffic isn't prone to disruption, but you can't really call it "reliable," either.

Figure 1 shows the fields of a TCP segment, the part of an IP packet that follows the IP header information. The first 16 bits identify the source port, and the second 16 bits identify the destination port. Port numbers provide a way for IP hosts to multiplex numerous types of concurrent connections at a single IP address. The combination of a 32-bit IP address and a 16-bit port address identifies a socket in most modern operating systems. The combination of a source socket and a destination socket defines a TCP connection. There are 216 or 65,536 possible ports. The lowest 1,024 ports are called well-known ports; these are set aside by default for particular Application layer protocols. For example, HTTP uses port 80 by default, while POP3 uses port 110. Other applications can use the higher port numbers.

The next two fields, the sequence number and the acknowledgement number, are the keys to TCP's reliability functions. When a TCP connection is established, the initiating host sends an arbitrary initial sequence number to the initiatee. The initiatee adds 1 to the sequence number and returns it to the initiator in the acknowledgement field, thereby indicating the next byte that should be sent. Once data begins to flow, the sequence and acknowledgement numbers keep track of which data bytes have been sent and which data bytes have been acknowledged. Because each field is 32 bits, it can have 232 values, so each field ranges from 0 to 4,294,967,295 and wraps around to 0 when it passes the upper limit.

The 4-bit data offset field simply indicates how many 32-bit words the TCP header has. This information is necessary because there are optional header fields, and the data offset marks where the header ends and the data begins.

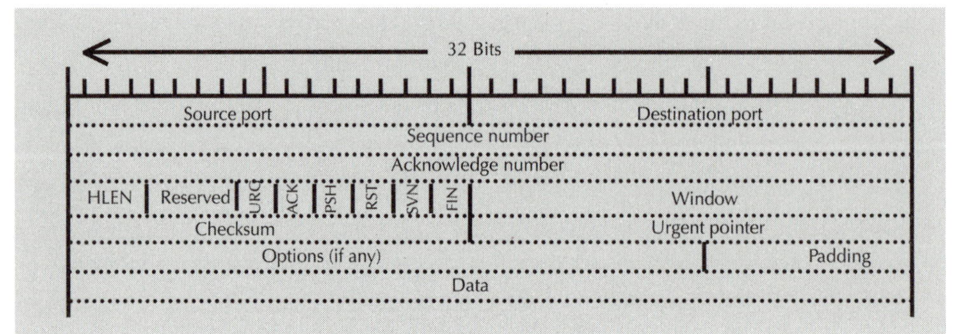

■ **Figure 1:** This diagram illustrates the fields of a TCP segment.

TCP designers set aside the next 6 bits, just in case they might be needed for future development. Since RFC793 (Transmission Control Protocol) dates to 1981, and no one since has established a good reason to use these reserved bits, Jon Postel and his colleagues must have been overly cautious.

Each of the following 6 bits is a flag. An URG (urgent) flag with a value of 1 indicates that the data in the urgent pointer field, farther along in the header, is significant. An ACK (acknowledgement) flag with a value of 1 indicates that the data in the acknowledgement number field is significant. (Note that an initial setup or SYN packet has a meaningful sequence number, but not a meaningful acknowledgement number, since it isn't acknowledging anything.) The PSH (push) flag prevents data from waiting to be sent and from waiting to be processed by the receiving process. The RST (reset) flag shuts down a connection. The SYN (synchronization) flag indicates that the sequence number is significant. The FIN (finish) flag indicates the sender has no more data to send.

The Window field, 16 bits long, indicates the size of a so-called sliding window, which tells the sender how many bytes of data it's prepared to accept. TCP controls flow and congestion by adjusting the window size. A window equal to 0 tells a sender the receiver is overwhelmed and can't accept anything more without further notice. Large window sizes enable as many as 65,536 unacknowledged bytes to be in transit at any given time, but congestion—indicated when the retransmission timer expires without an acknowledgement—cuts the window size in half, effectively slowing the transmission rate.

The 16-bit checksum field protects the data payload's integrity, the TCP header, and certain fields of the IP header. The sender calculates the checksum value and inserts it in this field, and the receiver recalculates the value based on the received packet and compares the two. If they match, the data is probably intact.

The urgent pointer is a 16-bit offset value indicating the last byte that must be expedited when the urgent flag is set. The options field can hold 0 or more 32-bit words, extending TCP's capabilities. The most commonly used option supports window sizes greater than 65,536 bytes, reducing the time spent waiting for acknowledgements, especially at high date rates.

TCP processing entities have multiple timers. The retransmission timer begins when a segment is sent and stops when the acknowledgement is received. If it times out without an acknowledgement, the segment is sent again. One tricky problem is setting the value for the timeout period. If it's too long, unnecessary waiting occurs when the network drops or garbles numerous segments. If it's too short, the network will have too many duplicate segments when response slows down. Modern TCP implementations set the retransmission timer value dynamically in response to conditions.

The persistence timer is necessary to prevent a particular deadlock condition. If the network receives a 0-size window acknowledgement and loses the subsequent acknowledgement that restarts the flow, the persistence timer expires and sends a probe. The response indicates the window size (which may still be zero, in which case the timer starts over.)

The keepalive timer checks whether there is still an active process at the other end of the connection after no activity. The timer shuts down the connection if no response occurs.

A closing connection timer also provides for a period of twice the maximum packet lifetime when a connection is shut down. This timer makes sure the traffic is flushed through the connection before it's closed.

No matter how efficiently the retransmission process is implemented, however, a small number of dropped packets can seriously undermine the throughput of a TCP connection. Each packet, or fragment of a packet, that isn't received will only be missed when the retransmission timer expires. The receiving process must deliver the byte stream in order, so retransmission stops the flow of data until the missing bytes can be replaced. These retransmissions account for the sometimes herky-jerky performance of TCP-based links.

If you compare a UDP segment's structure (see Figure 2) to that of TCP, it's apparent that UDP doesn't have TCP's complex reliability and control mechanisms. UDP's source and destination port numbers support multiplexed applications on a host, just as TCP's do. The content of a 16-bit UDP length field is equal to the length of the 8-byte header plus the length of data, while the checksum field enables integrity checking. (Many applications commonly employing UDP, such as streaming media, derive no added value from data integrity, and wouldn't retransmit corrupted packets even if they identified them.)

TCP is clearly the protocol of choice for data transactions where performance must give way to integrity, controllability, and reliability. UDP is the best choice when performance matters more than perfect data integrity, as in voice and multimedia applications. UDP is also a good choice for transactions so short that connection setup overhead is a large fraction of total traffic, for example, in DNS exchanges. The decision to base SNMP on UDP was made partly because designers thought that UDP would have a better chance of delivering management data when networks were distressed or congested because of UDP's lower overhead. TCP's rich functionality sometimes results in unpredictable performance, but reliable end-to-end connections are likely to support most networked applications for the foreseeable future.

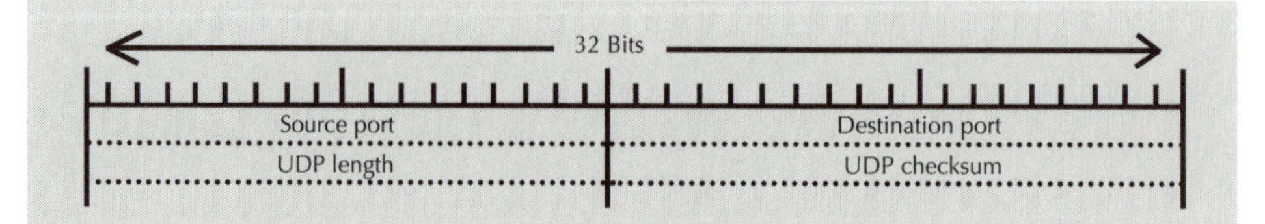

■ **Figure 2:** The UDP Leader is much simpler than the TCP Leader.

RESOURCES

The horse's mouth for TCP is RFC793, which dates to 1981. Among the many Internet RFC repositories is www.freesoft.org/CIE/index. htm. This Internet Encyclopedia site also includes further descriptions of TCP operation.

The preeminent discussions of TCP for application developers and others needing implementation details can be found in:

TCP/IP Illustrated, Volume 1: The Protocols by W. Richard Stevens (1994, Addison Wesley, ISBN: 0201633469)

TCP/IP Illustrated, Volume 2: The Implementation by Gary R. Wright and W. Richard Stevens (1995, Addison Wesley, ISBN: 020163354X)

Internetworking with TCP/IP Vol. I: Principles, Protocols, and Architecture by Douglas E. Comer (1995, Prentice Hall, ISBN: 0132169878)

Internetworking with TCP/IP, Vol.II, ANSI C Version, Design, Implementation, and Internals by Douglas E. Comer, David L. Stevens (1998, Prentice Hall, ISBN: 0139738436)

This tutorial, number 151, by Steve Steinke, was originally published in the February 2001 issue of Network Magazine.

IP Quality of Service

Not long ago, IP was used primarily in Unix environments or for connecting to the Internet; other protocols, like SNA and IPX were used for other purposes. Now, however, many companies have begun using IP for everything—from sharing information within the company to running voice and other real-time applications across their global enterprise networks.

The rise of IP as a foundation for a universal network raises several issues for both enterprise IT departments and ISPs, not the least of which is how to guarantee that applications will receive the service levels they require to perform adequately across the network. For example, network managers might need a way to define a low level of latency and packet loss to ensure that a large file or a business-critical traffic flow gets to its destination on time and without delay. Or, they may need to ensure that a real-time session such as voice or video over IP doesn't look choppy or out of sequence.

The problem with IP is that, like Ethernet, it is a connectionless technology and does not guarantee bandwidth. Specifically, the protocol will not, in itself, differentiate network traffic based on the type of flow to ensure that the proper amount of bandwidth and prioritization level are defined for a particular type of application. By contrast, the cell-based ATM standard incorporates such service requirements in its specifications.

Because IP does not inherently support the preferential treatment of data traffic, it's up to network managers and service providers to make their network components aware of applications and their various performance requirements.

THE QOS DILEMMA

Quality of Service (QoS) generally encompasses bandwidth allocation, prioritization, and control over network latency for network applications. There are several ways to ensure QoS, no matter what type of network you're talking about—Ethernet or ATM, IP or IPX. The easiest one is simply to throw bandwidth at the problem until service quality becomes acceptable. This approach might involve upgrading the backbone to a high-speed technology such as Gigabit Ethernet or 622Mbit/sec ATM. If you have fairly light traffic in general, more bandwidth may be all you need to ensure that applications receive the high priority and low latency they require.

However, this simplistic strategy collapses if a network is even moderately busy. In a complex environment—one that has a lot of data packets moving in many paths throughout the network, or that has a mixture of data and real-time applications—you could run into bottlenecks and congestion.

Also, simply adding bandwidth doesn't address the need to distinguish high-priority traffic flows from lower-priority ones. In other words, all traffic is treated the same. In the network realm, such egalitarianism is not good, because network traffic is, by its nature, unpredictable. For instance, on some days, you'll see traffic bursts occurring at 8 a.m., while on other days you'll see them at noon or at the end of the day. These traffic bursts can move around too. One day, your Internet gateway or one of your switches is the bottleneck; another day, it's your intra-campus video conferences or heavy voice traffic causing the congestion.

As you can see, additional bandwidth can solve some of your short-term problems, but it's not a viable long-term solution, particularly if you already have enough bandwidth to accommodate all but the most highly sensitive network applications.

So how can you flag special traffic as high priority on an IP network? Options like Resource Reservation Protocol (RSVP), multiple flows, and tagging fields can help you give sensitive applications the resources they need.

MAKE A RESERVATION

RSVP, which enables non-QoS technologies such as Ethernet and IP to make QoS requests of the network, is an IETF Internet-Draft that has received a lot of attention. Specifically, the protocol lets end stations request specific QoS levels for QoS-enabled applications. For example, for a video conferencing application, a request could include information that defines the maximum frame transmission rate, the maximum frame jitter, and the maximum end-to-end delay.

When an end station makes an RSVP request for an application, each router situated between it and the source makes note of the QoS request and attempts to honor it. As the request travels through the routers to the source, it merges with requests from other end stations (see Figure 1). If a router can't comply with the request—that is, if it can't guarantee the bandwidth or performance—the requesting station receives an error message, the equivalent of a busy signal on the voice network.

As you can see from all of these conditional statements, nothing is absolutely guaranteed with IP-based networks, even with an additional protocol like RSVP. For example, if a router refuses an RSVP request because it has already allocated its RSVP bandwidth, the packets associated with the request get dumped.

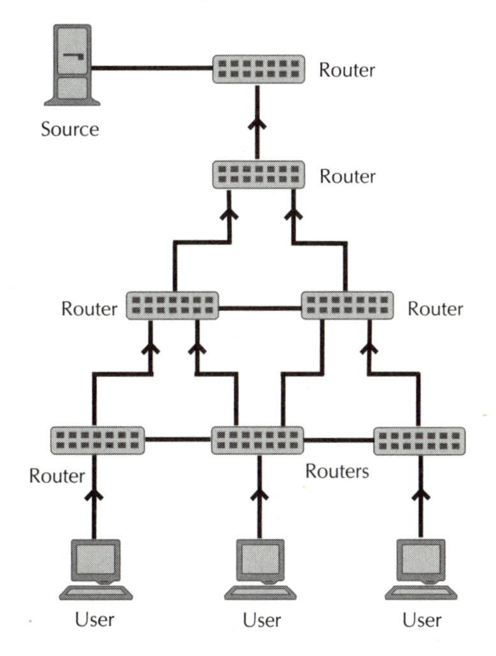

■ **Figure 1:** End-user stations send RSVP requests on behalf of an application. These requests go through each RSVP-enabled router up to the source. A router that can't honor the request sends an error message to the end user.

Another potential downside of RSVP is cost. Because the protocol requires each router to understand and support QoS requests, you might have to invest in firmware or even hardware upgrades to enable it.

RSVP could also lead to a perceived degradation of some network services. In a non-RSVP network, an application might run slowly or badly—but at least it'll run. In an RSVP network, the application might not run at all if there's not enough bandwidth to support it. Similarly, if high-priority requests consume all the allotted bandwidth, a router could dump nonprioritized flows. If you end up with an "arms race," where many nonbandwidth-sensitive applications claim to be high priority anyway, you'll see no real benefits.

Another potential caveat is that routing protocols such as OSPF and Border Gateway Protocol (BGP) do not currently support RSVP—and, ideally, they should, because a QoS request is actually made *after* a route is chosen for a particular data flow. It would be far better for an RSVP priority level to be taken into account *before* a route was chosen. Once the IETF's QoS Routing Working Group finishes its Internet-Draft for RSVP, it will begin working with the Working Groups for OSPF and BGP to integrate the routing protocols with RSVP.

Another criticism of RSVP is its inability to scale. Some industry watchers have warned that the protocol might not be suitable for large enterprise networks, specifically because each and every router along a particular path must support it. Given that requirement, using RSVP successfully over the public Internet could prove challenging.

Another router-based method for handling QoS requests is multiple queues. Routers that support this scheme identify flows that are tagged as high priority and move them to fast queues to expedite transmission.

Along these lines, the IETF's Integrated Services Working Group is debating how to use the 8-bit Type of Service (ToS) field in an IP packet header to tag traffic with different service levels. Currently, this field prioritizes traffic according to which end station initiated the flow, rather than the type of flow itself. If used in this new way, the TOS field would tell network devices that certain types of packets should receive a certain level of service. [Editor's note: Ultimately the Differentiated Services (diffserv) Working Group elaborated new standards for the TOS field (now renamed the differentiated services or DS field) in RFC2474 and RFC2475.]

POLICY MAKERS

RSVP, multiple queues, and tagging can be valuable prioritization methods, but they aren't the only ways to address QoS.

A fairly new idea in the QoS realm is to make the network itself more intelligent by incorporating something called policy-based management. With this method, network managers could define policies that spelled out what levels of service certain types of traffic should have and what particular routing paths they should use. For example, with voice traffic, a path with a minimal number of hops would reduce delay and maintain the highest possible quality.

Policy-based management systems, which major networking vendors are starting to support, give IT managers and service providers a way to take a predefined policy and apply it across the board, rather than have to configure each network device with QoS support.

The Directory-Enabled Networks (DEN) initiative, which was launched in September 1997, is an interesting concept along these lines. Initially endorsed by Microsoft and Cisco, DEN allows network managers and service providers to create policy-based rules so that they can provide services based on users and applications. With DEN, directory services and networks are integrated, which means that network elements and services are represented in a directory. This scheme allows a centralized policy for network services to be associated with particular end-user stations and applications.

Because network devices and services are integrated with a directory, the network can be customized to provide a predictable level of service for users no matter where they are. For example, you might approve certain users for video conferencing applications, regardless of whether they log in to the network from their desktop or from a conference room with video conferencing equipment.

NOT OUT OF REACH

Although many of the QoS standards and protocols mentioned here are still in their infancy and not yet widely deployed, the growing need for QoS in IP-based networks will soon drive them to the top of any IT manager's or service provider's to-do list.

In corporate intranets or extranets, where all routers are typically part of a particular enterprise and therefore subject to the same policies, methods such as RSVP and policy-based networking shouldn't be difficult to use. The public Internet is another story, because you have no idea which routers are out there and whether or not they will honor QoS requests.

The IETF hopes to change that with IPv6, the next version of IP, which includes inherent provisions for QoS. For instance, IPv6 will allow applications to request different levels of service, and will guarantee these service levels even when a request goes over the WAN.

This inherent QoS will be a big boost for real-time applications such as voice and video. Currently, the only viable way you can ensure a specific QoS without adding on support for other protocols is through the use of ATM's User-to-Network (UNI) signaling and Private Network-to-Network Interface (PNNI) routing mechanisms. UNI allows sending and receiving stations and the network to work together to ensure that a particular traffic flow gets the QoS it needs. PNNI selects the most appropriate route through the network for traffic.

IP'S BRIGHT FUTURE

As you can see, IP quality of service isn't quite the oxymoron it once was. By tweaking your existing routers, applications, and other network components, you can ensure that users and applications will have the bandwidth guarantees and low latency they require—all while reaping the management and cost benefits that IP offers.

RESOURCES

For information on the IETF Working Group addressing RSVP and an index to the Internet Drafts, go to www.ietf.org/html.charters/rsvp-charter.html

For several informative reports on policy management and quality of service, go to The Burton Group's Web site www.tbg.com

To learn more about the directory-enabled network initiative, go to www.cisco.com/warp/public/734/den

This tutorial, number 119, by Anita Karvé, was originally published in the June 1998 issue of Network Magazine.

Dynamic Host Configuration Protocol

Automatic assignment of IP numbers to network clients on boot-up eliminates some of IP's administrative complexity.

As an open, standards-based protocol that is robust and WAN efficient—not to mention the fact that it's *the* protocol of the Internet—TCP/IP offers many advantages over other network protocols. But like anything else, TCP/IP has a downside. Its biggest drawback is administrative complexity; great effort is required to keep a TCP/IP network running well.

The problem stems from the protocol's need to have a valid IP address for every device on the network. If you've worked with TCP/IP networks, you're probably aware of several important rules that network administrators must follow when assigning IP addresses:

- The IP address must consist of four bytes (four eight-bit numbers).
- Each IP address must be unique.
- All devices on a segment must have the same network number and subnet number.
- Each subnet number must be unique.

Communicating with nodes on a different subnet or a different network requires a router. (For more information on TCP/IP addressing, see the Tutorial "IP Addresses and Subnet Masks.")

Assigning an IP address to a node—and doing it correctly—can be a significant administrative hassle. In particular, you need to track the addresses that have been assigned in order to avoid duplication. Moreover, if a node is moved to a different subnet, the subnet portion of the address must be changed to that of the new subnet. You also need to check that the node-specific portion of the IP address (the remainder, after you exclude the subnet bits) doesn't conflict with any other node on the new subnet.

AUTOMATING THE JOB

Keeping track of IP address assignments and making new assignments sounds like the perfect kind of mind-numbing task to delegate to computers—and it is! The Dynamic Host Configuration Protocol (DHCP) was developed to accomplish that exact job. DHCP is fully described in the Internet Requests for Comments (RFC) 1541, but I'll summarize the highlights.

DHCP is based upon the Bootstrap Protocol (BOOTP), a system for automatically delivering configuration information from a BOOTP server to BOOTP clients on boot-up—the point when clients first connect to the network.

Under DHCP, a computer is designated as the DHCP server. All of the other computers on the network—at least, those that need an IP address—will be DHCP clients (computers that already have a permanently set IP address don't need to participate). The network administrator needs to initially configure the DHCP server. Part of that configuration process involves assigning the DHCP server a block of IP address numbers that it can dispense to nodes that need IP addresses.

When a new node comes onto the network—assuming it is capable of being a DHCP client—it will broadcast a request for an IP address. Simply put, the DHCP server will respond by checking its table of address assignments, selecting the next available address, and sending a response back to the requesting node.

The actual process is more complex than that, as the requesting client must first find a DHCP server. Also, the protocol is constructed so that a client may negotiate with more than one DHCP server.

Here's how it works: A DHCP client that is in need of configuration broadcasts a dhcpdiscover packet in search of servers (see Figure 1). This packet will contain the hardware address of the requesting client, for example, its Ethernet or Token Ring address. It might also contain a suggested IP address. Next, one or more DHCP servers will evaluate the request and respond with a dhcpoffer packet, which contains an offer of a specific IP address, together with a "lease period" (the length of time the client may use the address).

The client then selects one from among the dhcpoffer packets it receives. (The client's selection will depend on its design; it may seek the longest lease, for example.) The client then issues a dhcprequest packet, which contains the address of the server that issued the dhcpoffer it prefers.

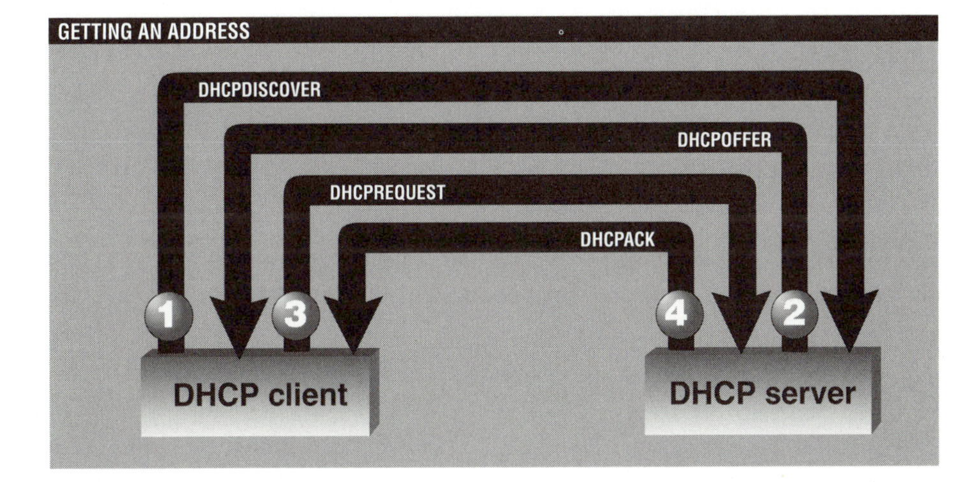

Figure 1: A DHCP client negotiates IP address "lease" through this sequence of packet exchanges.

The chosen server then issues an acknowledgment packet (dhcpack), which closes the negotiations. The dhcpack packet contains the IP address and lease period that have been agreed upon. The server now marks this address as committed; it cannot be assigned to any other client for the duration of the lease. The client configures itself to use its new assigned address and begins normal network operation.

Note that more than one DHCP server may respond to a dhcpdiscover with a dhcpoffer. The client must pick one offer, responding with a dhcprequest packet containing the server identifier for the server it has chosen. The other servers monitor the dhcprequest packet and infer from the server identifier that their offers were not chosen; they then know that the IP addresses they offered are still available for assignment to another client.

As mentioned, the selected server finalizes the offer by issuing the dhcpack, but in the event that the server cannot commit the configuration, it will issue a dhcpnak (negative-acknowledgment) packet. In this case, the client must start the whole request process over again, from the top.

DISPENSING IP ADDRESSES

DHCP provides for IP addresses to be allocated in three different ways: With *automatic allocation*, the DHCP server assigns a permanent IP address to a DHCP client requesting an address. Using *dynamic allocation*, the DHCP server would assign an IP address for a limited period of time (the "lease period") or until the DHCP client specifically relinquishes it, whichever comes first. The third method is *manual allocation* in which the IP address is chosen by the network administrator, but the DHCP server is used to convey the assignment to a DHCP client.

Dynamic allocation is particularly useful for computers that will connect to the network only occasionally. When a DHCP client is about to disconnect from the network, and thus no longer needs an IP address, it can notify the DHCP server, which can then reassign the number to the next node that needs an address. Dynamic allocation makes more efficient use of a limited number of IP addresses. A group of occasional users, for example, can share a smaller pool of IP addresses than if each required a permanent address of its own.

Dynamic allocation won't work for every node, however. If the IP addresses of network servers were to change frequently and randomly, network clients would have a hard time finding servers. Thus, as a general rule, it's best to use static addresses for servers and dynamic addresses for clients.

As mentioned earlier, DHCP is an offshoot of BOOTP. Where DHCP differs from BOOTP is that the BOOTP server merely stores a preset configuration for a BOOTP client and delivers it on boot-up. It doesn't eliminate the need to set up a configuration for the client. DHCP, by contrast, automatically configures DHCP clients, using rules pre-established by the administrator.

When setting up DHCP on a network, you must decide where to locate the DHCP servers. You could put one server on each subnet, or you could have one central DHCP server to which all the clients must connect in order to get an IP configuration. Because much of the communications between DHCP clients and servers take place via broadcasts, the central DHCP server approach requires that routers be capable of forwarding DHCP packets or that some other type of "relay agent" forwards the packets. BOOTP uses relay agents, and DHCP takes advantage of this. The format of DHCP packets is specifically designed to be nearly identical to BOOTP packets, so that BOOTP relay agents can forward DHCP packets.

Figure 2 shows a simple two-subnet network, in which a router is acting as a relay agent. When a DHCP client on Subnet 1 issues a broadcast in order to discover a DHCP server, the router forwards the packet onto Subnet 2 where the DHCP server resides, using the unicast address of the DHCP server as the destination address. The router also notes from which subnet the discovery packet was issued and encodes that information into the discovery packet. When the DHCP server issues a reply packet, the router will forward it to Subnet 1, using IP unicast or broadcast, depending on the client's capabilities. A DHCP client on Subnet 2 doesn't need the services of a relay agent, because it can interact directly with the DHCP server (both are on the same subnet).

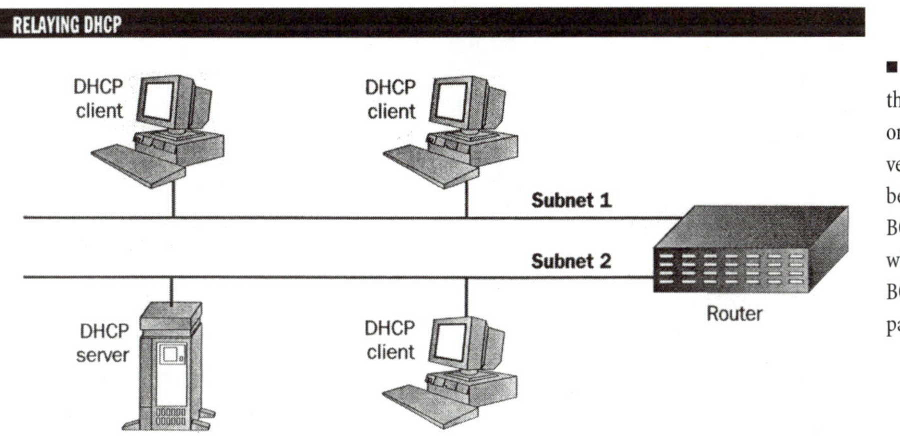

■ **Figure 2:** Unless there's a DHCP server on each subnet, intervening routers must be capable of being BOOTP relay agents, which can forward BOOTP and DHCP packets.

WHO'S USING IT

The first RFCs for DHCP were issued in 1993. Most vendors have adopted it and developed products that support it. Most prominently, perhaps, is Microsoft, which built DHCP server capability into Windows NT 3.5 and higher. Windows for Workgroups 3.11 and Windows 95/98 both have DHCP client capability. Thus, you could set up one or more Windows NT Servers to be DHCP servers. Other networked nodes running Windows 95, Windows for Workgroups, Windows NT Workstation, or Windows NT Server could be DHCP clients.

Apple's Open Transport (the successor to MacTCP) includes DHCP client capability. SunSoft's (Mountain View, CA) SolarNet PC Admin network management software includes a DHCP server. Many of the third-party TCP/IP packages for Windows have DHCP client capability. On Technology (Cambridge, MA) announced a DHCP server NLM for NetWare 3.11 and 4.x file servers, and subsequent versions of NetWare servers and NDS have native support for DHCP.

THE SIMPLE LIFE

DHCP can simplify life dramatically for managers of TCP/IP networks. It automates the process of assigning an initial IP address to a client, easing the task of adding new clients to a network. Moreover, if a client moves from one subnet to another, DHCP can make the appropriate adjustments to the client's IP configuration. Lastly, dynamic allocation lets you time-share a block of IP addresses among many clients, reducing the total number of IP addresses required.

This tutorial, number 91, by Alan Frank, was originally published in the March 1996 issue of LAN Magazine/Network Magazine.

Multicasting

Most traditional IP applications, such as Web browsing and e-mail, employ unicasting, in which a separate connection is set up between a server and each of its clients. In cases where every client has unique needs—looking at a different Web site, reading a particular message—unicasting makes sense. There is no wasteful duplication of data.

But consider cases in which multiple clients all want the same data at the same time. Here, a server once again is connected to each of its clients, but this time it sends an identical data stream to each client. This is a waste of both server and network capacity. For example, suppose that a server is offering a live video stream that requires 1Mbit/sec for each client. With a 100Mbit/sec NIC on the server, its interface would be completely saturated after 90 or so clients had connected.

Worse, replicated unicast transmissions are extremely wasteful of network capacity. Figure 1 demonstrates this graphically, suggesting how wide the data pipe must be to get 90 copies of the same information from the server to all of its clients. With 90 clients separated from the server by two router hops and two switch hops, as shown in the figure, a 1Mbit/sec unicast channel winds up consuming a total of 180Mbits/sec of router bandwidth and 180Mbits/sec of switch bandwidth.

This is where multicasting comes in. It permits our server to get away with transmitting only a single stream, regardless of how many clients have requested it. Whenever this stream traverses a multicast-enabled switch or router, it is simply copied—but only to branches of the tree where clients that requested the stream are located. As Figure 2 illustrates, this makes a dramatic difference to overall bandwidth consumption.

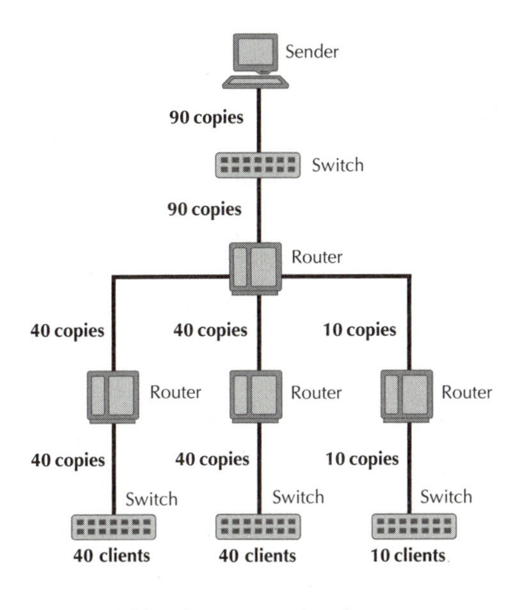

■ **Figure 1:** With unicast propagation of a stream, data must be wastefully replicated.

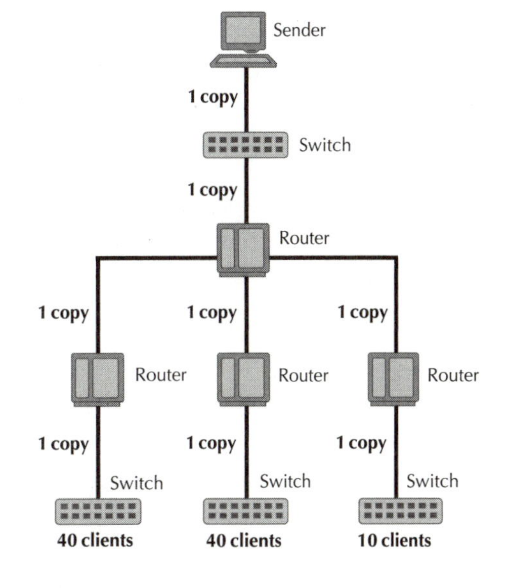

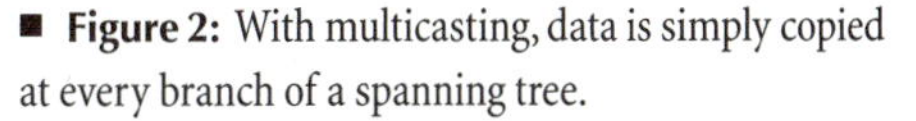
■ **Figure 2:** With multicasting, data is simply copied at every branch of a spanning tree.

WHAT MULTICASTING IS GOOD FOR

It is important to concede that multicasting can work its magic only when all clients want the identical data simultaneously. It is ideal for live audio or video streams or for applications such as near-real-time stock tickers. Multicasting has also become popular for distributing software updates.

Unfortunately, multicasts are not practical on the Internet because, as we shall see, they cannot work unless every router or switch between the receiver and the sender is multicast-enabled. Most ISPs have not modified their networks to support multicasting because (among other reasons) there has been relatively little public demand for it—a classic chicken-and-egg problem.

Multicasting is being promoted by the IP Multicast Initiative (IPMI), an industry consortium of more than 100 companies, including Cisco Systems, 3Com, Lucent Technologies, and Stardust Technologies (Campbell, CA).

Meanwhile, numerous multicast-enabled subsets of the Internet already exist. To list them all would be beyond the scope of this tutorial, but I must briefly mention the Multicasting Backbone (MBone). The first part of the MBone was created as long ago as the summer of 1988, when a multicast tunnel was established between Stanford University and Bolt, Beranek, and Newman (Cambridge, MA). For more information on the history of multicasting, see "IP Multicast Streams to Life," *Network Magazine*, October 1997.

WHAT YOU NEED TO USE IT

Multicasting is easiest to implement over a single LAN, as Ethernet readily supports unicasting, broadcasting, and multicasting at the Data-link layer. To expand multicast traffic to a WAN, however, the Network layer must be involved.

The first issue to be dealt with at layer 3 is addressing. When a sender initiates an IP multicast, it assigns to it an IP address defined by RFC 1112, called "Host Extensions for IP Multicasting." This Class D address, falling within the space from 224.0.0.0 to 239.255.255.255, specifies a multicast host group. A host group has a dynamic membership, as end users (hosts) can tune in or out at any time. In fact, a host group's IP address is much like a radio station's frequency, with the exception that hosts are able to be members of multiple groups simultaneously (few of us voluntarily listen to two radio stations at the same time!)

The second issue is registration, a method for letting the WAN know that any given host wants to be part of a group. Registration is accomplished via the Internet Group Management Protocol (IGMP), which was also defined by RFC 1112. A user's host application, such as the Windows Media Player or other multicast-enabled client, sends an IGMP report—encapsulated in an IP datagram—to the LAN router. From there, the report traverses other routers, if necessary, until it reaches the sender.

Periodically, each multicast-enabled router must send an IGMP query to all the hosts on its subnet. Each host must then respond with a further report or reports—one for each of the host groups it wants to continue being a member of. (A host does not need to send a report for a given host group if it has already seen a report for that group on its subnet. This helps eliminate network congestion locally.)

PRUNING THE TREE

If no IGMP reports are received from a given subnet, then no multicast traffic needs to be sent to that subnet. A simplistic method for shutting off transmissions where they are not required would be for the sender to maintain a table identifying all participating subnets, and then to send a copy of its data to each. However, that would negate some of the advantages of multicasting, since many of the data streams would follow the same path throughout most of the network.

So we now come to the third layer-3 issue—routing. IGMP specifies that routers interact with one another to exchange information about neighboring routers. A single router is selected as the Designated Router for each physical network; these routers then construct a spanning tree to connect members of any given IP multicast group.

Thanks to the spanning tree, there is just one path between every pair of routers. Multicast-enabled routers create new branches in the tree only when they need to, copying multicast datagrams to each as required. When a branch no longer leads to any hosts, it is pruned from the tree.

Many different IP multicast routing protocols and algorithms have been proposed. Dense-mode protocols are most appropriate to networks where bandwidth is relatively plentiful and there is at least one multicast group member in each subnet. They initially assume that all hosts are part of a multicast group, sending out traffic until they are informed otherwise (at which point the tree is pruned). Examples of dense-mode protocols include Distance Vector Multicast Routing Protocol (DVMRP), Multicast Open Shortest Path First (MOSPF), and Protocol Independent Multicast-Dense Mode (PIM-DM).

Sparse-mode protocols begin with the assumption that few routers in the network will be involved in any multicast. They minimize network traffic by adding branches to the tree only when explicitly requested to do so. As you can imagine, sparse-mode protocols, which include Core-Based Trees (CBT) and Protocol-Independent Multicast-Sparse Mode (PIM-SM), are better suited to WANs than are dense-mode protocols.

TRANSPORTS OF DELIGHT

Now that we have our multicast-enabled network, we still have to decide which protocol to use to give us adequate QoS at the Transport layer. Most of the Internet, of course, uses TCP, which establishes a feedback loop between the sender and the receiver. As mentioned in the Tutorial, "H.323", TCP is highly reliable, but it's also prone to delay and to sending data in a bursty manner.

Most multimedia applications therefore improve performance, even when unicasting, by switching to UDP and settling for its "best-effort" services. TCP would not be suitable for multicasting in any case, since any server would gag when faced with simultaneous acknowledgements from dozens or hundreds of clients.

Though many multicast applications feature loss-tolerant data such as audio or video, some kind of feedback mechanism is still desirable. Real-Time Transfer Protocol (RTP) and its feedback-loop companion Real-Time Transport Control Protocol (RTCP) provide one answer to this problem. RTP is both a proposed IETF standard (RFC 1889) and an ITU standard (H.225.0). It typically runs on top of UDP, adding to it the timing information necessary to synchronize and display audio and video data. RTCP works in conjunction with RTP, providing feedback from receivers about the QoS they are experiencing.

RTP and RTCP can keep feedback traffic to 5 percent or less of overall session traffic. When any receiver detects that it has failed to hear a sequenced packet, it waits for a random interval, after which it sends a report back to the sender and to all other receivers. If, while waiting to send a report, it hears one from another receiver, it will cancel its own report.

Another protocol relevant to multicasting is the Real Time Streaming Protocol (RTSP), first published as a proposed standard (RFC 2326) in April 1998 by the IETF. RTSP does not deliver data by itself, but instead uses RTP as a transport protocol. RTSP acts as a framework for controlling multiple data delivery sessions, helping switch between TCP, UDP, and RTP sessions as required. One significant aspect of RTSP is relevant only to unicasting: It can provide VCR-style controls such as pause, fast-forward, reverse, or seek a particular spot in a previously recorded stream.

However, RTSP also helps with firewall configuration, historically a problem in both multicasting and unicasting. Application-level proxies need to be aware of, and to translate, specific client and server ports; similarly, network address translators must remap specific client ports. If incoming protocols and ports are unrecognized, the traffic winds up getting rejected—to the frustration of would-be recipients. RTSP helps prevent this by including information about transport mechanisms and port numbers in an initial setup transaction.

IN THE FUTURE

This tutorial has touched on just a few of the many protocols and technologies involved in multicasting. It is a technology that is rapidly evolving, both to enhance interoperability and to help scale performance for use on the Internet.

The main development that multicasting awaits, however, is simply public demand. This will materialize—perhaps sooner rather than later—as users migrate to high-speed Internet connections.

RESOURCES

For a primer on IP multicasting, links to information about the Multicasting Backbone, and implementation notes, see www.cisco.com/ warp/public/ 732/ multicast/.

For a list of current Internet drafts relevant to IP multicasting, see www.ipmulticast.com/techcentral/drafts.htm.

To learn more about Real Time Streaming Protocol (RTSP), see the RTSP FAQ at www.real.com/devzone/library/fireprot/rtsp/faq.html.

See also Dave Kosiur's book *IP Multicasting: The Complete Guide to Interactive Corporate Networks* (Wiley and Sons, 1998, ISBN 0471243590).

This tutorial, number 127, by Jonathan Angel, was originally published in the February 1999 issue of Network Magazine.

Internet Protocol Version 6

To many, IPv4 is like a beloved but aging pair of jeans. The knees are patched, the cuffs are fraying, and you sure don't need a belt to keep them up anymore, but they fit—more or less. Getting rid of them means a lot of hassle: going to the mall, trying on a bunch of new pairs, and washing them half a dozen times before they feel right. Who needs that?

The fact is, the Internet is expanding. Besides the routers, servers, PCs, and laptops coming online, devices such as PDAs, cell phones, copier machines, and automobiles are clamoring for IP addresses. And even if the wild-eyed prophecies of Internet-enabled toasters don't pan out, eventually the buttons are going to pop on version 4's 32-bit address space.

Yet address scarcity is just one argument for the next generation of IP. The standard's architects tout other benefits of version 6: improved routing efficiency, simplified administration, and the opportunity to tailor the protocol to the new demands of modern global communications.

This tutorial examines the basics of the next generation of IP. It reviews the number of new addresses, explains the shape of an IPv6 packet, describes how IPv4 and IPv6 devices and networks will interoperate, and briefly examines IPv6's new security features.

WHAT'S YOUR ADDRESS?

IPv4, with its 32-bit address space, allows for approximately 4.2 billion addresses. While this is a huge number, the protocol's developers didn't anticipate how quickly the Internet would grow. Techniques such as Network Address Translation (NAT) and Classless Interdomain Routing (CIDR) will extend IPv4's wearability for years to come, but sooner or later the Internet will outgrow IPv4.

IPv6's 128-bit address space promises more legroom than the mind can comfortably comprehend. Just how many addresses is that? Approximately 3.4x1038. If that's too abstract, try this on for size: The new version allows for 6.5x1023 (or 655,570,793,348,866,943, 898,599) addresses for every square meter of the Earth's surface.

Besides having a far greater quantity of addresses, IPv6 also departs somewhat from IPv4's familiar "dotted quad" format (for example, 193.10.10.154). Instead, IPv6 uses hexadecimal notation, and colons replace the periods. A mock example shows the structure of the address: FEDC:BA98:7654:3210:FEDC:BA98:7654:3210.

CHECK YOUR HEAD

While the IPv6 header has twice as many bytes as an IPv4 header (40 bytes vs. 20 bytes), IPv6 boasts a streamlined header format. As Figure 1 shows, the IPv6 header has eight fields to IPv4's 14. IPv6 removes and repurposes several of the header fields in IPv4 to make packet processing more efficient.

Here's how the two headers stack up. The Version field is unchanged between the two protocols. IPv4's Internet Header Length, Type of Service, Identification, Flags, Fragment Offset, and Header Checksum fields are dropped. The Total Length, Time to Live (TTL), and Protocol fields have new names and slightly redefined functions in IPv6. The Option field in IPv4 has been removed from the header and repurposed as an extension capability. Lastly, IPv6 includes two new fields: Class and Flow Label. The following section examines each of the header fields in an IPv6 packet.

Version: The Version field is still 4 bits long and identifies the version number of the protocol.

Traffic Class: This 8-bit field permits packets to be assigned different classes or priorities. It's similar to the Type of Service field in IPv4 and should allow for differentiated services.

Flow Label: The Flow Label field is new to the IPv6 header. A source node uses this 20-bit field to request special handling (that is, better than best-effort forwarding) for a specific sequence of packets. Real-time data transmissions such as voice and video can use the Flow Label field to help ensure QoS.

Payload Length: This 16-bit field tells the length of the payload. Unlike the Total Length field in an IPv4 packet, the value in this field doesn't include IPv6's 40-bit header; only the extensions and data that follow the header are tallied. Because the field is 16 bits, it can indicate a data payload of up to 64Kbytes. For larger payloads, a "jumbogram" extension is available.

Next Header: This 8-bit field approximates the Protocol field in IPv4, but with some differences. In IPv4 packets, a Transport-layer header such as TCP or UDP always follows the IP header. In IPv6, extensions can be inserted between the IP and Transport-layer headers. Such extensions include authentication, encryption, and fragmentation capabilities. The Next Header field indicates whether a Transport-layer header or an extension follows the IPv6 header.

Hop Limit: This 8-bit field replaces the TTL field in IPv4. It prevents a packet from being forwarded forever by discarding the packet after a specified number of router hops. Each router that a packet traverses reduces value in the Hop Limit field by one. IPv4 uses a time value, subtracting one second from the TTL field for each router hop. IPv6 simply swaps out a time value for a hop value.

Source Address: This field identifies the source of the originating host. It is 128 bits.

Destination Address: This field identifies the recipient of the transmission. It is 128 bits.

Networkers may be surprised to see the checksum and fragmentation fields missing from IPv6's header. The packet header checksum was discarded to improve router efficiency. While errors in the packet header are still possible, the new protocol's designers decided that the risk was acceptable, especially considering that the Data-Link and Transport layers (just below and above the IP layer) check for errors.

As for fragmentation, IPv6 does allow packets to be split, but the process happens in a header extension, rather than in the header itself. In addition, an IPv6 packet can be broken up only by the source node and reassembled at the destination node—intervening routers aren't permitted to break up or reassemble a packet. This fragmentation feature is intended to reduce processing overhead in transit, and assumes that the frame sizes of today's networks are large enough that most packets won't require fragmenting.

If an IPv6 packet must be divided, the source node determines the Maximum Transmission Unit (MTU) of each link. One way it does this is by sending a test packet to the destination address. If the test packet is too large for a particular link, that link will return an Internet Control Message Protocol (ICMP) message to the source node, which will scale down the packet size.

The extension mechanism that enables fragmentation (among other options) represents a significant redesign feature of IPv6. It replaces the Options field in IPv4, which allows enhancements such as security capabilities and source routing choices in the IPv4 packet.

Rather than pack such enhancements into the IPv6 header, the architects designed extensions to be inserted between the IP header and the higher-layer protocol header (see Figure 2). This enables packets without extensions to be processed more quickly, and also provides a host of extensible options, such as encryption, authentication, fragmentation, source routing, and hop and destination options, among others. As mentioned ealier, such extensions count toward the packet's overall payload length.

IPv4 Header

Version	IHL	Type of Service	Total Length	
Identification			Flags	Fragment Offset
Time to Live		Protocol	Header Checksum	
Source Address				
Destination Address				
Options				Padding

IPv6 Header

Version	Traffic Class	Flow Label	
Payload Length		Next Header	Hop Limit
Source Address			
Destination Address			

■ **Figure 1:** The IPv6 packet header has fewer fields than its predecessor for more efficient processing. (Field lengths are not drawn to scale.)

IPv6 Header Next Header= TCP	TCP Header and Data

IPv6 Header Next Header= Routing	Routing Header Next Header= Fragment	Fragment Header Next Header= TCP	Fragment of TCP Header and Data

■ **Figure 2:** IPv6 uses extension headers to provide a host of options, such as fragmentation and source routing. The extension headers, which replace the Options field in IPv4, are inserted between the IP header and the Transport-layer header.

INTEGRATING V4 AND V6

While it's pleasant to imagine one grand and all-encompassing wardrobe change from version 4 to version 6, both versions must interoperate, and it will likely be years—if ever—before IPv6 entirely replaces IPv4. Thus, transition methods have been put forward to ensure that IPv4 packets aren't stiffed by IPv6 devices and vice versa. Those methods are dual stacks and tunneling.

The dual stack proposition is simple. New IPv6 devices will be backward-compatible with IPv4, and IPv4 devices can be programmed with both IPv4 and IPv6 stacks to process respective packets appropriately. Hosts with dual stacks can send and receive both IPv4 and IPv6 data, and routers with dual stacks can forward either kind of packet.

The tunneling mode sends packets between two IPv6 domains over an IPv4 network. To do so, a dual-stack node encapsulates an IPv6 packet with an IPv4 header. This encapsulated packet can then be routed over an IPv4 network—the tunnel—until it reaches the second IPv6 domain. A second node strips away the IPv4 header and processes the packet accordingly.

Tunnels can be configured manually or automatically. A manual configuration requires a network administrator to define IPv4-to-IPv6 address mappings at the tunnel endpoints. While IPv6's 128-bit address can normally be used on either side of the tunnel, the router at the tunnel entry point must be manually configured to define which IPv4 address will cross through the tunnel.

Automatic configuration uses an IPv4-compatible address, which is a 32-bit address padded out with zeros to reach 128 bits. If an IPv6 node uses the compatible address, the router at the tunnel entrance simply removes the extra zeros to reveal the true IPv4 address. Once the packet passes through the tunnel, the node at the tunnel exit removes the IPv4 header to reveal the actual IPv6 address.

PACKET PROTECTORS

IP datagrams can be intercepted, and the data read and modified. Thus, IPSec protects IP packets through authentication and encryption. Because IPSec was developed some time after IPv4, support for IPSec has to be grafted onto IPv4 packets. By contrast, IPv6 supports IPSec by design, through extension headers.

Authentication header extensions can help administrators verify that packets are indeed coming from the source address in the header. This should help prevent address spoofing, a technique in which an attacker forges the source address to make the packet appear as if it's coming from a legitimate or trusted location.

The encryption extension, known as the Encapsulating Security Payload (ESP) service, renders the packet's payload data illegible unless the recipient has the proper key to unscramble the data. Encryption provides a measure of confidentiality and data authentication. Administrators can choose to encrypt only the transport and data payload of a packet or the entire packet, including headers and extensions. If the entire packet is encrypted, an additional unencrypted header must be appended to the packet so that it can reach its destination.

RESOURCES

The IETF has a host of RFCs on IPv6 implementation. RFC 2460 is the most current commentary on general IPv6 specifications. Go to www.ietf.org.

IPv6: The New Internet Protocol, Second Edition, by Christian Huitema, is a concise and informative guide to the next-generation protocol.

The 6bone network at www.6bone.net is an international testbed for IPv6. It has links to other IPv6 related sites.

For white papers, news, training, and information on IPv6 for Linux, Berkeley Software Distribution (BSD), Microsoft, and other operating systems, go to www.hs247.com.

For more information on integrating IPv4 and IPv6 environments, see a white paper entitled "Connecting IPv6 Routing Domains over the IPv4 Internet." The paper is available at www.cisco.com/ipv6/. Click on the "Technical documents" link at the bottom of the page.

This tutorial, number 159, by Andrew Conry-Murray, was originally published in the October 2001 issue of Network Magazine.

Mobile IP

Though the Internet can seem anarchic, IP routing depends on a well-ordered hierarchy. At the Internet core, routers aren't concerned with individual users. They look only at the first few bits of an IP address (the prefix) and forward the packet to the correct network. Routers further out look at the next few bits, sending the packet to a subnet. At the edge, access routers look at the final parts of an address and send the packet to a specific machine.

The hierarchy depends on devices that remain fixed to one subnet and on subnets that don't move between larger networks. If a computer is unplugged from one subnet and connected to another, its IP address must be altered. Likewise, an enterprise that changes its ISP might have to renumber its entire network. The result is that many client machines don't have a permanent IP address, but acquire a new one each time they log on to a network. Most laptops, for example, have an IP address on the employer's network while docked at the office, but one from the employee's ISP while at home.

This isn't a problem if users don't often switch between networks and are willing to log off and on again whenever they do. However, it's a problem if users need to stay connected while moving between networks. Higher-level protocols, such as TCP, use the IP address to identify users, so a user can't maintain a TCP connection if the IP address changes. The solution to this is mobile IP, an IETF standard enabling users to keep the same permanent IP address no matter how they're connected.

Mobile IP is still used rarely, partly because there's little need for it and partly because present implementations waste bandwidth and require at least two precious IP addresses per user. However, mobile IP is expected to become more important as wireless networks and IPv6 become ubiquitous. Cellular vendors are pushing it hard, as a way to allow seamless roaming between third-generation (3G) networks and higher-bandwidth hot spots based on Bluetooth or Wi-Fi (802.11b).

Mobile IP will allow an employee to unplug a handheld computer from its Ethernet cable, then continue to download a file or conduct a Voice over IP (VoIP) conversation while the connectivity is transferred, first to the office's Wireless LAN (WLAN), then to an outside cellular network, and finally to a home DSL line.

TUNNEL VISION

Every type of mobile IP depends on giving the mobile node two IP addresses: a permanent address on its home network, and a care-of address on another network. The permanent address is the one that higher-level protocols use, while the care-of address signifies the node's actual location within a network and its subnets.

Whenever the node moves to a new network, it must acquire a new care-of address on the network it's visiting. In IPv4, this means requesting one from a special mobility agent—essentially a DHCP server, with some Authentication, Authorization, and Accounting (AAA) functionality added—on the foreign network. IPv6 has so many addresses available that the mobile node can make up its own by combining the visited network's prefix with an identifier unique to the device, such as its MAC adress. This eliminates the need for a mobility agent, speeds up the process, and ensures that a care-of address is always available.

Back at the home network, another mobility agent, usually an edge router with some AAA functions, keeps track of all the mobile nodes with permanent addresses on that network, associating each with its care-of address. The mobile node keeps the home agent informed of its whereabouts by sending a binding update via the Internet Control Message Protocol (ICMP), whenever its care-of address changes. These updates can incorporate a digital certificate, to ensure that they're actually sent by the mobile agent, rather than an attacker seeking to impersonate him or her.

When another machine on the Internet needs to correspond with the mobile node, this machine sends packets via the home network. The home agent must intercept these packets and forward them to the visited network, a process known as tunneling. This allows correspondent nodes to use the permanent address and remain unaware of the mobile node's movements.

The next step depends on which type of mobile IP you're using. In IPv4, all packets intended for the mobile node are tunneled via the home network, where the home agent intercepts and forwards them to the care-of address. This is the simplest way to enable mobility, but it adds extra routing hops which use more bandwidth and increase latency. The latter is particularly important for wireless networks, whose main application is still latency-sensitive voice, and where latency is already high and unpredictable.

In the original version of mobile IPv4, standardized in 1996, mobile nodes were supposed to send replies directly to correspondents (see Figure 1). For compatibility with higher-level protocols, the "source" address field in these packets had to be the permanent address on the home network, even though routers on the Internet would see that the packets were actually coming from the care-of address on the visited network. This wasn't a problem in 1996, but it is now.

Thanks to Denial of Service (DoS) attacks, where malicious packets often claim to be from fake IP addresses, routers on the Internet began to incorporate ingress and outgress filtering. Routers would only allow a packet through if its source address field was consistent with its origin. To get around these filters, mobile IPv4 was updated in 2002 to include reverse tunneling. Instead of taking a triangular path, all packets travel via the home network in both directions (see Figure 2). Unfortunately, this step wastes even more bandwidth and adds further latency, making it unsuitable for wireless networks running VoIP.

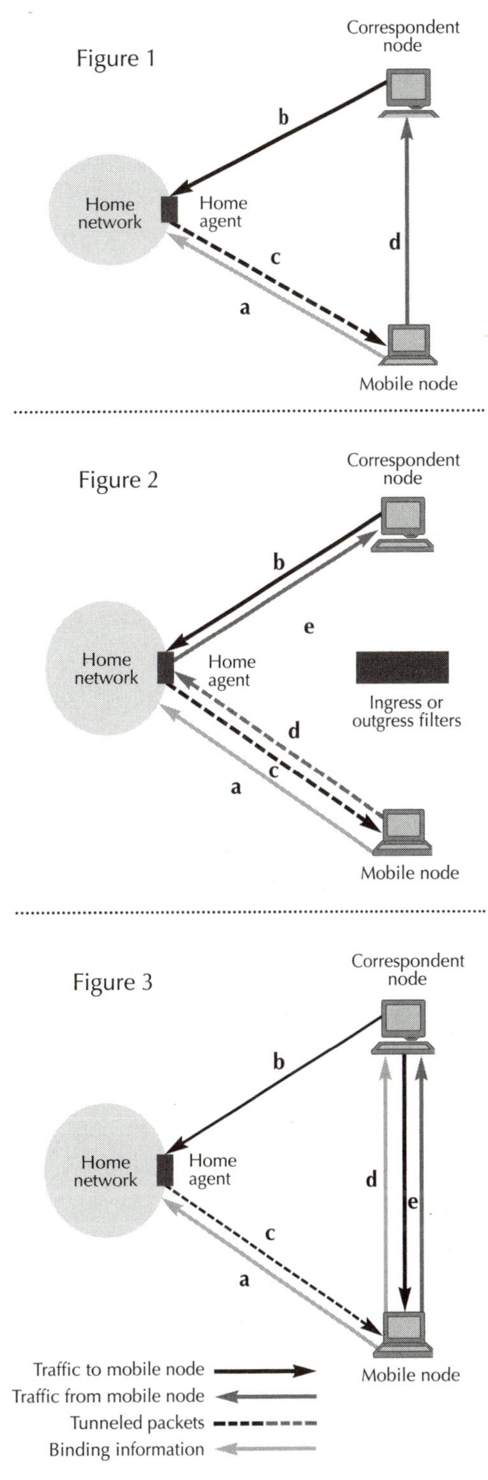

ADDRESSING THE ISSUE

Mobile IPv6 tries to solve the bandwidth and latency problems by avoiding tunneling as much as possible. Though the first few packets of every session are still tunneled via the home agent, the mobile node also sends binding updates to every correspondent. Future packets can be sent directly, just as if the mobile node belonged on the network it was visiting. You can apply the same principle to entire mobile subnets, such as a WLAN inside a moving vehicle.

You can accomplish this with extensible headers, a feature allowing IPv6 packets to contain extra protocol information to deal with issues such as QoS and prioritization. In mobile IP, you use extensible headers to make each packet contain both the permanent and the care-of address, satisfying both higher-level protocols and Internet routers. (See Figure 3.)

The extra bandwidth taken up by this information can be significant, especially for small packets such as those used in VoIP, so headers are compressed using a standard called Robust Header Compression (ROHC). By taking advantage of the fact that consecutive packets often have identical headers, ROHC can reduce header size by around 95 percent.

Because a mobile node can move rapidly, it might have several care-of addresses at any one time. These addresses include the primary one, representing the network the node is attached to, and several older ones on networks the node previously passed through. Packets sent to these older care-of addresses must be tunneled by agents on the previously visited network, just as if they were sent through the home network. To prevent a node from accumulating too many old care-of addresses, binding updates in mobile IPv6 always include an expiry time for a care-of address.

All of this would seem to require extra functionality within every device connected to the network. Edge routers must be able to tunnel packets not just to their own mobile nodes, but also to other nodes that have previously used a care-of address on their network. TCP/IP stacks on individual devices must be able to understand the difference between a permanent and a care-of address.

However, this functionality is standard in the IPv6 specification, whereas the ability to act as a home or foreign agent has to be retrofitted to IPv4 devices. This fact, rather than the larger address space, is why the wireless industry is so keen to promote IPv6 adoption.

RESOURCES

The mobility schemes for IPv4 are well-defined in several RFCs, available at www.ietf.org. In particular, the protocol is covered in RFC 3220, authentication in RFC 2977, header compression in 3095, and reverse-tunneling in RFC 3024.

Cellular vendor Nokia has some informative whitepapers explaining how IP mobility works in 3G wireless networks at its IPv6 site, www.nokia.com/ipv6/.

The European Commission's Information Society Technologies Programme oversees many research projects into the wireless Internet, at www.cordis.lu/ist/. Three of these projects deal specifically with the future direction of mobile IP.

Wireless IP Network as a Generic platform for Location Aware Service Support (WINE-

GLASS), at http://domobili.cselt.it/wineglass/, looks at IP roaming between cellular networks and Wireless LANs (WLANs).

Broadband Radio Access for IP Networks (BRAIN), at www.ist-brain.org, attempts to optimize the protocol for wireless networks.

Mobile IP Network Developments (MIND), at www.ist-mind.org, tries to integrate mobility with QoS and other features necessary if all traffic is to be carried over IPv6.

This tutorial, number 166, by Andy Dornan, was originally published in the May 2002 issue of Network Magazine.

H.323

IP-based networks have become indispensable, thanks to their reliable transmission of data packets. However, they are not optimized for multimedia traffic, which requires careful attention so that packets make it to their destinations on time and in sequence.

If an e-mail message gets pulled apart and reassembled, you'll still wind up with a readable missive. When the packets of a videoconferencing session get jumbled, however, its participants will unwittingly resemble Max Headroom.

H.323, approved by the International Telecommunications Union (ITU) in 1996, addresses this problem. The specification's components define how audio and video should be transmitted, thereby providing real-time applications with the Quality of Service (QoS) they need.

UNDER THE UMBRELLA

H.323 grew out of the H.320 spec, which was approved in 1990 to support videoconferencing over ISDN and other types of circuit-switched networks. If you've ever used a room-based videoconferencing system, it probably supported H.320.

But because ISDN is still expensive and never took off as many people hoped it would, companies began turning to the much more economical Internet for their communications needs. When that happened, a new protocol was needed to address this inherently unreliable network.

The H.323 standard does not address the network type or Transport layer being used, nor is it dependent on any particular hardware or operating system. The umbrella is big enough to cover plenty of other things, such as specifications for audio and video compressor/decompressor (codec) devices, standards for call setup and control, an interface that supports data conferencing, and provisions for real-time transmissions.

On the audio front, H.323 supports several standards. One of the key components is G.711, a voice standard for digital encoding. Under G.711, voice is usually sent at either 56Kbits/sec or 64Kbits/sec, speeds that can be supported by most network environments. Other audio standards supported within H.323 are G.722, G.723, G.728, and G.729.

As for video, H.323 supports H.261 and H.263, two key video codec standards. H.261 is also used by H.320, H.321 (broadband ISDN, ATM), and H.324 (PSTN analog phone system). H.261 specifies fully encoding some video frames in some instances, and encoding only the changes between a frame and the previous frame in other instances.

The other video codec, H.263, is considered an optional component. It is a newer codec that's backward compatible to H.261 and produces better picture quality.

Important as audio and video standards are, a number of other components within H.323 are also essential. For example, a subgroup usually referred to as call/conference control, or system control, provides the all-important functions of call setup, flow control, and message transmission that oversee the operation of H.323.

The key component in this subgroup is a standard called H.245—a control channel protocol that transmits a variety of information through a channel during an H.323 communication. H.245 includes information about flow control, preference requests, and other general commands that need to be sent back and forth during a call. It also defines separate send and receive capabilities and the means to send these details to other devices that support H.323.

Another part of the control subgroup is the Q.931 signaling protocol, which sets up and terminates a connection between two H.323 devices. (Note that while Q.931 uses a fixed IP port, other elements of H.323 use ports that are dynamically allocated. Firewalls must therefore snoop the control channel to determine which dynamic sockets are in use for H.323 sessions, and allow traffic as long as the control channel is active.)

Also within this subgroup is the Registration/Admissions/Status (RAS) channel, a signaling channel that provides a variety of communications among devices in an H.323 call. For example, RAS monitors status, changes in bandwidth, and other functions between H.323 devices and gatekeepers, which I'll look at later on.

Although H.323 is geared toward audio and videoconferencing, it does support data conferencing; the standard supports the ITU's T.120 spec, which defines point-to-point and multipoint data conferencing sessions.

H.323 also includes QoS. It utilizes TCP as the Transport layer for communications that involve data—including T.120 and H.245 information. TCP is highly reliable, and the fact that it is prone to delay doesn't affect data information negatively.

But for audio, video, and information going over the RAS channel, H.323 makes use of UDP, which routes packets in sequence and with best effort. When sending this type of information, H.323 supports the Real-Time Transfer Protocol (RTP), an IETF specification for delivering packetized audio and video over the Internet.

Related to RTP and also supported by H.323 is the Real-Time Control Protocol (RTCP), which keeps track of QoS and sends information to the participants about what's happening in the session. Not defined by H.323 is support for RSVP. Currently, only some H.323-compliant products support this method of requesting bandwidth, but this support is expected to increase.

H.323 also supports IP multicast, a way of sending UDP packets to multiple parties without repeating the information being sent out.

BASIC ARCHITECTURE

Building upon the protocols discussed in the previous section, H.323 defines four major network components. See Figure 1 for a sample diagram.

The H.323 terminal is the LAN endpoint that allows users to communicate with each other in real time. Examples of terminals are videoconferencing or audio conferencing clients, which in many cases are PCs. H.323 stipulates that endpoints must conform to certain standards. They must support G.711 speech compression, H.245 for controlling the media between the clients, the Q.931 signaling protocol, the RAS channel, and RTP/RTCP.

Optionally, terminals can support video codecs, additional audio codecs, and T.120 data conferencing.

Another component of an H.323 network is a gateway, which connects H.323 terminals to other endpoints that do not support the standard—generally on circuit-switched networks

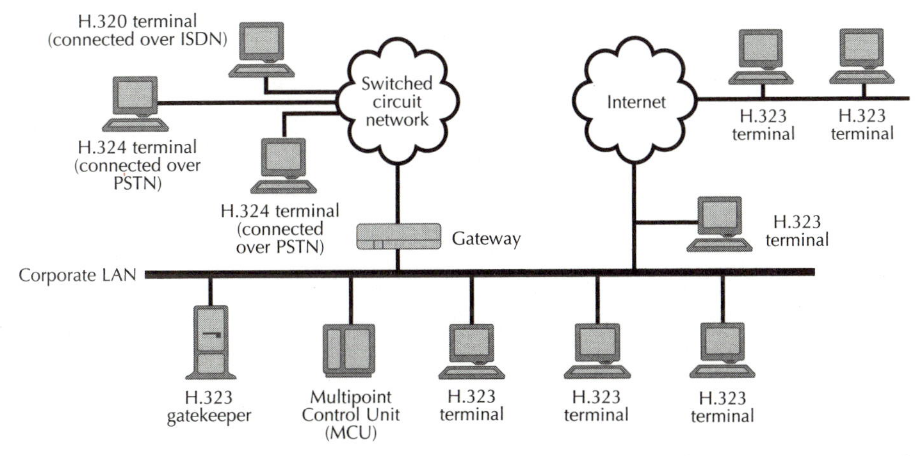

■ **Figure 1:** This sample network architecture shows roughly where the four major H323 components—terminals, gateways, Multipoint Control Units (MCUs), and gatekeepers—would be installed, and how they would communicate with one another and with other network resources.

such as ISDN and the regular phone system. Gateways do this by translating protocols such as audio and video codecs between an H.320 (ISDN) or an H.324 (PSTN) system and clients in an H.323 call, allowing all parties to carry on a meaningful conference. Gateways are also responsible for transferring information between the different networks. Terminals send information to gateways using the H.245 and Q.931 protocols.

Typically, a gateway consists of several parts. One is a switched-circuit network interface, incorporating T1 or ISDN PRI interface cards. Gateways usually also contain NICs for communication with devices on an H.323 network. Other components include digital signal processors, which take care of voice compression and echo cancellation, and a control processor that oversees all other gateway functions.

Gateways are actually an optional component of an H.323 network. If you're communicating directly over a LAN, and not to endpoints on other networks, you won't need a gateway.

On the other hand, if your calls involve three or more endpoints, you need a Multipoint Control Unit (MCU). This consists of a Multipoint Controller (MC), which processes H.245 requests among terminals and controls conference resources, and multipoint processors, which process audio, video, and data.

The fourth and most important piece of any H.323 network is the gatekeeper, which acts as the central point for all calls within its zone. A gatekeeper's zone is defined as the H.323 terminals, translation gateways, and multipoint units over which it has jurisdiction.

Gatekeepers are required to perform four functions. First, they must translate terminal and gateway LAN aliases to IP or IPX addresses. Second, gatekeepers perform bandwidth control, which involves allocating bandwidth during a call; they can refuse to create more connections once a pre-established upper limit for a number of simultaneous conversations has been reached.

A third gatekeeper task is admissions control, which uses RAS messages to authorize network access. The fourth required function is zone management, which involves performing the previous three tasks for all terminals, gateways, and MCUs within its zone.

Gatekeepers can also perform several optional functions. One is call-control signaling, which permits the gatekeeper to process Q.931 signaling messages.

A gatekeeper may also perform bandwidth management, an extension of bandwidth control, which means it can determine when there is no available bandwidth for a call, or if there is no more available bandwidth when a call in progress requests more.

Other optional gatekeeper services include call authorization, which involves the acceptance or rejection of calls based on criteria such as time of day, type of service, and lack of bandwidth. Gatekeepers also may perform call management, which involves keeping track of H.323 calls in progress to know which terminals are busy. This helps gatekeepers redirect calls or save call-setup time by not trying to reach a terminal already in use.

THE NEXT STEP

In January 1998, the ITU ratified the second version of H.323, which includes a number of enhancements as well as several new features.

Major enhancements include the ability for endpoints to set QoS through RSVP; gatekeeper redundancy; support of URL-style addresses; the ability for gateways to let a gatekeeper know of its present resource availability; and increased audio and video capabilities.

The second version of H.323 also includes faster call setup and new security features. Supporting a standard called H.235, this latest version offers authentication, integrity for data packets, privacy, and nonrepudiation.

Since 1997, the number of products supporting H.323 has increased dramatically. Not only do desktop videoconferencing products support it, but so do IP telephony gateways and whiteboarding products. And, as is the case whenever a critical mass of a particular type of product hits the market, interoperability tests have been performed on H.323-compliant products, making the decision on what to buy that much easier.

While most of us have yet to participate in our first videoconferencing session or place our first telephone call over IP, standards such as those defined within H.323 will make it easier for vendors to create products that work well with others, and that deliver real-time video, audio, and data to a much wider audience.

RESOURCES

For a primer on H.323, see www.databeam.com/h323/h323primer.html.

For an in-depth look at the standards efforts behind H.323, visit the ITU's Web site at www.itu.ch/.

To see more detailed information about H.323 gatekeepers, gateways, and standards, visit the H.323 center at www.elemedia.com.

This tutorial, number 126, by Anita Karvé, was originally published in the January 1999 issue of Network Magazine.

Inside the Session Initiation Protocol

Network managers getting ready to roll out IP telephony networks had better think twice about the gear they're going to deploy. A new IETF protocol may change the way next-generation phone networks are built. The Session Initiation Protocol (SIP) could provide the underlying mechanisms for establishing calls between users on an IP telephony network.

That might not seem like a big deal. After all, the H.323 suite of protocols has long provided comparable functionality. However, SIP's smaller footprint makes the protocol more scalable and faster than existing H.323 implementations. The catch? The protocol is still in its early stages, making products hard to come by.

Until recently, network managers looking to roll out intelligent networks have relied heavily on the H.323 suite of protocols. With H.323, a compliant client, such as Microsoft NetMeeting, queries an H.323 gatekeeper for the address of a new user. The gatekeeper retrieves the address and forwards it to the client, which then establishes a session with the new client using H.225, one of the H.323 protocols. Once the session is established, another H.323 protocol, H.245, negotiates the available features of each client.

It may sound simple, but H.323 suffers from some key problems. At the top of the list is call setup time. Since H.323 first establishes a session and only then negotiates the features and capabilities of that session, call setup can take significantly longer than an average PSTN call (see Figure 1).

Just how long depends on the particular network and the distance between locations, but the total time for someone to answer the call can reach up to 8 seconds, according to Pauli Saksanen, project manager for IP telephony at Sonera (www.sonera.com), the former Postal, Telegraph & Telephone (PTT) administration of Finland. The delays are even worse on international calls, he says, with lag times as long as several seconds. Saksanen should know. He runs Sonera's nationwide IP telephony network, the first of its kind anywhere in the world.

The network is currently based around H.323, but Saksanen says it will ultimately be able to accept clients using other protocols as well. By working with existing component suppliers, Saksanen says he's been able to get the delay down to between 100ms to 200ms, but the

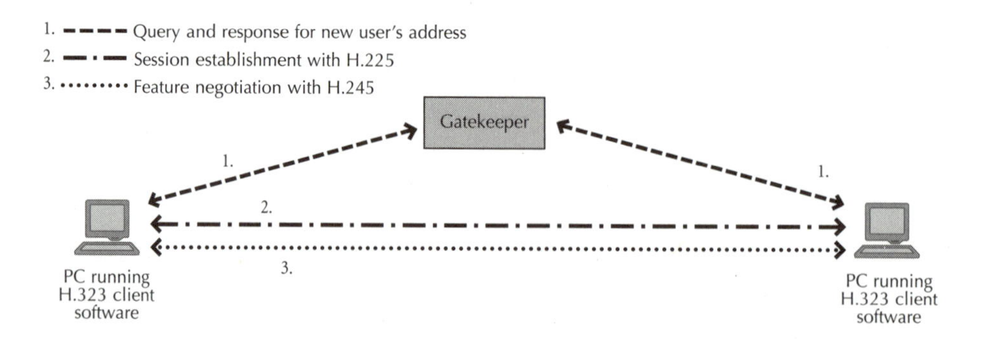

■ **Figure 1:** The time needed to set up a call is the key problem with H.323. Since the features of the call, such as whether or not to invoke video, are only negotiated after the call is established, setup times are much longer with H.323 than with SIP.

problem still exists. A new version of H.323, dubbed "H.323 fast," will address the problem, but H.323 fast isn't widely accepted yet.

What's more, opponents say H.323 doesn't scale well. A case in point is H.323 addressing. Creating separate phone-numbering schemes complicates interconnecting carrier networks. Critics also charge that the H.323 standard itself is too large and complex to make deployment easy. "H.323 is built in a telecom manner," says Hans Eriksson, chief technology officer at Telia Network Services (www.telia.com), a division within Telia, the Swedish PTT. Eriksson has evaluated H.323 as a way of rolling out telephony services over an IP network. Finally, H.323 doesn't provide a simple way for connecting two circuit-switched networks across an IP network.

ENTER SIP

All of these problems are addressed by SIP. With SIP, each user is identified through a hierarchical URL that's built around elements such as a user's phone number or host name (for example, SIP:user@company.com). The similarity to an e-mail address makes SIP URLs easy to guess from a user's e-mail address.

When a user wants to call another user, the caller initiates the call with an invite request. The request contains enough information for the called party to join the session. With a unicast session, this includes the media types and formats that the caller wants to use and a destination for the media data (see Figure 2). A session might include, for example, sender requests to employ H.261 video and G.711 audio.

The request is sent to the user's SIP server. The SIP server may be a proxy server, which receives the request and, using its own internal algorithms, determines the user's location. Alternatively, the SIP server may be a redirect server that returns to the client the appropriate SIP URL, which the user then queries. In either case, the server's address is learned by querying the DNS, the distributed database that matches high-level host names with the underlying IP address.

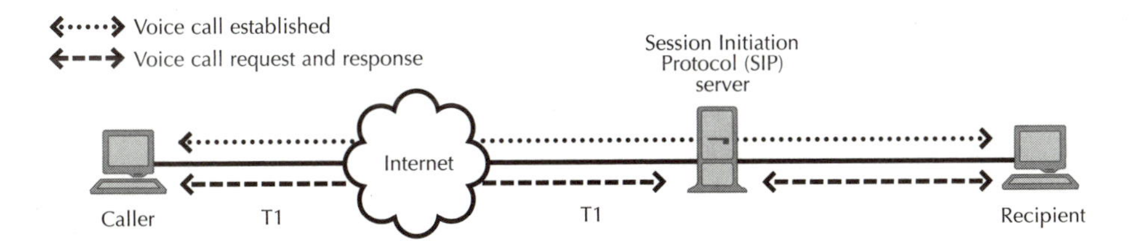

■ **Figure 2:** Session Initiation Protocol (SIP) speeds up call setup by bundling all of the configuration information in the request. The request is either sent directly or via the proxy server to the recipient, who accepts the call and immediately initiates the session.

Once found, the request is sent to the user, and from there several options arise. In the simplest case, the request is received by the user's telephony client—that is, the user's phone rings. If the user takes the call, the client responds to the invitation with the designated capabilities of the client software, and a connection is established. If the user declines the call, the session can be redirected to a voice mail server or to another user.

"Designated capabilities" refer to the functions that the user wants to invoke. The client software might support videoconferencing, for example, but the user may only want to use audio conferencing. Regardless, the user can always add functions—such as videoconferencing, whiteboarding, or a third user—by issuing another invite request to other users on the link.

SIP has two additional significant features. The first is SIP's ability to split, or "fork," an incoming call so that several extensions can be rung at once. The first extension to answer takes the call. This feature is handy if a user might be working between two locations (a lab and an office, for example), or where someone is ringing both a secretary and a boss.

"You could extend H.323 to offer that feature, but under the existing standard, forking just isn't possible," says Henning Schulzrinne, associate professor in the departments of computer science and electrical engineering at Columbia University in New York, and one of the original authors of the SIP standard.

The second significant feature is SIP's unique ability to return different media types. Take the example of a user contacting a company. When the SIP server receives the client's connection request, it can return to the customer's phone client via a Web Interactive Voice Response (IVR) page, with the extensions of the available departments or users provided on the list. Clicking the appropriate link sends an invitation to that user to set up a call.

BORDER PATROL

With the basics covered, it's easy to see where SIP fills in some of H.323's holes. First there is the issue of call setup time. By including a client's available features within the invite request, SIP negotiates the features and capabilities of the call within a single transaction. The upshot, says Schulzrinne, is that SIP can set up a call within about 100ms, depending on the network.

SIP also scales better than H.323. One of the attractive features for providers like Telia is SIP's simplicity. It doesn't purport to solve the whole telephony equation, says Eriksson. He notes SIP can be used just to identify an end user, relying on other protocols and applications to manage the call. The result, says Eriksson, is that the protocol is much easier to implement than H.323 and much easier to adapt.

A case in point is SIP's addressing scheme, which leverages the existing DNS system instead of recreating a separate hierarchy of telephony name servers. Then there's the way SIP handles connectivity with circuit-switched networks. A new SIP draft will also extend the protocol to address a number of other problems. Chief among them is the need to develop a standard way to map the telephony service parameters onto SIP packets using the MIME standard. By using MIME for signaling, the data ends up being passed along on a SIP transaction in much the same way that e-mail attachments are transported with a mail message.

The protocol's light weight lets it be enhanced in other ways as well. The draft standard is expected to add several special-purpose functions to SIP. Among these options is an ability to negotiate security and QoS. There's also the ability to let callers indicate their various preferences. A user calling a company, for example, might want to speak only with someone who speaks Spanish. This option can be embedded in the Invite command so that the user will be routed to the correct company contact.

Are there drawbacks to SIP? No question. The biggest issue today is availability. While H.323 is widely accepted and deployed today, SIP products are, for the most part, nowhere to be found. That will soon change, however. Currently, as many as 20 vendors are working on SIP implementations. 3Com announced last October that it's including SIP support within its CommWorks IP telephony solution for service providers. Cisco Systems is doing the same by including SIP support in its Architecture for Voice, Video and Integrated Data (AVVID) architecture. And while Nortel Networks has not announced SIP support directly, the company is collaborating with Telia Mobile on developing next-generation telephony services utilizing the protocol.

That's a pretty strong vote of confidence, but it may not be enough to compensate for the Microsoft factor. "Users already get an H.323 client today with every version of Windows," says Sonera's Saksanen. "Will they be willing to go out and purchase a SIP client as well?" That's the question.

RESOURCES

The actual Session Initiation Protocol (SIP) standard (RFC 2543), submitted by Mark Handley, Henning Schulzrinne, Eve Schooler, and Jonathan Rosenberg, is at ftp://ftp.isi.edu/ in-notes/rfc2543.txt/.

Henning Schulzrinne, one of the original authors of the SIP standard, has an excellent site at www.cs.columbia.edu/~hgs/sip/sip.html. There's an overview of SIP, a schedule of the interoperability tests (bake-offs) of SIP gear, plenty of technical details on the protocols, and links to associated sites. If you're worried about grammar (SIP grammar, that is), the site provides all the gory details on how to format SIP requests.

SIP implementers can stay current with the latest information by subscribing to the list at majordomo@cs.columbia.edu. Send mail to the address with the line "subscribe sip-implementers" in the body.

This tutorial, number 138, by David Greenfield, was originally published in the January 2000 issue of Network Magazine.

The AppleTalk Protocols

The Apple Macintosh operating system, as its many supporters will tell you, is extremely easy to learn and use. Apple's AppleTalk network system brings the same kind of simplicity of use to Macintosh connectivity.

Although not an official LAN standard, AppleTalk can be considered a de facto standard: With AppleTalk connectivity options built into every Macintosh, millions of Macs possess ready-made networking capabilities. This has not been lost on Mac aficionados, who have used AppleTalk to link thousands of Macs into efficient, cost-effective LANs.

NETWORKING THE MAC

When you consider the kind of work performed by the typical Macintosh user, it's not surprising that Mac users have readily accepted networking. For example, take the desktop publishing environment where the Mac prevails: Few writers are good artists, and vice versa. The nature of their jobs, however, demands that they combine their diverse efforts into a single product.

The ability to share and combine files online means those producing documents with PCs can easily merge graphics and other images with text without having to swap diskettes or "cut and paste" hard-copy images. This means the job gets done faster and more efficiently. Networked Macs are thus the rule rather than the exception in these situations, and AppleTalk is Apple's solution to Mac connectivity.

Apple calls AppleTalk "a comprehensive network system" made up of hardware and software components. An AppleTalk network can consist of many different kinds of computer systems and servers and a variety of cabling and connectivity products. Because it was designed to support a variety of machines, Apple developed a suite of proprietary protocols that permits communication between the varying devices that users might need to attach to an AppleTalk network.

However, AppleTalk is not a network operating system, a media-access control (MAC) method such as Ethernet, or a cabling system (LocalTalk is a trade name of Apple's cabling system). Rather, AppleTalk is a nonstandard suite of protocols that while not fully compliant, still provides most of the functions spelled out by the International Standards Organization's Open Systems Interconnection (OSI) reference model.

As Figure 1 illustrates, the six-layer suite of AppleTalk protocols supports numerous connectivity options, including LocalTalk, Ethernet, and Token Ring. AppleTalk also supports Northern Telecom's Meridian, a now defunct 2.5Mbps twisted-pair network. This set of protocols allows connections of virtually any computing device to an AppleTalk network. Here's how it works.

AT THE PHYSICAL LAYER

To many network users, the media (or cabling system) that connects PCs into a network is the network—that's all they ever see of it. Their NOS software operates transparently, having been set up by their network administrator, and their network interface card (NIC) is installed inside their computer, out of sight and mind. In the case of AppleTalk, the original (and only) media users see are Apple's own LocalTalk products.

This scheme, driven by Apple's data-link layer, LocalTalk Link Access Protocol (LLAP), uses proprietary modular plugs and wiring to link Macs and LaserWriter printers into a network.

Since AppleTalk's 1984 release, Apple and other third-party vendors have developed data-link protocols to support Ethernet, Token Ring, and ARCnet networks, which exchange data at 10Mbps, 4Mbps, and 2.5Mbps respectively, all faster than LocalTalk's 230.4Kbps rate.

Despite its relative lack of performance, LocalTalk offers one major benefit these technologies lack: Every Macintosh computer that Apple has manufactured (prior to the iMac) contained the built-in LocalTalk connection; Apple LaserWriters and Apple IIgs computers, as well as many other Apple peripherals, also contained this built-in connection. Apple's Quadra computers and subsequent models come with Ethernet built in.

This ready-made networking option makes LocalTalk an ideal connectivity option for Mac users, particularly those who don't require the better data-exchange performance delivered by Ethernet, Token Ring, or ARCnet. LocalTalk users get most of the benefits of networking—that

APPLETALK PROTOCOL ARCHITECTURE AND THE ISO-OSI REFERENCE MODEL

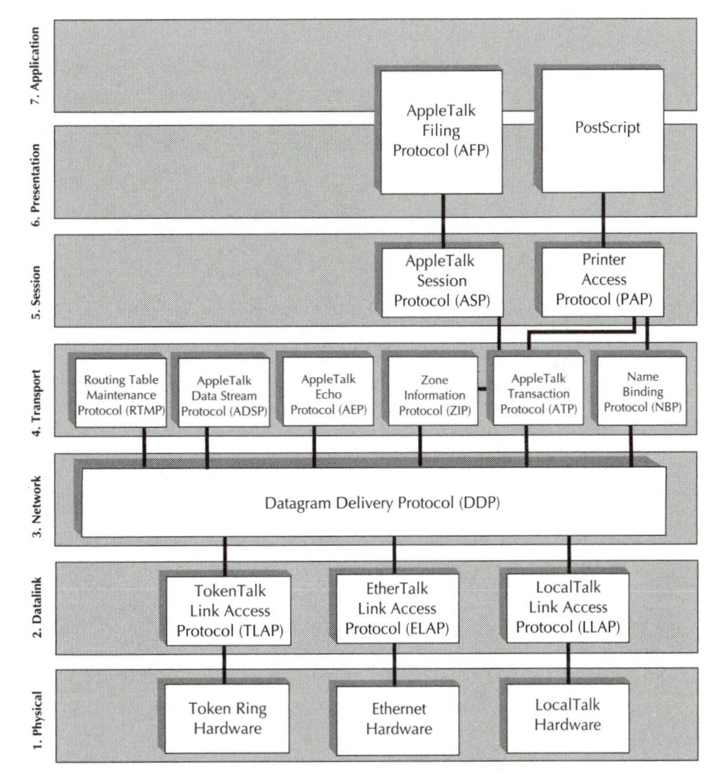

■ **Figure 1:** Apple Computer's six-layer AppleTalk protocol suite, although not fully compliant with the seven-layer OSI reference model, provides many of the capabilities and functions defined by OSI. Here, the two protocol suites are compared side-by-side.

is, file and printer sharing, access to electronic mail, and other shared resources—without the added costs associated with a network adapter board.

LOCALTALK'S ACCESS METHOD

LocalTalk, like Ethernet, uses a Carrier-Sense, Multiple-Access (CSMA), media-access scheme to place data packets on the network wire. It does not rely on collision detection (CSMA/CD), as does Ethernet. It uses CSMA/CA, for Carrier-Sense, Multiple-Access with Collision Avoidance.

Stations on a CSMA/CA network, rather than sensing collisions between data packets sent by multiple stations, send out a small (three-byte) packet that signals their intent to place data on the wire. This packet tells all other stations on the wire to wait until the signaling node's data has been sent before they attempt to send data. If collisions between packets are going to occur, they will occur between the preliminary packets, not the actual data packets.

This best effort packet-delivery system, managed by LLAP, does not guarantee that the packet reaches its destination, but it does ensure that all packets delivered are free of errors. The LLAP provides the data-link access specifications and uses a dynamic address-acquisition method that enables AppleTalk's plug-and-play capabilities over twisted-pair wiring.

LOCALTALK'S LIMITATIONS

LocalTalk, though convenient, suffers from other limitations besides its slow data-transfer rate. For example, LocalTalk workgroups are limited to 32 nodes over a 1,000-foot cable run. Ethernet and Token Ring both support substantially greater numbers of nodes.

The EtherTalk, TokenTalk, and ARCnet Link Access Protocols (ELAP, TLAP, and ALAP, respectively) manage AppleTalk network access to Ethernet, Token Ring and ARCnet networks. Apple developed EtherTalk and TokenTalk as extensions of the two protocols' industry-standard data-link processes. Standard Microsystems developed ALAP. One of the key responsibilities of ELAP, TLAP, and ALAP is mapping AppleTalk addresses into the standard data-link Ethernet, Token Ring, or ARCnet address required for proper routing of data.

Because the Ethernet, Token Ring, and ARCnet addressing schemes are incompatible with LLAP, AppleTalk node addresses must be translated into the appropriate format; the AppleTalk Address Resolution Protocol (AARP) handles this translation.

One layer up in the AppleTalk stack is the Datagram Delivery Protocol (DDP). The DDP works with the Routing Table Maintenance Protocol (RTMP) and AppleTalk Echo Protocol (AEP) to ensure data transmission across an Internet.

END-TO-END SERVICES

The DDP exchanges data packets called datagrams. Datagram delivery is the basis for building other value-added AppleTalk services, such as electronic mail. The DDP permits running AppleTalk as a process-to-process, best-effort delivery system, in which the processes running in the nodes of an interconnected network can exchange packets with each other.

The DDP provides these processes with addressable entitles called sockets, and processes can attach themselves to one or more sockets in their nodes. Once associated with a socket, a process can exchange packets with other nodes via these sockets. Once linked to a socket, the process becomes accessible from any point on the AppleTalk network. It is then called a network-visible entity.

The RTMP provides the logic that routes datagrams through router ports to other networks; it permits routers to dynamically learn routes to other AppleTalk networks in an Internet. The AEP lets nodes send datagrams to any other nodes and to receive a copy, or "echo," of the datagram sent. This confirms the existence of a node and helps measure round-trip delays.

RELIABLE DATA DELIVERY

The data-delivery group of protocols—the AppleTalk Transaction Protocol (ATP), Printer Access Protocol (PAP), AppleTalk Session Protocol (ASP), and the AppleTalk Data Stream Protocol (ADSP)—guarantee the delivery of data. These protocols can be further broken into two groups, one offering transaction-based services, the other data-stream-based faculties.

Transaction-based protocols use the request-response model typically found in server-workstation interactions. Data stream protocols deliver bi-directional data flow between two communicating nodes.

The ATP directs the AppleTalk transaction processes, in which sockets issue requests that require response (typically, status reports). ATP binds the request and response to guarantee a reliable exchange. The PAP sets up a connection-oriented service that sends print requests to AppleTalk-compatible printers.

ASP opens, maintains, and closes transactions during a session, while ADSP provides a full-duplex, byte-stream service between any two sockets on an AppleTalk Internet.

At the highest level of AppleTalk are the AppleTalk Filing Protocol (AFP) and the PostScript protocol. The AFP, built on top of ASP, permits users to share data files and applications on a shared server, while PostScript, a programming language understood by Apple's LaserWriter and numerous other output devices, provides a standard way of describing graphics and text data.

The AFP, which conducts the dialog between a user's computer and an AppleShare server, is one of the key AppleTalk protocols. AFP was designed to provide the tools that allow supporting different types of computers—that is, Macs and IBM PCs—over an AppleTalk network.

The AFP is also important because any network operating system that is fully compatible with it can operate transparently on any AppleTalk network. In turn, this means that such an NOS can support all AppleTalk-compatible applications.

This tutorial, number 25, was originally published in the August 1990 issue of LAN Magazine/Network Magazine.

The Resource Reservation Protocol

End-to-end QoS over the Internet has long been a goal of those desiring optimized services over this public medium, but the fact that systems on a public network cannot be as tightly controlled as those on a private one has made this quest daunting.

One of the more promising approaches to this problem is the IETF's Resource Reservation Protocol (RSVP). Its version 1 functional specification is found in RFC 2205 (1997); RFC 2750 (2000) contains updates to the standard.

RSVP was designed to signal QoS requirements across Internet connections. RSVP is an IP-based, hop-by-hop protocol that provides devices throughout the path of a data flow with information on the QoS requirements of that particular flow.

The protocol's ultimate goal is to reserve resources so that applications can get the amount of bandwidth they need. In this resource reservation approach, bandwidth management policy can be applied via policy objects, further increasing the efficiency of resource allocation. In effect, RSVP is a method for overlaying a circuit-switched network onto a packet-switched IP network. Due to IP's design, networks based on this protocol place the responsibility for maintaining the state of connections solely on end devices, whereas RSVP requires that each intermediate router maintain connection state information.

The RSVP protocol is part of the IETF's Integrated Services (IntServ) architecture (described in Informational RFC 1633, drafted in 1994), which enables heterogeneous devices to communicate about QoS requirements.

IntServ provides a protocol for signaling the QoS requirements of applications, and incorporates specifications for describing service requirements and related functionality of devices and other network elements that support QoS. IntServ is geared toward supporting real-time data such as voice and video.

PRIORITY MAIL

RSVP can be used with unicast and multicast traffic. In multicast traffic, one sender (source of a data flow) forwards a copy of each data packet to multiple receivers. One rationale for making the reservations receiver-based lies in the importance of support for multicast traffic. This type of traffic often involves reservation requests from receivers with very different characteristics—thus, a protocol was needed that would support diverse requests.

Due to the complexity of multicast flows, reservations must be merged. In multicast traffic, packets that must be delivered to different next-hop nodes are replicated. In RSVP, reservation requests must be merged at each replication point. In effect, multiple reservation requests are combined at this merge point and then forwarded as a single reservation request.

This relates to the concept of distinct reservations and shared reservations, which are both supported by RSVP. In distinct reservations, a single reservation is made for each upstream sender in a session. In shared reservations, a number of senders in a multicast flow use the reservation. (Receiver or destination nodes or endpoints are referred to as downstream; sender or source nodes or endpoints are referred to as upstream.)

There are subcategories of reservation styles. In the Wildcard-Filter (WF)-style reservation request, a single reservation is generated and shared among all data flows from all upstream senders. In the Fixed-Filter (FF)-style reservation, a distinct reservation is generated for each sender. In the Shared-Explicit (SE)-style reservation, selected upstream senders share a single reservation. These reservation styles help to optimize reservation handling from different senders in the same session.

THE MESSAGE IS THE MEDIUM

RSVP is executed via a series of messages. A standard RSVP packet header consists of a 4-bit Version field denoting the protocol's version number, accompanied by an unspecified 4-bit Flags field. Eight-bit-long fields include a Reserved field, a Send TTL field indicating the message's sent time-to-live value, and a Message Type field containing the message's function. The Checksum and Length fields, both 16 bits, yield a standard TCP/UDP checksum, and the total length of the common header and the variable-length objects that follow, respectively (see Figure 1).

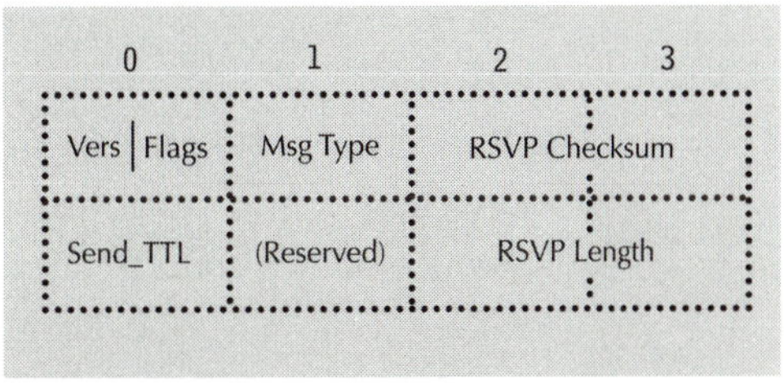

0	1	2	3
Vers | Flags	Msg Type	RSVP Checksum	
Send_TTL	(Reserved)	RSVP Length	

■ **Figure 1:** The RSVP message header includes various fields that denote characteristics such as the message's function, its time-to-live value, the protocol's version number, a TCP/IP checksum, and the total length of the common header.

The header is followed by a set of objects, which includes the information needed to describe and characterize that particular message. These objects are divided into classes, and each class of objects can contain multiple types that characterize the data's format in greater detail.

There are two strains of RSVP: Native and UDP-encapsulated. In Native RSVP, the header and payload are encapsulated in an IP datagram, with the protocol number 46. UDP-encapsulated RSVP can enable end systems to communicate with first-hop and last-hop routers, if these systems aren't RSVP-enabled.

In RSVP, a Path message is transmitted from a sender downstream to a receiver (or multiple receivers). Path messages store path information in every node along a traffic path; at a minimum, a Path message contains the IP address of each previous hop along that path. These IP addresses establish the path for transmitting subsequent reservation request (Resv) messages.

The Path message contains a Session object, which includes destination address and port information, and a Previous Hop (PHOP) object, which identifies the previous router in the direction of the traffic flow.

In addition, the Path message contains a Sender Template and a Sender traffic specification (Tspec) object. The Sender Template contains filter specifications (Filter Specs) and identifies the format of the traffic that the sender will originate. (The Filter Spec, in combination with the session information, defines the set of packets that will receive the QoS specified in the flow specification, or Flowspec. The Flowspec defines the QoS to be applied to a particular data flow.)

The Sender Tspec characterizes the traffic flow that the sender intends to generate, in terms of attributes such as bandwidth parameters, jitter, and delay.

The Path message may also contain an Adspec object, which describes the type of services, service-specific performance characteristics, and amount of resources available for a particular reservation. The Adspec is sent to the local-traffic-control process at every node or router in a path for updating; the updated version is then sent downstream in the Path message.

The Path message may also contain a Policy Data object, which includes policy information pertaining to the source or the previous hop of a specific data flow. The policy control process uses policy data to establish authorization and usage feedback for a particular data flow.

Upon receiving the Path message, downstream RSVP-enabled routers invoke a Path state, which, at a minimum, includes the IP address of the PHOP node. This information is used to send Resv messages in the reverse direction on a hop-by-hop basis. The Path state informs devices along the path of adjacent RSVP nodes on a particular flow.

RETURN TO SENDER?

The receiver makes the reservation request by transmitting a Resv message in the upstream direction (ultimately to the data source). The Resv message includes a reservation specification or Rspec, that specifies what type of service is required—Guaranteed or Controlled Load. Guaranteed service provides a connection like a virtual circuit, whereas Controlled Load service exceeds a best-effort service but isn't up to par with a Guaranteed service.

Resv messages include information about the reservation styles, as well as a Flow Descriptor, which consists of the Flowspec object and the Filter Spec object.

The Flowspec establishes the desired QoS and related parameters for the packet scheduling process. The Filter Spec performs the same function in the packet classification process.

Upon receiving the Resv message, RSVP-enabled routers or nodes invoke an admission control process to indicate whether those nodes can provide the requested QoS. If this process is successful, the Resv message is sent to the next router.

RSVP-enabled routers along the path schedule and prioritize packets according to the request. These systems send incoming data packets to a packet classifier; the messages are then queued in a packet scheduler. A packet filter assigns or maps packets to specific service classes, defining the route and QoS class for these packets. The packet scheduler enforces resource allocation and selects packets for transmission (see Figure 2).

After the last router along the path grants the request, a reservation confirmation (Resv-Conf) message notifies the receiver that the request was successful. RSVP is a soft-state protocol, so the reservations must periodically be refreshed. This soft-state characteristic helps to accommodate changes in routing, and changes in multicast group membership.

Sessions are terminated by teardown messages that cancel out the path or reservation states from nodes or devices upon receipt. PathTear messages are created by the senders (or

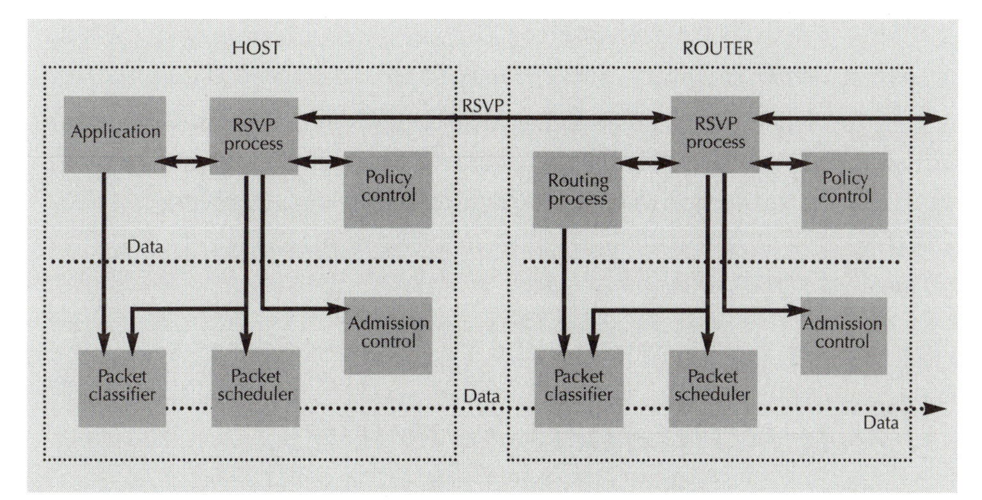

■ **Figure 2:** This diagram represents the implementation of RSVP in hosts and routers. On each end of this equation, stringent policy control and admission control are crucial to deliver the desired level of service. Packets must be classified and scheduled, and queued in the packet scheduler as necessary.

via RSVP's time-out function) in any node on the path, and sent to all receivers. Sent by the specific node that created it, the message then cancels out path state in every node along that path. ResvTear messages cancel out reservation states in these nodes or devices.

In the event of admission control failure, an error message is sent back to the originator of the request. If any device along a path can't execute a Path state, it sends a Path Error (PathErr) message back to the sender. If a reservation state can't be invoked, the device sends a Reservation Error (ResvErr) message to the sender.

MIXED DELIVERY

RSVP can be combined with other QoS protocols and technologies, such as Differentiated Services (DiffServ) and Multi-protocol Label Switching (MPLS). While DiffServ marks and prioritizes traffic, RSVP secures the resources necessary to transmit that traffic. Also, RSVP contains very specific provisions for policy support. It's also compatible with MPLS, in that MPLS is capable of assigning labels in accordance to RSVP Flowspecs.

On the downside, RSVP's sophistication and complexity can impair performance on backbone routers. Thus, it's simply too involved to reasonably apply to certain applications, and is often best substituted with other protocols, such as DiffServ, on the network backbone.

RESOURCES

The following books provide detailed overviews of RSVP and its relationships with other network elements and protocols:

Inside the Internet's Resource ReSerVation Protocol, by David Durham and Raj Yavatkar, John Wiley & Sons (ISBN: 0-471-32214-8)

Internet Performance Survival Guide, by Geoff Huston, John Wiley & Sons (ISBN: 0-471-37808-9)

A white paper called "QoS Protocols and Architectures," located at www.stardust.com/qos/whitepapers/protocols.htm, contains an excellent explanation of protocols such as RSVP, Integrated Services (IntServ), and Differentiated Services (DiffServ).

This tutorial, number 157, by Elizabeth Clark, was originally published in the August 2001 issue of Network Magazine.

Differentiated Services

The previous tutorial explored the issue of delivering QoS over IP-based networks ("Lesson 157: The Resource Reservation Protocol"). Now, we'll look at another approach to this goal: Differentiated Services (DiffServ).

DiffServ was designed to provide a simpler, more coarse approach to establishing differentiated classes of service for Internet data. In the Integrated Services (IntServ) model, resources are allocated to individual flows, which can lead to scalability limitations. In DiffServ, traffic is divided into a small number of forwarding classes, and resources are allocated on a per-class basis. The desired performance levels are achieved through the proper mix of provisioning, prioritization, and admission control. This is in contrast to techniques such as end-to-end resource reservation, which is the foundation of RSVP.

DiffServ was designed to provide service to aggregate forwarding classes comprising multiple traffic flows. Simplicity is fostered by the fact that these flows are grouped into a relatively small number of aggregates that receive a limited number of differentiated treatments (defined via policies) throughout the network.

One of DiffServ's goals was to eliminate the need for per-flow resource reservation state, as well as signaling in each router along a data path. Most classification and policing is done at the network edge. In the core, routers inspect only one field in the IP header—the DiffServ field—to determine where to send the packet next, as opposed to storing information about each individual flow. (The DiffServ field is also referred to as the DS field or DS byte.)

The ultimate aim of the DiffServ architecture is to simplify forwarding in the core and to place the processing burden that accompanies traffic classification and profiling at the network edge. This architecture is more conducive to facilitating the levels of scalability required for today's networks than many possible alternative approaches. (The DiffServ architecture is described in detail in the IETF's RFC 2475.)

MOVING TO THE HEAD OF THE CLASS

When traffic enters the DiffServ network's ingress interface, it's classified and subjected to a preconfigured admission process, and then conditioned to meet policy requirements in accordance with a specific classification. (Conditioning is also referred to as shaping or policing.)

The data stream is then assigned to a behavior aggregate. This is done by marking the IDS fields of the packets' IP headers with the corresponding Differentiated Services Code Point (DSCP). The DSCP value initiates a specific Per-Hop Behavior (PHB) in devices in the network and classifies the packet service level. (The term PBH refers to specific forwarding treatments that occur at a particular node.)

In DiffServ, the 8-bit Type of Service (ToS) field in the IPv4 header is supplanted by the DS field, which contains a value that DiffServ-enabled routers use to determine a specific forwarding treatment (PHB) at each node along a traffic path (see figure). The first six bits of the DS field comprise the DSCP. This is mapped to the PHB, which is received by the packet containing this field at each DiffServ-aware node. The values within the DSCP field are called code points.

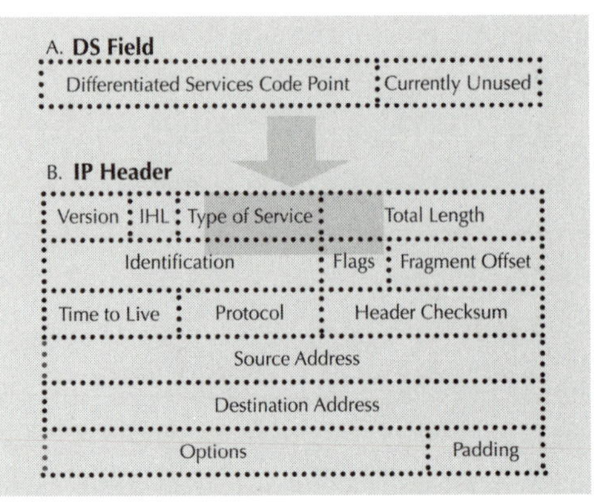

■ **Figure 1:** In Differentiated Services (DiffServ), the DS field (or DS byte) shown in part A of this diagram replaces the 8-bit Type of Service (ToS) field in the IPv4 header (shown in part B). The DS field contains a 2-bit Currently Unused (CU) field and the Differentiated Services Code Point value, which triggers certain treatments of the packet in devices within the network.

The last two bits in the DS field are designated Currently Unused (CU), and support legacy (non-DiffServ-enabled) devices that use the original ToS byte to determine forwarding treatment. The DS field can have up to 64 different values. (For more information on the DS field, see RFC 2474.)

The DS field is ultimately a fundamental ingredient in ensuring that Service Level Agreements (SLAs) between network subscribers and service providers are met. On a more granular level, Traffic Conditioning Agreements (TCAs, basically subsets of an SLA) are designed to help achieve this goal. The TCA may include parameters regarding traffic profiles, performance metrics (such as latency, throughput, and drop priorities), and instructions on how out-of-profile packets will be handled. The TCA can also describe additional traffic-handling services supplied by the service provider. In addition to what the TCA specifies, an SLA may also include specific availability levels, monitoring provisions, and accounting and billing agreements.

TERMS AND CONDITIONS

Two fundamental processes in DiffServ implementation are classification and conditioning (see Figure 2). In DiffServ, traffic is classified and conditioned at the network ingress interface based on the TCA parameters that exist between the service provider and the network subscriber.

In DiffServ, a classifier selects packets based on information in the packet header correlating to preconfigured admission policy rules. There are two primary types of DiffServ classifiers: the Behavior Aggregate (BA) and the Multi-Field (MF). The BA classifier bases its function on the DSCP values in the packet header. The MF classifier classifies packets based on one or more fields in the header, which enables support for more complex resource allocation schemes than the BA classifier offers. These may include marking packets based on source and destination address, source and destination port, and protocol ID, among other variables.

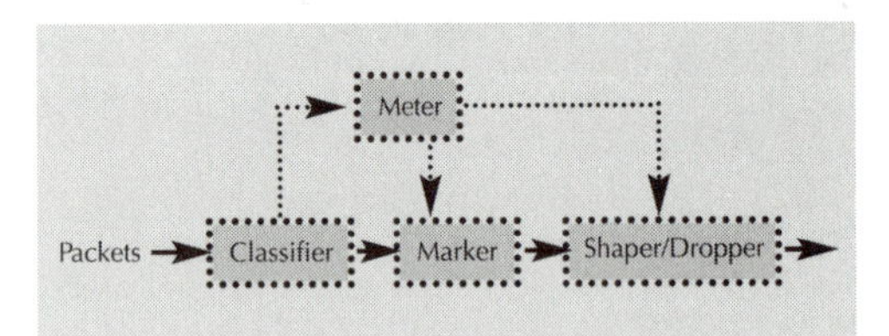

■ **Figure 2:** In the DiffServ process flow, packets are classified and then conditioned to ensure that they conform to the classification's particular policy requirements. Classification is based on Traffic Conditioning Agreements (TCAs) between service providers and network subscribers. Conditioning, also called traffic shaping or policing, involves metering, marking, shaping, and dropping.

Conditioning involves metering, marking, shaping, and dropping. A meter monitors traffic based on how the traffic is classified. It determines whether classified traffic is meeting the specified traffic profile, and can also gather statistics on flows for accounting and billing functions.

A marker sets the packet's DS byte at a specific value. Packets may be marked for a particular flow to help ensure the proper sequence of PHBs for that flow. The marker can also be used to re-mark packets (for example, to change the existing value of a DS field), or to demote packets that fall outside the traffic profile. Packets that travel through a series of different domains may need to be re-marked repeatedly to ensure that they're in compliance with the various traffic profiles located at multiple domain boundaries. (A domain is a set of nodes that operates under a common set of service provisioning policies and PHB definitions.)

The shaper function keeps packets in a queue to ensure that traffic is in accordance with the traffic specification, often storing large bursts of packets until they can be safely released into the network.

When a flow violates the traffic specification, excess packets may simply be dropped from that flow. Alternatively, these packets may be delayed or reduced in priority level. Such actions are taken when the flow exceeds the negotiated transmission rate, or when a burst exceeds a maximum limit.

A key aspect of DiffServ implementation is determining which packets should receive forwarding priority. There are two primary types of forwarding: Expedited Forwarding (EF) and Assured Forwarding (AF). EF provides minimal delay, jitter, and packet loss, and assured bandwidth. In EF, the arrival rate of packets at a node must be less than the output rate at that node. Packets that violate the traffic profile are dropped or delivered out of sequence. EF is suitable for delay-sensitive applications such as voice and video.

In AF, there are four classes, each containing three drop precedences and allocated certain amounts of buffer space and bandwidth. Drop priorities are assigned to help determine which packets should be dropped during congested periods. Out-of-profile packets are dropped according to drop precedence; those with the highest drop priority are dropped first, and those with lower drop priorities follow.

ENTERING THE DIFFSERV DOMAIN

In DiffServ, SLAs between network subscribers and service providers establish policy criteria and define traffic profiles. Traffic is typically classified and conditioned at the network ingress interface, but when traffic spans multiple domains, some of these processes may be performed at the network egress interface, or at interior DiffServ-enabled nodes. In some cases, out-of-profile packets may be forwarded at an extra charge to the sender. Policy criteria may include source and destination addresses, port numbers, and time of day.

The Internet and some enterprise networks contain multiple domains. Bandwidth provisioning across multiple domains can be challenging, if the desired level of end-to-end QoS is to be achieved. The profile of traffic that crosses domain borders is specified in the SLA that exists between two domains. But as traffic between the two domains increases, the need to adjust traffic profiles often also increases, creating the need for more flexible resource allocation.

Enter the Bandwidth Broker (BB). The BB starts an end-to-end call setup to other BBs across the desired path, enabling resource negotiation among multiple domains. BBs perform admission control, policy control, and reservation tracking and aggregation. BBs can facilitate reservation requests between an enterprise network and its users, or between the network subscriber and service provider.

DOUBLE DUTY

DiffServ has the potential to support multicast traffic, but some traffic estimation issues must be resolved before this can occur on a widespread basis. The fact that multicast group membership can change so frequently and quickly makes it difficult to gauge the amount of traffic potentially involved in a multicast session—a must for ensuring reliable multicast transmissions. In addition, a multicast distribution tree may have one ingress point but many egress points, which also complicates traffic estimation. (For an overview of some of the IETF's efforts in this area, see http://search.ietf.org/internet-drafts/draft-bless-diffserv-multicast-01.txt.)

Work is also underway to determine how DiffServ could best be used with RSVP. One promising approach is using RSVP at the network edge and DiffServ in the core.

The two technologies have some very complementary properties that could make this a compelling combination. RSVP excels at per-flow resource management, but isn't highly scalable. Also, because it's inherently more complex than DiffServ, it's recommended that RSVP not be used on most backbones, as it can impair performance there.

DiffServ, on the other hand, has limited resource management capabilities but is more scalable than RSVP. Thus, using DiffServ in the network backbone and RSVP at the network edge could represent a very efficient approach in the quest to sustain desired levels of QoS across the network.

The IETF has also proposed measures for refining the use of DiffServ with Multiprotocol Label Switching (MPLS; see http://search.ietf.org/internet-drafts/draft-gant-mpls-diffserv-elsp-00.txt). DiffServ traffic can be mapped to MPLS processes, but this can involve some relatively complex resource allocation schemes and label assignment procedures.

Despite these issues, DiffServ remains a promising candidate for use in combination with other QoS architectures. The primary challenge will lie in achieving interoperability with other technologies, particularly when it comes to specifying and maintaining SLAs and TCAs.

RESOURCES

The following books contain useful overviews of Differentiated Services (DiffServ) and other QoS architectures:

Internet Performance Survival Guide, by Geoff Huston, John Wiley & Sons (ISBN: 0-471-37808-9)

Designing Quality of Service, by Eric D. Siegel, John Wiley & Sons (ISBN: 0-471-33313-1)

A white paper entitled "Internet QoS: A Big Picture," by Xipeng Xiao and Lionel M. Ni, contains an informative section on DiffServ. See www.cs.columbia.edu/~zub/myloral/qos/netmg/ qos.pdf.

This tutorial, number 158, by Elizabeth Clark, was originally published in the September 2001 issue of Network Magazine.

SECTION V

Application Layer Protocols

Providing Internet and World Wide Web Services

Getting connected isn't as difficult as you might think.

The growth of the Internet has been an interesting reflection of the growth of networking in general. The first networks to be deployed in most companies were workgroup networks—islands of connectivity. They were of various types, and they weren't connected to each other.

As networking technologies matured, and networking took on greater importance in many organizations, the workgroup networks grew and often became interconnected. The next step was enterprise-wide networking, and it wasn't long before companies began to deploy e-mail across those enterprise networks.

Today, many companies have full internal networks in place, and the growth of the Internet signals a continuance of the networking trend. Companies are now connecting via the Internet to their trading partners and prospective customers, much as the early workgroup networks interconnected to form a larger corporate network where different departments could collaborate on projects.

The Internet is also mirroring another trend: Just as we've seen microcomputers and workstations move from text-oriented, command-driven operating systems to graphical user interfaces, so too have services on the Internet shifted from the terse command-line types to the graphical World Wide Web.

A consequence of this shift toward the Internet is that network managers are often being asked to set up Internet connections and World Wide Web sites. This Tutorial is the first in a series designed to introduce network managers to the Internet and Web technologies.

INTERNET SERVICES

The World Wide Web steals the lion's share of attention lately, but there is actually a wide variety of services available on the Internet. "Internet Services," gives a brief description of some of the key services available on the Internet.

In the early years, the Internet was mostly used for electronic mail and for exchanging files between computer systems. These applications tended to be textual and command-line oriented, which means that the Internet was, at that time, mostly used by the "initiates." The Internet didn't really open up to the masses until just a few years ago, when the World Wide Web—an application with a graphical interface—was deployed.

THE WORLD WIDE WEB

What is the Web? It has many aspects, which makes it difficult to describe in just a sentence or two. I'll give you a sentence, but then I'll need several paragraphs to elaborate: The World Wide Web is a client-server system for delivering information in hypermedia form.

The medium of the Web is the Hypertext Markup Language (HTML). HTML is essentially a page description language, similar to Adobe Systems' Postscript or Hewlett-Packard's PCL (Printer Control Language). HTML tells the Web browser on the user's PC how to display the text and graphics that represent the content of a particular Web site. A *Web browser* is an HTML interpreter that requests and receives HTML-coded documents from a Web server and displays the information according to HTML commands embedded in the code.

The server component of this client-server system is a computer running software that operates according to the hypertext transport protocol (http). The Web server responds to users' Web browsers by sending the files the browsers request.

In most cases, a Web server delivers a document one page at a time. (Of course, that page can be much longer than the height of your display screen—you may have to scroll through several screens to see the entire page). These documents are *hypertext*, much like the Windows Help system. Certain key words are *hyperlinks*. Usually, the browser will indicate hyperlink text by underlining it and displaying it in a different color than the rest of the text. (Images can also be hyperlinks.) When you click on a hyperlink, it causes the browser to issue a request for the HTML document associated with that link. The Web server will then service that request. In Web lingo, each request for a file (text document or graphic image) is called a *hit*.

One difference between hyperlinks in HTML and those in other hypertext systems, such as Windows Help, is that HTML hyperlinks can take you to an entirely different server. These hyperlinks, in effect, make the Web one giant document management system, which explains how the World Wide Web got its name. Published on a Web server, other Web sites or Web documents referenced within this article could be made into hyperlinks; a reader could jump to each reference with just a click of the mouse.

Web servers are attractive electronic publishing systems. In the past, the only way to publish something on the Internet—and ensure that everyone could read it—was to present it in plain ASCII text. Richer formats, such as text displayed in a particular font, size, or style (italics, for example) were word processor-specific. Graphics, too, require specific viewer programs, which the reader may or may not have had. These factors hindered the presentation and effectiveness of electronic publishing. Enter the World Wide Web.

The Web has given us a level of platform independence. It's somewhat similar to having a videocassette that can be played on a wide variety of videocassette recorders, regardless of the vendor. Standardization in http and HTML means that any Web browser can read any Web document (at least, in theory). As HTML develops, vendors tend to add extensions that add new features to HTML or make life easier for its coders. Not every Web browser can read every proprietary extension, so certain features might not work with all browsers. It's still true, though, that if you stick with base-level HTML and avoid proprietary extensions, almost any browser will be able to read and display your documents. Of course, you need a Web browser running on your computer, and it's safe to say that there are now browsers for almost every type of computer.

Web servers were developed to reside on the Internet, but there's no reason you can't use one on any other TCP/IP network, large or small. This has given rise to the idea of corporate *intranets*—networks that are completely contained within the organizations they serve. Figure 1 shows an example of an intranet, as well as a connection to the Internet. Everything behind the firewall (that is, everything within the dashed lines) is the corporate intranet.

The concept of the Web as a platform-independent, client-server system is tantalizing to developers. Not only is platform independence a nice feature to have on the Internet, it's effective for the intranet as well. Companies such as IBM's Lotus Development are bringing out Internet interfaces for their client-server systems (Lotus Notes, in this case).

Typically, whenever you revise (rev) a client-server system, you have to develop both a new client piece and a new server piece. The amount of work needed on the client side is multiplied several times if you're trying to provide clients for several different operating systems. However, if you use a Web browser as the client, there's no work to be done on the client side whatsoever—you can simply let companies such as Netscape Communications (Mountain View, CA) or Spyglass (Naperville, IL) provide the browsers.

WAN links—at least those that most companies can afford—are typically very restrictive in terms of data throughput when compared with LAN links. Most people consider a T1 line (1.544Mbps) a high-speed link, but it crawls in comparison to 10Mbps Ethernet. For this reason, you must carefully plan the graphic design of your Web pages. Keep graphics small, and never put more than a few on each page, or else your readers will be staring at the Windows hourglass icon for minutes at a time.

As bad as this problem can be in the wide area, it disappears for intranets due to the tremendous throughput of local area networks. If you're going to strictly dedicate a Web site as an intranet server, you can afford to go hog wild with graphics. Ironically, Web servers, which were born on the Internet, seem to be realizing their full potential on the intranet.

If there's a fly in this soup, it's HTML, which is essentially a document publishing system. HTML is read-only and as such, it is not interactive, although you can request new pages by clicking on hyperlinks. There are ways around this, as we'll explore later in the series, but Web designers must really bend over backward to compensate for the one-way nature of HTML.

Internet Services

Many ways to surf the Net.

The World Wide Web is only one of the many services available on the Internet. Here's a brief synopsis of ten Internet services:

archie Archie servers catalog the names of files residing on many Internet *ftp* sites and index keywords about those files. Using archie, you can obtain a list of files that match your keyword, as well as the *ftp* server where each file is located. Once you know which file you want, you use *ftp* to fetch it. An archie search can save you a tremendous amount of work because you don't have to log in to hundreds of hosts and search each one individually.

Electronic mail (e-mail) Internet mail uses the Simple Mail Transport Protocol (SMTP) to transport e-mail messages across the Internet.

file transfer protocol (*ftp*) *Ftp* lets you copy files from one computer to another or across a network (the Internet, for example). In most cases, you're required to log in to the remote computer before you can obtain access to any of the files. Some systems, however, are meant to offer files to the public. For this purpose, anonymous *ftp* exists, wherein you log in with the user name "anonymous," and your IP address serves as your password.

gopher Gopher is an easy-to-use, menu-oriented search tool. Gopher servers catalog information by subject area, and the menu structure lets you "drill down" to successively more specific topics. Gopher includes a plain-text viewer, which enables you to view individual files (if they're text-only) so you can determine whether those files are what you're looking for. Gopher will fetch the file for you, saving you from the need to use *ftp* to retrieve the file. Gopher sites are interconnected, such that selecting a particular menu item may leapfrog you to a different gopher server. Gopher was developed at the University of Minnesota, where the "mother gopher" still resides.

Network news You can post a message on a particular topic, and it will be widely disseminated to a distribution list of subscribers. These topic-oriented BBSs are known as newsgroups. The underlying messaging protocol used is the Network News Transport Protocol (NNTP).

telnet This is a terminal-emulation program that runs on your PC and emulates a terminal for some host computer. A key difference between telnet and earlier terminals is that while

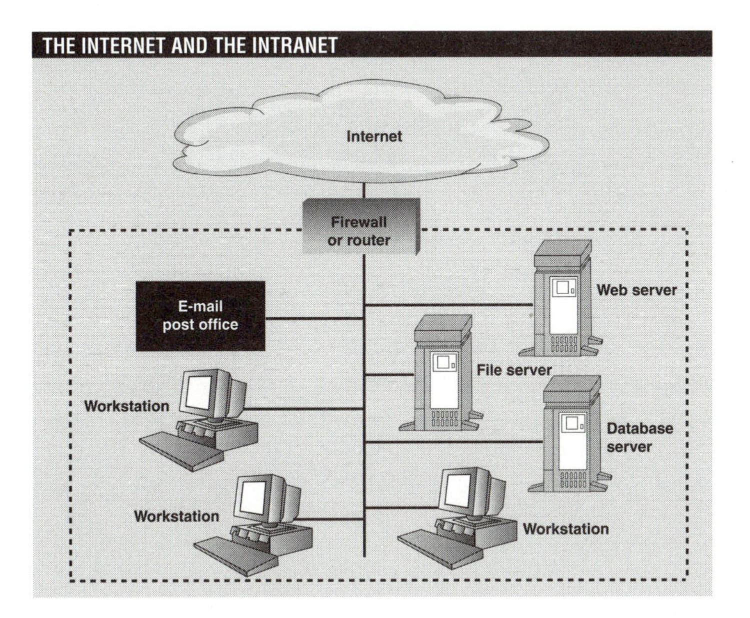

■ **Figure 1:** A typical corporate Internet connection. Everything behind the firewall is part of the corporate intranet which is the organization's private network. The dashed line encompasses this particular intranet.

terminals originally used RS-232 serial connections or some other type of terminal cable to connect to the host computer, telnet uses the network to make the link.

veronica The veronica system indexes the menus of all of the gopher servers. The collection of all the menus in all the gopher servers is known as "gopherspace," and veronica gives you a powerful way to search all of gopherspace for the subject in which you're interested.

Wide Area Information Service (WAIS) Instead of indexing file names (as archie does, for example), WAIS indexes the text within the files, allowing you to find information that might not be stored in file names.

World Wide Web The World Wide Web is a networked, graphically oriented, hypermedia system. It uses the hypertext transport protocol (http) and the Hypertext Markup Language (HTML).

This tutorial, number 94, by Alan Frank, was originally published in the June 1996 issue of LAN Magazine/Network Magazine.

Hypertext Markup Language

Writing HTML code isn't as difficult as you might think.

"Providing Internet and World Wide Web Services" was the first of a multipart series explaining Internet and World Wide Web services. In this tutorial, we'll take a closer look at HTML and examine some actual HTML code from a sample Web page.

There are specialized HTML editors, but HTML code is plain ASCII text and, therefore, can be created using any text editor that can save a file in ASCII text format. Although Web browsers are typically used to access HTML pages from a Web server, most Web browsers also load files from your local disk drive, which allows you to test some simple Web pages without having a Web server up and running.

When creating HTML files, remember to name them using the .html extension. If your Web server is based on Windows 3.x and can only work with file names in the MS-DOS "8.3" format, use the extension .htm.

ROCKET SCIENCE

Listing 1 displays HTML code for "a sample home page" of a mythical company called Retro Rocket, which manufactures the RetroRocket product line. I named the document retro.htm.

HTML uses tags to tell the Web browser how the text should be displayed. In Listing 1, the code begins with the tag <HTML>. Place this tag at the beginning of your HTML pages; it lets the Web browser know that the following code is HTML and should be rendered accordingly. Note also that Listing 1 ends with the tag </HTML>. Most HTML tags are used in pairs, with an opening tag and an ending tag to delineate which text you want handled in a particular way. Ending tags are the same as opening tags, but with the addition of a forward slash (/).

Web browsers can identify an HTML file by the file extension, and many browsers will display pages that don't carry the <HTML> tag. When viewing your document with your own browser you might be able to get away without the tag, but you can't be sure that everyone who wants to view your page will be using the same browser you use. It's best to stick with good programming habits and to avoid the temptation to take shortcuts.

Following the opening HTML tag in Listing 1, there are three comment lines delineated by <! and >. The browser doesn't do anything with comment lines—comment lines are not displayed. A comment's only purpose is to enable you to document your code.

HTML pages consist of a head and a body portion. In Listing 1, the head is that portion of text between the <HEAD> and </HEAD> tags. In this example, there's one item in the head, and that's the title. The title is delineated by (you guessed it) the <TITLE> and </TITLE> tags. The title tags indicate what will be displayed in the title bar of the Web browser.

Everything that shows up on the Web page itself is found in the body of the HTML document. (I'll let you figure out which tags delineate the main body of the HTML page.)

Any text in an HTML document that doesn't have specific HTML tags bracketing it will be displayed by the Web browser as body copy. In Listing 1, the paragraph that begins with "Welcome to the RetroRocket World Wide Web site" is body copy and is displayed in the font and point size that has been determined for the browser.

Web developers can't live by text alone, so I included a graphic image—the RetroRocket Co. logo—on the sample page. Web browsers can display files in the .gif and jpeg file formats. I created the sample logo using the Windows Paintbrush program, then converted the .bmp file to .gif format with a graphics conversion program.

In Listing 1, I referenced the image with the HTML statement <IMG ALIGN=bottom SRC="retro.gif">. This statement is one that does not use beginning and ending tags—it's all one statement. The SRC="retro.gif" portion of the statement tells the Web browser which file to request. In this case, the .gif file (retro.gif) is located in the same directory as the retro.htm file. If it were in another directory, I would need to enter the file's full path name. If the file were on another Web server, I would need to give the full URL, including path extensions, such as SRC="http:// www.lanmag.com/images/retro.gif".

In addition to body text, HTML lets you have several subhead text styles, ranging (in descending order of point size) from H1 to H6. I used the H1 headline style for the main headline (Retro), which I positioned next to the graphic. If you look at the HTML image statement in Listing 1, you'll note that it is bracketed by the <H1> and </H1> tags, and the word *Retro* immediately follows the image statement. The H1 tags cause all the text between those tags (the word *Retro* is all there is) to be rendered in Headline 1 style—the largest headline size. The ALIGN=bottom portion of the image statement aligns the bottom of the text with the bottom of the graphic. (You can also use ALIGN=top or ALIGN=middle if you want a different text position. ALIGN=top aligns the top of the text with the top of the image, while ALIGN=middle aligns the middle of the text with the middle of the image.)

The line following the image statement ("Your best buy in a luxury spacecraft.") is set as Headline 6—the smallest head.

LISTING AND LINKING

I have already mentioned the body copy ("Welcome to the RetroRocket World Wide Web site."), so I won't belabor it, except to point out the <P> tag I used at the end of the paragraph. This tag tells the Web browser that it has reached the end of a paragraph. The Web browser will insert a carriage return followed by a blank line wherever you place a <P> tag in the document.

You can use carriage returns and blank lines when writing your HTML code to make it easier to read and understand, but Web browsers will ignore them. This means you will have to use HTML tags to force carriage returns and blank lines to appear in the displayed pages. The Web browser will automatically enter a carriage return at the end of a headline, but in most cases, you need to specifically call for carriage returns or blank lines in your HTML code.

Following the body text, you'll notice three bulleted items ("RetroRocket Product Line," "Dealer Listings," and "History of Retro Rocket Co."). I decided to make these three items hyperlinks to other documents, but the list feature is not necessarily tied to hyperlinking; I simply decided to combine the two features, as you will likely do in many cases.

Let's discuss lists first. In general, there are two broad categories of lists in HTML: ordered lists and unordered lists. Ordered lists are numbered lists. You would most likely use an ordered list when items must be displayed in a specific order. One example might be steps you need to follow to assemble some product: Step 1 comes first, then Step 2, and so on. You create ordered lists by surrounding them with the <OL> and </OL> tags. Then, at the beginning of each item in the list, insert the <LI> tag (for "list item"). The Web browser will automatically take care of numbering the ordered list, starting with "1" for the first list item.

The second type of list is the unordered list, which is what I used in the sample. This list is not numbered ("ordered"). It is also called a bulleted list, as each list item is preceded by a bullet. You create an unordered list by surrounding the collective list items with the <UL> and </UL> tags. As before, you need to place the <LI> tag at the beginning of each list item.

As I mentioned, each list item is also a hyperlink, so now let's look at the creation of hyperlinks. A hyperlink, also known as an "anchor," takes the following form: <A HREF= *"file name"*>*displayed text*</A>. Here, "file name" is the name of the document to which you want to link and Displayed Text is the underlined wording that you want the reader to see. (I should note here that by convention, text representing hyperlinks is usually displayed in a different color from regular text, and the hyperlink text is underlined. It is the Web browser that does this displaying, and these displays are usually configurable options, so the end result will vary according to the browser and user choices.)

If you look at the HTML code for the first of the three list items, you'll note that the opening tag (<A HREF="rockets.htm">) gives the file name of the document to which we are linking, while the second portion ("RetroRocket Product Line") is what the reader sees when looking at the hyperlink via his or her Web browser. In this case, I didn't need to reference a complete URL, or even a path, as the file rockets.htm is in the same directory as the current document (retro.htm).

Just below the three list items in our sample document, you will see the tag <HR>; it calls for a horizontal rule.

Finally, I added a line that asks for readers' comments, and I included the e-mail address of the Webmaster at retrorocket.com. There's even an <ADDRESS> tag in HTML, which my browser (Netscape Navigator 2.0) converts to italics.

In the tutorial "HTML and CGI," we'll continue our discussion of HTML. In the meantime, fire up those text editors and Web browsers and try a few experimental Web pages!

LISTING 1—A SAMPLE WEB PAGE

```
<HTML>
<! This is a sample HTML file, to show you what's behind a Web>
  <! page. This line, the one above, and the line below this>
  <! are comments. Comments don't show up in the displayed page.>
<HEAD>
  <TITLE>Retro Rocket Company </TITLE>
  </HEAD>
<BODY>
<H1><IMG ALIGN=bottom SRC="retro.gif">Retro</H1>
  <H6>Your best buy in a luxury spacecraft.</H6>
  Welcome to the RetroRocket World Wide Web site. This is the place
  to check out the latest in personal spacecraft technology.<P>
<UL>
  <LI><A HREF="rockets.htm">RetroRocket Product Line</A>
  <LI><A HREF="dealers.htm">Dealer Listings</A>
  <LI><A HREF="history.htm">History of Retro Rocket Co.</A>
  </UL>
<HR>
Comments? Please e-mail to:
  <ADDRESS>webmaster@retrorocket.com</ADDRESS>
</BODY>
  </HTML>
```

This tutorial, number 95, by Alan Frank, was originally published in the July 1996 issue of LAN Magazine/Network Magazine.

HTML and CGI, Part One

The common gateway interface is the key to interactive Web sites.

This series of tutorials, which began with the "Providing Internet and World Wide Web Services," has examined services commonly available on the Internet and has explored what you, as a network manager, need to do to make such services available within your company (via your corporate intranet) and to others (via the Internet).

In "Hypertext Markup Language" we focused on HTML and took a first look at hypertext links. This month, we'll continue to explore HTML, and we'll take a look at the common gateway interface (CGI).

Using simple tags, you can make any word or phrase a hyperlink. Hyperlinks are not limited to text, however. They can also be images. The following code fragment shows how to make an image into a hyperlink:

```
<A HREF="http://www.amphibian.com/frog.htm"><IMGSRC="frog.gif"></A>
```

This sample code displays the graphic image frog.gif, and—if the user clicks on that image—causes the Web browser to request the Web page frog.htm from the Web server www.amphibian.com.

THE COMMON GATEWAY INTERFACE

HTML, along with HTTP, is the enabling technology of the World Wide Web. HTML's most compelling feature is hypertext linking. When you look for information in a conventional book you might search the index for a key word, then flip through the book to find the referenced page, and hope to find the information you seek. By contrast, when searching for information on the Web, simply clicking on a hyperlink and immediately jumping to the referenced page is gratifyingly quick and easy. HTML is, however, essentially a read-only format. It's not interactive except when using the mouse to navigate. HTML pages are static documents that don't change until someone (the Webmaster, in many cases) accesses the Web server and edits them.

From the start, Web developers have been looking for ways to make Web servers interactive—to allow users to input information as well as retrieve information. The standard way of delivering information in this manner is now known as the common gateway interface. Using CGI, the Web server, in response to a user request for information, runs a program to search for that information and returns the results to the user. In order to achieve these results, the Web server typically generates an HTML page by inserting a few HTML statements into a preexisting page. A Web server using CGI in this manner could potentially run an infinite variety of programs.

If you've used a Web search engine to find an item of interest, you've used CGI. But CGI itself isn't the search engine. Rather, it's a way to interface programs, such as search engines, with Web servers.

HTTP (Web) servers are designed primarily to serve up HTML documents. CGI files, however, aren't documents, they're programs. Therefore, to store CGI programs, most Web servers use a special directory, commonly named cgi-bin. The Web server knows that files stored in the cgi-bin directory are to be executed rather than simply sent to the user's Web browser for display. CGI programs can be written in a wide variety of languages, including DOS batch files, BASIC, C, and scripting languages such as Perl. It's the job of the CGI to activate the CGI program at the proper time and pass the program to any necessary user- or operating environment-generated data. The CGI program then processes whatever data is generated. Once the program accomplishes this process (which should take only seconds—the goal here is interactivity), it must return some output to the user via the user's Web browser. To make the output appropriate for browser display, the CGI program must format its output in the form of an HTML document.

A FORM FOR EVERYTHING

A CGI program has to somehow get from the user any data to be processed. That's where HTML forms come into play. HTML forms are similar to paper forms: They provide specific spaces in which to enter specific data items. A simple form is shown in Figure 1. This sample form has one text input field: product name. The form also has some multiple-choice fields, in which, for example, the user can check off the type of product, as well as what networking protocols the product supports. The form also has a button the user can click on to submit the entry. (A CGI program will only act upon the information in a form when the form data has been submitted to the CGI program.) Clicking on the form's Clear Form button resets the form back to its default values. The HTML code for this form is shown in Listing 1. Note that to aid in discussion, Listing 1 includes line numbers, which you wouldn't use in actual HTML code.

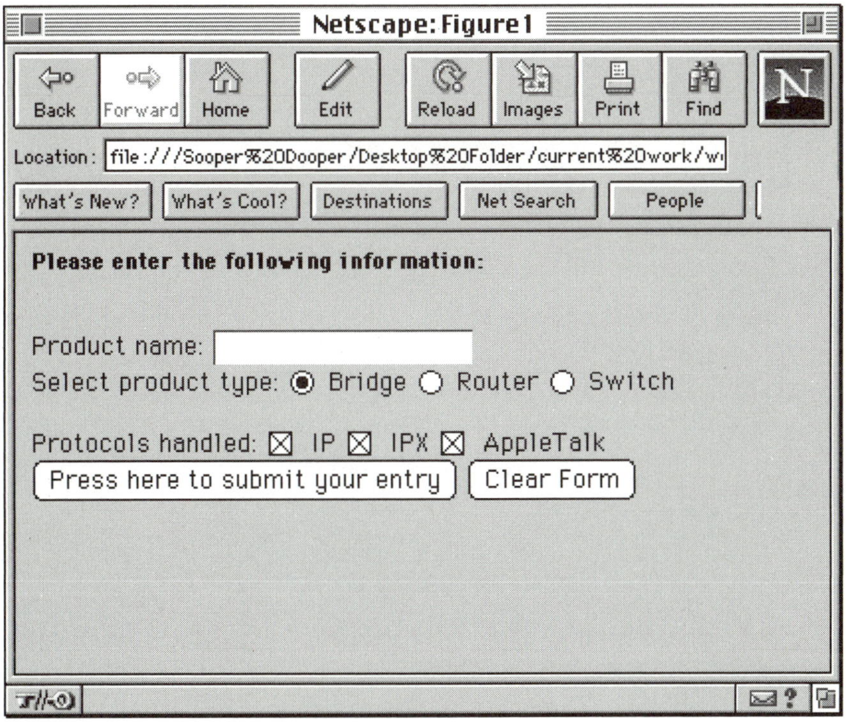

■ **Figure 1**

LISTING 1. HTML CODE FOR FIGURE 1

```
1   <FORM ACTION="URL" METHOD="GET">
2   <H5>Please enter the following information:</H5><BR>
3   Product name:
4   <INPUT TYPE="TEXT" NAME="NAME" SIZE="25" MAXLENGTH="30"><BR>
5   Select product type:
```

```
6  <INPUT TYPE="RADIO" NAME="ProdType" VALUE="Bridge" CHECKED> Bridge
7  <INPUT TYPE="RADIO" NAME="ProdType" VALUE="Router"> Router
8  <INPUT TYPE="RADIO" NAME="ProdType" VALUE="Switch"> Switch<P>
9  Protocols handled: <INPUT TYPE="CHECK BOX " NAME="IP" CHECKED> IP
10 <INPUT TYPE="CHECK BOX " NAME="IPX" CHECKED> IPX
11 <INPUT TYPE="CHECK BOX " NAME="AppleTalk" CHECKED> AppleTalk
12 <INPUT TYPE="SUBMIT" VALUE="Press here to submit your entry">
13 <INPUT TYPE="RESET" VALUE="Clear Form">
14 </FORM>
```

The source code for a form is bound by the <form action="url"> and </form> tags (see line 1 and line 14 of Listing 1). The use of action specifies the action to be taken when the form is submitted. In other words, action determines which CGI program should be run in order to process the form, and you specify either a full or partial URL for that program. Depending on the Web server, you may be able to set a default directory for all CGI programs. As long as the program you want to run is stored in that directory you need only specify the program's file name. Otherwise, you will probably have to provide a complete URL for the file.

Use method to move input data from the form into the CGI program that's going to process that data. We'll explore the methods by which this is done in more detail in the tutorial "HTML and CGI, Part II," but for now, just be aware that there are two general methods you can use: get and post. If you're going to use the get method to transfer data from the form to the CGI program specified in the action URL, be sure that the CGI program is written to accept data in this manner. By the same token, if you use post to send data from your form, be sure the CGI program is expecting post data input.

Line 2 and line 3 of Listing 1 prompt the user to enter data. After the prompts comes the code that puts a text field into the form (line 4). Entering input type="text" tells the Web browser that it is a text field. Specifying name="name" tells the browser that the name of this field is name. (This example asks for the name of a product; if you want to ask for the user's address, you might name this field address.) size="25" indicates that the text box will be large enough to see 25 characters at a time. maxlength="30" means that the field can hold up to 30 characters of text. In this case, the length of the field exceeds the length of the box, so a scroll bar will appear if the user types a long text string.

Next on the form are radio buttons, followed by check boxes. These are both multiple-choice, but they differ slightly in how they work in that, using check boxes, you can check more than one box at a time, but you can select only one radio button. Clicking on a radio button clears the radio button that was previously selected. By comparison, selecting one check box does not deselect another.

Let's discuss radio buttons first. You determine the button style by setting the type for the input equal to radio. There are three statements for radio buttons (line 6 through line 8)—one for each of the possible selections. Each statement contains name="prodtype." This creates a logical field named prodtype. If the first button is checked, the field prodtype takes on the value bridge. (You can determine the value by using the statement "value='bridge'"). Note also the keyword checked in line 6. When the form first appears, the radio button for bridge will be checked; in other words, we've made this setting the default. The user can override this default setting by clicking on the Router or Switch button.

Next comes a series of three check boxes that allow the user to indicate which network protocols the product supports (represented by lines 9 through 11of the sample code in Listing 1). When a user clicks in a box, "X" appears in that box to indicate that it is checked. Clicking again on a checked check box removes the "X," or unchecks the box.

You create a check box field with the statement input type= "check box ." You also need to specify the name of the input field. In Listing 1, the name of the input field for the first check box is set to IP. The sample code uses the keyword checked, indicating that this box is selected by default, which the user can clear by clicking on the check box.

When a check box is selected, by default the field with that name has its value set to "on." Although it is not included in the example, you can optionally use the value attribute to set the value, for example, to "green" when the box is selected. When the check box is not selected, the field has no value. In this example, the default setting for all three boxes is checked. Thus, unless the user clears the boxes, the form returns ip=on, ipx=on, and appletalk=on. If only the IP box is selected, the form returns only ip=on. The IPX and AppleTalk boxes will be disregarded.

I don't have quite enough space here to finish the discussion of our sample form, so we'll have to conclude that in the tutorial "HTML and CGI, Part Two." Also, so far, I've discussed gathering data to be passed to a CGI program, but I haven't said much about how that data is transferred or how the CGI program sends output back to the user. I'll cover that (and more) in the tutorial "HTML and CGI, Part Two."

RESOURCES

Using HTML, by Neil Randall (Que, ISBN: 0-7897-0622-9)

This book is an excellent overall resource for those who are just learning HTML. Written in an engaging, easy-to-comprehend style, it is liberally sprinkled with examples of HTML code and the screen images they generate.

CGI Programming on the World Wide Web, by Shishir Gundavaram (O'Reilly & Associates, ISBN: 1-56592-168-2)

Gundavaram's book is the most thorough treatment of CGI programming I've found. It's augmented by a prodigious number of examples, which are mostly written in the Perl scripting language (the most commonly used language for CGI programs).

One of the best ways to get information about Internet services such as the World Wide Web is the Web itself. I've listed a few URLs here, but there are plenty more to be found. Submitting the phrase "common gateway interface" to Digital's AltaVista search engine (http://www.altavista.digital.com) yields a huge number of potential links.

http://hoohoo.ncsa.uiuc.edu/cgi

http://www.ncsa.uiuc.edu/demoweb/url-primer.html

http://www.ncsa.uiuc.edu/SDG/Software/Mosaic/Docs/fill-out-forms/overview.html

This tutorial, number 96, by Alan Frank, was originally published in the August 1996 issue of LAN Magazine/Network Magazine.

HTML and CGI, Part Two

Sending user input to a CGI program.

This is the fourth installment in a series of tutorials about providing Internet and intranet services. In the tutorial "HTML and CGI Part One," we began to explore HTML forms, which make it possible to collect information from users and route it to a common gateway interface (CGI) program. I showed a simple HTML form, along with the code that makes it work. I didn't get all the way through my discussion of the sample form before running out of column space, so let's conclude that now, before moving on to CGI.

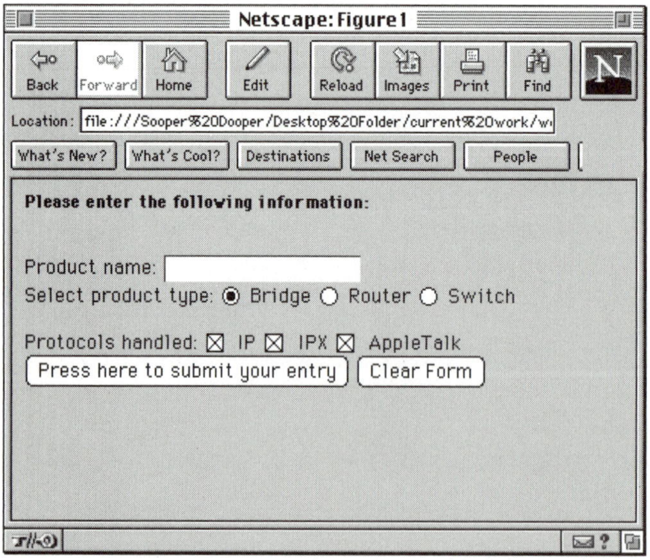

For your reference, I've reproduced the code listing for the sample in Listing 1. If you'd like to see how this listing looks when displayed by a Web browser, see Figure 1.

I need to correct a statement I made about the listing last month. In discussing the text field, line 4 of Listing 1, I mentioned that if the maximum length you specify for the field (the maxlength option) exceeds the displayed size of the text box (the size option), a scroll bar will appear when the user types a long text string. After further testing, I've found that what

■ **Figure 1**

you actually get is a scrolling text box, not a scroll bar. For example, if you have set size to 25 and maxlength to 30, the text begins to scroll once you type in more than 25 characters. Once you exceed 30 characters, the text box will no longer accept keystrokes, and you'll get beeps instead.

Also in Listing 1 last month, the statements for creating check boxes used the syntax input type="check box" with a space between check and box. To get this to work correctly, you must specify the input type as one word, in this case checkbox. Listing 1 as displayed here is correct.

Another modification I've made to Listing 1 is to place a specific URL into the form's action statement (line 1). This lets me show more detail about how information gathered by the form gets turned over to a CGI program for processing.

FINISHING OFF THE FORM

Last month's discussion covered line 1 through line 11 of Listing 1, so let's take up with line 12, which creates a Submit button. When the user finishes filling out the form, he or she then clicks the Submit button to submit the completed form to a CGI program running on the Web server. As you can see from line 12, you create a Submit button by setting the input type to submit. You apply a label to the button by setting value equal to a text string.

When the user clicks on the Submit button, the Web browser takes whatever action you specified in the action statement for the form (for example, line 1 in Listing 1). In this example, the Web server will execute the Perl script named example.pl, located in the cgi-bin directory.

How does the Web server know that it should execute the referenced program, rather than simply delivering it to the requesting Web browser? It knows by virtue of the directory that the program is stored in cgi-bin. On most Web servers, this directory is reserved specifically for executable files. (You may have other directories that hold executable files, but cgi-bin should hold all of the files that can be remotely executed via the Web.)

Keeping all executable files in cgi-bin is also a good security measure. As Webmaster, you should allow the Web server to execute programs only if they are located in this directory. Also, thoroughly scrutinize and test any programs you place in this directory to ensure that they cannot be used to damage or replace files or circumvent your security measures. If, for example, someone can use one of your programs to read system password files, your system security is obviously at risk.

In my example, I mentioned Perl, the Practical Extraction and Report Language. For those unfamiliar with Perl, it's an interpreted language that was originally developed for use on Unix systems, but has since been ported to many other operating systems. Perl interpreters are available for Macintosh and Windows NT systems, for example. Perl programs are usually referred to as scripts, as they are relatively quick and easy to dish up compared with C or other "full-blown" programming languages. Perl is much closer to DOS batch file programming than it is to the more classical programming languages. One area where Perl excels is in string manipulation, which—as we'll see—is very important in CGI programming.

Line 13 of Listing 1 creates a Reset button. If the user clicks on this button, all values will be reset back to their original default settings. In this example, the radio button for Bridge is checked, as are all three of the check boxes (IP, IPX, and AppleTalk), so the form would revert back to these settings.

When the user clicks on the Submit button, the information will be transmitted to the Web server using pairs of field names (also called keys) and values. For example, suppose the user had entered the name SuperDuper as the product name in our sample form, checked the Router radio button, and checked IP and IPX for protocols handled. In this case, when the user clicks the Submit button, the form will be submitted with name=SuperDuper, ProdType=Router, ip=on, and ipx=on. (If the AppleTalk check box is unchecked, you won't get "AppleTalk=off"—AppleTalk will simply not be returned as a field.)

We've seen how HTML forms collect information from a Web user and submit it to a CGI program as key/value pairs. Because the action specified in the opening <form> tag references a file located in the cgi-bin directory, the Web server knows that the referenced file should be executed as opposed to being displayed by the requesting Web browser, as it would be with a typical HTML document.

GETTING TO KNOW CGI

Writing CGI programs is only slightly different from writing other types of programs. At the risk of oversimplifying, I'll characterize programming as obtaining input from a user or from a data file, storing that input in program variables, manipulating those variables to achieve some desired purpose, and sending the results to a file or video display. If you're a programmer, you know that programs typically get their input from the logical device known as standard input (STDIN, for short), and send output to the device standard output (STDOUT). STDIN and STDOUT typically represent the computer console (keyboard and video display), but most operating systems support redirection, so STDIN and STDOUT could be disk files or other devices.

Many operating systems and programming languages also support the use of environment variables—variables that can be set in the operating system and read by programs, or vice versa. Environment variables allow information to be passed between the operating system and running programs or between programs written in different languages.

CGI programs are similar to regular programs. They typically get their input from STDIN or from environment variables and send output to STDOUT.

To understand how a user's Web browser and a CGI program interact, we need to take a step back and examine how a browser submits simple HTML requests and how a Web server responds.

Suppose you embed the following hypertext link in an HTML document:

```
<A HREF="TEST.HTML">
```

If you were to click on this link, the browser would issue the following request to the Web server:

```
GET /TEST.HTML HTTP/1.0
Accept: text/plain
Accept: text/html
```

Each of these lines is referred to as a header. The first one is the get header, which tells the Web server that the browser wants to get the document test.html, and that it's using version 1.0 of the Hypertext Transport Protocol. Because only the file name was specified in this case, the Web server defaults to looking for the file in the server's Web-document root directory. If you want to obtain a file that is located in a subsidiary directory, your hypertext link must specify the complete path name to the file, relative to the server's root directory for Web documents. If you want to reference a file located on another Web server, your hypertext link has to specify the complete URL for the new file.

Following the get header are two Accept headers, which state that the browser can accept plain text or HTML-formatted text files. If the browser can accept more data types, there will be more Accept headers, detailing each type in terms of Multipurpose Internet Mail Extensions (MIME).

The Web server's response to this request would look like this:

```
HTTP /1.0 200 OK
Date: Monday, 24-May-96
11:09:05 GMT
Server: NCSA/1.3
MIME-version 1.0
Content-type: text/html
Content-length: 231

<HTML>
<HEAD>
<TITLE>This is the document title</TITLE>
</HEAD>
This is a test HTML page.
</HTML>
```

The server's header gives the Web server name and version number, and the version of HTTP used. Other headers describe the content type (HTML-formatted text, in this case) and content length (231 bytes) of the material being sent. The Web browser then reads and executes the HTML portion of the file.

With CGI, things are not much different, except that the file being requested will be in the cgi-bin directory, which tells the Web server that the requested file is to be executed, instead of merely sent to the Web browser for display as an HTML document.

In the Tutorial "HTML and CGI, Part I", I mentioned that information gathered from users or their Web browsers can be sent to the Web server using one of two methods: the get method or the post method.

With the get method, all the form data is included in the URL in what's known as a query string. As an example, suppose we have a simple form that has only two fields, named color and size, and that the user typed sky blue and large, respectively, in response. Let's also assume that the CGI program that's going to process the data is a Perl script, named example.pl and located in the cgi-bin directory. When the user clicks on the Submit button, an HTTP request will be generated and sent to the Web server.

The code for our HTML form must contain an action statement, as well as tell the Web server which method (get, in this case) is being used to send data. Thus, the first statement for our form must read:

```
<FORM ACTION="/CGI-BIN/ EXAMPLE.PL" METHOD="GET">
```

This lets the Web server know the complete path name of the program to be executed ("/cgi-bin/example.pl"), and that the get method will be used. As mentioned earlier, the get method uses a query string to pass data to the CGI program. In this example, when the user clicks on the Submit button, his or her Web browser will send the following request to the Web server:

```
GET /CGI-BIN/EXAMPLE.PL? COLOR=SKY%20BLUE&size= LARGE HTTP/1.0
```

The continuous string of text that follows the question mark represents the query string. In response to this request from the Web browser, the server executes the script example.pl and places the string color=sky%20blue&size=large in the query_string environment variable. Your CGI program will then be able to read the query_string environment variable.

In the tutorial "CGI and Web Servers," I'll discuss what your CGI program must do to process the query string. I'll also cover the post method of submitting data to CGI programs.

LISTING 1-HTML CODE FOR A SAMPLE FORM

"Sample Form"

```
1   <FORM ACTION="/CGI-BIN/EXAMPLE.PL" METHOD="GET">
2   <H5>Please enter the following information:</H5><BR>
3   Product name:
4   <INPUT TYPE="TEXT" NAME="NAME" size="25" MAXLENGTH="30"><BR>
5   Select product type:
6   <INPUT TYPE="RADIO" NAME="ProdType" VALUE="Bridge" CHECKED> Bridge
7   <INPUT TYPE="RADIO" NAME="ProdType" VALUE="Router"> Router
8   <INPUT TYPE="RADIO" NAME="ProdType" VALUE="Switch"> Switch<P>
```

```
9  Protocols handled: <INPUT TYPE="CHECKBOX" NAME="IP" CHECKED> IP
10 <INPUT TYPE="CHECKBOX" NAME="IPX" CHECKED> IPX
11 <INPUT TYPE="CHECKBOX" NAME="AppleTalk" CHECKED> AppleTalk
12 <INPUT TYPE="SUBMIT" VALUE="Press here to submit your entry">
13 <INPUT TYPE="RESET" VALUE="Clear Form">
14 </FORM>
```

This tutorial, number 97, by Alan Frank, was originally published in the September 1996 issue of LAN Magazine/Network Magazine.

CGI and Web Servers

Fundamentals of the Common Gateway Interface.

In the tutorial "HTML and CGI, Part Two," I discussed HTML forms and how to use them to obtain input from Web users. I also began to describe how to use the common gateway interface (CGI) to turn the information supplied in those forms over to a program running on the Web server that can process the data and produce a response.

In this tutorial I'll provide a basic understanding of how CGI works and how it fits almost hand in glove with HTML forms techniques to make a Web site interactive. But before I proceed, a few disclaimers:

1. This tutorial is not an in-depth, how-to discussion of CGI. CGI is quite complex and one of the tougher Web server technologies to implement correctly. Complete coverage of the subject would require an entire book.
2. If you're new to Web servers, you might want to limit your first few Web site projects to using HTML only. The learning curve for HTML is much gentler than that of CGI. Trying to make your first Web site interactive is a little like having an introductory lesson in rock climbing, then deciding to tackle Yosemite's El Capitan as your first climb. There's a lot to explore in HTML, so wait until you have some experience under your belt before embarking on a CGI excursion.
3. This is not to paint CGI as an impossibly difficult subject: You don't have to be a gonzo C programmer to use CGI. If and when you feel ready for the challenge, you have several programming options from which to choose. A lot of CGI programming can be done with scripting-type languages, such as Perl or Tcl (tool command language). I experimented a bit with Santa Monica, CA-based Quarterdeck's WebServer 1.0, a Web server that runs on Windows 3.x and lets you use MS-DOS batch files as the scripting language. (Although simple to use, the DOS batch language has serious limitations as a CGI scripting language, so I don't recommend using DOS scripts for CGI.)

With so many caveats to consider, why would I even attempt to cover CGI in this brief format? If you run a Web site or if you are considering doing so, it's useful to be familiar with how CGI works—at least in general terms. For example, if you want to put together an interactive Web site for your organization, but you are repelled by the idea of programming, you might decide to outsource the development work. Knowing the broad outlines of how CGI works will make you a well-informed decision-maker.

If, by chance, this discussion whets your appetite and you want to learn more about CGI programming (I know, that may be stretching it), I recommend you take a look at the books and Web documents listed in "For More Information" and "In Cyberspace."

WHICH METHOD?

Disclaimers aside, let's dive into our subject. In the tutorial "HTML and CGI, Part II," I discussed two basic methods by which HTML forms can submit user data to a Web server's Common Gateway Interface: the get method and the post method. Let's pick up where I left off.

The get method passes user data to the gateway by adding the data to the requested document's path. (I use the term "document" loosely here, because most CGI programs create an HTML-formatted response on the fly. The response therefore is a dynamically generated page, or *virtual document.)*

When a user clicks the Submit button in an HTML form, the HTTP request generated by the form contains a string of text known as the *query string.* This string contains the user data that is submitted to the CGI program (see the tutorial "HTML and CGI, Part II" for an example). The other portion of the request contains the file name of the CGI program to be executed; this program processes the submitted data.

The gateway extracts the query string and places it in an environment variable named, appropriately enough, query_string. Now your CGI program can read that environment variable and process it accordingly. In general, you place that value into a temporary variable for further processing. Following the example from the tutorial "HTML and CGI, Part II,", a user might respond sky blue to the prompt for color and large for size. The value for the environment variable query_string would thus be, color= sky%20blue&size=large. All the data—including the names of the returned information fields and their values—is jumbled together, along with a cast of weird characters, such as ampersands (&) and percent signs (%). Placing the text string into a temporary variable lets you parse the information into the appropriate variables.

As you can see, parameters are passed to the CGI program as key/value pairs (such as color=sky%20blue or size=large). These key/value pairs are separated by ampersands. Because spaces and several other characters are not permitted in URLs, these characters are replaced by the percent sign followed by the hexadecimal ASCII value of the character. Thus, a space becomes %20, for example.

The order of the key/value pairs in the query string may vary depending on the user's Web browser; the number of returned pairs can vary, as well. As I pointed out in the tutorial "HTML and CGI, Part II," if the check box for AppleTalk is not checked, no key/value pair is returned. As a result, your CGI program should not assume that a set number or order of key/value pairs will be returned. Once your program has read the query_string environment variable and placed it into a temporary or working variable, you must parse the string for key/value pairs and convert special characters (such as spaces) from their hexadecimal representation to their original characters. You can then assign each value to a corresponding variable in your program.

There are some drawbacks to using the get method. First of all, it limits the amount of information that can be passed to the CGI program. Because the get method simply appends all the information to be passed to the URL, it can create an extremely long URL. As a result, the information you are attempting to send may be truncated by the Web server. Another drawback is that with the get method it is hard to set any rules that determine a "safe" length for the string because where the string is truncated depends on the Web browser and server you use. For these reasons, you may want to use the post method.

POST YOUR MESSAGES

If a Web browser uses the post method, the request will look similar to the following:

```
POST /cgi-bin/example.pl HTTP/1.0
Accept: www/source
Accept: text/html
Accept: text/plain
User-Agent: Content-type: application/ x-www-urlencoded
Content-length: 28

COLOR=SKY%20BLUE&SIZE=LARGE
```

Note the content-type header (`application/x-www-urlencoded`). HTML forms typically use this content type—when the form creates a request, it automatically sets this type.

With the post method, the server passes the information contained in the submitted form as standard input (STDIN) to the CGI program. Your CGI program needs to know how much data to read from STDIN. Fortunately, the browser's request includes a content-length header. When form data is transferred via CGI, the gateway sets an environment variable named `content_length` to report the amount of data being transferred. Thus, before it can read from STDIN, your CGI program must first read the environment variable `content_length`.

Although in this example we're transferring only a small amount of data (28 bytes), the advantage of using the post method is that it lets you send an unlimited amount of data.

So how does the Web server know whether data is being sent via the get method or post method? The first header in the browser's request specifies the type of request, which, in turn, sets another environment variable, `request_method`, accordingly. However, if you're developing an application using CGI, you'll likely develop both the client and server ends of the application, so you'll know (or will decide) which method both the requester and the server use. So, for example, if you program your HTML forms to submit data using the get method, you should write your CGI program to accept data using that method.

In his book *CGI Programming on the World Wide Web,* Shishir Gundavaram shows a short Perl routine that can sample the request_method environment variable, obtain the data from either the query_string environment variable or STDIN (depending on whether the request is get or post), and place the information into a local variable named query_string (not to be confused with the environment variable of the same name). You can then use the data in the query_string local variable without concerning yourself as to whether it came in via a get request or a post request.

CREATING A VIRTUAL PAGE

Assuming you have the data in your CGI program for processing, how do you generate the response? It's quite simple: You send your response to standard output (STDOUT)—which you can usually do with print statements. The gateway then directs the HTTP server (the Web server) to return the response to the user's Web browser as a document—typically, an HTML document.

A simple response might look similar to this:

```
HTTP/1.0 200 OK
Date: Monday, 24-May-96
11:09:05 GMT
Server: NCSA/1.3
MIME-version 1.0
Content-type: text/html
Content-length:

<HTML>
<HEAD><TITLE>Simple CGI Test</TITLE></HEAD>
<BODY>
This is a test.
</BODY>
</HTML>
```

Notice the blank line between the last header line and the beginning of the HTML code. It has to be there, as it serves as the delimiter between the header and the main part of the document. In the tutorial "Creating 'Virtual Documents' with CGI," we'll look at a Perl Script that could be used to create this sample virtual page.

RESOURCES

A column, even a series of columns, can serve only to familiarize you with such dense content matter as CGI programming. But if you have found your interest sparked and you would like to learn more about CGI programming, then I would recommend the following list of books and Web sites.

CGI Programming on the World Wide Web

by Shishir Gundavaram
O'Reilly & Associates
ISBN: 1-56592-168-2

This book provides excellent coverage of CGI. It includes numerous examples, most of which are written in Perl. It is a good introduction to CGI, and also dives deeper into the subject.

Introduction to CGI/Perl

by Steven Brenner and Edwin Aoki
MIS Press
ISBN: 1-55851-478-3
I came across this reference on a Web page. I haven't read this book, so I can't comment on it, but it's one bookstore possibility.

The WWW Common Gateway Interface 1.1

http://www.ast.cam.ac.uk/~drtr/draft-robinson-www-interface-00.html
This document, written by David Robinson of the University of Cambridge, is an Internet draft describing CGI. The opening paragraphs stress that "it is inappropriate to use Internet drafts as reference material or to cite them other than as 'work in progress.'" So, if you bear in mind that this draft is subject to change or replacement at any time, you will find it has much useful information about the interface in its current state.

The Common Gateway Interface

http://hoohoo.ncsa.uiuc.edu/cgi/
The University of Illinois' National Center for Supercomputing Applications (NCSA) is where Mosaic was developed. The document listed here offers a good introduction to CGI.

Perl for CGI

Practical Extraction and Reporting Language

http://jumpgate.acadsvcs.wisc.edu/publishing/cgi/perl.html
This document provides information on the Perl scripting language.

This tutorial, number 98, by Alan Frank, was originally published in the October 1996 issue of LAN Magazine/Network Magazine.

Creating 'Virtual Documents' with CGI

Using the common gateway interface to send 'dynamic documents' to users' Web browsers.

This column is the fifth in a series of tutorials covering Internet and World Wide Web services. I began the series with "Providing Internet and World Wide Web Services," which provided an overview of Internet services. With "HTML and CGI," I began discussing Web servers. This lesson concerns the use of the common gateway interface (CGI)—the key technology enabling Web servers to respond to user input.

On the client side, an important feature related to CGI is HTML forms capability. In the tutorials "HTML and CGI, Part II" and "CGI and Web Servers," I delved more deeply into the nuts and bolts of how HTML forms pass data to the gateway, discussing the get and post methods of submitting the request—as well as the form data—to the Web server.

In "CGI and Web Servers," I also briefly discussed how CGI programs send responses back to the user. In short, the CGI program uses print statements to send responses to standard output (STDOUT); the gateway then directs the Web server to deliver the output—usually in the form of an HTML page—to the user's Web browser. I gave an example of a simple response packet, which contained a basic HTML page called "Simple CGI Test" and the body text "This is a test."

The example in "CGI and Web Servers" showed you how a response packet might look. Now, let's discuss how you might generate such a response in HTML format. One way to do so is to use the Perl script shown in Listing 1.

The first line of Listing 1 calls the Perl interpreter, which is located in the /usr/local/bin directory (you might keep the interpreter in a different location on your server). The second line is a print statement that generates a header line. The server can generate most of the header lines, saving you from doing the work, but you must still create the content-type header yourself. The "\n" is an escape sequence that calls for a new line. Some operating systems, such as MS-DOS and Windows, use a carriage-return/line feed combination to indicate a new line, while others, such as Unix, use a single newline character at the end of each line. How you handle line feeds will depend on which operating system your Web server uses.

Note that the print statement for the content-type header has two newline commands at the end of it. These commands create a blank line immediately following the header. It's mandatory that a blank line exist between the last header line and the beginning of the document. This blank line is a delimiter that enables the Web server and the browser to distinguish the header from the document. (If you're trying out a CGI program for the first time and it isn't working, a missing blank line is the first thing you should look for.)

The remaining lines, except for the last, contain print statements that output the closing HTML tags for the Web browser to interpret. The very last line exits the program.

CREATING DYNAMIC DOCUMENTS

Listing 1 illustrates output from a CGI program. The service CGI performs is redirecting print statements (which usually go to the printer) to the Web server. The server, in turn, forwards the statements to the user's Web browser as a "virtual HTML document." In this example, only one HTTP header is generated (the content-type header). The Web server fills in the rest of the headers. Alternatively, your CGI program can generate all the necessary header lines, in which case the server delivers the output directly to the requesting Web browser without attempting further processing. This second approach is called a *nonparsed header.* For most applications, you can let the Web server do the work of creating the headers.

LISTING 1

```
#!/usr/local/bin/perl
print "Content-type: text/html", "\n\n";
print "<HTML>", "\n";
print "<HEAD><TITLE>Simple CGI Test</TITLE></HEAD>", "\n";
print "<BODY>", "\n";
print "This is a test.", "\n";
print "</BODY>", "\n";
print "</HTML>", "\n";
exit (0);
```

You've probably noticed that this example creates only "canned" HTML pages. What about taking the user's input, processing it, and producing a response based on the input? The Perl script in Listing 2 does this.

In this example, I'm assuming three things: that you have created a simple HTML form to submit a request, that the request is submitted via the get method, and that the form contains one or more information fields. (The information fields are consolidated into the query string, which is appended to the URL). If you're unfamiliar with how HTML forms submit information to a CGI program, please review "HTML and CGI, Part One,", "HTML and CGI, Part Two," and "CGI and Web Servers."

The first line of this script calls the Perl interpreter. The second line generates the content-type header, followed by the requisite blank line between the header and the rest of the document. The third and fourth lines create the beginning of the HTML code and the document title.

The fifth line samples the environment variable HTTP_USER_AGENT. When a Web browser sends a request to a Web server, the request typically includes a good deal of information in the header. One of the items of information is the browser type, which CGI stores into the HTTP_USER_AGENT environment variable. The Perl program shown in Listing 2 creates a local string variable called "$browser_type" and loads it with the value it found in the environment variable HTTP_USER_AGENT.

The sixth line operates in similar fashion to the previous line, sampling the environment variable QUERY_STRING and loading the text string it found into a local variable called $query_string. (Note that QUERY_STRING and $query_string are not the same variable; QUERY_STRING is an *environment* variable, while $query_string is a *local* variable, used only within our Perl script. What they share in common is the value that we're shuttling between them, but otherwise, they are two separate entities.)

The seventh line prints an HTML tag, while the eighth creates a sentence informing the user what Web browser he or she used in submitting the request. The next line echoes back to the user the contents of the query string, while the remaining lines send the closing HTML tags.

While this program is a good learning tool, it's not very useful in practice, because all the information we want is still packed together into a single environment variable (QUERY_STRING). Recall from the tutorial "HTML and CGI, Part Two" that when a user clicks on the Submit button on an HTML form, the browser returns a query string that includes key/value pairs, separated by ampersands (&). Furthermore, plus signs (+) are substituted for spaces between words, and any other characters that are not acceptable in a URL (or on the shell's command line) are replaced by the percent sign, which is followed immediately by the hexadecimal value of the character, as represented in the ASCII code.

When writing a CGI program, you'll have to parse the query string, reversing-out the URL encoding, breaking the string up into individual variables, and loading the appropriate values into each one. Discussing the details of how to parse strings is beyond the scope of this tutorial; however, if you know your way around one or two programming languages—and I'm assuming you do if you're planning to write CGI programs—you should have a pretty good idea of how to do so. Incidentally, Perl's strength in string manipulation makes it the preferred language for CGI programming. You may also find a library of routines that can parse strings for you. Search the Internet for companies selling programming libraries, to see if there are any libraries of CGI routines available for your language of choice.

GETTING YOUR FEET WET

As I said in the tutorial "CGI and Web Servers," the learning curve for CGI is considerably steeper than that for HTML. As such, I present this discussion as a general overview of how CGI works, as opposed to a hands-on tutorial about implementing it. Following are a few tips to help you get started on your journey.

CGI can be tricky to set up. When first testing it, choose the simplest type of CGI program. (The Perl scripts contained in Listings 1 and 2 would be good initial test scripts.) How smoothly your setup goes depends a lot on proper configuration of your Web server and any interpreters (such as Perl) you might be using. Since many of these factors are likely to be server-specific, you should consult the documentation for your server or contact the tech support department at your Web server supplier. If you happen to be using a freeware Web server, you are, of course, much more on your own in terms of support. However, you may find an Internet newsgroup or online forum that covers your Web server, and these groups can be a good source of technical tips. Another option is to hire a Web server consultant to configure your CGI gateway correctly and show you how to pass parameters back and forth between the user's Web browser and your CGI programs.

While the initial setup of CGI can be tricky, the concept of how the technology works is quite simple. You submit requests to the Web server with either the get or post methods. The gateway passes user-supplied information to your CGI program, using environment variables. Your program massages the information obtained, adds other information from the server itself (for example, you might have a database that holds product availability data), and spits out to the user a "virtual document" in HTML format.

You've probably noticed that I've focused all of this discussion on getting information into your CGI program and then getting it back to the user. I've done so because—except for the input and output process—writing a CGI program is not much different from writing any other program. The only difference that you must keep in mind is the nature of Web browser access. Since the user is waiting for something to happen, you must always design your CGI programs to deliver a response in a few seconds' time.

LISTING 2

```
#!/usr/local/bin/perl
print "Content-type: text/html", "\n\n";
print "<HTML>", "\n";
print "<HEAD><TITLE>Regarding Your
Request...</TITLE></HEAD>", "\n";
$browser_type = $ENV{'HTTP_USER_AGENT'};
$query_string = $ENV{'QUERY_STRING'};
print "<BODY>", "\n";
print "The browser you're using is ", $browser_type,
"<BR>\n";
print "This is the query string that was sent along with
your request: ", $query_string, "\n";
print "</BODY>", "\n";
print "</HTML>", "\n";
exit (0);
```

This tutorial, number 99, by Alan Frank, was originally published in the November 1996 issue of LAN Magazine/Network Magazine.

Web Server Image Maps

How to link parts of an image to different Web pages.

Over the course of our tour of Web technologies, we've seen how HTML offers hypertext links, where clicking on a hyperlink causes a jump to the page to which the anchor points. We've also seen how images can be hyperlinks. If an anchor is created for a given image and a reader clicks on that image, the browser will request the hypertext page referenced by the anchor.

Each image can have one hypertext link. When you click on a given image, you'll jump to a given document. You might, for example, have three or four small images (icons, in other words), with each representing a different choice to the reader.

It's time now to talk about another technique: *image mapping*. With image mapping, where you jump to depends upon which *part* of an image you click on. A single image can thus be a vector to many possible paths. Image maps require that your Web browser somehow communicate with a process on the Web server; it must pass coordinates that describe exactly where on the screen the user clicked.

FIX UP YOUR IMAGE

This example shows how image maps work. The image below is a graphic in .gif format. Try placing the cursor over one of the shapes in the image and clicking the mouse button.

The figure is an image map. In this example, I've created a fairly large (almost full-screen)

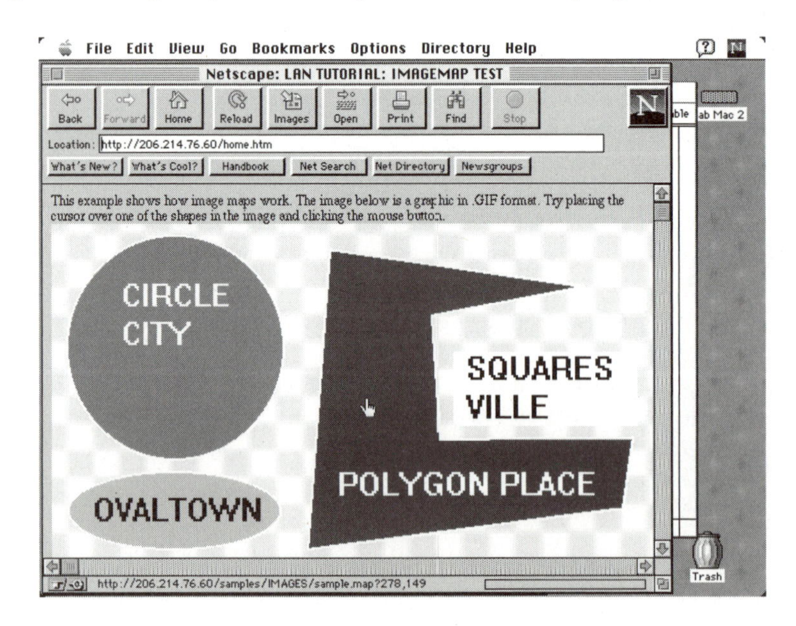

graphic image and saved it as a .gif file called sample.gif. Note that the graphic is divided into several regions—Circle City and Squaresville, for example. There's also a pale-yellow background area, which is not marked as a region per se.

Using the Web browser, a reader can position the cursor over any one of these regions, then click the mouse button. Because this is an image map, clicking on the circle will result in a hyperlink jump to the HTML document circle.htm. If, instead, the user clicked on the polygon, the document poly.htm would be fetched. This particular example is set up so that if the user clicks on the graphic—but not on one of the marked regions—a default document will be displayed. (I named my default document misc.htm.) Also, you're not required to choose destination document names according to the shape of the region they correspond to—I just happened to do so for this example.

LISTING 1—HOME.HTM

```
<HTML>
<HEAD>
<TITLE>LAN TUTORIAL: IMAGEMAP TEST</TITLE>
</HEAD>

<P>This example shows how image maps work. Figure 1 is a
graphic in .gif format. Try placing the cursor over one of
the shapes in the image and clicking the mouse button.</P>

<A HREF="/images/sample.map">
<IMG SRC="/images/sample.gif" ALIGN="BOTTOM" ismap
border=0></A>

</HTML>
```

Listing 1 shows the HTML code necessary to make this image map work. If you've been following this series of tutorials, which began with "Providing Internet and World Wide Web Services," you're probably familiar with most of the HTML code shown here, so I'll just focus on the line that reads as follows:

```
<AHREF="/images/sample.map"><IMG
SRC="gifs/sample.gif"ALIGN="BOTTOM" ismap border=0 ></A>
```

This is the key line for this image map example.

The most important word in that line is the key word ISMAP, which tells the user's Web browser that this is an image map. The first part of the line, <A HREF="/images/sample.map">, is the name of the map file, which will be used to interpret which hypertext link to take, based on where the user clicked on the image. The second part of the line, <IMG SRC="/images/sample.gif" ALIGN="BOTTOM" ismap border=0>, gives the browser the name of the graphic that it should display. In this case, the file name is sample.gif. As always, you can reference a file on a completely different Web server, if you give the complete URL, including server name and full directory path, to the file.

If the user clicks the mouse while the cursor is positioned over the graphic, the Web browser sends a request to the Web server. The URL referenced in the request is the file name of a map file, to be used by the image map program running on the Web server. Map files are plain ASCII text files that list cursor coordinates which bound certain regions of the image with the URLs to be fetched if that particular region is selected. These coordinates are in pixels (picture elements), with the upper-left corner of the graphic referenced as location (0, 0).

Also appended to the URL are the coordinates of the cursor at the moment of the mouse-click. The image map program (which is really a Common Gateway Interface, or CGI, program running on the Web server) takes the coordinates reported by the Web browser and looks in the map file to see if the cursor coordinates fall into a region defined in the map file. If they do, it looks (also in the map file) for the URL of the document to be fetched, fetches it, and sends it to the browser, fulfilling the browser's request.

LISTING 2—SAMPLE.MAP

```
default /samples/misc.htm
circle (110,107) (204,107) /samples/circle.htm
oval (15,215) (197,274) /samples/oval.htm
poly (248,23) (462,54) (334,78) (342,184) (510,184)
(502,244) (226,278) /samples/polygon.htm
rect (356,110) (516,173) /samples/rect.htm
```

MAPPING MADE EASY

Listing 2 shows the map file I used for the sample program. I named this file sample.map. The first line in sample.map gives the default URL—this is the URL of the document to be fetched in those cases where the user clicks on the graphic, but doesn't click on a defined region. In my case, I decided to call a document named misc.htm.

Format specifics for the map file depend on the specific Web server you're using and its image map program, but in general, there is one line of text for each region you define. On each line, the first item of information is the type of region, which includes rectangles, polygons, circles, and ovals.

After describing the type of region, you need to give the coordinates for the boundary. For a rectangle, you need to supply the (x, y) coordinates of the upper-left and lower-right corners. For a polygon, give the coordinates for each of the polygon's points. For a circle, give the coordinates of the center, plus a second set of coordinates that describe a point on the perimeter. For an oval, give upper-left and lower-right coordinates.

The last item of information you need to give for each line in the map file is the URL of the document you want delivered if that region is selected. Either give a full or partial URL, as appropriate.

READING MAPS

Image maps are used frequently on the World Wide Web, so if you surf around, you will find many examples to explore. Most Web browsers have a field somewhere in their display that, if you place the cursor over a hyperlink, shows the URL of the document that will be fetched if you click on that link. This also applies to image maps. The difference is that for image maps, you see an extended URL that includes not only a file name, but one that's followed by a question mark, plus a pair of numbers separated by a comma. Those two numbers represent the current cursor coordinates, and they change as you move the mouse. (Don't click the mouse button just yet—simply move the mouse around to explore the image map area.) If you move the mouse around, you can determine the boundaries of the overall image map area. (You won't be able to see which regions have been defined within that area—that information is in the map file on the Web server. But the underlying graphic usually gives you plenty of clues—there might be fancy buttons on the screen, for example.)

The file name to the left of the question mark is the name of the map file that defines the regions and the URLs they point to. When you click somewhere on an image map, the entire line you see displayed—including file name, the question mark, and cursor coordinates—is sent as a request for a document.

Image map files are stored in a particular directory (the location of the directory may vary, depending on the Web server). When the request for this URL comes in, the server knows it is an image map request because of the directory cited. The server then hands the URL-encoded data to the image map program (a CGI program), which is also running on the server. The image map program examines the data to see which map file to request and to determine the cursor coordinates for the mouse-click. It then gets the map file and compares the user-generated cursor coordinates with the entries in the map file to see which region the mouse-click was in and which URL corresponds with that region. Knowing that, it directs the Web server to fetch the document that resides at that URL, and the Web server sends it to the browser.

MAKING MAPS

Once you have a graphic to display, you need to create a map file that is keyed to that graphic. There are various commercial and shareware programs available that can help you with this task—though I haven't tried them out, so I can't vouch for them. For this example, I took these steps:

First, create the graphic by using a paint program or other image editor. Size the graphic according to how you want it to look on the user's browser. I always make sure my screen's resolution is set to 640dpi by 480dpi before doing any image editing. If you were to edit with 1,024dpi by 768dpi resolution, for example, you could easily create a graphic that's far to big to fit on the screen of a user with 640dpi by 480dpi resolution.

Second, create the HTML code for the document in which you're going to place the image map. (This would be comparable to Listing 1.)

Third, create a map file. You won't know the exact coordinates to enter, but for the moment, you can enter just about anything.

Load the HTML document, the graphic, and the map file into the appropriate directories on your Web server.

Load the document, using a Web browser. When you place the cursor over the image map, the browser will display the cursor coordinates. Don't click the mouse at this point—just move the mouse around and note the (x, y) coordinates for the various regions you want to define. Also, decide what shapes (circle, rectangle, and so on) are most appropriate for each region. Write your figures down on a scratch pad, then load the map file into a word processor and edit it to reflect the correct cursor coordinates for each region. Be sure to double-check the accuracy of each URL.

Finally, test the finished map file by firing up your Web browser and clicking on the image map. Does clicking on each region take you to the document you had intended? If not, go back and make minor adjustments. If it works as advertised, pat yourself on the back, map-maker!

For the example shown here, I used Microsoft's Internet Information Server 2.0, running on Windows NT Server 4.0. Not every Web server supports image maps (but the majority seem to), so if you're planning to roll your own image maps, consult the documentation for your server to see if you do have the required support. Different Web servers use different default directories to store certain files, such as the map files. Again, consult your documentation to see which directories you should use. The syntax for image map anchors and map files may vary, so this is something else you'll want to verify in your documentation.

This tutorial, number 100, by Alan Frank, was originally published in the December 1996 issue of LAN Magazine/Network Magazine.

Web Tools

What exactly is a Web site? Technically speaking, this question isn't difficult to answer. A Web site can be defined as a collection of related files, usually written in HTML, that are delivered to users via HTTP.

Beyond these basics, of course, there's much more technology that a Web site can employ. Server- and client-side scripting are common techniques for making Web pages dynamic and interactive; links to databases are increasingly important; ActiveX controls and Netscape plug-ins have extended the range of data types that Web browsers can interpret; and HTTP is no longer the only protocol that matters (for example, UDP audio and video streams are increasingly important).

But for people involved with the Web—consumers, publishers, developers, those who evaluate their efforts—mere technology doesn't define a Web site. Rather, most of us evaluate a site in terms of its former physical equivalent, whether that be paper or bricks and mortar.

For example, a publisher experienced in producing traditional print media is likely to see a Web site as a magazine, designed to serve readers with interesting articles and, along the way, lure them into reading some paid advertising. A retailer, on the other hand, is likely to see a Web site as a store, designed to showcase products and herd customers toward a metaphorical checkout counter.

To those of you who are Web site developers, or merely interested in their predicament, I would like to propose a less commonly encountered metaphor. Think of a Web site as if it were an airplane: The goal is to carry as many passengers as possible, get each traveler to his or her destination safely and efficiently, and leave customers pleased and eager to return for their next journey.

Clearly, any successful trip involves extensive preflight checking; only this can ensure that nothing in the plane is broken and that there is adequate support for every foreseeable requirement during the journey. The trip also requires in-flight monitoring. Error conditions should be tracked, and preferably responded to automatically via redundant systems.

Finally, after a successful landing (a user leaving the Web site) there is still plenty of postflight work to be done. Since on this particular airline we are both the engineering and the marketing department, we must do everything from keeping track of how many seats were full to surveying customers to find out what they thought of the food and the movie, to doing essential maintenance on the aircraft itself.

The metaphor is useful because Web tools, the subject of this tutorial, parse rather neatly into preflight, in-flight, and postflight categories. Read on, and I will provide an overview of the first two categories, then look at postflight analysis in greater detail.

PREFLIGHT TASKS

If a Web site is brand-new, then its developers face a task fortunately unfamiliar to any airline employee: engineering it from the ground up. The first preflight task is, then, creating the text and graphics that will be incorporated into the site.

The next step is conversion into HTML. Many productivity programs—such as both Microsoft Word and Excel in the new Office 2000 versions—can produce HTML-formatted output as a matter of course. However, many developers still favor simple ASCII text editors such as Notepad or BBEdit, using them either to code HTML by hand or to tweak the output of some other program.

Some programs maintain their classic what-you-see-is-not-what-you-get format, while adding menus that insert the most commonly used HTML tags. Others, such as Softquad's HotMetal Pro, let the user switch between a text editor view and a what-you-see-is-what-you-get view.

Many editors also offer HTML syntax checking, plus additional Web features such as cascading style sheets, dynamic HTML, and Extensible Markup Language (XML). (For more information on XML, see "Lesson 124: XML and XSL," November 1998.)

Whether out of necessity or just the desire to stand out in a crowded marketplace, many editors offer some degree of site management. For example, when used with complementary server extensions, Microsoft's FrontPage lets multiple users author parts of a site simultaneously—from remote clients if necessary. It also simplifies creating threaded discussions and site indexes.

Whatever editor you use, the next essential preflight check is to test Web pages using multiple browsers—different releases of Netscape Navigator and Microsoft Internet Explorer at a bare minimum, and preferably others. The purpose of this is merely to evaluate the appearance of the new pages as rendered by the different browsers.

The final preflight step, checking a site to make sure all of its links work, used to be laborious and boring. Now it is easily performed by link-checking software. For detailed information on this, see "Control Web Site Content," June 1999.

IN-FLIGHT MONITORING

Once a Web site has been devised, checked out, and placed on a server, some developers consider their job done—until and unless they get e-mail from a user complaining that some feature doesn't work. However, with the critical nature of today's sites, this might seem the moral equivalent of waiting for a passenger to press the flight attendant call button in order to find out that an engine is on fire.

High-availability Web sites may require redundant servers that are teamed via simple round-robin DNS or perhaps a dedicated hardware unit such as Cisco Systems' Local Director, which watches TCP connections, detecting when one server fails to respond and switching clients to another.

With or without the aid of such hardware, in-flight software tools can monitor Web transactions for a bevy of error conditions. They can then automatically alert an administrator, distribute traffic around a failed server or a broken link, perform load balancing, and monitor the performance of back-end databases.

In fact, these tools include Web-specific versions of Application Performance Management (APM) software (see "Application Performance Management Software," May 1999). As such, they can operate via monitoring SNMP traps, sniffing packets, collecting data from specially designed server extensions, or a combination of all of these.

BACK DOWN ON THE GROUND

While not every Web site requires active, in-flight monitoring, each—to abuse my metaphor one last time—certainly needs a postflight "ground crew." Scrutinizing server logs has always been important. Now, given the commercial importance of today's Web sites, it is vital.

Automatically created as users access a server, log files were originally devised simply for the benefit of system administrators. As well as providing a crude way of tracing broken links and other errors, they gave an ongoing indication of exactly when demand for a site was highest. Today, logs are being used as a primary source of demographic information about visitors. What ad banners do they click on most? What pages do they read often, and how long do they linger before clicking through to another? What companies do they come from?

To answer questions such as these, every Webmaster needs a log file analyzer. I will devote the remainder of this tutorial to this type of program and what it can (and cannot) do.

Over the years, HTTP servers have used a variety of log file formats. The baseline standard, however, is the Common Log Format (CLF), originally created by the National Center for Supercomputing Applications (NCSA) for its HTTPd server software.

The access log in CLF records the bulk of the information. Every successful (or merely attempted) file access results in a new line being added to this log. Each line contains seven data fields that are separated by spaces and further delineated, when necessary, by brackets or quotation marks.

To provide you with a random example, I used my desktop computer at Network Magazine to access a personal home page that is hosted externally (see Figure).

Only the first seven fields in this table would be found in a CLF access log. Servers using CLF record information about what Web page visitors came from last and what type of browser they are using, but they place this in separate referrer and agent log files. Since juggling these files can be cumbersome, most Web servers today use Extended Log Format (ELF), which incorporates referrer and agent information into the access log itself (as seen on the bottom line in the figure).

Common and Extended Log Formats

An Extended Log Format (ELF) access log entry, stored on a single line, typically looks like this:

```
alto2.mfi.com - - [06/May/1999:11:44:44 -0700] "GET /logo.gif HTTP/1.0" 200 4305 "http://www.angel.org/"
"Mozilla/4.51C-Caldera [en] (X11; I; Linux 2.2.5 i686)"
```

An ELF's nine fields—Common Log Format (CLF) access logs have only the first seven—mean the following:

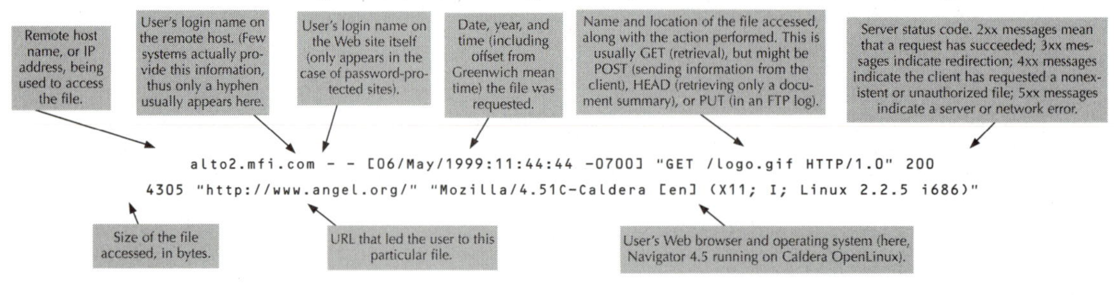

COLLECTING MORE INFORMATION

Other log file formats have been designed to collect more extensive information. For example, Microsoft's Internet Information Server (IIS) adds metrics such as the name or IP address of a server, the amount of time it takes the server to process a request, the number of bytes received from a client, and Windows NT-specific status information.

Since the majority of visitors to any Web site do not have permanent, routable IP addresses, one visitor may tend to look much like another. If, for example, 20 or 30 users from AOL are accessing your site at the same time, your log file will record plenty of AOL activity but will not help you understand the movements of any individual. To get around this problem, cookies were invented. A cookie is a unique identifying code that a Web server can give to a client to store locally when a site is first visited. Later, the server can read the cookie—only the one it created, not others—to see if the user is a repeat visitor. Naturally, the cookie can be linked to a name, password, and other demographic information that you provide on first visiting the Web site.

Most users now realize that cookies are not a serious privacy threat. They can, however, cause problems in cases where many users share the same machine—in a public library, for example. Conversely, cookies have difficulty keeping track of individuals who habitually use several machines to do their Web browsing.

When you import a log file into analyzer software, the program may start by converting raw IP addresses into domain names, assuming reverse DNS lookup has not already been performed by the Web server. It may then convert file names to actual page titles, making all subsequent reports easier to read.

Beyond that, an analyzer can then parse the log file to provide a wide variety of reports, both tabular and graphical. High-end analyzers, known as Web mining solutions, can even link to other databases in an enterprise, answering such questions as "How many visitors to the site purchased something from our online store?"

Download one of the many demos available and try it with your own log files. This will let you evaluate the software with all the appreciation, and skepticism, it deserves.

This tutorial, number 132, by Jonathan Angel, was originally published in the July 1999 issue of Network Magazine.

SSL and S-HTTP

Secure Communication over the Internet.

Most companies with connections to the Internet have implemented firewall solutions to protect the corporate network from unauthorized users coming in—and unauthorized Internet sessions going out—via the Internet. But, while firewalls guard against intrusion and unauthorized use, they can guarantee neither the security of the link between a particular workstation and a particular server on opposite sides of a firewall nor the security of the actual message being conveyed.

To create this level of security, two related Internet protocols, Secure Sockets Layer (SSL) and Secure hypertext transfer protocol (S-HTTP), ensure that sensitive information passing through an Internet link is safe from prying eyes.

SECURE CONNECTION

SSL was developed by Netscape Communications (Mountain View, CA) for such instances as when sensitive, private information is sent from client to server through a TCP/IP connection—for example, during an electronic commerce transaction. The protocol is application independent and operates at the Transport layer. This means application protocols such as HTTP, ftp, telnet, gopher, Network News Transport Protocol (NNTP), and Simple Mail Transport Protocol (SMTP) are easily and transparently layered on top of SSL, and TCP/IP is layered beneath it (see Figure 1).

When both a client (usually in the form of a Web browser) and a server support SSL, any data transmitted between the two becomes encrypted. Netscape Navigator and Microsoft's Internet Explorer both support SSL at the browser level, and Netscape's Commerce Server supports the protocol at the server level. Other Web servers that support SSL include Safety-Web from CompuServe (Columbus, OH) and WebSite from Sebastopol, CA-based O'Reilly & Associates. When an SSL-compliant client wants to communicate with an SSL-compliant server, the client initiates a request to the server, which in turn sends an X.509 standard cer-

■ **Figure 1:** SSL creates secure communications links across the Internet at a very basic level, making it possible for use in a variety of Internet sessions, including telnet, ftp, gopher, and SMTP.

INTERNET PROTOCOLS

HTTP · *Telnet* · NNTP · *FTP* · SMTP · *Gopher*	**Application layer**
SSL	**Transport layer**
TCP/IP	**Network layer**

tificate back to the client. The certificate includes the server's public key and the server's preferred cryptographic algorithms, or ciphers.

The client then creates a key to be used for that session, encrypts the key with the public key sent by the server, and sends the newly created session key to the server. After it receives this key, the server authenticates itself by sending a message encrypted with the key back to the client, proving that the message is coming from the proper server. After this handshake process, which results in the client and server agreeing on the security level, all data transferred between that client and that server for a particular session (whether HTTP, telnet, ftp, and so on) is encrypted using the session key (see Figure 2).

When using the Netscape Navigator browser, a user can tell that the session is encrypted by the key icon in the lower-left corner of the screen. When a session is encrypted, the key, which is usually broken, becomes whole, and the key's background becomes dark blue. Another way to detect a secure session is by the URL. When a secure link has been created, the first part of the URL at the top of the browser will change from http:// to https://. The process I just described involves server authentication. In addition, a second phase, client authentication, may take place for extra security measures. SSL supports several cryptographic algo-

■ **Figure 2:** SSL enabled sessions require server authorization, which consists of the server sending its public key to the client, which then generates a master key to be used to encrypt all communication for that particular session.

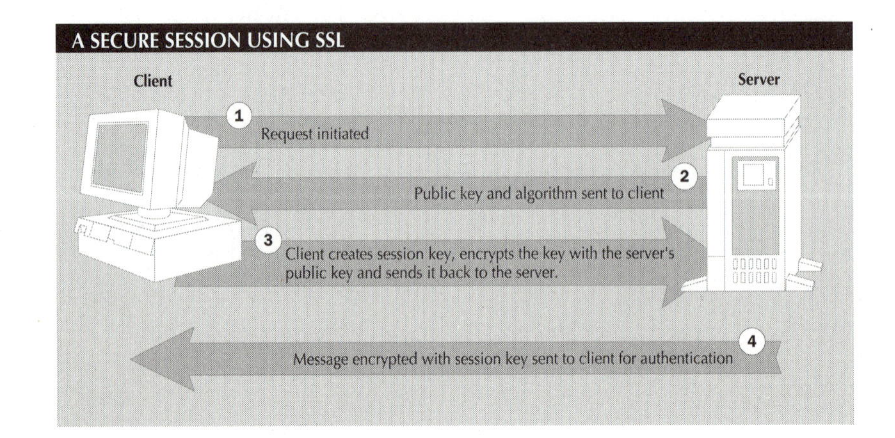
A SECURE SESSION USING SSL
Client
Server
1
Request initiated
2
Public key and algorithm sent to client
3
Client creates session key, encrypts the key with the server's public key and sends it back to the server.
4
Message encrypted with session key sent to client for authentication

rithms to handle the authentication and encryption routines. One algorithm, which was developed by RSA Data Security (now part of Security Dynamics), is RC4, a variable key-size cipher. Other cryptographic algorithms supported by SSL include Data Encryption Standard (DES), which works on 64-bit blocks of data with a 56-bit key, and triple-DES, which runs an encrypted message through DES three times.

SECURE MESSAGES

S-HTTP, as you might have guessed, is simply an extension of HTTP, the communications protocol of the World Wide Web. S-HTTP is created by SSL running under HTTP. The protocol was developed in 1994 by Enterprise Integration Technologies (EIT, Menlo Park, CA) as an implementation of the RSA encryption standard. While SSL operates at the Transport layer, S-HTTP supports secure end-to-end transactions by adding cryptography to messages at the Application layer, which means that while SSL is application independent, S-HTTP is tied to the HTTP protocol. Also, while SSL encrypts the entire communications link between a client and a server, S-HTTP encrypts each message on an individual basis.

S-HTTP, like SSL, can be used to provide electronic commerce without customers worrying about who might intercept their credit card number or other personal information. According to the Internet Engineering Task Force's (IETF's) latest S-HTTP Internet draft, an S-HTTP message consists of three parts: the HTTP message, the sender's cryptographic preferences, and the receiver's preferences. The sender integrates both preferences, which results in a list of cryptographic enhancements to be applied to the message.

To decrypt an S-HTTP message, the recipient must look at the message headers, which designate which cryptographic methods were used to encrypt the message. Then, to decrypt the message, the recipient uses a combination of his or her previously stated and current cryptographic preferences and the sender's previously stated cryptographic preferences.

S-HTTP doesn't require that the client possess a public key certificate, which means secure transactions can take place at any time without individuals needing to provide a key (as in session encryption with SSL).

An S-HTTP message includes a request line or status line and style headers, all of which precede the message's content. According to the IETF draft for version 1.2 of S-HTTP, all S-HTTP requests should include a request line that looks like this:

```
Secure*Secure-HTTP/1.2
```

This request line distinguishes an S-HTTP request from existing HTTP implementations. Servers should respond to this request with:

```
Secure-HTTP/1.2 200 OK
```

regardless of whether the request succeeded or failed.

An S-HTTP message header consists of several pieces of information, including Content-Type and Content-Privacy-Domain. An optional component outlined in the IETF draft is something called a Prearranged Key-Info header, which allows the use of session keys for return encryption when one party doesn't have a public and private key pair. However, it can also be used when both parties choose another method of encryption in place of the public and private key pair method, such as a one-time key.

The Prearranged Key-Info specification defines three ways to exchange keys: Kerberos, Inband, and Outband. Kerberos and Inband show that the session key has already been exchanged prior to the current communication. In the case of Kerberos, this can be accomplished through access to a trusted Kerberos server. Outband indicates that the sending and receiving parties have external access to keys through a database or other means. This part of the message header should look like this:

```
Prearranged-Key-Info: <Hdr- Cipher>','<CoveredDEK>','
   <CoverKey-ID>
   <CoverKey-ID> := <method>':'
   <key-name> <CoveredDEK> :=<hex-digits>
   <method> :='inband' |'krb-' <kv> |'outband'
   <kv> :='4' | '5'
```

In this example, the `<Hdr-Cipher>` field should be filled with the name of the cipher used to encrypt the session key. This cipher could be DES, triple DES, IDEA (International Data Encryption Algorithm), or RC2. IDEA is a block cipher similar to DES, but with a larger key; RC2 is a variable key-size block cipher that encrypts blocks of 64 bits.

`<CoveredDEK>` is the Data Encryption Key, which is the key under which the message was encrypted. The draft specifies that this key should be randomly generated by the sender and then encrypted by the session key using the header cipher.

CALLING A TRUCE

Not surprisingly, when Netscape announced SSI, and EIT announced S-HTTP, the industry and users quickly divided into two camps. Thankfully, vendors have woken up and realize the two protocols don't have to compete. Rather, the two can complement one another very well because SSL secures an Internet connection while S-HTTP secures HTTP-based messages.

Terisa Systems (Menlo Park, CA), an Internet security company formed by RSA Data Security and EIT, shipped the first security toolkits, which developers can use to build a Web server that supports both SSL and S-HTTP. The SecureWeb Client Toolkit 2.0 and Server Toolkit 2.0 allow companies to build secure Web servers for such applications as electronic commerce and other applications requiring the exchange of sensitive information. The products support RSA public key cryptography, as well as other cryptographic methods.

In the past year, two major credit card companies have teamed with the likes of Netscape, RSA Data Security, and Microsoft to provide yet another protocol to deliver secure commerce over the Internet. Secure Encryption Transaction (SET) is being supported by MasterCard and Visa as a means for customers to conduct electronic payment transactions.

While one can safely transmit credit card information through an SSL-encrypted session or within an S-HTTP message, SET differs from these two protocols by proposing to supply users with an X.509 digital certificate once their credit card numbers are authorized. The certificate would be attached to a user's browser and then used for all future purchases.

Technically, however, SET can work with both S-HTTP and SSL. And, with cooperation among the various companies to bring secure electronic transactions to the masses, final standards shouldn't be too far away.

RESOURCES

For more information on Secure Sockets Layer, including a link to the most recent IETF Internet draft, access Netscape's Web site at http://www.netscape.com/newsref/std/ SSL.html.

To view the latest S-HTTP Internet draft from the Internet Engineering Task Force, access ftp://www.ietf.org/internet-drafts/draft -ietf-wts-shttp-03.txt.

This tutorial, number 101, by Anita Karvé, was originally published in the January 1997 issue of LAN Magazine/Network Magazine.

Lightweight Directory Access Protocol

An offshoot of X.500, LDAP may be the simple, universal directory of the future.

This lesson relates closely to "Directories" in Section VII of this book, which discusses directories in general. This month I will focus on a specific directory technology: the Lightweight Directory Access Protocol (LDAP).

As "LDAP" suggests, it started life as an access method, not a directory. By establishing a protocol, LDAP standardizes how an application can "talk" to a directory. LDAP was originally designed to be a lightweight front end to X.500 directories. X.500 is the ISO's robust and comprehensive directory standard.

However, at least in its early stages, X.500 took quite a bit of computing power to implement. Researchers at the University of Michigan conceived LDAP as a simplified replacement for X.500's native access protocol, known as the Directory Access Protocol. An X.500 directory could be run on a mainframe, minicomputer, or other high-horsepower machine, while individuals with PCs and other computationally challenged machines could use LDAP to access the directory.

Despite LDAP's beginnings as a front end to X.500 directories, work at the University of Michigan soon turned to developing a standalone LDAP-based directory that could be accessed by any LDAP client. Since then, IETF standards efforts have also been directed toward standalone LDAP directories.

THE LDAP STANDARD(S)

Efforts to make LDAP an Internet standard resulted in RFC 1487, which describes LDAP 1. This document has since been superseded by RFC 1777, which covers LDAP 2. The IETF working group responsible for Access, Searching, and Indexing of Directories (ASID) has since developed a proposal for LDAP 3 and submitted it for approval in the fall of 1997. For all practical purposes, it has been approved. The draft document for LDAP 3 can be found at ftp://ftp.ietf.org/internet-drafts/draft-ietf-asid-ldapv3-protocol-08.txt.

In theory, any LDAP-compliant application should be able to access any LDAP-compliant directory. The LDAP-compliant directory would then play the role of a server, while the application would function as an LDAP client. LDAP can be used to get information from a directory (such as finding out a user's e-mail address), as well as to store information (such as creating a new account).

HOW IT WORKS

The goal of a directory is to keep track of some entity, which might be a person, but could also be a printer or other resource, and to store relevant and needed information about that entity. An LDAP directory is organized into entries; there is an entry for each entity you want to keep track of. Each entry has attributes (for example, a name or e-mail address) that hold information about the entity. Each attribute has a name, a type, and one or more associated values. Take, for example, the network user John Doe. The entry for John in the directory may have an attribute named "mail," which stores John's e-mail address. The attribute type of the mail attribute might be String, which means that the information is stored as an ASCII text string, and the attribute value of the mail attribute for John Doe's entry might be jdoe@xyz.com.

Since you need to keep track of different attributes for a person than for a printer, entries have a type associated with them. An entry's type is known as its object class. An entry's object class defines what attributes it will have. How do we know what an entry's type is? That's held in a special attribute named objectClass.

What attributes are stored for a given object type is known as an object's schema. The schema is enforced according to an entry's object class, but one of the major extensions contained in LDAP 3 is the provision for an extensible schema. The LDAP 3 proposal defines a special object class called extensibleObject. If an entry's objectClass attribute contains the value extensibleObject, anything goes as far as what attributes are allowed. In other words, schema enforcement is overruled.

LDAP 3 also provides a method for clients to discover the schema, which is a very important, much-needed capability. For example, is a Social Security number or address information in the directory? If so, what's the attribute name? Schema discovery gives the LDAP client a means to find this out. In LDAP 1 and LDAP 2, a client had to know what the schema was—a major limitation.

NAVIGATING THE DIRECTORY

How do you find a particular entry in an LDAP directory? LDAP supports a hierarchical organization of the information in the directory. It's based on concepts borrowed from X.500, so if you've worked with X.500 or Novell Directory Services, which bases its naming scheme on X.500, you're probably familiar with LDAP's naming convention. LDAP entries are named according to their Distinguished Name, known as DN. For example, the DN for the John Doe entry might be "CN=John Doe, O=XYZ, C=US." This specifies the entry as having the common name "John Doe" in the U.S.-based organization "XYZ."

The complete description of LDAP's naming scheme can be found in RFC 1779.

LDAP OPERATIONS

An LDAP client uses TCP/IP to communicate with an LDAP server. By default, LDAP servers listen to port 389, so LDAP clients need to direct their requests to that port.

Communications between LDAP clients and servers are session-oriented, and the first thing a client must do when initiating a session is send a bind request. This establishes the session.

As part of the session setup procedure, LDAP directories require authentication of the client. LDAP 1 and LDAP 2 provide for a simple password-based authentication, but the password is sent in plain text. This is one of the issues spurring LDAP 3 work. (LDAP 1 and LDAP 2 can also use Kerberos authentication.) The LDAP 3 proposal allows for the use of something called Transport Layer Security (TLS), which is based on Secure Sockets Layer (SSL) 3.

LDAP defines several operations that an LDAP server can perform for a client. These include bind (which was already mentioned), unbind (which is issued to terminate the session), search (search for an entry), modify (modify an entry), add (add an entry for a new entity), del (delete an existing entry), modifyRDN (modify the RDN of an entry), compare (which I will explain), and abandonRequest (abandon a request that's pending).

Searches can, and typically are, limited in scope and by search criteria (known as search filters). A search filter is a means of determining what you're searching for. For example, you may be looking for any entries where "CN=Elvis." Another alternative is to do a substring search. A substring search for common names beginning with "e" might return entries for Edward Jones, Edna Edwards, and Eloise Smith, for example.

The ability to set filters is a very powerful aid in searching the directory, and LDAP's filtering possibilities are very flexible. You can require either exact or approximate matches for your filter criteria, as well as use greater-than-or-equal-to or less-than-or-equal-to comparisons. You can also use the And, Or, and Not functions to look for matches for various combinations of attributes and values.

When submitting a search request, the client must provide a parameter called baseObject, which is the DN of the entry, from which point the search will begin. For example, if you're looking for someone in company XYZ, you might specify "O=XYZ, C=US" as the baseObject.

With baseObject as the starting point, how far should the search extend? That's defined by the scope parameter. Scope can have one of three values. The first is a subtree search, which means everything "below" the baseObject will be searched (in other words, baseObject will be the root of the search operation).

The second possibility for scope is a one-level search. This will search all entries for which the specified baseObject is the immediate parent. For example, if there was an entry for the accounting department in company XYZ, you could search for all the entries that are immediately subordinate to the accounting department entry by setting the base-Object to "OU=Accounting, O=XYZ, C=US," and setting the scope to one-level. ("OU" stands for the object type Organizational Unit.) Note that if you wanted to find every single entry that falls under the organizational unit "Accounting," you would need to use the subtree scope, not one-level scope.

The third option for scope is a base-object scope, which will only return the entry for the base object itself. This is handy when you want to look up a certain attribute (or collection of attributes) for a specific entry, and when you know exactly which entry you want. For example, if you just wanted to find out John Doe's e-mail address, you might specify a baseObject "CN=John Doe, O=XYZ, C=US," and set the scope to a base-object search.

When an LDAP server gets a request from a client, it will execute that request, if it can, and return a response message. The response might be an error code (in cases where the request could not be honored) or the results of the operation requested by the client. In the case of a search request, for example, the LDAP server will return all entries that meet the submitted scope and filter criteria.

When submitting a search request, you also specify how much information should be returned. For example, you might specify that only the first 30 entries that are found should be returned. You can also specify how many seconds to spend on the search.

Whenever the server finds an entry that matches your search criteria, it will return the entire entry (subject to whatever access permissions you have). But you can elect to have only certain attributes returned, such as a telephone number. Your request message must include a list of all the attributes you wish to have returned; if no attributes are specified, that is, if it's an empty list, you will get the entire entry.

The compare operation lets you check to see if the value of an attribute for a specific entry matches some value that you're testing for. You could, for example, verify that John Doe's e-mail address is indeed jdoe@xyz.com by issuing a compare request for the entry "CN=John Doe, O=XYZ, C=US," and specifying the attribute/value assertion "mail= jdoe@xyz.com." The LDAP server will look up the requested entry and return a response code of compareTrue or compareFalse. If it can't execute the operation, it will return an error code.

REFERRALS

In LDAP 1 and LDAP 2, no provision was made for LDAP servers returning referrals to clients. However, in the case of failed queries (where, for example, a search is made for an entry that does not exist in one LDAP directory, but might exist in another), LDAP 3 provides a method for the LDAP server that's getting the initial request to return a referral to another LDAP server. The LDAP client can then issue the request to the server identified in the referral.

ENABLING DIRECTORY ACCESS

LDAP sets forth a standard for how network applications can access directories. It's also simple and easy enough to implement that it's been enthusiastically embraced by many vendors, so it's rapidly becoming the lingua franca of directory operations. In a world where each application and operating system has its own proprietary directory, LDAP holds forth the promise of directory interoperability. But for now, that interoperability extends for the most part only to simple browsing operations.

RESOURCES

A good overall description of LDAP and LDAP concepts can be found in the book *LDAP: Programming Directory-Enabled Applications with Lightweight Directory Access Protocol,* by Tim Howes and Mark Smith (1997, Macmillan Technical Publishing). Other good sources of information are *An LDAP Roadmap & FAQ*, by Jeff Hodges of Stanford University (www.leland.stanford.edu/ group/ networking/ directory/ x500ldapfaq.html), and LDAP information pages maintained by Critical Angle, a developer of LDAP servers and related technology (critical-angle.com/ldapworld/ldapv3.html).

This tutorial, number 113, by Alan Frank, was originally published in the January 1998 issue of Network Magazine.

X.400 Messaging

E-mail is an essential application, one that can often justify the installation of a network all by itself. E-mail provides a vehicle for people to communicate regardless of their physical location and time zone. A form of e-mail exists for nearly every type of operating system, from ones running on personal computers to large hosts. Mail-enabled applications can be constructed on the mail system, allowing greater productivity.

A corporation's departments are likely to use a wide variety of systems. IBM's PROFS and SNADS are commonly used on IBM mainframes; All-In-1 is popular on VAX/VMS systems; WangOffice is used on Wang VS machines; HPOffice is used on Hewlett-Packard minis; SMTP is used on Unix and TCP/IP systems; Lotus' cc:Mail, DaVinci's eMail, and Microsoft Mail are among those used on PC LANs; and CE Software's QuickMail is often used on Mac LANs.

As corporations build enterprise networks, the dissimilar e-mail systems must be connected via a common transport. Very often, that platform is X.400. X.400 is vendor-independent, able to run on a wide variety of computer systems, and internationally accepted and used.

The International Organization for Standardization (ISO) X.400 Message Handling Systems (MHS) is a suite of protocols that define a standard for store-and-forward messaging. X.400 provides message creation, routing, and delivery services. It runs over an X.25 packet-switched network or asynchronous dial-up lines.

X.400 software is available for a variety of computers, either as native implementations or as gateways.

There are two versions of the X.400 standard: the 1984 and the 1988 specifications. [Note: the specification was further upgraded in 1992.] Most X.400 mail systems implement the older specification. The new spec offers key additions to X.400's flexibility and usefulness. It permits the use of asynchronous lines, a useful tool for PC or laptop users not connected to an X.400 network. Directory services and security features have been added. As more 1988 X.400 implementations become available, native X.400 usage should rise.

X.400 COMPONENTS

X.400 divides an electronic mail system into a client, called a *User Agent* (UA) in ISO parlance, and a server, called a *Message Transfer Agent* (MTA). Essentially, the User Agent is a mail box; it interfaces directly with the user. It is responsible for message preparation, submission, and reception for the user. It also provides text editing, presentation services, security, message priority, and delivery notification. The User Agent is an interface, not an end-user application, so it does not define the specifics of how it interacts with the user. The product developer decides those issues.

The Message Transfer Agent routes and relays the messages. Its responsibilities include establishing the store-and-forward path, ensuring channel security, and routing the message through the media. An MTA's operation is relatively straightforward. The User Agent sends its message to the local Message Transfer Agent. The MTA checks the message for syntax errors, then delivers the message to a local User Agent, or if the message is not local, it forwards it to the next MTA. That MTA repeats the process until the message is successfully delivered.

A collection of MTAs is known as *Message Transfer System* (MTS). The MTS is usually specialized to a particular vendor's product.

X.400 also uses *Distribution Lists* (DLs), which are like routing lists commonly used in offices.

The *Message Store* (MS) provides a facility for message storage, submission, and retrieval. It complements the User Agent for devices that are not always available, such as PCs or terminals. Essentially, it is a database of messages.

The *Access Units* (AUs) provide connections to other types of communications systems, such as telex and postal services. AUs defined in the 1988 spec are the Telematic Agent for Teletex terminals, Telex Agent for Telex service, and Physical Delivery Agent for connection to the traditional postal service.

A management domain is a collection of at least one Message Transfer Agent and zero or more User Agents that is administered by a single organization. Management domains may be *private* (PDMD) or *administrative* (ADMD). An ADMD is managed by an administration such as a PTT or telephone company, and a Private Management Domain is managed by any other type of organization. A hierarchy of management domains enables the configuration of a worldwide X.400 system with unique addressing.

SENDING X.400 MAIL

An X.400 mail system follows the metaphor of a post office, and X.400 messages follow the metaphor of the letter. Messages are packaged into envelopes, and an envelope describes the control information necessary to deliver the message's content, including the body type, syntax, and semantics. X.400 can deliver messages to other X.400 users via a *message transfer* service or to other communications facilities such as Telex via an *interpersonal messaging* service.

As with the postal services, users and distribution lists must have unique addresses to deliver the message anywhere in the world. X.400 uses two kinds of names, a *primitive name,* which identifies a unique entity such as an employee number, and a *descriptive name,* which identifies one user of the X.400 system.

A name is typically looked up in a directory to find the corresponding address, but X.400 allows a machine name to also be a directory name, which makes it easier for humans to interpret. X.400 addresses consist of attributes that describe a user or distribution list or locate the user distribution list within the mail system. Attributes are personal (such as last name or first name), geographical (street name and number, town, or country), organizational (name or unit within organization), and architectural (X.121 addresses, unique User Agent identifier, or management domain). In practice, X.400 names are lengthy and complex and should be hidden from end users by offering them aliases.

In addition to messages, the message handling system and the message transfer system use probes and reports. A probe contains only an envelope—no content—and it is used to determine if messages can be delivered. For instance, a probe may be sent to test out a path, asking the receiving MTS if it can accept a particular message type. By testing the waters with a probe, a lengthy message is less likely to be rejected by the recipient. A report is a status indicator that relates the progress or outcome of a transmission to users.

WAYS OF HANDLING MESSAGES

X.400 defines several protocols for handling messages among the different system components. Two Message Transfer Agents may communicate directly with each other, without the intervention of a User Agent. They do so by using the P1 protocol. If a User Agent wants to communicate with a service outside the X.400 domain, it uses P2, the interpersonal messaging protocol. The 1988 implementation of P2 defined additional body types for messages, so beyond supporting Teletex and Group III fax, P2 supports externally defined body types such as word-processing formats. This specification paves the way for electronic document interchange.

P3 defines the conventions for transferring a message from the User Agent to the Message Transfer Agent. Initially defined in the 1984 spec, P3 assumes that the User Agent is online and ready to accept messages from its Message Transfer Agent. In 1984, X.400 did not provide for a User Agent that would be online intermittently. In practice, most User Agents are implemented on personal computers, and therefore will not always be online.

To remedy the situation, Message Store was added in the 1988 spec, and P7 was defined for the communication between the User Agent and the Message Store. The Message Store, always connected to the Message Transfer System, stores messages for the User Agents. User Agents submit messages through the Message Store as well as retrieve, list, summarize, and delete messages from the Message Store database. P7 support is crucial for anyone using laptops.

DIRECTORY AND SECURITY ISSUES

The 1988 X.400 recommendations suggest using ISO's X.500 directory service for naming, storing distribution lists, storing profiles of User Agents, and user authentication. By using X.500, users do not have to contend with ungainly machine-oriented names, but can use more intuitive names. However, X.500 is still under development in ISO. Pilot projects are under way, but there are no commercial implementations. No clear-cut solution exists for a directory service that's suitable for a heterogeneous network.

Security is another key issue. The 1988 spec provides facilities to authenticate who originated the message, verify who originated a delivery or nondelivery notice, check that the message or its contents were not altered and that all recipients received a copy. It also provides return receipt and registered mail services.

Security must be improved beyond these capabilities. Dangers lurk in one user masquerading as another. For example, no facility ensures that a user does not impersonate and misuse a Message Transfer System or falsely claim to originate a message. No facility ensures that message sequence is preserved. If a message were resequenced, messages could be replayed, reordered, or delayed.

X.400 is just one option for building the backbone of the enterprise messaging network; however, it has international acceptance and vendor independence. With the products based on the 1988 specification emerging, X.400 can serve a more useful role.

This tutorial, number 47, by Patricia Schnaidt, was originally published in the June 1992 issue of LAN Magazine/Network Magazine.

XML and XSL

The Extensible Markup Language (XML) is a World Wide Web Consortium (W3C) standard, approved in February 1998, for describing the content of Web pages. The Extensible Stylesheet Language (XSL) is a draft W3C standard, released in August 1998, for describing how to present XML pages within a Web browser. But before I talk about what XML does, consider what the ubiquitous Hypertext Markup Language (HTML) doesn't do. (In this tutorial, I'm assuming that you have some knowledge of HTML and can read simple HTML code.)

HTML BASICS

Say you wish to display the contents of this tutorial on a Web page. You might put the word "Tutorial" in headline style 2, the lesson number (Lesson 124) and title (XML and XSL) in headline style 1, the author's name in headline style 3, then the text in normal body style. In simple form, and with apologies to *Network Magazine's* long-suffering Webmaster, the code would look like:

```
<html>
<h2>Tutorial</h2>
<h1>Lesson 124: XML and XSL</h1>
```

```
<h3>by Alan Zeichick</h3>
The Extensible Markup Language (XML) is a World Wide Web
Consortium (W3C) standard ...
</html>
```

So far, so good—except there's no logical rhyme or reason behind the coding. If you want to search a site (or a long document) for tutorials, for certain lesson numbers, or for everything by a particular author, you'd have to know that the site usually uses h2 for the article type, and h3 for the author name. Not particularly efficient, and rather ad hoc. You'd probably prefer to design the code to read as:

```
<html>
<article_type>Tutorial </article_type>
<lesson_number>124 </lesson_number>
<article_title>XML and XSL </article_title>
<author>by Alan Zeichick</author>
<article_text>The Extensible Markup Language (XML) is ...
</article_text>
</html>
```

That's the principle behind XML, which lets Web site developers use meaningful tags based on the content of a Web page. Furthermore, XML allows site creators to define new tags as needed, rather than rely on a fixed set of generic HTML tags blessed by the W3C, or "embraced and extended" by a particular browser manufacturer.

Before you get carried away, please note that the second code fragment above is nonsensical: It's not HTML, and it's not really XML (but it's close). So, next I will look at what makes up an XML document, and then rewrite the example in genuine XML.

XML DOCUMENTS

XML isn't a page-description language—it's a structured data-definition language. It describes the content of a Web page using tags. (XML can actually describe any arbitrary data, but for this tutorial assume XML code is being written for the Web.) An XML document must start with an XML prolog, which begins with a declaration that the document is written in XML; it may optionally include a Document Type Definition (DTD) that describes the elements, tags, attributes, and other elements of the document. (DTD will be discussed in more detail later.) The prolog is followed by a single tag that encapsulates the entire document. That tag usually is named root or document.

Further, in XML, unlike HTML, all tags must be closed, and although tags may be nested, they may not overlap.

So, here is the example rewritten in XML (save it as tutorial.xml). The XML prolog, which begins with <?xml, contains two attributes; one describes the version of XML used, and the

other states that the document is complete in this one file. An XML document that adheres to these rules (and a few others regarding reserved characters) is said to be well-formed and should be interpretable by any XSL processor.

```
<?xml version="1.0" standalone="yes"?>
<document>
<article_type>Tutorial </article_type>
<lesson_number>124</lesson_number>
<article_title>XML and XSL </article_title>
<author>Alan Zeichick</author>
<article_text>The Extensible Markup Language (XML) is ...
</article_text>
</document>
```

XSL: DISPLAYING XML

XML, as stated earlier, is a way to describe the meaning of a document. Unlike HTML, it does not describe how to display a document in a Web browser. No browser can understand arbitrary tags like <author> and <lesson_ number>. That's why you need the Extensible Stylesheet Language (XSL), which describes the intended physical appearance of an XML document. You will also need software, called an XSL processor, to read the XML document, apply the XSL style sheet to its tags, and produce standard HTML as output. I will talk about where to find XSL processors, but first I will create a sample XSL style sheet.

An XSL style sheet consists of a text file contained within <xsl> and </xsl> tags. Between those tags are a series of rules describing the XML tags within an XML document and telling how to format them in HTML. Remember I wanted the article_type above to be displayed in HTML headline style 2? The XSL rule for that XML tag would be:

```
<rule>
<target-element type="article_type"/>
<h2><children/></h2>
</rule>
```

The special element <children/> tells the XSL processor to apply the XSL rule to the contents of the tagged item. Thus, the XML text <article_type>Tutorial</article_type> will be processed into <h2>Tutorial</h2>. Note that any HTML tag can be included as part of an XSL rule. Also, XML tags can be nested; in that case, XSL rules are applied recursively as needed. This capability allows Web site designers to exercise very fine control over the appearance of pages—far more than is illustrated using this simple example.

Here is the complete XSL document tutorial.xsl for the tutorial.xml file. The final rule, which doesn't explicitly mention a target-element type, is a catch-all that applies a default formatting to all tags not specifically defined.

```
<xsl>
<rule>
<target-element type="article_type"/>
<h2><children/></h2>
</rule>
<rule>
<target-element type="lesson_ number"/>
<h1>Lesson<children/></h1>
</rule>
<rule>
<target-element type="article_title"/>
<h1><children/></h1>
</rule>
<rule>
<target-element type="author"/>
<h3>by <children/></h3>
</rule>
<rule>
<target-element type="article_text"/>
<p><children/></p>
</rule>
<rule>
<target-element/>
<p><children/></p>
</rule>
</xsl>
```

GENERATING HTML

So, you'd like to see the HTML code generated by the XSL style sheet? Well, it's not as easy as loading it on a browser, as no generally available Web browser currently contains XSL processing capabilities. Microsoft includes some rudimentary XML parsing capability in Internet Explorer 4, and Netscape has demonstrated some XML functionality in Mozilla 5 (the core of its next-generation browser), but for now you'll need external software to apply the XSL style sheets to an XML document. You can find links to a number of freeware XSL processors at www.w3.org/XML/. [Note: Newer browser versions fully support XSL.]

The simplest processor to play with is Microsoft's MSXSL.EXE, downloadable from www.microsoft.com/xml/xsl/ msxsl.asp/. This command-line utility can apply XSL style sheets to an XML document, and produce an HTML output file. It's simple enough to run: from a DOS prompt, run MSXSL –i xmlfilename –s xslfilename –o htmlfilename.

If there is an error in the XML or XSL code, the MSXSL processor will let you know roughly where the problem occurred, and it may even guess at what went wrong (such as an argument mismatch: In my sample XSL file above, I had initially terminated the <h1> command with

</h2>). But if all goes according to plan, you'll end up with the HTML file:

```
<div><h2>Tutorial</h2>
<h1>Lesson 124</h1>
<h1>XML and XSL</h1>
<h3>by Alan Zeichick</h3>
<p>The Extensible Markup Language (XML) is ...</p></div>
```

If this example represented the best that XML could do, the technology would have died a swift death; after all, many of those capabilities can be handled using straightforward HTML with Cascading Style Sheets (CSS). The real payoff will come from using XML's more advanced features, such as data validation using the DTDs and the Extensible Linking Language.

DOCUMENT TYPE DEFINITIONS

Earlier, I discussed the prolog section of an XML document. The prolog must contain the <?xml statement, but it may optionally include either DTDs or a link to another file containing DTDs for application to the XML file. DTD validates a well-formed (that is, syntactically correct) XML document; all tags used in the body of the XML document must be defined in the DTD.

The DTD for this Tutorial, for example, would need to define the article_type field. It could simply define it as a random string of characters, but that wouldn't be of much benefit. A better strategy might be to predefine all of the possible article types, such as "Tutorial," "Feature," or "NT Techniques." DTDs can further specify that the article_type field must occur once, and only once, within an XML document, but that the author field can occur more than once, so articles with multiple authors can be supported. It could further specify that a new field, author_email, is valid only if the "@" symbol is included within it exactly once, or if the author_phone field can contain only digits. You get the idea.

An XML file can be validated only with an XML validating parser. (For a current list of validating parsers, nearly all of which are Java classes a developer can include in custom applications, rather than ready-to-run programs, see www.w3.org/XML/.) In general, if you're using only XML to build pages for displaying within a browser, you need not worry about DTDs. Those rigid definitions are required, however, to use XML as a domain-specific data-definitional language—to pass e-commerce data between servers, for example, or to standardize a way to describe chemical data, astronomical readings, or consumer credit reports. In those cases, having a strict definition of an XML document's allowable fields, and each field's allowable values and format, will make it easy to implement Web pages that enable the automated transfer of data between applications or organizations. One group doing just that is the XML/EDI Group (www.xmledi.com), which submitted an e-commerce-oriented DTD to the W3C in August 1998.

RESOURCES

The World Wide Web Consortium's site at www.w3.org/ XML/ should be your first stop. This page provides many links to standards proposals and other papers, newsgroups and forums, and XML software and demo code.

Microsoft maintains a fairly extensive collection of documents about XML and XSL at www.microsoft.com/ xml/. The **MSXSL.EXE** command-line processor and instructions can be found at www.microsoft.com/ xml/ xsl/ msxsl.asp. There are also additional documents at www.microsoft.com/standards/ xml/default.asp.

Netscape, which also supports XML, has posted developer information at http://developer.netscape.com/ tech/ metadata/ index.html, but it's not as extensive as Microsoft's library.

The best book I've found on XML is *XML: Extensible Markup Language,* by Elliotte Rusty Harold (IDG Books, 1998); if you're looking for one book on the topic, this is it.

This tutorial, number 124, by Alan Zeichick, was originally published in the November 1998 issue of Network Magazine.

Megaco and MGCP

Though IP telephony may well be the springboard for the kinds of new, enhanced voice and data services that carriers crave, deployment has been slowed by lack of (or too many!) Voice over IP (VoIP) standards. The latest call control protocol, Megaco (an evolution of Media Gateway Control), adds to the problem while seeking to reduce the number of protocols in use.

Megaco addresses the relationship between the Media Gateway (MG), which converts circuit-switched voice to packet-based traffic, and the Media Gateway Controller (MGC), sometimes called a call agent or softswitch, which dictates the service logic of that traffic (see Figure 1). Put another way, Megaco is designed for intradomain remote control of connection-aware or session-aware devices, such as VoIP gateways, remote access servers, Digital Subscriber Line Access Multiplexers (DSLAMs), Multiprotocol Label Switching (MPLS) routers, optical cross-connects, PPP session aggregation boxes, and so on.

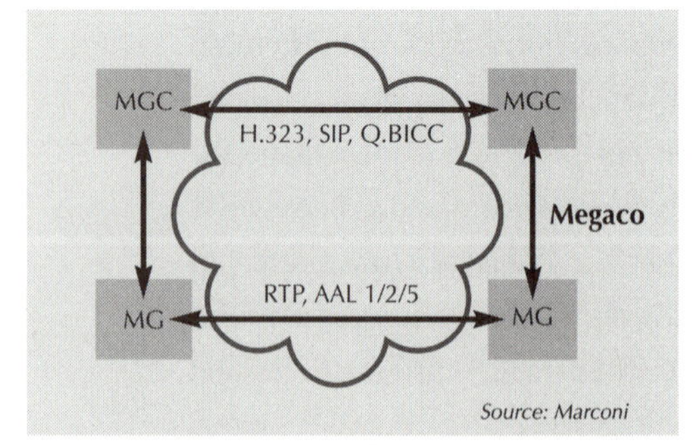

■ **Figure 1:** Megaco links the Media Gateway (MG) and Media Gateway Controller (MGC) for intradomain remote control of connection-aware or session-aware devices.

For carriers and vendors, the decision to implement Megaco means answering the question "How much signaling—the protocols that make connections between endpoints (such as telephones, PBX uplinks, and videoconference stations)—should be in the gateway?" From time to time, this debate has swung from "Endpoints are where the action is; they should be smart," to "Interaction is king; best to centralize."

For those who believe that end devices should be intelligent, signaling terminates in the MG itself. An example is a Session Initiation Protocol (SIP) phone, as the call control signaling runs directly on the end device. For those who believe there is merit in leaving the signaling on a general-purpose computer, leaving adaptation of media on and off the network to a specialized device, there is need for a protocol between the media handling part (the MG) and the signaling part (the MGC).

That's where MGCP and the newest kid on the block, Megaco, (also known by its ITU designation, H.248) come in. These are relatively low-level device-control protocols that instruct an MG to connect streams coming from outside a packet or cell data network onto a packet or cell stream such as the Real-Time Transport Protocol (RTP). Megaco is essentially quite similar to MGCP from an architectural standpoint and the controller-to-gateway relationship, but Megaco supports a broader range of networks, such as ATM.

For example, MGCP typically conditions the endpoint to look for an off-hook indication (when a person lifts the receiver to make a call). When the MG detects the off hook, it tells the MGC, which might respond with a command to instruct the MG to put dial tone on the line and listen for DTMF tones indicating the dialed number. After detecting the number, the MGC determines how to route the call and, using an inter-MGC signaling protocol such as H.323, SIP, or Q.BICC, contacts the terminating MGC. The terminating MGC might instruct the appropriate gateway to ring the dialed line. When the MG detects the dialed line is off hook, both MGs might be instructed by their respective MGCs to establish two-way voice across the data network. Thus, these protocols have ways to detect conditions on endpoints and notify the MGC of their occurrence; place signals (such as dial tone) on the line; and create media streams between endpoints on the MG and the data network, such as RTP streams.

Right now, many vendors consider it more practical to build large gateways that separate the signaling from the media-handling because of the density of the interconnections (which may have OC-3 or even OC-12 connections). Removing the signaling to a fast server is more practical than trying to integrate it into the MG. Also, by removing the signaling from a residential gateway, network operators retain a higher degree of control, which many believe will result in more reliable networks—vital if VoIP systems support lifeline/emergency services.

SO WHAT'S IN MEGACO?

There are two basic constructs in Megaco: terminations and contexts (see Figure 2, page 38). Terminations represent streams entering or leaving the MG (for example, analog tele-

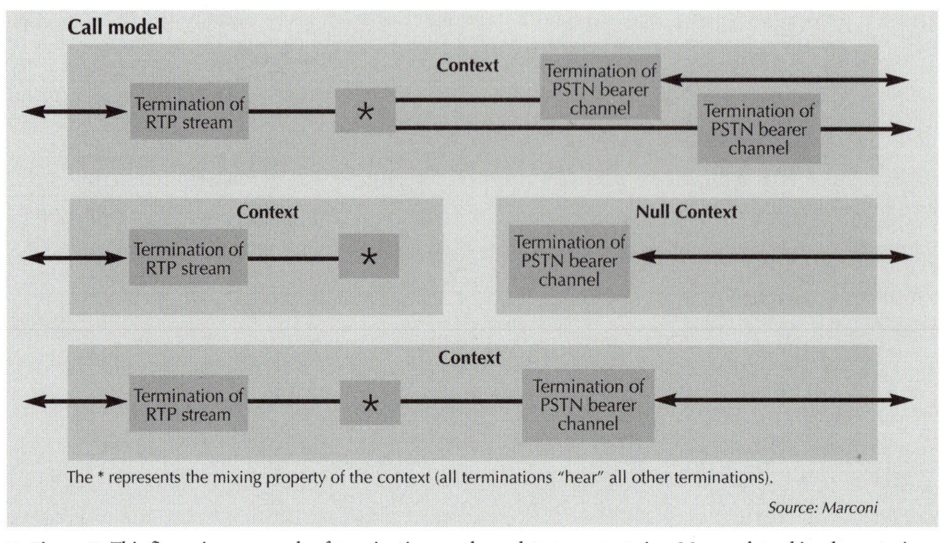

■ **Figure 2:** This figure is an example of terminations as they relate to contexts in a Megaco-based implementation.

phone lines, RTP streams, or MP3 streams). Terminations have properties, such as the maximum size of a jitter buffer, which can be inspected and modified by the MGC. A termination is given a name, or TerminationID, by the MG. Some terminations, which typically represent ports on the gateway, such as analog loops or DS0s, are instantiated by the MG when it boots and remain active all the time. Other terminations are created when they are needed, get used, and then are released. Such terminations are called "ephemerals" and are used to represent flows on the packet network, such as an RTP stream.

Terminations may be placed into contexts, which are defined as when two or more termination streams are mixed and connected together. The normal, "active" context might have a physical termination (say, one DS0 in a DS3) and one ephemeral one (the RTP stream connecting the gateway to the network). Contexts are created and released by the MG under command of the MGC. Once created, a context is given a name (ContextID), and can have terminations added and removed from it. A context is created by adding the first termination, and it is released by removing (subtracting) the last termination.

While a simple call may have two terminations per context (the termination representing the end device, such as the telephone, and the termination representing the RTP connection to the network), a conference call might have dozens, each one representing one leg of the conference. It's also possible to have a context with only one termination (say, in a three-way call). At any one time, two of the terminations are in one context (and are therefore connected together), while the third termination is in a context all by itself. When the user indicates that he or she wants to switch from the active to the standby caller (by using "flash," for instance), one of the terminations is moved to the other context.

A termination may have more than one stream, and therefore a context may be a multi-stream context. Audio, video, and data (for example, T-120 shared whiteboard) streams may exist in a context among several terminations.

Besides making basic connections by placing terminations in contexts, MGs may also be able to generate tones, announcements, ringing, and other signals that the user can hear. Megaco includes facilities to apply signals to terminations and control them. This enables gateways that have Interactive Voice Response (IVR) functionality to be controlled with Megaco, and also provides normal call-progress indications.

Asynchronous events, such as the hookswitch on a phone or a DTMF keypress, can be detected on the MG and reported to the MGC. By using a shorthand notation called a "digitmap," MGs may be programmed to look for an entire phone number or feature invocation and send the "dial string" to the MGC as one event. Statistics may be kept by the MG, and reported to the MGC, for such quantities as bytes sent or received, packets lost, and so on.

Megaco uses a series of commands to manipulate terminations, contexts, events, and signals. The Add command adds a termination to a context and may be used to create a new context at the same time. The Subtract command removes a termination from a context and may result in the context being released if no terminations remain. The Move command moves a termination from one context to another. Modify changes the state of the termination. The commands AuditValue and AuditCapabilities return information about the terminations, contexts, and general MG state and capabilities. ServiceChange creates a control association between an MG and an MGC and also deals with some failover situations. All of these commands are sent from the MGC to the MG, although ServiceChange can also be sent by the MG. The Notify command, with which the MG informs the MGC that one of the events the MGC was interested in has occurred, is sent by the MG to the MGC.

Commands are grouped into transactions, which are a series of commands that are executed in order, acting as the unit of transfer from the MG to the MGC and back again. Transactions are sent by a variety of transports between MGs and MGCs, including UDP and TCP (other transport options are coming). The MGC then sends an acknowledgement, along with replies for each command, back to the MG.

A package is a set of properties, events, signals, and statistics that are defined in a document and realized on a set of terminations in a gateway. Megaco defines a base set of packages for very common capabilities, such as analog and digital loops, DTMF detection and generation, and RTP. Both the IETF and ITU are working on additional package definitions. By defining a package, all the specific characteristics of a termination that realizes how that package is defined, and all gateways that implement that package can be controlled by an MGC that understands the package. The analog loop package has events for hookstate, signals for ring, and statistics for bytes sent and received, for example. A termination can have more than one package implemented on it. Packages can also be defined by any organization; a vendor could even define its own package.

ARE WE THERE YET?

With approximately a dozen independently developed implementations expected at a late-August 2000 interoperability-testing event at the University of New Hampshire's Interoperability Laboratory (see www.tsemantics.com/megaco-interop), Megaco is off the starting block. Several carriers have begun asking their vendors for Megaco support. There are, of course, a number of MGs already on the market. These devices use older, de facto protocols, most notably IPDC and MGCP. Of concern to many is whether Megaco will supplant MGCP, or whether the on-the-ground reality of MGCP will stymie acceptance of Megaco as the international standard for media applications.

"I can't predict if Megaco will take root, at the expense of MGCP or IPDC, or [if] the reverse will happen," says Ike Elliott, vice president of softswitch services at Level 3 Communications. "In truth, the protocols bear a strong resemblance to one another, and for many applications, it won't matter much which protocol is used. However, I can say that Megaco is more closely coupled with media applications than MGCP because the base protocol includes semantics for conferencing. Because of that, MGCP may be a better base for non-media-centric applications, such as MPLS-based session control.

RESOURCES

This article was largely taken from notes by Brian Rosen, the chief technology officer at Marconi (www.marconi.com), who put together the lion's share of this article's material.

Learn more about Megaco at the IETF's home page (www.ietf.org/html.charters/megaco-charter.html). The current draft is available at ftp://standards.nortelnetworks.om/megaco/docs/latest/megah248_protocol.pdf/.

This tutorial, number 147, by Doug Allen, was originally published in the October 2000 issue of Network Magazine.

SECTION VI

Network Hardware

Cabling

Cable is the medium that ordinarily connects network devices. Cable's ability to transmit encoded signals enables it to carry data from one place to another. These signals may be electrical as in copper cable or light pulses as in fiber-optic cable.

A few networks don't use cable at all. Instead, data is carried through the air as microwave, infrared, radio frequency, or laser-produced visible light signals. These wireless networks are often expensive, and may require licenses from the Federal Communications Commission. When the cost of running cable is prohibitively high or a network must be mobile or temporary, a wireless LAN can make sense.

THREE CHOICES

Network users have three basic cable choices: coaxial, twisted-pair, and fiber-optic. Coaxial and twisted-pair cables both use copper wire to conduct the signals; fiber-optic cable uses a glass or plastic conductor. Before the Ethernet standards for unshielded twisted-pair installations were approved in 1992, the majority of LANs used coaxial cable, but a high proportion of subsequent installations have used the more flexible and less costly unshielded twisted-pair medium. The use of fiber-optics in local networks is growing, albeit slowly. Fiber is most often used on the backbone network and is not commonly run to the desktop.

Originally, access protocols were tied to cable type. Ethernet and ARCnet ran on coaxial cable only. (Most of the installed coaxial cable is there because Ethernet and ARCnet have been around for so long.) However, these protocols have since been modified to run on shielded and unshielded twisted-pair, and fiber-optic cable. Cable type is no longer tied to the access method. ARCnet and Ethernet run on coaxial cable, unshielded twisted-pair, and fiber-optic cabling. Token Ring runs on unshielded and shielded twisted-pair and fiber-optic cabling.

A tradeoff between speed and distance exists, especially with copper cable. It is possible to increase the speed of data transmission, but this reduces the distance that data can travel without regeneration. Signal regenerating products like repeaters and amplifiers can help, but the physical properties of cable impose certain limitations.

COAXIAL CABLE

Coaxial cable, or coax, has a long history. If you have cable television in your home, you have coaxial cable. Broadband transmission uses the same principles as cable TV and runs on coax. Broadband and cable TV take advantage of coax's ability to transmit many signals at the same time. Each signal is called a channel. Each channel travels along at a different frequency, so it does not interfere with other channels.

Coax has a large bandwidth, which means it can handle plenty of traffic at high speeds. Other advantages include its relative immunity to electromagnetic interference (as compared to twisted-pair), its ability to carry signals over a significant distance, and its familiarity to many cable installers.

Coax cable has four parts (see Figure 1). The inner conductor is a solid metal wire surrounded by insulation. A thin, tubular piece of metal screen surrounds the insulation. Its axis of curvature coincides with that of the inner conductor, hence the name coaxial. Finally, an outer plastic cover surrounds the rest.

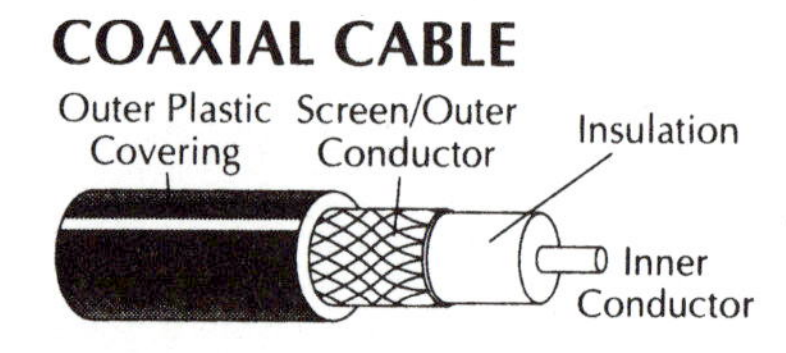

■ **Figure 1:** Coaxial cable, also called coax, is the oldest network cable. It is proven, easy to use. It has a large bandwidth and can support transmission over long distances.

Coax comes in several sizes. Standard Ethernet cable, the yellow stuff called thick Ethernet, is about the diameter of a man's thumb. Thin Ethernet, the black cable, is about as thick as a woman's pinky finger. ARCnet uses RG/68 coax cable. Thicker coax is more robust, harder to damage, and transmits data over longer distances. It's also more difficult to connect.

Standard Ethernet requires a "vampire tap" and drop cable to connect a LAN device. This combination is bulky and expensive. Thin Ethernet uses a biconic (or BNC) connector, which is easier to install than vampire taps.

TWISTED-PAIR

Twisted-pair cable has been around a lot longer than coaxial, but it has been carrying voice, not data. Unshielded twisted-pair is used extensively in the nationwide telephone system. Practically every home that has telephones is wired with twisted-pair cable.

In the past few years, vendors have been able to transmit data over twisted-pair at reasonable speeds and distances. Some of the first PC LANs, such as Omninet or 10Net, used twisted-pair cable but could only transmit data at 1Mbit/sec. Token Ring, when it was introduced in 1984, was able to transmit data at 4Mbits/sec over shielded twisted-pair. In 1987, several vendors announced Ethernet-like technology that could transmit data over unshielded twisted-pair, but computers can only be about 300 feet apart, not the 2,000 feet allowed by thick coax. Recent developments in technology make it possible to run even 16Mbit/sec Token Ring and 100Mbit/sec FDDI traffic over unshielded twisted-pair.

Twisted-pair offers some significant benefits. It's lighter, thinner, more flexible, and easier to install than coax or fiber-optic cable. It's also inexpensive. It is therefore ideal in offices or work groups that are free of severe electromagnetic interference.

Although there are a variety of types of twisted-pair cable types, shielded (see Figure 2) and unshielded (see Figure 3) are the two most important. Shielded twisted-pair has an RF-insulating material wrapped around the two twisted wires. Unshielded twisted-pair (or ordinary telephone wire) does not. Shielded twisted-pair is more immune to interference, which usually translates into higher transmission speeds over longer distance—it is more expensive, however.

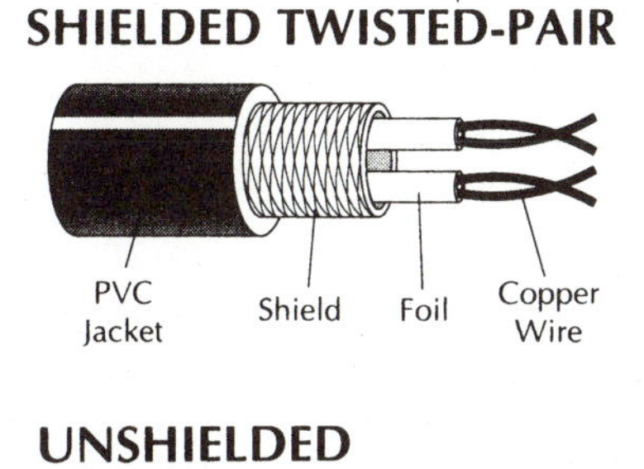

■ **Figure 2:** Shielded twisted-pair's shield increases its immunity to electromagnetic interference which allows it to transmit data over longer distances than unshielded twisted pair.

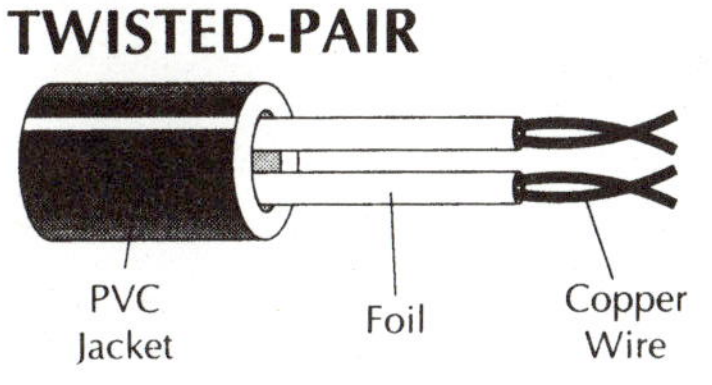

■ **Figure 3:** Unshielded twisted-pair is installed nearly everywhere. Besides being inexpensive and readily available, it is flexible and familiar to cable installers. It has become the cable of choice for the departmental network.

Unshielded twisted-pair is fast becoming the media of choice. By 1993, the market research firm Dataquest projected that 78 percent of Ethernet connections would be made via twisted-pair cables. Unshielded twisted-pair also gained in popularity for Token Ring networks, which were traditionally wired with shielded twisted-pair.

The most important result of the telephone industry's use of twisted-pair is modular cabling. A modular cabling system, built with patch panels, wiring closets, and connector jacks, makes it easier to move computers from one place to another without rewiring the LAN. A modular cabling system allows a company to prewire a building for its phone and data services. Once the wire is in place, people can move from office to office, and new cabling does not have to be run.

FIBROUS DIET

Fiber-optics has been touted as the answer to all the problems of copper cable. It can carry voice, video, and data. It has enormous bandwidth and can carry signals for extremely long distances. Because it uses light pulses, not electricity to carry data, it is immune to electromagnetic interference. It is also more secure than copper cable, because an intruder cannot eavesdrop on the signals, but must physically tap into the cable. To get at the information inside, a device must be attached, and the light level will subsequently decrease.

Despite its many advantages, fiber-optic's deployment in the LAN has been slow. According to Dataquest figures, by 1993, fiber-optics held only 1.4 percent of the LAN market. Cable installer's experience and fiber's high cost is holding back its widespread installation. Very simply, installing fiber-optic cable is very difficult. Splicing fiber-optic cables together is even more difficult. Putting connectors on the fiber-optic cable is also harder than for copper cable. The expense of diagnostic tools is another problem. Time domain reflectometers, ohmmeters, voltmeters, and oscilloscopes can be easily connected to any type of copper cable. But such tools must be specifically designed or adapted for fiber-optics use.

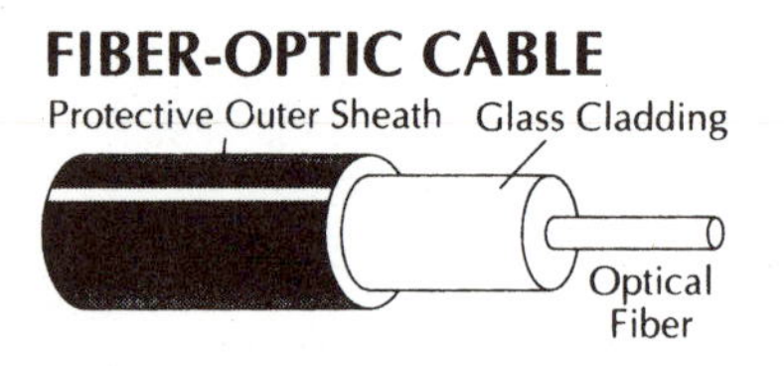

■ **Figure 4:** Fiber-optic cable offers tremendous bandwidth, tight security, immunity to electromagnetic interference, and can carry data over long distances. It is mostly used in backbones.

Fiber-optics has enjoyed its greatest success as a backbone medium for connecting sub-networks. Its properties make it ideal for the heavy traffic, hostile environments, and great distances that characterize backbone networks. Its immunity to electrical interference makes it ideal for the factory floor, another popular application.

Fiber-optic cable itself is a core fiber surrounded by cladding (see Figure 4). A protective covering surrounds both. LEDs or light emitting diodes send the signals down the cable. A detector receives the signals and converts them back to the electrical impulses that computers can understand. While the bits are encoded into light in a number of ways, the most popular method is to vary the intensity of the light.

Fiber-optic cable can be multimode or single-mode. In single-mode cable, the light travels straight down the fiber, which means data can travel greater distances. But since single-mode cable has a smaller diameter than multimode cable, it is harder (more expensive) to manufacture. In multimode cable, the light bounces off the cable's walls as it travels down, which causes the signals to weaken sooner, and therefore data cannot travel great distances. Single-mode cable is most often used in the nationwide telephone system, and multimode cable is most often used in LANs, since data is not required to travel across the country.

Standards for fiber-optic LANs have been developed. ANSI's Fiber Distributed Data Interface (FDDI) describes a network that can transmit data at 100Mbits/sec. It also specifies a dual, counter-rotating ring, which makes it fault tolerant. The IEEE has also developed standards for fiber-optic Ethernet.

Imaging applications and the proliferation of networks will force installation of high capacity LANs. Fiber-optics has enormous potential. Its capacities are tremendous. When wiring a new building, the best strategy is to run fiber-optic backbones, with twisted-pair to the desktops.

This tutorial, number 7, was originally published in the February 1989 issue of LAN Magazine/Network Magazine.

Lay of the LAN—Cabling Basics

If you're on a network, your machine has to be connected to at least one other machine, whether through metal cabling, fiber-optic light guides, or radio waves. Even as new technologies, such as radio wave-based networks and infrared light-based networks, produce new ways for connecting nodes to each other, the least expensive and most popular medium for networking computers is still copper wire.

Networks based on copper cabling generally use one of two main cable types: coaxial or twisted-pair. ARCnet started with RG-62/U and RG-59/U coaxial cable, but now includes twisted-pair and even fiber-optic cabling. Likewise, Ethernet, which was defined in the IEEE 802.3 standard, was originally implemented only on thick coaxial baseband cabling. But the Ethernet standard has expanded to include broadband coaxial, fiber-optic cabling, and twisted-pair.

ETHERNET CABLING

The thick Ethernet cabling standard was designed to serve as a network backbone. In this design, machines are connected to a thick, 0.4-inch double-shielded coaxial segment, or bus, through transceiver cables, using 15-pin transceiver or AUI cables. Each AUI cable can be up to 50 meters long—which is a blessing given the difficulty of maneuvering the thick, trunk cabling.

The thick Ethernet cabling system has been standardized by the IEEE as 10Base5. The "10" in 10Base5 refers to its 10Mbit/sec transmission rate, while the term "Base" refers to baseband cabling and the "5" refers to the maximum length for a segment in hundreds of meters. The coaxial transceivers are connected directly into the coaxial trunk either by piercing the cabling (called a vampire tap, appropriately enough) or by in-line connection with N connectors. From there, an AUI cable connects each transceiver to a network interface card on a node. Each end of the 10Base5 segment must have a 50-Ohm terminating resistor installed, and only one terminator should be grounded.

ETHERNET SLIMS DOWN

10Base2, also known as thin Ethernet—or in slang as cheapernet, was developed to lower the cost of installing an Ethernet network. In fact, this thin coaxial cabling can still be the most cost-effective solution for companies that want to connect a few PCs within a relatively small area.

However, with 10Base2, the Ethernet bus connects directly to a T connector on the back of each node. Consequently, the design is much more prone to disaster: If any user breaks the chain (for example, by accidentally disconnecting the wrong part of the T connector), either the entire network will go down or, at the very least, the side of the network that doesn't have a server will be isolated.

The 10Base2 standard has a 10Mbit/sec transmission speed and it uses thin, flexible RG-58A/U coaxial cable, typically 0.2 inches in diameter, with a stranded conductor. Like 10Base5, the coaxial bus must be terminated at each end, and one end must be grounded. The maximum length of a 10Base2 segment is 185 meters, which is not the 200 meter limit you might expect based on its name.

WITH A TWIST

Partly in response to complaints about the difficulty of troubleshooting coaxial Ethernet and partly to take advantage of the existing wiring in many offices, the 10BaseT standard was developed to work with Category 3 twisted-pair copper cabling. (The "T" in 10BaseT stands for twisted pair.)

Common in newer buildings, Category 3 cabling is often used as phone wiring. But because it has low noise and crosstalk characteristics, it can be used to support a network installation that can deliver 10Mbit/sec performance.

In some older buildings, the cabling may not be up to Category 3 specifications, and subsequently, may not be able to support a network installation. In this situation, you may still want to install twisted-pair cabling rather than coaxial despite its higher price, because it has an easier installation.

The 10BaseT specification calls for unshielded twisted-pair cabling, commonly called UTP, to use two of the four pairs of conductors in a typical Category 3 cable (see Figure 1). One pair transmits data, while the other is meant to receive data. The standard doesn't specify

10BASET WIRING CONFIGURATION

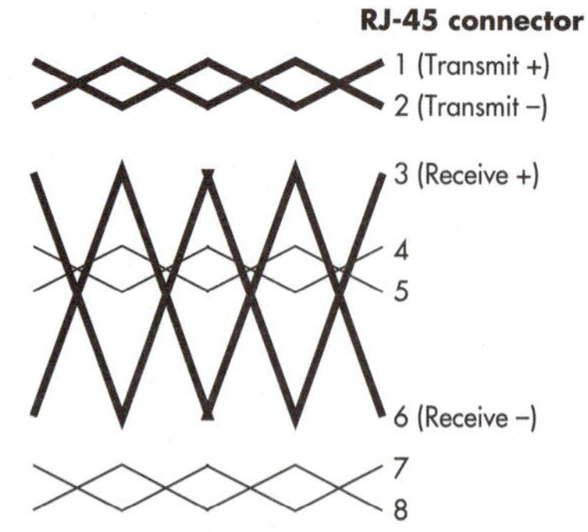

■ **Figure 1:** UTP cabling typically has four pairs of wire in one sheath, which connects to an RJ-45 modular jack. For 10BaseT, only the 1-2 pair and the 3-6 pair are wired, but not used with Ethernet.

TOKEN RING PHYSICAL TOPOLOGY

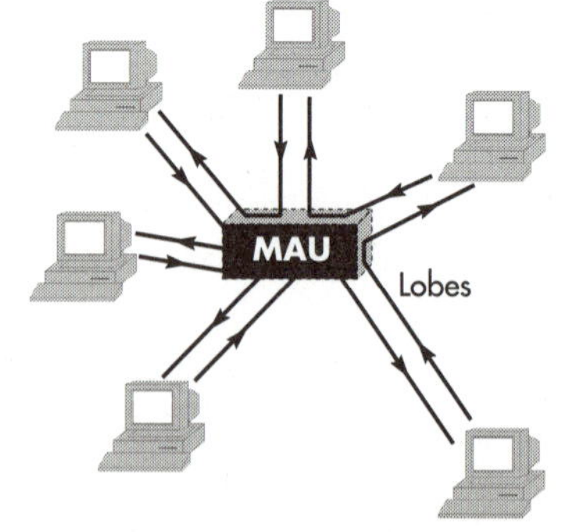

■ **Figure 2:** In Token Ring networks, packets are forwarded from station until they arrive at the correct address. Although Token Ring uses a star configuration, each station is still connected, via the MAU, to the next station on the ring.

exactly what the remaining two pairs of conductors can be used for, so many cable installers use them to create a second 10BaseT data connection. However, this practice may complicate future conversion to faster transmission speeds, because the standards for 100Mbit/sec transmission over Category 3 cabling require the use of all four pairs of cable.

Unlike 10Base2 and 10Base5, 10BaseT uses a star topology in which each node connects to a central concentrator or multiport repeater, which is typically located in a central equipment room or wiring closet. This topology meshes well with the existing cabling layout in most buildings.

FAST ETHERNET CABLING

With many users outgrowing the 10Mbit/sec data rate of 10BaseT, vendors are responding with new products and standards to wring more speed from copper cabling. Two physical cabling strategies are emerging: using all four pairs of conductors in existing Category 3 cabling, or using two pairs of conductors in the faster and more noise resistant Category 5 cabling. Category 5 specifications permit signaling rates of up to 100MHz, compared to Category 3, which allows rates of up to 16MHz.

Two emerging standards, 100BaseTX and 100BaseT4, conform to the IEEE 802.3 standard, while a third, 100VG-AnyLAN conforms to the new IEEE 802.12.

TOKEN RING CABLING

Based on a token-passing logical ring topology, Token Ring networks tend to show predictable performance degradation curves when under heavy loads, as opposed to Ethernet networks, which degrade less predictably when stressed.

Although it's based on a logical ring topology, a Token Ring network looks more like a star, with each node connected directly to a central MAU. Think of this cabling arrangement as being a collapsed star, in which the middle of each segment of the ring has been pulled back to the central hub (see Figure 2). The internal circuitry of the MAU is configured as a ring. As a node is inserted into the ring and activated, a relay in the MAU closes to include the node (called a lobe) in the electrical ring.

Token Ring networks come in two "flavors": 4Mbit/sec and 16Mbit/sec. Telephone-grade UTP cabling can be used for 4Mbit/sec Token Ring equipment. IBM calls this cabling Type 3, and it is roughly equivalent to Category 2 cabling. Connecting a 4Mbit/sec Token Ring adapter to Type 3 cabling requires a media filter to reduce electrical noise. Like 10BaseT, Token Ring networks use two pairs of conductors, but in a different configuration (see Figure 3).

IBM Type 1 cabling is required for 16Mbit/sec Token Ring networks. This requires the same wiring configuration as UTP cabling, but with solid, thicker, and heavier conductors. Each pair of wires is shielded with foil, and the resulting four-pair cable is further protected

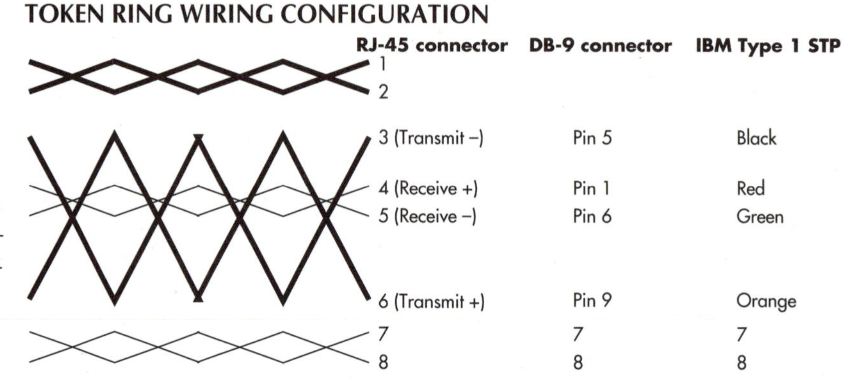

TOKEN RING WIRING CONFIGURATION
RJ-45 connector
DB-9 connector
IBM Type 1 STP
1
2
3 (Transmit –)
Pin 5
Black
4 (Receive +)
Pin 1
Red
5 (Receive –)
Pin 6
Green
6 (Transmit +)
Pin 9
Orange
7
7
7
8
8
8

■ **Figure 3:** Token Ring initially used IBM Type 1 STP wiring. Now it typically uses a 9-pin D-subminiature (DB-9) connector, although it also has been implemented over UTP using RJ-45 connectors. As shown, wiring configurations depend on which type of cabling you're using.

with braided shielding. IBM developed special hermaphroditic connectors to allow Type 1 cables to attach to either workstation patch cords or to other Type 1 cables as extensions.

PLAN FOR SUCCESS

Whether you're using Token Ring or Ethernet, or if you're just starting to network your company, the cabling you choose and how you install it can make a big impact on network performance. So take the time to carefully plan your wiring layout. For example, make sure your cable distances don't exceed the recommended standard length, and double-check the cabling and connections to ensure they meet established specifications.

Such steps can do more than make your network faster and more reliable; they can free you from the chore of constantly reacting to problems caused by the physical wiring plant, which are generally hard to trace. Additionally, if you plan right, you can use the same wiring when you upgrade to a faster network, saving your company money.

This tutorial, number 80, written by Dave Fogle, was originally published in the April 1995 issue of LAN Magazine/Network Magazine.

Cable Testing

Understanding the electrical impulses of UTP cabling sheds light on cable performance.

With the exception of wireless systems, networks rely on cable to conduct data from one point to another. In the case of copper-wire-based, unshielded twisted-pair (UTP) cabling, data is conveyed in the form of electric, digital signals. Because these signals are essentially bursts of electricity, the electrical characteristics of the cable itself greatly affect the integrity of the signal being transmitted. A bad length of cable or a poor cable installation can result in signal loss or distortion, and consequently, network failure.

To minimize such occurrences, cable vendors test their cables to guarantee performance. However, this doesn't make their products fault-proof; bad cabling does exist. In some cases, the error lies in improper cable installation. Network managers can use cable testers to ensure that a cable can conduct signals correctly. They can also use cable testers to verify if a cable is properly installed and to troubleshoot faulty cable.

A solid grounding in the electrical properties of UTP is a good way to learn how cable can affect the performance of a network.

THE ELECTRICAL CIRCUIT

A network can be broken down in simplistic fashion into an electrical circuit metaphor. In this case, a network essentially comprises energy sources, conductors, and loads. An energy source is a network device that transmits an electrical signal (data). The conductors are the wires that the signal travels over to reach its destination, which is usually another network device. The receiving device is known as the load. In its entirety, the connected network is a completed circuit (see Figure).

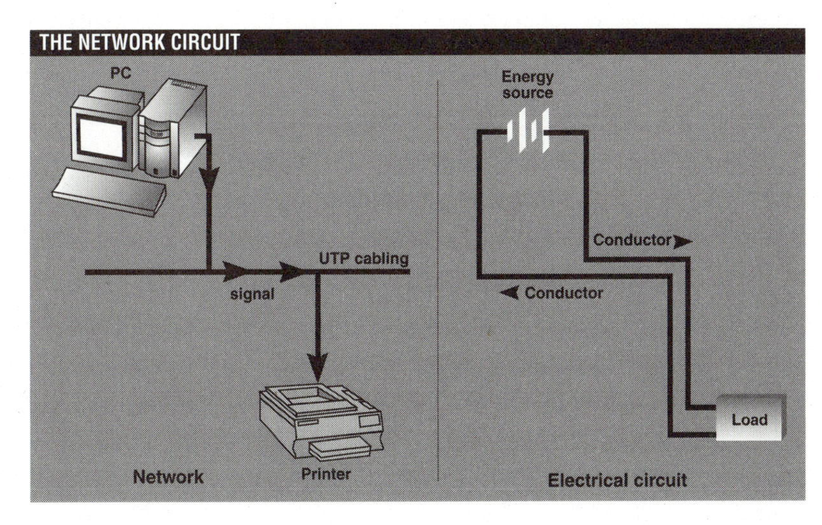

■ **Figure 1:** Simple Circuitry: When two devices communicate over a network, they form an electrical circuit. Cables serve as the transfer medium and are bound to the same electrical properties as normal conductors.

When an energy source transmits a signal, it is outputting an electric charge onto the conductor by applying voltage to the completed circuit. Voltage is measured in volts. The voltage propels the charge across the cable, and the flow of the charge is known as a current, which is expressed in amperes, or amps.

In the computer world, the electric signal transmitted by an energy source is a digital signal known as a pulse. Pulses—in the form of a series of voltages and no voltages—can be used to represent a series of ones and zeros. Digital pulses form bits, and a series of eight bits creates the almighty byte.

The key to a successful signal transmission is that when a load receives an electrical signal, the signal must have a voltage level and configuration consistent with what had been originally transmitted by the energy source. If the signal has undergone too much corruption, the load won't be able to interpret it accurately.

In short, a good cable will transfer a signal without too much fudging of the signal, while a bad cable will render a signal meaningless.

PROPERTY LIMITS

Due to the electrical properties of copper wiring, the signal will undergo some corruption during its transit. Obviously, signal corruption within certain limits is acceptable. Once the electrical properties exceed the limits prescribed to a certain cable type, the cable is no longer reliable and must be replaced or repaired.

As a signal propagates down a length of cable, it loses some of its energy. So, a signal that starts out with a certain input voltage will arrive at the load with a reduced voltage level. The amount of signal loss is known as attenuation, which is measured in decibels, or dB. If the voltage drops too much, the signal may no longer be useful.

The table lists the attenuation values allowable at the end of 100 meters of Category 3 through 5 UTP.

(Please note that the attenuation and near-end crosstalk [NEXT] values in the table are performance specifications detailed in Telecommunications Systems Bulletin [TSB] -67, written by the Electronics Industry Association/Telecommunications Industry Association [EIA/TIA]. All other values are suggested limits, not standards. In addition, the table shows limits for certain frequencies, although different frequencies can operate on each category of cabling. The limits for some properties vary according to frequency.)

TABLE 1—STANDARD CABLE LIMITS

Cable	Frequency	Impedance	Capacitance	Attenuation	NEXT
Cat 3	10MHz	100ohms +/-15%	20pF/ft.	11.5dB	22.7dB
Cat 4	16MHz	100ohms +/-15%	17pF/ft.	9.9dB	33.1dB
Cat 5	100MHz	100ohms +/-15%	17pF/ft.	24dB	27.1dB

Attenuation has a direct relationship with frequency and cable length. The higher the frequency used by the network, the greater the attenuation. Also, the longer the cable, the more energy a signal loses by the time it reaches the load.

A signal loses energy during its travel because of electrical properties at work in the cable. For example, every conductor offers some resistance to a current. Resistance, which is measured in ohms, acts as a drag on the signal, restricting the flow of electrons through the circuit and causing some of the signal to be absorbed by the cable. The longer the cable, the more resistance it offers.

Due to its electrical properties, a cable not only resists the initial flow of the current, it opposes any change in the current. The property that forces this reaction is called reactance, of which there are two relevant kinds: inductive reactance and capacitive reactance.

In an inductive reaction, a current's movement through a cable creates a magnetic field. This field will induce a voltage that will work against any change in the original current.

Capacitance is a property that is exhibited by two wires when they are placed close together. The electrons on the wires act upon each other, creating an electrostatic charge that exists between the two wires. This charge will oppose change in a circuit's voltage. Capacitance is measured in farads or picofarads (see table).

Reactance can distort the changes in voltage that signify the ones and zeros in a digital signal. For example, if the signal calls for a one followed by a zero, reactance will resist the switch from voltage to no voltage, possibly causing the load to misidentify what the voltage represents.

IMPEDING PROGRESS

When you combine the effects of resistance, inductance, and capacitance, the result is the total opposition to the flow of the current, which is known as impedance and is measured in ohms.

It's important for components of a circuit to have matching impedance. If not, a load with one impedance value will reflect or echo part of a signal being carried by a cable with a different impedance level, causing signal failures. For this reason, cable vendors test their cables to verify that impedance values, as well as resistance and capacitance levels, comply to standard cable specifications.

It's also important for the impedance of a cable to be uniform throughout the cable's length. Cable faults change the impedance of the cable at the point where the fault lies, resulting in reflected signals.

Cable testers use this trait to find cable faults. For example, a break in a wire creates an "open circuit," or infinitely high impedance at that point. When a high frequency signal emitted from a cable tester encounters this high impedance, it will reflect back towards the tester like an ocean wave bouncing off a seawall. Similarly, a short circuit represents zero impedance, which will also reflect a high frequency signal, but with an inverted polarity.

The cable testing device can then tell you approximately how far down the cable the fault lies. The formula for this feature uses a cable value known as nominal velocity of propagation (NVP), which is the rate at which a signal can flow through the cable, expressed as a percentage of light speed. The cable tester multiplies the speed of light by the cable's NVP and by the total time it takes the pulse to reach the fault and reflect back to the tester, and divides it by two, for the one-way distance.

The same concept is used to check the electrical length of a cable installation. In this case, you must make sure not to terminate one end of the cable. The open end will register as infinite impedance and reflect a pulse back to the tester. Again, this response time is plugged into the formula to estimate the overall electrical length of the wire.

As an aside, some cable testers can't check the first 20 feet or so of a cable. The reason for this blind spot is that a pulse transmitted by the tester will be reflected back to the device before it is entirely transmitted. Thus, the tester can't get an accurate reading.

MIXING SIGNALS

Finally, the successful transmission of a signal can be jeopardized by noise, which can introduce false signals, or noise spikes, at different frequencies on a wire. A load may interpret a noise spike as part of a digital signal, distorting the original content of the signal. Common sources of noise spikes include AC lines, telephones, and devices such as radios, microwave ovens, and motors. Some cable testers test for noise, running tests at different frequencies.

Another type of interference is called crosstalk, or more specifically, near-end crosstalk (NEXT). As mentioned, when a current moves through a wire, it creates an electromagnetic field. This field can interfere with signals traveling on an adjacent wire. To reduce the effect of NEXT, wires are twisted—thus the name twisted pair. The twisting allows the wires to cancel each other's noise.

The risks of NEXT are highest at the ends of a cable because wire pairs generally don't have twists at their ends, where they enter connectors. If the untwisted end length is too long, NEXT levels can rise to distorting levels.

Also, due to attenuation, signals are strongest when they are transmitted, and weakest when they arrive at their destination. So, the magnetic field of a signal being transmitted from a device on one wire may overwhelm a signal arriving at the same device on the wire's pair.

NEXT is measured in decibels, which represent a ratio of a signal's strength to the noise generated by crosstalk (see table). The stronger the signal and weaker the noise, the higher the NEXT value. For this reason, a high NEXT reading is good. Low NEXT readings, which indicate high crosstalk interference, can mean the cable is terminated improperly.

THE NOT-SO-FINAL WORD

In the past, the topic of cable testing and performance has been a contested one, due mostly to the absence of accepted testing and performance standards. Recently, the EIA/TIA finalized TSB-67, which defines what it calls "transmission performance specifications for field testing of unshielded twisted-pair cabling systems." These specifications cover Category

3, 4, and 5 UTP. TSB-67 also details specifications for cable tester performance—something that has been noticeably absent.

Although the document is doing much to calm some pretty turbulent waters, other questions concerning cable performance and testing loom on the horizon. For example, TSB-67 addresses specifications for Category 5 cabling up to 100MHz. How does this relate to 155MHz ATM, which is supposed to run over Category 5?

This tutorial, number 93, by Lee Chae, was originally published in the May 1996 issue of LAN Magazine/Network Magazine.

Basic Electricity Boot Camp

This crash course in electricity will help you with some of the terms and technology.

The circuitry in computer systems is powered by direct current (DC), the type of electricity produced by batteries. A battery produces a voltage at a constant level. Electric utilities, however, supply power with alternating current (AC), in which voltage constantly varies. To power a computer system, the computer's circuitry converts AC into DC of the proper voltage.

GENERATOR X

When an electrical current flows through a conductor, a magnetic field (or "flux") develops around the conductor. The highest flux density occurs when the conductor is formed into a coil having many turns. In electronics and electricity, a coil is usually known as an inductor. If a steady DC current is run through the coil, you would have an electromagnet—a device with the properties of a conventional magnet, except you can turn it on or off by placing a switch in the circuit.

There's reciprocity in the interaction between electron flow and magnetism. If you sweep one pole of a magnet quickly past an electrical conductor (at a right angle to it), a voltage will be momentarily "induced" in the conductor. The polarity of the voltage will depend upon which pole of the magnet you're using, and in which direction it sweeps past the conductor.

This phenomenon becomes more apparent when the conductor is formed into a coil of many turns. Figure 1 shows a coil mounted close to a magnet that is spinning on a shaft. As the north pole of the magnet sweeps past the coil, a voltage is induced in the coil, and, if there is a "complete" circuit, current will flow. As the south pole of the magnet sweeps past, a voltage of opposite polarity is induced, and current flows in the opposite direction.

This relationship is the fundamental operating principle of a generator. The output, known

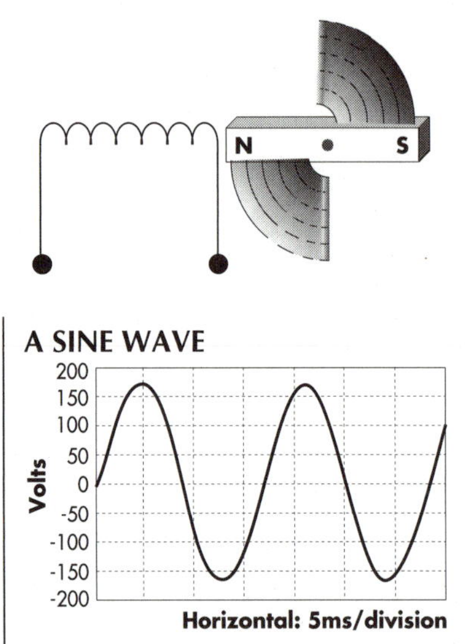

■ **Figure 1:** To generate electrical power, a coil is mounted close to a magnet that is spinning on a shaft. As the poles of the magnet sweep past the coil, voltages of alternating polarity are induced in the coil.

■ **Figure 2:** A 120-volt, 60-Hz generator produces power output that cyclically varies from 169.7V to -169.7V.

as alternating current, is the type of power that electric utility companies supply to businesses and homes. A practical generator would likely have two coils mounted on opposite sides of the spinning magnet and wired together in a series connection. Because the coils are in a series, the voltages combine, and the voltage output of the generator will be twice that of each coil.

Figure 2 is a graph of the voltage produced by such a generator as a function of time. Let's assume that this happens to be a 120-volt, 60-Hz generator. The voltage at one point in the cycle momentarily passes through 0 volts, but it's headed for a maximum of 169.7 volts. After that point, the voltage declines, passing through 0 volts, then reverses its polarity, and has a negative "peak" of -169.7 volts.

This curve is known as a sine wave since the voltage at any point is proportional to the sine of the angle of rotation. The magnet is rotating 60 times a second, so the sine wave repeats at the same frequency, making the period of a single cycle one-sixtieth of a second.

CURRENT EVENTS

In a direct-current system, it's easy to determine voltage because it is nonvarying or varies slowly over time. You can simply make a measurement with a DC voltmeter. But in an AC circuit, the voltage is constantly changing.

Electrical engineers state the voltage of an AC sine wave as the RMS (root-mean-square), a value equal to the peak value of the sine wave divided by the square root of two, which is approximately 1.414. If you know the RMS voltage, you can multiply it by the square root of two to calculate the peak voltage of the curve. If you were to power a light bulb from 120V(RMS) AC, you would get the same amount of light from the bulb as you would by powering it from 120V DC. Yet another device uses electromagnetic induction: the transformer.

Remember that a coil (inductor) develops a magnetic field when current flows through it. If alternating current is sent through the coil, it will produce an undulating magnetic field that reverses its polarity whenever the current reverses direction. If a second coil is wound around the first (but the two are electrically insulated from each other), the magnetic field of the first coil will induce a voltage in the second coil. In effect, the first coil sets up the same type of alternating magnetic field that is produced by the spinning magnet of a generator.

Just as an iron core improves the inductance of a coil, it has the same positive effect in a transformer, and most power transformers are wound on iron cores. As Figure 3 shows, a transformer is made up of two coils (usually referred to as windings) that are electrically insulated from each other. The two parallel lines separating the windings indicate that this is an iron-core transformer. The winding on the left, termed the primary, is connected to the electric utility's AC power grid. On the right is the secondary winding, in which the magnetic flux induces an AC voltage.

A TRANSFORMER

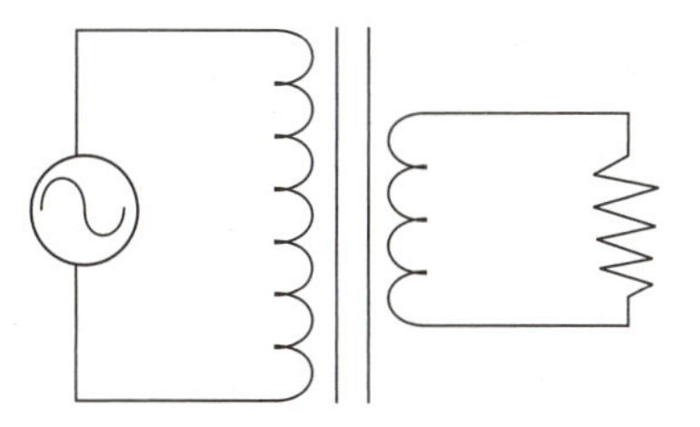

■ **Figure 3:** In a transformer, two coils (usually referred to as windings) are electrically insulated from each other. The left-hand winding, which is connected to the utility AC power grid, is called the primary winding. The secondary winding, on the right, is connected to a load.

CONNECTING DEVICES TO GROUND

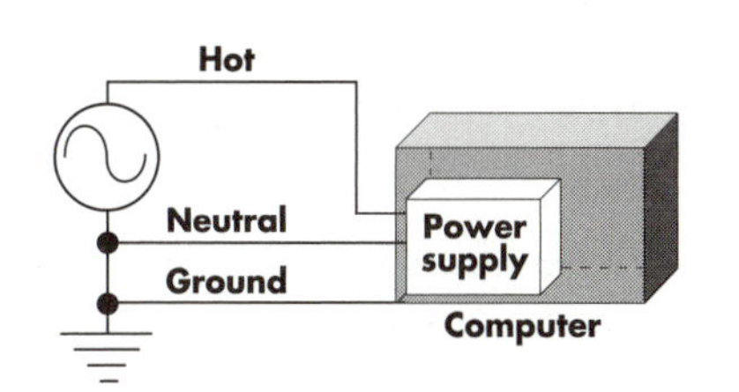

■ **Figure 4:** Power-using devices such as computers are connected to "ground" through the facility wiring.

The "turns ratio" of the transformer describes how many turns are on the primary and secondary winding relative to each other. A turns ratio of 5-to-1 means that there are five times as many turns in the primary winding as there are in the secondary winding.

A transformer "transforms" AC voltages in direct proportion to the turns ratio. With a 5-to-1 turns ratio, the voltage on the secondary will be one-fifth that on the primary. This would be a "step-down" transformer, since it steps down (reduces) the voltage. A transformer with a 1-to-3 turns ratio, on the other hand, would be a "step-up" transformer, with the voltage on the secondary being three times that on the primary. The current, however, will be reduced by the same factor; the current on the secondary will be one-third that on the primary.

Step-up transformers at utility power plants increase the voltage for transmission over long distances. Then, nearer the delivery points, utility substations use step-down transformers to bring the voltage down to a lower level for distribution through neighborhoods. Finally, numerous transformers on utility poles step the power down again before it goes to commercial and residential customers. Energy losses in power distribution are caused by the resistance of the conductors, and forcing large currents through a resistance produces high power losses. By stepping up the voltage in long distance transmission lines, then, utilities are able to minimize the current, and thereby minimize transmission losses.

The world's earliest commercial electric power systems were DC, but AC eventually won out, primarily because transformers, which only work with AC, could easily step voltages up or down, as needed.

WELL-GROUNDED

One further point we need to address: How is voltage measured? What is the frame of reference? On virtually all computers (and many other devices), the chassis is connected to "ground." This means that the chassis is at the earth's potential, and that is the reference point for measuring the voltages in the system's various circuits.

The block diagram in Figure 4 shows how grounding works. The symbol of a sine wave in a circle represents the power source. For our purposes, this is the building's main switch panel. Below the power source, the stacked series of lines represents ground. This normally consists of a heavy-gauge wire or cable that connects the switch panel to a thick copper ground rod driven into the earth. (In large, steel-framed buildings, one of the building's steel columns will serve as the ground point.)

Note that one side of the AC power wires is also connected to ground at the switch panel. This line is known as the "neutral" line, because it is at essentially earth potential. The other power-carrying wire is usually referred to as the "hot" line. It is this line that will rise and fall in voltage over the 360-degree cycle of a complete sine wave. (The neutral line can't vary significantly from 0 volts, because it is grounded at the main switch panel.)

On the right side of the figure is a block diagram of the computer, including its power supply. The computer is connected to the main switch panel via the building's wiring, with its three-wire power cord plugged into a wall outlet. Of the three wires, the two power-carrying wires (hot and neutral) go into the power supply—and nowhere else. The computer's chassis is connected to the grounding wire. The grounding wire serves two purposes: safety and signal reference. Grounding the chassis ensures that it will be at the same potential as the earth, which eliminates any possibility of electrocution if someone should happen to touch the chassis. The second function performed by grounding the chassis is to set up a reference (of 0 volts) to which all other voltages in the system can be compared.

This tutorial, number 74, written by Alan Frank, was originally published in the October 1994 issue of LAN Magazine/Network Magazine.

Electrostatic Discharge

Electrostatic discharge (ESD) isn't an everyday topic of discussion, so it's easy to put it out of mind. However, careless work habits or the right—in this case, maybe the wrong—environmental factors can create a lot of trouble for you.

Recognize this experience? You walk across a carpeted room on a day when the humidity is very low, reach for a doorknob, and get zapped. Here's how it happens. Electrostatic charges, also known as static electricity, build up when two insulating materials are rubbed together. Electrons are rubbed off of one surface and onto another. So, one surface will now have an excess of electrons and thus be negatively charged, while the other will have a relative deficiency of electrons and be positively charged.

The magnitude of these charges is impressive—as much as tens of thousands of volts. Specific magnitudes of electron flow, or current, expressed in amperes, pose an electrocution hazard to people. Fortunately, the current in the situation just mentioned is very low. Although the shock that we get when we touch a doorknob or other conductive material is startling, it doesn't do any damage.

Even though people aren't harmed by such ESD incidents, transistors and the integrated circuits built from them are extremely sensitive to minute currents and any voltages that exceed the 5 volts to 12 volts from which they typically are powered.

THE UNSEEN THREAT

Manufacturers of computers and other electronic devices, such as telephones, copiers, fax machines, and so on, take great pains to design their systems so they're armored against ESD hazards. In general, they're mostly successful in that regard. Most computers are enclosed in a grounded metal chassis that will shield the internal electronics from any discharge. The chassis is further surrounded by an insulating plastic outer casing. About the only vulnerable points of most computers are serial and parallel ports, and users are unlikely to be touching these in everyday use.

The part of a computer most likely to get zapped is the one you come into contact with most frequently: the keyboard. 3M (St. Paul, MN) has come up with a self-adhesive static dissipative strip that you can apply to a keyboard, just below the space bar. It includes a ground wire that you connect to a known electrical ground (the screw on the faceplate of a grounded AC wall outlet, for example).

By touching the grounding strip before touching anything else on your computer, you'll dissipate any electrostatic charge that you may have built up by walking across the carpet. I've also seen other types of keyboard wrist rests and related devices that incorporate some type of anti-static feature.

Perhaps the situation where you must take the most precautions against ESD is when you're servicing a computer or handling circuit boards. Electronics manufacturers are keenly aware that unmounted integrated circuits (ICs) and circuit boards are extremely vulnerable to ESD hazards, and they take elaborate precautions in their manufacturing operations.

Most do not install carpeting on shop floors, preferring linoleum or other floor tiles, polished with electrically conductive floor waxes. Assembly tables are covered with conductive laminates. Workers wear cotton anti-static shop coats and eschew nylon or other nonconductive synthetic textiles. Parts bins and trays are molded from special conductive plastics. And finished circuit boards are shipped in anti-static plastic bags.

MINDING THE SHOP

A completed circuit board—a network adapter or modem, for example—is most vulnerable to ESD right when it arrives in your shop, ready for installation in a computer. But you eliminate ESD hazards by taking a few simple precautions and investing in a couple of low-cost anti-static components.

A good start is to obtain conductive anti-static table mats for your service bench. These have a ground wire for connection to a reliable ground point. It's a good idea to place any computer that you're going to service on such a mat before opening it up. You can also get anti-static wrist straps. These come with a coiled cord that connects to ground through a 1-megohm resistor. The resistance is for personal safety, in the event you should come into contact with 110-volt power. The coiled cord shouldn't interfere with your work. And, most use a snap attachment to the wrist strap, so if you need to walk away from the bench to fetch something, it's a simple matter to unsnap the coiled cord, then quickly snap it back onto the wrist strap when you return.

If you refuse to be chained to your workbench, you can get elastic grounding straps that go around your ankle and under your shoe. (Perhaps the pinnacle in geek attire, they could make a great conversation starter if you wear one out to lunch. Maybe you'd like to rethink that wrist strap now.) For the grounding straps to work, you need to be standing on a grounded, conductive surface. A grounded anti-static mat is best.

For those times when you need to work on a computer at the user's site rather than in the shop, you can get static dissipative field service kits that consist of a wrist strap and a roll-up flexible plastic mat.

If you need to work on a computer, but don't have an anti-static mat and wrist strap, there are still some precautions you can take. For example, before opening up the computer, make sure the power is turned off, but leave the power cord plugged in. This will ensure that the computer's chassis is grounded through the power cord.

Alternatively, use a wire terminated with alligator clips to connect the chassis to an earth ground. Touch some point on the metal chassis, the back panel, for example, before opening the computer. This will drain off any static charge that may have been on your body. When you've opened up the system— and before you reach in to remove any circuit boards—touch the chassis again.

Before opening the anti-static bag on any new board you're installing, hold the board in its protective bag in one hand while you touch the chassis with your other hand. By doing this, you'll drain off any static charge that may exist on the outside of the bag. You can then open the bag and remove the board. It's good practice to handle circuit boards by their edges or mounting brackets, and avoid touching ICs or circuit traces. Obviously, if the board has jumpers that need to be set, you're going to have to touch the board. But it's still a good idea to avoid handling circuit boards to the extent that you can.

Walking around is a good way to build a static charge, so if you need to step away from the system, make sure you again ground yourself by touching the chassis upon your return.

By using these techniques, you can work on a system with very little likelihood of doing any ESD damage. Clearly, though, using an anti-static wrist strap is a far more positive way to guarantee that you aren't harboring a static charge.

BE SENSITIVE

If a circuit board arrives in an anti-static plastic bag, you should probably conclude that the component is static-sensitive. So, don't pull it out of the bag until you're ready to install it in a computer. First, place the computer and all the boards you're going to install (still in their conductive plastic bags) on the grounded anti-static mat. Then, put on your wrist strap before pulling any circuit boards out of their bags. Any boards you remove should be put into conductive anti-static bags. (It's a good idea to hang onto anti-static bags that come with new boards.)

Handling polystyrene cups, polystyrene peanuts and packing material, and the plastic shrink-wrap film that new software comes in can expose you to a high-voltage static field that can zap components, so keep these things out of your server rooms, wiring closets, and service shop.

To understand why nonconductive plastics are such an ESD hazard, consider the polystyrene cup. Let's assume it has a high-voltage static charge on its surface. Touching the cup to a grounded electrical conductor will drain off the charge only at the point of contact. Because the plastic in the cup acts as an insulator, the charge that exists over the rest of the cup's surface remains in place.

Ionized-air blowers are available for service benches. These produce a stream of ionized air (both positive and negative ions) that neutralizes any static field that may exist on nonconductive plastic surfaces. The blowers can be a nice insurance policy, in case nonconductive materials come into contact with sensitive electronic components. Even if you take this precaution, however, it's still good practice to have a policy of keeping plastic cups and other materials away from the service bench.

ON THE CARPET

You'll also want to give some attention to the type of floor covering that's being used in your offices. Many synthetic carpets are complete insulators and promote static build-up. If you're outfitting a new office or putting new carpet into an existing one, look for carpets that are conductive—either by nature of the material they're made of, or from some coating that's been applied. You can also treat existing carpeting with anti-static sprays. (Repeated carpet cleaning tends to remove anti-static treatments and sprays, so you'll need to periodically reapply them.)

Many people have trouble rolling their office chairs across carpeting (especially if it's deep pile), so they put a hard plastic or rubber chair mat on top of the carpet, to get a nice, smooth rolling surface. The only problem with this is that most of these mats are electrical insulators, which again means there's potential for static charges to build up. An alternative is to buy special static-dissipative chair mats. These are made of conductive plastic and come with a ground wire that you can connect in the same fashion as the keyboard anti-static strip mentioned previously.

Also, avoid low relative humidity. Humid air conducts electricity, to a certain extent. By contrast, very dry air is an insulator. Heaters and furnaces not only heat air, they reduce the relative humidity as they do so. You may want to add a humidifier to your heating system to add moisture back into the air. Obviously, you don't want to go overboard with this—relative humidity in excess of 90 percent can result in condensation that can short-circuit electronics. It's ultra-low relative humidities (15 percent or less) that you want to avoid.

HOW BAD CAN IT BE?

Am I going overboard with this ESD stuff? Somewhere between paranoia and oblivion lies a happy medium. Some organizations don't have much of an ESD problem. Others, particularly those in dry or cold climates, can have severe problems. What's so insidious about ESD is that an organization may have a problem and not even know it. It takes fairly high voltages (up in the thousands of volts) to produce a noticeable spark when you touch a metallic object. You can be carrying a static charge of a few hundred volts, which is enough to damage a circuit board, and not know it, because you don't feel the discharge when you touch metal.

One way to determine whether hazardous ESD levels exist in your environment is to use a static field meter. These devices, which typically sell for $300 to $500, can measure the static charge on a surface, when placed within a few inches of the surface.

Making changes to your office building's heating and air conditioning equipment is not a trivial expense. If you're not willing to take that plunge, there are still a number of easy, inexpensive steps you can take. The logical move is to eliminate the points of greatest vulnerability. So, focus your efforts on those areas where computers are being worked on and circuit boards are being handled.

For the most part, it doesn't cost a fortune to purchase the equipment and supplies needed for ESD protection. Anti-static wrist straps cost $25 or less. Most table mats are priced at less than $100. Wrist strap resistance testers are in the $100 to $300 range. The cost of all of this probably adds up to less than that of a single computer. (I can hear it now: "If it saves even one computer's life, it's worth it." But it's true.)

For anti-static materials and equipment, check out suppliers of electronic assembly equipment and computer furniture. If you don't know of any sources, here are a couple you can use as a starting point. This is by no means an exhaustive list, however.

Contact East, a distributor of test, repair, and electronic assembly products has a catalog that lists numerous static protection products, including anti-static mats, wrist straps, parts trays, static-shielding bags, anti-static sprays and floor waxes, and ionized-air blowers. It even carries a few videotapes and books on ESD awareness and techniques. (I haven't reviewed the tapes or the books, so I can't vouch for their merit. But they may well prove to be quite informative).

Misco, a distributor of computer peripherals, supplies, and furniture, carries anti-static chair mats and the 3M keyboard strip previously mentioned. Jensen Tools, which distributes a broad range of electronics toolkits, carries anti-static wrist straps, floor mats, table mats, and roll-up field service kits.

SOURCES OF SUPPLY

Here are some vendors of anti-static equipment and supplies:

Contact East
335 Willow St.
North Andover, MA 01845
(508) 682-2000
Fax: (508) 688-7829

Jensen Tools
7815 S. 46th St.
Phoenix, AZ 85044
(800) 426-1194
(602) 968-6231
Fax: (800) 366-9662

Misco
1 Misco Plaza
Holmdel, NJ 07733
(800) 876-4726
Fax: (908) 264-5955

This tutorial, number 85, written by Alan Frank, was originally published in the September 1995 issue of LAN Magazine/Network Magazine.

Fiber Optics for Networks

In networking applications, fiber-optic cable offers several advantages over copper wiring. For a given length of cable, fiber has far less signal attenuation than copper, which allows you to use longer cable runs.

A copper circuit can pick up unwanted electrical "noise" (spurious signals unrelated to the desired source signal). This noise is electrically induced into the circuit by virtue of the cable running close to AC power wires, fluorescent lights, or other noise sources. A copper circuit can also pick up stray signals from adjacent data circuits. What's more, if the circuit is close enough to be picking up outside data signals, it's also likely to be inducing its own signals into the other circuits. This problem is known as "crosstalk."

A further disadvantage of copper data links is that they are susceptible to transient overvoltages (or impulses), which can momentarily reach hundreds or thousands of volts. Transient overvoltages can damage computers and networking equipment and, at the very least, they can be erroneously interpreted as data, leading to problems similar to those caused by electrical noise contamination.

Finally, the numerous copper-wire network links between computers can potentially cause electrical ground loops that interfere with signal reception.

With optical fiber networks, these types of problems can be avoided. Fiber-optic links use light as the medium for data signaling. An optically clear glass fiber functions as a light guide, carrying a light signal from source to destination. The fact that the fiber does not use or carry electrical signals makes it immune to induced electrical signals and impulses. And since glass is an electrical insulator, using fiber-optic links is the best way to avoid network ground loops. Even with these advantages, fiber optics is still a new technology to the LAN world, and it may seem foreign to many network managers. This Fiber Optics for Networks "tutorial," then, is aimed at enlightening you (if you'll pardon the pun) about this relatively new medium.

BOUNCING ALONG

Figure 1 shows how light travels along a fiber-optic light guide. The light source could be almost any type of light, but for networking applications where very high-frequency signaling is desired, a specialized LED is typically used. At the other end of the light guide is some type of light receptor—often a photodiode.

A fiber-optic cable consists of several layers (see Figure 2). The "core" is glass of extremely high optical clarity—a tiny filament typically 62.5 microns in diameter. The core is surrounded by another layer of glass, known as the cladding, which typically has an outer diameter of 125 microns. The entire fiber (core and cladding) is usually encased in a plastic jacket, for mechanical protection. The material used for the jacket will depend on building codes and other mechanical requirements of a particular network installation.

REFLECTIONS

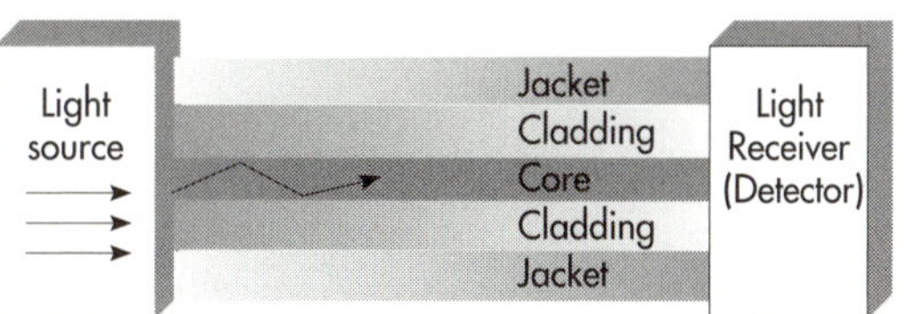

■ **Figure 1:** Lan travels along a fiber-optic light guide (or cable) from a special LED light source to a photoreceptor. The cladding layer reflects back any stray light waves traveling along the core.

Only the core and cladding are involved in the light transmission. Cables are typically specified according to the diameters (in microns) of these two layers. Thus, the cable just described would be listed as 62.5/125 micron optical fiber.

CABLE CROSS-SECTION

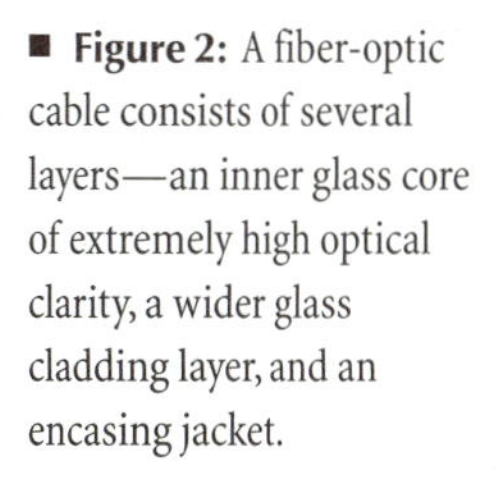

■ **Figure 2:** A fiber-optic cable consists of several layers—an inner glass core of extremely high optical clarity, a wider glass cladding layer, and an encasing jacket.

Physicists classify light-passing media according to their index of refraction, a measurement that correlates with the speed at which light travels through it. The kinds of glass used for the fiber's core and cladding have slightly different optical indices of refraction. As a result, light rays are reflected at the boundary between the two layers. As shown in Figure 1, light rays traveling along the core will be reflected back inside, should they stray toward the cladding. Thus light can't get out of the core, and that makes the core an excellent light guide.

A COMMON THREAD

One type of fiber-optic cable can serve virtually all local area networking needs, says N. D'Arcy Roche, president and CEO of Raylan (Palo Alto, CA). This is the 62.5/125 micron multimode fiber that's specified for Fiber Distributed Data Interface (FDDI). "It is a dual-fiber cable that can be either a two-fiber duplex cable (like an electrical lamp cord) or two fibers in a round jacket," Roche says. "This cable can accommodate 10Mbit/sec Ethernet, up through OC-12 (622Mbit/sec) speeds. This eliminates the question of what fiber to choose for a given installation."

Roche says that this fiber can serve for either of two commonly used transmission wavelengths—1,300nm and 850nm. FDDI, as currently specified by the IEEE, uses the 1,300nm transmission window, but Roche says that several vendors (Raylan included) have introduced 850nm options for FDDI. The LED light sources and photodiode receivers for 1,300nm operation are expensive, while 850nm devices are substantially lower in cost.

The 62.5/125 micron multimode fiber has a bandwidth of 160MHz over a length of 1km, with 3.75dB of signal attenuation (light reduction) when operating in the 850nm band. At 1,300nm, the same fiber has a bandwidth of 500MHz, over a distance of 1km, with 1.5dB of attenuation. Over shorter distances, bandwidth increases and attenuation decreases, but you can see that fiber cable can comfortably accommodate cable lengths approximately an order of magnitude greater than UTP copper wiring—with greater bandwidth and far less attenuation.

GOING THE DISTANCE

Telephone companies and other long-distance carriers typically use single-mode optical cable, which has a much smaller core—in the range of 8 microns. It offers a wider bandwidth than multimode fiber, and substantially lower attenuation per kilometer. These factors make it the ultimate medium for city-to-city communications, as signal repeaters can be spaced many kilometers apart, reducing costs and maintenance requirements proportionately. But taking advantage of single-mode fiber, typically, also means using rather exotic (and expensive) laser light sources and photo receivers. For in-building networking (as well as campus cable runs of less than 1km), multimode fiber is the cable of choice.

Thomas-Conrad (Austin, TX) is another major player in optical fiber-based networking, with its Thomas-Conrad Networking System (TCNS), a 100Mbit/sec network that competes with FDDI. (Unlike FDDI, TCNS is proprietary.)

Peter Rauch, director of product marketing for TCNS, says that Thomas-Conrad specifies multimode cable with core diameters of 50 microns to 140 microns, but that 62.5-micron cable is the most frequently used. TCNS uses the straight tip (ST) connector developed by AT&T. This bayonet-type connector fastens with a quarter-turn and looks similar to the BNC connector commonly used with thin Ethernet cabling.

LIGHT STANDARDS

It's common to associate fiber-optic cabling with high data rate applications such as FDDI and TCNS, but optical fiber can also be used for Ethernet and Token Ring. The Fiber Optic Inter-Repeater Link (FOIRL) standard is used to connect Ethernet hubs. FOIRL doesn't boost the data rate for Ethernet—it's still 10Mbits/sec—but it does allow you to connect hubs with cable runs of up to 1km. More recently, the IEEE has come out with a 10BaseFL specification that's comparable to FOIRL, but allows 2km cable runs.

The IEEE's 802.5j standard addresses Token Ring over fiber-optic cable, but this specification only covers "drops" (hub-to-end-node). There is no standard covering the use of fiber for the connections between hubs (using the Ring Out and Ring In ports). While various vendors offer solutions for fiber connections between hubs, these are proprietary, so you can't count on interoperability among them.

The fact that all commonly used networking technologies can use 62.5/125 micron multimode fiber means you can install fiber now, run Ethernet or Token Ring initially, then move up to higher speeds—100Mbit/sec FDDI or TCNS, or 155Mbit/sec Asynchronous Transfer Mode (ATM)—in the future.

You can't count on all systems using the same termination at the ends of the cables, however. The ST connector is perhaps the most commonly used. If you should move from one type of network to another, you may have to change connectors accordingly, but both Roche and Rauch say that this is seldom a problem. For example, if you have in-the-wall cabling that uses ST connectors, and network adapters that use a different connector, you can use patch cords with the appropriate connectors at each end to make the transition.

Who are the major suppliers of fiber-optic cable and connectors, and how do you choose the specific type you need? Rauch recommends that you seek the advice of your local cable installation contractor, who will know the local building codes and which cable is best for specific applications.

Fiber-optic cabling is gaining popularity as its cost comes closer to parity with copper. Its transmission properties are unquestionably desirable, particularly for high-speed networks or in cases where long distances must be spanned. If you're looking for more illumination on the subject, consult vendors of fiber-optic systems for background materials. I'd also recommend The Technician's Guide to Fiber Optics by Donald J. Sterling (Albany, NY: Delmar Publishers, 1987 and 1993; ISBN 0827358350). The first edition lacks the latest information—on Ethernet 10BaseFL, for example—but it offers an excellent discourse on the physics of optical cabling, light sources, and detectors. I haven't had the opportunity to review the second edition, but it should be considerably more up-to-date on the latest standards.

This tutorial, number 75, written by Alan Frank, was originally published in the November 1994 issue of LAN Magazine/Network Magazine.

Wireless Networking

Dealing with cabling problems is one of a network manager's primary jobs. Troubleshooting cables for breaks and bad connections and overseeing the relocation of computers are not only time-consuming tasks, but frustrating as well.

Installing new cable can be an expensive proposition, costing hundreds or even thousands of dollars per network connection. The cable itself is not that expensive—unshielded twisted-pair wire, for example, costs only a few cents per inch. But paying union technicians to pull new cable is costly, especially in old buildings that may contain asbestos or other hazardous materials.

As a result, some network managers turn to wireless networking, rather than cable, when they expand existing LANs or build new ones. Wireless LANs give network designers a level of flexibility unavailable to wire-based systems. Wireless networks—which send and receive data through a transmission and reception device attached directly to the network adapter card—make moving PCs connected to wireless networks a much simpler process than with wire-based systems.

Network managers can merely move the PC, without adding cable or testing its associated connections. This capability makes wireless networks a tempting alternative to cable-based ones.

GOING WIRELESS—PROS AND CONS

Wireless LANs offer network managers a host of advantages over traditional hard-wired network technologies. In addition to simplifying the move of networked PCs, they can also be less expensive to buy and build—over the lifetime of the LAN—than cabled LANs. The most obvious cost savings come from not needing to remove and install and test the cable. Their initial installation costs, however, are higher, since you must purchase the wireless transmission and reception electronics. Some wireless connections won't require building permits, as can be the case with large-scale cable installations. This feature can save additional time and money.

There are tradeoffs, however. By nature of the medium, wireless networks transmit at slower data rates than wire-based networks. They also impose limitations on the number of connected nodes and how far apart those nodes can be placed from each other.

Some wireless networks—infrared ones—require line-of-sight communications. This limitation restricts their usefulness in many situations, such as in so-called hard-walled offices and multi-story buildings.

Manufacturers use three basic technologies to carry data over their wireless networking products: infrared light signals, narrow-band Radio Frequency (RF) signals, and spread-spectrum RF signals. Here's a look at the basics of how each operates and what their operational characteristics mean to network managers and users.

MAKING LIGHT OF DATA

Infrared networking products transport data via light waves that are invisible to humans. Infrared light, falling in the 1,000 gigahertz (GHz) and higher range, shares all the properties of visible light: It can be reflected off, but cannot penetrate, solid objects such as walls.

The primary advantage of infrared technology is its great bandwidth, allowing it to carry hundreds of megabits of data per second. In addition, because no government body regulates use of light frequencies, infrared data transmission is unlicensed.

Infrared networks share the technology used with the remote-control units that come with home electronics equipment, such as TVs and stereos. Receivers for wireless networks, like the channel changer and TV, must be visible to the transmitter, either directly or via reflection.

Infrared-based networks can be implemented with mirrors that focus the light signal to an extremely tight beam. Because focusing delivers essentially all of the transmitted signal to the receiver, it permits high-speed communication that can equal or surpass that of 16Mbit/sec Token Ring networks. Mirror-based systems are well suited for point-to-point applications, such as building-to-building connections where transceivers are seldom moved. They are less satisfactory for in-building networks, however: The receiver must have a direct, unobscured view of the sending unit, and any movement in either unit can break the connection.

WIRELESS NETWORKING

To work around this line-of-sight requirement, some infrared networks spread, or "flood" light around an area, which allows a single transmitter to reach multiple receivers, while reducing the effects of the transmitter or receiver being moved. A more diffused signal, however, reduces data rates and shortens the distance over which the signal can be reliably sent.

Another disadvantage of infrared networks is their susceptibility to interference from other light sources, including the sun and some lighting fixtures. Focused systems, because they produce a stronger light beam, offer greater immunity to light interference than unfocused ones.

UP IN THE AIR

Radio frequency transmission, while not limited to line-of-sight environments, offers its own set of thorny technical issues, including available bandwidth, signal reflections and interference, and Federal Communications Commission (FCC) regulations.

Manufacturers have taken two tacks in developing RF network products. Narrowband transmission requires licensing by the FCC because it needs a "clear" communications channel—one that is uninterrupted by other narrow-band transmitters. Spread-spectrum transmission uses special unregulated frequencies that require no FCC licensing.

Narrow-band networking products transmit data directly on a center frequency, much like a radio broadcast, so the transmitter and receiver must be tuned to the same bandwidth. Like the signals from radio and TV stations, narrow-band signals are subject to interference from signal reflections. This interference is caused when signals reflected off walls and other objects arrive at an antenna at different intervals.

Such "ghosts" make data communication unreliable. Unlike the human eye, communications equipment is not sophisticated or intelligent enough to discern the difference between reflections and the "real" transmission.

Vendors of narrow-band wireless products must thus be able to guarantee their customers a clear channel. A clear channel can only be ensured by carefully allocating each available frequency band to make sure that two nearby networks do not share the same frequency.

SPREADING DATA AROUND

Spread-spectrum transmission distributes, or "spreads," a radio signal over a broad frequency range. To do this, spread-spectrum networks use what is called a predetermined pseudo-random sequence to transmit data. This pseudo-random sequence is actually a predetermined digital signal pattern that places data on a combination of frequencies from across the entire spread-spectrum band.

A receiving device must thus know the specific signal pattern used by the transmitting device to decode data. This technique makes spread-spectrum LANs secure and reliable. In fact, the spread-spectrum technology was developed by the U.S. military during World War II for secure voice communications.

One of spread-spectrum's main benefits is that it allows multiple networks to share a single frequency as long as different pseudo-random sequences transfer data. In these situations, the signals from one network are interpreted by another as random noise and are ignored. Another advantage of spread-spectrum networks: They don't require line-of-sight communications, making them suitable for hard-wall offices as well as open office environments.

WIRELESS TOPOLOGY ISSUES

Like their cable-based counterparts, wireless networks must provide clearly defined methods of accessing the transmission channel. Vendors use two basic techniques for granting this access: One, called a peer-level system, lets every node on the network communicate directly with every other node. The second uses a dedicated server architecture, with all network devices communicating through a central control station.

WIRELESS NETWORKING

Peer-level wireless networks typically cost less to build than those with central control units, primarily because they don't require the central hardware. But because there is no central point of control in peer-level systems, managing the network, including providing security, collecting network statistics, and performing network management and diagnostics, can be difficult.

Because they do feature a dedicated server environment, controller-based wireless networks permit centralized management, security, and maintenance capabilities. The central control unit also provides access to other services, including local and wide area networks.

THE WIRELESS MARKETPLACE

A number of vendors, including NCR, Motorola, Windata, and Photonics, market wireless networking products. These offer a variety of features and benefits, including varying degrees of performance, transmission characteristics, and network operating system support. Additionally, the IEEE has formed a committee, 802.11, to study and standardize wireless networks.

Many of the new pen-based, hand-held PCs include wireless networking capabilities as an option. These products can give end users even greater flexibility in their networking choices.

This tutorial, number 40, written by Jim Carr, was originally published in the November 1991 issue of LAN Magazine/Network Magazine.

802.11 and Spread Spectrum

A close look at the MAC and Physical layers of wireless LANs.

In a world of cellular telephones and pagers, it's no surprise that the wireless phenomenon has also infiltrated the world of networks. Wireless LANs (WLANs) have found a prominent place in vertical markets such as healthcare, retail, and manufacturing, where workers are often away from a desk, yet they still need to access the wired network.

Although WLANs have practically become staples in these niche areas, a lack of vendor-neutral standards may have hindered those companies that wanted to deploy wireless capabilities for mobile users.

In June 1997, the IEEE 802.11 committee approved a WLAN protocol that defines both the Physical layer (layer 1 on the OSI model) and the MAC layer (the lower part of the Data-Link layer). (See Figure 1 for a graphical representation of the IEEE 802.11 layers.)

In this tutorial, we will take a closer look at the 802.11 standard and two methods of delivering wireless network capability.

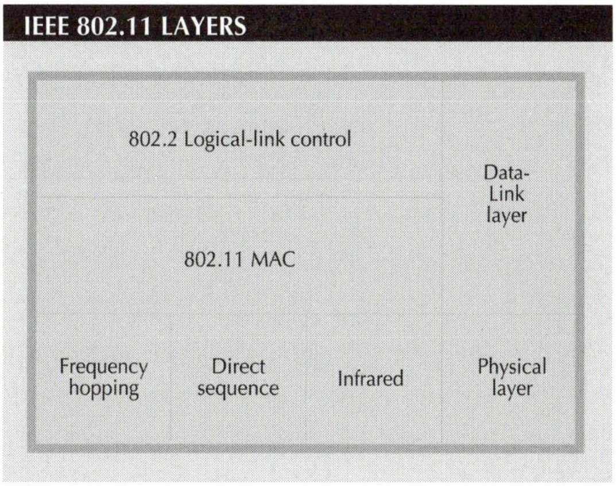

IEEE 802.11 LAYERS

802.2 Logical-link control

802.11 MAC

Data-Link layer

Frequency hopping

Direct sequence

Infrared

Physical layer

Figure 1: The IEEE 802.11 protocol covers the MAC and Physical (PHY) layer specifications of wireless networking. The standard consists of one MAC that works with three PHYs: two radio frequency and one infrared.

THE BIG MAC ATTACK

The Data-Link layer (layer 2) of the OSI model handles a variety of transmission functions, including ensuring that data is packaged properly before it's sent over cabling and performing flow and error control. The upper part of the Data-link layer is called the *Logical Link Control sublayer;* it ensures that data is reliably sent over the physical link.

The MAC-or Media Access Control-layer, which resides at the lower portion of the Data-link layer, does just as its name states; additionally, it controls access to the physical transmission medium. It's this layer that received a lot of attention from the 802.11 committee.

The 802.11 standard specifies that Carrier Sense Multiple Access with Collision Avoidance, or CSMA/CA, should be used as the method for transmitting information in a WLAN. This may look familiar to anyone with a basic working knowledge of the Ethernet standard; Ethernet uses CSMA/CD (CSMA/Collision Detection) as its transmission protocol.

The CSMA part of both of these methods determines whether the transport medium is currently busy with another transmission. But in cases where two or more end stations hear a quiet network and start to blast information at about the same moment, collision is inevitable. With Ethernet, the CD part of the equation allows packets to be re-sent in the event of collision.

It makes sense for 802.11 to share MAC layer aspects with Ethernet, since in many cases a wireless network will be tied into a wired Ethernet network. But as you can imagine, CD doesn't quite work on a WLAN. For one, CD would require that wireless radios be able to send and receive at the same time—a requirement that would increase the price of products and make them more complex. For another reason, on a wireless network it's not always a given that each station can hear all the other stations, as is the case with a wired Ethernet LAN. Because all stations can't necessarily hear each other, a sending station that is free to transmit has no way of knowing if the receiving station is not busy as well.

Instead, 802.11 supports CSMA/CA along with something called *positive acknowledge,* which differs a bit from CSMA/CD. With the collision avoidance method, a station that wants to transmit first checks the medium to see if it's free. If it is free, then the station is allowed to send. The station on the receiving end of the transmission then dispatches an acknowledgment to inform the sending station that a collision did not occur. If the sending station doesn't receive an acknowledgment packet, it will assume the original packet did not make it through and will resend until an acknowledgment is received.

To minimize the possibility of collision due to stations not being able to hear each other, 802.11 defines a virtual-carrier sense feature. This allows for a station that wants to send something to first send a Request to Send (RTS), which is a short packet that contains the source and destination addresses, as well as the duration of the transmission. If the medium is free, the receiver station will then reply with a short packet called Clear to Send (CTS), which will include the same duration-of-transmission information (see Figure 2).

The total duration is the time in microseconds that it takes to send the next information packet, plus a CTS frame, an acknowledgment packet, and three Short Interframe Space (SIFS) intervals. The SIFS separates transmissions in the same dialogue.

Stations that receive the RTS, the CTS, or both packets will set their network allocation vector, which is a virtual carrier sense indicator, to the designated duration, which will then be used when sensing the medium. By sending out these short RTS and CTS packets, the likelihood of collision drops because stations that may not normally be able to hear each other will know to consider the medium busy until the end of the transmission.

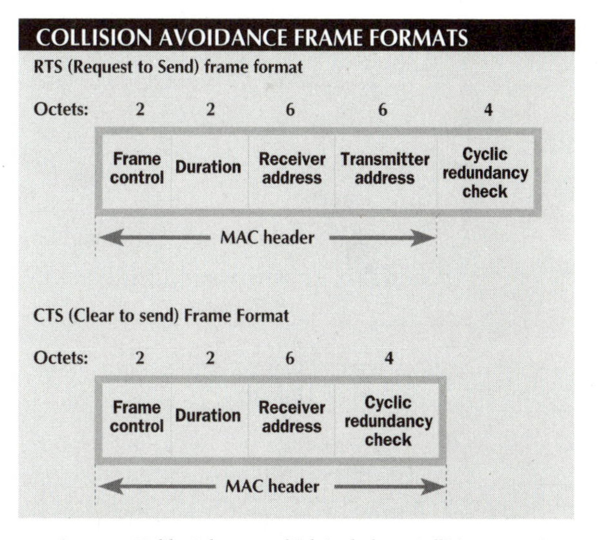

■ **Figure 2:** Unlike Ethernet, which includes a Collision Detection (CD) feature, WLANs use Collision Avoidance (CA). Before sending packets, the sending station transmits a Request to Send (RTS); the destination station then sends a Clear to Send (CTS) packet. Only when this occurs will the medium be reserved for the duration of the transmission.

This is a brief description of some of the technical details behind a WLAN's layer 2 functions. But as with any traditional wired network, like Ethernet and Token Ring, the Physical layer is necessary to get raw information from one place to another.

PHYSICAL ACTIVITY

The MAC layer, as we've seen, is crucial to getting information from one place to another safely and reliably. The 802.11 protocol defines one MAC that interacts with three different Physical layers (PHYs). The PHY of the OSI model brings an interface to the network medium and provides the actual signaling function across the network.

802.11 defines two radio frequency PHYs, both of which are variations of spread spectrum, a technology that's been around since World War II. Spread spectrum radio technology was used during the war because it was largely immune to enemy interference and jamming. The transmission signal is, as the name suggests, spread over a wide range of the radio spectrum. These qualities are still desirable for companies that trust the airwaves to transmit what could often be sensitive information.

The IEEE defines two specific types of spread spectrum technologies: direct sequence and frequency hopping. Both methods operate at the 2.4GHz to 2.4835GHz ISM (Industrial, Scientific, and Medical) band—an unlicensed range opened by the FCC.

Direct Sequence Spread Spectrum (DSSS) has been defined by the IEEE to operate at either 1Mbit/sec or 2Mbit/sec speeds. DSSS works by spreading a signal over a wide range of the 2.4GHz frequency band. Initially, DSSS enjoyed a very large installed base of products that operated at the 902MHz to 928MHz band. However, this frequency band was not available for DSSS in all parts of the world. In addition, the width of the frequency band was relatively small, and overcrowding soon became a problem. Today, the 2.4GHz range is available for DSSS usage worldwide, and it encompasses more bandwidth than what's available at the 900MHz band.

Frequency Hopping Spread Spectrum (FHSS) is the method adopted by the majority of vendors developing and shipping WLAN products. This subset of spread spectrum operates at a data rate of 1Mbit/sec, with an option to go as high as 2Mbits/sec.

Instead of spreading the signal over a wide band of frequency, FHSS transmits a short burst of data on one frequency, hops to another frequency and transmits for a short period of time on this frequency, then hops to a new frequency. The exact sequence of frequencies used is known as the *hop sequence.* This sequence must be synchronized between both sending and receiving stations, or they won't be able to communicate. Also, it's possible to have several communications occurring at the same time across the same frequency band—as long as each uses a unique hop sequence. In most cases, many overlapping channels are feasible. Because of this, and because the 2.4GHz band encompasses a lot of space, it's possible for many separate channels to be transmitted at different sequences.

The FCC requires that 75 or more frequencies be used with FHSS, and that a transmission dwell on a particular frequency for no more than 400 milliseconds. In the event of interference on one frequency, that data is retransmitted at the next frequency hop.

The third PHY defined by the IEEE is infrared, but so far most of the products tied to 802.11 fall into the radio frequency category. Probably because the 802.11 discussion dragged out for years and was mostly centered on radio as the wireless medium, many infrared vendors stopped going to 802.11 committee meetings, but the standard still addresses this issue.

WHAT'S THE DIFFERENCE

FHSS may appear to be the most popular PHY among both vendors and customers, but both spread spectrum methods of transmission have their place within companies using WLAN technology. Because FHSS signals constantly move around within the 2.4GHz band, anyone trying to listen in will find it just about impossible to pick up any of the signal. Also, the very nature of FHSS lends itself to areas with a heavy density of wireless users.

Technically, DSSS may not be as secure as FHSS, because its signals don't hop around, but anyone listening would only be able to pick up random slices of information. Because the signal is spread out, it becomes very difficult to pick up enough of the signal to do any serious damage. DSSS makes more sense at customer sites such as a warehouse, where users are more spread out and there is little concern for signal interference.

DSSS and FHSS cannot interoperate due to their different methods of transmission. In most cases, companies will settle on one method, or, in some cases, companies may use one method in certain installations and the other at additional sites.

Although companies have been shipping WLAN products—such as the access points that bridge the wireless with the wired network and the cards that reside within laptops and hand-held units—for several years, customers haven't had a 100 percent guarantee that products they purchased would conform to the eventual standard. Unless a company purchased all their gear from a single vendor, they couldn't have complete peace of mind that everything would work together.

As with any new standard, the step following approval is interoperability testing among products so customers can be assured that their products will work with those of other vendors.

RESOURCES

For more on the 802.11 protocol, visit the following Web sites:

http://stdsbbs.ieee.org/groups/802/11/index.html

www.breezecom.com/802.11links.html

www.proxim.com/proxim/apps/whiteppr/fh_vs_ds.htm

This tutorial, number 112, by Anita Karvé, was originally published in the December 1997 issue of Network Magazine.

Wireless LANs (802.11b)

We've covered wireless LANs before, as regular readers will have noticed (see "Wires Not Included," June 1999 issues of *Network Magazine*, and "802.11 and Spread Spectrum," the previous tutorial). Recent approval of the IEEE's 802.11b standard for 11Mbit/sec wireless networking, however, has been followed by a wave of new product announcements from Apple, Compaq Computer, Lucent Technologies, and others.

This tutorial will discuss what's new about 802.11b and provide some implementation examples, then explain how it compares with HomeRF (proposed for home-based wireless networks, at www. homerf.org) and Bluetooth (for Personal Area Networks, or PANs, at www.bluetooth.com). For information on wireless WAN technologies and standards, as opposed to the wireless LAN architectures this tutorial will discuss, see "Wide World of Wireless."

NOT ALL TRIPLETS ARE CREATED EQUAL

Every article on 802.11 (including our earlier two) inevitably points out that it is actually three standards in one. To be more specific, 802.11, ratified in 1997, gave birth to a single MAC standard for the lower portion of the Data-link layer, plus three possible Physical (PHY) layers.

There is nothing new and noteworthy about the MAC layer in 802.11b, so I gladly refer you to the December 1997 tutorial for in-depth information. Briefly, however, wireless networking uses Carrier Sense Multiple Access with Collision Avoidance (CSMA/CA). This modifies the Ethernet 802.3 standard so it can work via radio with a "virtual carrier sense" feature, based on brief Request to Send (RTS) and Clear to Send (CTS) packets.

The three PHY layers defined by the 802.11 standard are, confusingly, not interoperable. They are: infrared (never implemented by anybody); Frequency Hopping Spread Spectrum (FHSS) radio; and Direct Sequence Spread Spectrum (DSSS) radio.

FHSS was employed in most early 802.11 networks. Originally conceived during World War II by actress Hedy Lamarr and composer George Antheil (would I kid you about this?), it employs a narrowband carrier, changing its frequency in a pattern known only to both the sender and receiver. As intended, this makes information difficult to intercept.

While FHSS lives on in some products, as I will explain later, it is not part of 802.11b. The only triplet anointed as part of the new standard is DSSS. This type of signaling uses a broadband carrier, generating a redundant bit pattern (called a "chip") for every bit of data to be transmitted. While seemingly wasteful of bandwidth, DSSS copes well with weak signals: Data can often be extracted from a background of interference and noise without having to be retransmitted, making actual throughput superior.

Proponents of DSSS point to its superior range, plus its ability to reject multipath and other forms of interference. In fact, DSSS can reject noise from a microwave oven, for example, with relative ease, though it would still be swamped if deployed in the vicinity of a hospital's MRI scanner.

In any case, the 802.11b version of DSSS transmits data at a nominal 11Mbits/sec (actual rates vary according to distance from another transmitter/receiver). It is downwardly compatible with 1Mbit/sec and 2Mbit/sec wireless networking products, provided they also use DSSS and are 802.11-compatible.

With few exceptions, 802.11b is a worldwide standard. It uses the 2.4GHz to 2.48GHz Instrumentation, Scientific and Medical (ISM) frequency band, dividing this into as many as 14 different channels. In the United States, 11 channels are available for use.

Vendors must tailor their hardware access points (discussed later) to use legal channels in each country they ship to. Wireless NICs, however, can often adapt themselves automatically to whatever channels are being employed locally. Therefore, it is possible to travel with an 802.11b client and make connections in any country.

DEPLOYING A SYSTEM

It is difficult to plan a wireless network just by looking, or even by measuring distances. The antennas typically can, at the power levels permitted, transmit and receive for distances of about half a mile. This figure, however, only applies to outdoor, line-of-sight transmission.

Indoors, it is difficult to predict how a building's contour will affect propagation of radio waves. According to Harris Semiconductor, whose chipset is used in many 802.11b devices, range in an open plan "cube farm" may be from 200 feet to 500 feet. In a closed-wall office environment, it may be as low as 100 feet.

The metal found in an office building's floor can cut a signal by as much as 30 decibels (dB). Therefore, every floor in such a building will require one or more transmitters.

The simplest type of wireless LAN is a peer-to-peer setup that might be used in a conference room or at a trade show. Here, all stations are kept within a circle with a radius of approximately 300 feet, and direct communication between stations is possible.

To create this type of network, an administrator would install wireless NICs, setting their drivers to the ad hoc mode of operation, then selecting a radio channel for the workgroup. (In the United States, there is enough spectrum for three channels to coexist in one location, but channels must be 25MHz apart to avoid interference.)

In 802.11 lingo, this workgroup would be known as a Basic Service Set (BSS). A mechanism known as the Distributed Coordination Function (DCF), basically the "virtual carrier sense" function described earlier, provides best-effort delivery of data within a single, peer-to-peer BSS.

A more typical wireless network, however, is an "infrastructure" network—one that operates as an adjunct to a preexisting wired network. Here, Access Points (APs) are employed to act as a bridge (and usually a router), moving traffic between the wireless and wired networks (see "Gaining Access").

A hardware AP is a self-contained unit, typically featuring one or more Ethernet ports, plus either a built-in radio or a PC Card slot. (For the sake of versatility and easy upgrading, much 802.11b equipment employs PC Card-based transceivers, whether these will be installed in a portable computer or in a stationary piece of equipment.) A software AP is a functional, more affordable equivalent, using an existing computer that has been equipped with both wired and wireless NICs to perform bridging and routing.

As well as providing a gateway between network types, an AP has several other functions. For a start, the AP can provide Point Coordination Function (PCF), an optional connection-oriented mode. By broadcasting a beacon signal, the AP can temporarily silence ordinary terminals in order to provide point-to-point transmission of time-sensitive data, such as voice.

The primary functions of an AP, however, are authentication and association. The AP performs authentication to determine if a given wireless device is permitted to join the network, and can be based on MAC address, password, or some other parameter. Association is a handshaking relationship between the wireless device and the AP. It is designed to ensure that the client connects to only one AP at any given time.

ROAM SWEET ROAM

An Extended Service Set (ESS) is a logical collection of more than one BSS. Via an ESS, multiple APs can work together so that computers can roam from one to another while still staying in the same network.

To create this type of network, an administrator would install APs and wireless NICs, setting drivers to Infrastructure mode and making sure that all components are set to use the same ESS ID number (ESSID). To avoid interference, each AP should be set to a different channel.

Each 802.11 device associates with one AP initially, but a wireless network would be of limited use if stations were unable to roam. Fortunately, clients can switch from AP to AP in a way that is transparent to the user.

Logically, there are several ways roaming can take place, depending on the way APs have been set up. The simplest case is when different APs have the same ESSID and are on the same subnet of the same LAN. Slightly more complexity results when different APs have the same ESSID but live on different subnets. Here, DHCP re-registration is required, unless a Mobile IP solution is being used (see "Mobile IP Hits the Street," November 1999). Multiple APs can also form different logical networks on a single LAN via the use of different ESSIDs.

Given the nature of radio-based communications, eavesdropping is always a possibility. Therefore, the 802.11 standard includes a shared-key encryption mechanism known as Wired Equivalent Privacy (WEP). When a client tries to connect to an AP, the AP sends a challenge value to the station. Upon receiving this, the client uses the shared key to encrypt the challenge and send it to the AP for verification.

While useful, WEP only allows for 40-bit encryption; some vendors of 802.11b equipment offer optional 128-bit encryption or plan to make it available as a firmware upgrade. A few also sell wireless NICs that have been manufactured not only with a unique MAC address but also with a unique public/private key pair. Administrators can require that all allowable hardware address/public key combinations be entered into APs in advance. Alternately, they can simply configure APs to keep track of the combinations they encounter and subsequently reject any mismatches. This way, an attacker can be prevented from breaking into a network via MAC address masquerading.

CUTTING ACCESS COST

While valuable for the infrastructure network, features such as multiple ESSIDs, roaming, and 128-bit encryption increase the cost of hardware APs to around $1,000 each. Vendors such as Apple and Lucent, however, offer simplified APs without roaming for less than half that price. Not to be confused with prestandard wireless networking solutions, these "budget" APs have the advantage of being fully compatible with other 802.11b equipment.

This means users could purchase a single wireless NIC, then use it both in a corporate setting and in a home office. In order to popularize the advantages of 802.11b compatibility, the Wireless Ethernet Compatibility Alliance (WECA, at www.wirelessethernet.org) recently announced a labeling program known as "Wi-Fi."

Of course, 802.11b is not the only entrant into the 2.4GHz wireless networking melee. A rival of sorts is the HomeRF Shared Wireless Access Protocol (SWAP) system, which has been designed for consumers. It uses FHSS transmission and eliminates the more complex parts of 802.11 (such as PCF and RTS/CTS). An advantage here is that a single connection point can support both voice services via Time Division Multiple Access (TDMA) and data services via CSMA/CA.

Another contender, Bluetooth, uses the 2.4GHz band for localized connection between different devices on a PAN. These might include a PC and a handheld device, a phone and a headset, or a notebook computer and a printer. While there are grounds for concern about interference between 802.11b, HomeRF, Bluetooth, and the many other devices using the same spectrum (such as baby monitors and garage door openers), some observers seem to believe all these can coexist. A coexistence study group exists within the IEEE (P802.15) and presentations about this topic have been fairly encouraging.

Eventually, wireless LANs will migrate into the relatively wide-open spaces offered in the 5GHz band, where they will be able to exchange data at up to 54Mbits/sec. Just as portable computers have always lagged behind their desktop cousins in terms of speed and affordability, wireless networks will always lag behind what copper and fiber can offer. Once again, however, there's no question of which best suits the needs of a mobile worker.

RESOURCES

For information about the 802.11 standard and the P802.15 study group, see http://grouper.ieee.org/groups/802/11/ and http://grouper.ieee.org/groups/802/15/, respectively.

The Wireless LAN Alliance is at www.wlana.com, while the Wireless Ethernet Compatibility Alliance (WECA) is at www.wirelessethernet.org.

Other members of the nonexclusive 2.4GHz club are HomeRF (www.homerf.org) and Bluetooth (www.bluetooth.com).

The HiperLAN/2 global forum may be reached at www.hiperlan2.com.

This tutorial, number 137, by Jonathan Angel, was originally published in the December 1999 issue of Network Magazine.

Interface Cards

The Network Interface Card (NIC) provides the physical connection to the network. Every computer attached to a LAN uses some sort of network interface card or chip. In most cases, the card fits into the expansion slot of the computer, although some cards are external units that attach to the computer through a serial or parallel port. Internal cards are generally used for PCs, Macs, and some workstations. Internal interface cards can also be used in minicomputers and mainframes. External boxes are often used for laptops. In some cases, the network circuitry is integrated onto the computer's motherboard.

The interface card takes data from the PC, puts it into the appropriate format, and sends it over the cable to another LAN interface card. This card receives the data, puts it into a form the PC understands, and sends it to the PC.

The interface card's role can be broken down into eight tasks: host-to-card communications; buffering, frame formation; parallel-to-serial conversion; encoding and decoding; cable access; handshaking; and transmission and reception. These steps get the data from the memory of one computer onto the cable, and reversing the steps gets the data off the cable and into the memory of another computer.

BEFORE TRANSMISSION

The first step in transmission is the communication between the personal computer and the network interface card. There are three ways to move data between the PC's memory to the network interface card: I/O, direct memory access, and shared memory.

I/O is the simplest method. The most important types are memory-mapped I/O and program I/O. In a memory-mapped I/O transfer, the host CPU assigns some of its memory space to the I/O device, in this case the network interface card. Out of the possible 640KB of RAM that is available for DOS PCs, a few KB are allocated to the network card. This memory is then treated as if it were the PC's main memory. No special instructions in the CPU are needed to get data from the card since it is like taking data from one part of main memory to another.

With program I/O, the CPU is given a set of special instructions to handle the input/output functions. These instructions can be built into the chip or come with software. To send data, a request is sent from the network interface card to the CPU. The CPU then moves the data from the card over the bus to main memory. Because the CPU is required to handle the I/O process, it cannot perform other tasks while data is being transferred. This makes it slow. Also, I/O takes up PC memory.

Direct Memory Access (DMA) is another method. All Intel-based computers come with a DMA controller chip that takes care of transferring data from an input/output device to the PC's main memory so the PC's CPU does not have to get involved in the transfer. For DMA transfer, the controller or processor on the interface card sends a signal to the CPU indicating it wants to transfer information. The CPU then relinquishes control of the PC bus to the DMA controller.

Once the DMA controller has command of the bus, it takes the data from the card and places it directly in memory. The CPU had told it the appropriate memory address at which to begin putting data in memory. After all the data is in memory, the DMA controller returns control of the bus to the CPU and tells it how much data has been put in memory.

DMA is generally faster than I/O because the DMA controller removes work from the CPU, so the CPU can perform other functions while data transfer is taking place. The disadvantage is the CPU cannot access memory while the DMA controller is working.

In shared memory, part of host PC's memory is shared by the network interface card's processor. Shared memory is a very fast transfer method, since no buffering on the card is required. Because the card and the PC do their work on the data in the same place, no transfer is necessary. Although shared memory is the fastest method of moving data between the network interface card and the PC, it is more difficult to build than DMA or I/O. Shared memory takes more PC RAM than the other methods.

The second component of PC-to-NIC communication is buffering. The buffer is a storage place that holds data as it is moving into and out of the NIC. A buffer is necessary because some parts of data transfer are slower than others. For example, data comes into the card faster than it can be converted from a serial or parallel format, depacketized, read, and sent. This is true in both directions. To compensate for delays inherent in transmission, a buffer temporarily holds data either for transmission onto the cable or for transfer into the PC. While in the buffer, data may be acted on, such as put into frames, or it may simply wait while the NIC handles other things.

An alternative to buffering is to use PC RAM. This can be less expensive, but is usually slower and it requires memory.

The NIC's most important job is frame formation. Frames are the basic units of transmission. Files and messages are broken into frames for transmission. At the other end the frames are reassembled to form the file or message. A frame has three sections: header, data, and trailer. The header includes an alert to signal that the frame is on its way, the frame's source address, destination address, and other data. In some networks, headers also have preamble bits used for various purposes, like setting up parameters for transmission. They can also have a control field to direct the frame through the network, a byte count, and a message type field.

The data section contains the data being sent, for example, the numbers in a spreadsheet or words in a document. On some networks, the data section of a frame can be as large as 12KB. On Ethernet, it is 1,500 bytes. The upper size limits of most networks fall between 1KB and 4KB.

The trailer contains error checking information called a cyclical redundancy check (CRC). It is a number that is the result of a mathematical calculation the sending NIC does on the frame. When the frame arrives at its destination, the mathematical calculation is repeated. If the result is the same—all the ones and zeros are in the right place—no errors occurred in transmission. If the numbers don't match, an error has occurred and the frame must be retransmitted.

TRANSMISSION

Parallel-to-serial conversion is the next step in transmission. Data comes from the PC in parallel form, 8, 16, or 32 bits at a time, depending on the bus width. But it must travel over the cable in serial form, which is one bit at a time. Thus, the network interface card must convert between the two. A parallel-to-serial controller is responsible for this task. Once a frame is formed and changed from parallel to serial, it is nearly ready to be sent over the line. First it must be encoded, which means it must be converted into a series of electrical pulses that convey information.

Most network interface cards use Manchester encoding. Serial data is divided into bit periods. Each of these periods (or fractions of seconds) is divided in half, and the halves together represent a bit. From the first half to the second half of each bit period there is a change in the signal's polarity, from positive to negative, or vice versa. The change during each bit period represents the data. A change from negative to positive represents a one. A change from positive to negative represents a zero. Or vice versa depending on the network. Ethernet uses Manchester encoding. Token Ring uses another version called Differential Manchester encoding, in which the mid-bit transmission is used for clocking information. Either way, these ones and zeroes represent data.

Before data can be sent, however, the network interface card must have access to the cable. Token Ring and ARCnet use a token to grant network access. Ethernet lets any workstation transmit at will if it finds the cable unoccupied and senses collisions if two senders inadvertently transmit simultaneously. The access method protocol, circuitry, and firmware reside on the network access card.

After getting data from the PC, formatting it, encoding it, and getting access to the cable, the interface card for some types of networks has one more task before it can send data: handshaking. In order to send data successfully, a second NIC must be waiting to receive it. A short period of communication between two cards ensues before data is sent. During this period, the NICs negotiate the parameters for the upcoming communication. The transmitting card sends the parameters it wants to use. The receiving card answers with its parameters. Parameters include the maximum frame size, how many frames should be sent before an answer, timer values, how long to wait for an answer, and buffer sizes. The card with the slower, smaller, less complicated parameters always wins because more sophisticated cards can "lower" themselves while less sophisticated cards can't "raise" themselves.

Finally, everything is set. The only thing left is for the NIC's transceiver to put the data on the cable. The transceiver gives the data power to make it down the line. It actually puts the electrical signal out over the cable, making sure the data can get to the next NIC, repeater, amplifier, or bridge.

At the other end, a transceiver waits to accept the signal and begin the whole process in reverse, from modulated signal through decoding, serial-to-parallel conversion, and depacketizing the information into a format readable by the receiving device.

PICK A CARD

More than any other LAN component, the network interface card determines the performance of the LAN. The speed of the disk drives, file servers, and network operating system are important, but the speed of the interface card and its software driver determine the network speed. [Editor's note: As NIC throughputs have progressed from 10Mbits/sec to 100Mbits/sec and even 1Gbit/sec, NICs have ceased to be a significant bottleneck in most networks.]

Choosing network cards is a difficult process. Nearly every vendor claims to have the fastest cards. Benchmarking, even when done by independent sources, measures a myriad of parameters. Look at the bus width (a card using a 32-bit bus is normally faster than one that uses an 8-bit bus), the bus type (Extended Industry Standard Architecture bus or Micro Channel is faster than Industry Standard Architecture bus), the type of memory transfer (shared memory is faster than I/O and DMA), and whether the card can perform bus mastering. But most important, test the speed of the network card driver.

Performance is one, albeit critical, factor. Reliability is essential. Speed is irrelevant if the card causes errors, loses frames, drops the line, or just doesn't work. Nothing is more frustrating than having to isolate network hardware problems. Evaluate the vendor's reputation, longevity in the business, and technical support services.

This tutorial, number 5, written by Aaron Brenner, was originally published in the December 1988 issue of LAN Magazine/Network Magazine.

File Servers

A file server is a combination of computer, internal hardware, and software that allows network users to share computer programs and data. A file server usually has a significantly faster processor, faster network interface card, more memory, and more data storage than most PCs. It may also have a tape back-up unit, modems, and several printers attached.

SERVER HARDWARE

Server software is that portion of the network operating system that "serves" a computer's resources to other network users. This software accepts incoming requests from network users and gives back files and other resources. In the days of MS-DOS, users achieved this connection by mapping a local logical drive, drive F: for instance, to the server's physical drive. That is, using a designated drive letter from A to Z, users access the server's physical drive as if it were one of their own local drives. Application programs do the same. The network operating system takes care of routing traffic across the network to the proper file server hard disk and back.

A file server must find data quickly and get to the requesting workstation with minimal delay. Factors affecting the server's performance include the CPU's speed, the network interface card's speed, the amount of RAM available, the type of disk and controller, the cable type and length, the network software's efficiency, the application type being run, and the number of users on the network.

Most high-performance file servers include a very fast, very large hard disk. Access time is the time required to get data off the disk, and it is broken into seek and rotation times. Average seek time, which accounts for the bulk of the time, is the time the disk head takes to move to the correct track on the platter. Average rotation time is the time taken for the platter to turn to the sector where the data is stored. Most of the time, the average access time is all you need to know when buying a server hard disk.

One overlooked aspect of the hard disk is its controller. Standard ST, Enhanced Small Device Interface (ESDI), Small Computer System Interface (SCSI), and Integrated Drive Electronics (IDE) are four types. ST drives are practically obsolete, ESDI drives are faster, but less common with each passing month, SCSI drives are easily expandable, and IDE drives are often the default on workstations and low-end server models. The tradeoff depends on your application and financial situation. Today, most network drives are SCSI.

Disk capacity is crucial. A rule of thumb is to figure out how much disk space you will need, and then double it. LANs always grow, so expandability is important. The file server should be able to accommodate more disk drives as needed. A file server (and a NOS) should permit you to add storage beyond any practical limit you can foresee. Disk arrays, which are composed of multiple drives and controllers, will provide fault tolerance but won't necessarily improve performance.

MEMORY NEEDS

A cache can make the hard disk appear to work faster than it actually does. Cache is space in high speed memory that is set aside to hold the last data read from the disk. The server software and the disk controller take more data off the disk than the user actually requested. This adjacent data, stored in the cache, is available when the next request is made, saving the file server from going to the disk to get it. It doesn't speed the work of the disk, but it eliminates some disk access, which moves data faster.

Caching is effective because a file server usually makes several accesses to disk when retrieving or writing data. Caching works because the next information a user requests is generally stored sectors adjacent to the data just requested. Caching also works because the disk controller usually can't get all the data with a single access to disk. So the disk controller reads the next couple of sectors after what the user requested, because chances are that the data the user wants next will be in those blocks.

Some server software caches only disk reads. This way, if the file server crashes, no data is lost because the data in the cache is identical to what is on the disk. Other file server software caches both read and write.

The amount of RAM necessary for disk caching varies from vendor to vendor. In general, a bigger cache is better, although a point of diminishing returns sets in. Moreover, the more data kept in a cache, the more it is vulnerable to server failure. A power loss will wipe out data in the cache, although data written to disk will not be lost. Most vendors take precautions for this, writing and verifying data at specified intervals, thereby protecting as much data as possible while still increasing performance.

PROCESSOR, CLOCK, AND PORTS

File servers should be more powerful than workstations for the shared services they provide. In some small networks with nondedicated servers, a last-generation processor will perform fine. However, once data sharing passes beyond a handful of users or light word processing, more powerful file servers are a necessity. Servers that provide only file and print services are likely to be more dependent on the amount of RAM, the disk subsystem performance, and the performance of the system bus than they are on the processor itself. Applications servers, on the other hand, can often absorb all the CPU processing power they can get.

Every server should have plenty of expansion slots as well as multiple serial and parallel ports. More than one or two locally connected printers can take up all available ports, since few PCs have more than one of each type of interface on the motherboard. Keep this in mind if you are using the file server as a router or switch. Two or three network interface cards, one or two drive controllers, and extra serial or parallel ports may occupy more ports than are available in the file server.

After performance, the main concerns when choosing a file server are reliability and compatibility. A PC file server needs to have a good power supply so it does not experience power drops or outages. It must work with standard software drivers such as network drivers, disk drivers, and video drivers, which means its BIOS must be compatible with those of other major vendors. Most cheap IBM clones do not work well as file servers.

This tutorial, number 8, written by Aaron Brenner, was originally published in the March 1989 issue of LAN Magazine/Network Magazine.

PCI: New Bus on the Block

The PCI bus may have arrived just in time to carry the load that faster networks and multimedia put on the I/O subsystem.

As computer CPUs and memory subsystems advance in performance, the I/O subsystems are challenged to keep up. Video graphics controllers, network adapters, and hard disk controllers demand high levels of data throughput.

The Peripheral Component Interconnect (PCI) bus was developed to address these needs for higher I/O transfer rates. Like EISA, PCI is a 32-bit-wide bus, allowing data to be transferred four bytes at a time. But, while EISA has an 8.33MHz bus clock rate, the PCI 2.0 specification allows for a clock rate of up to 33MHz. The product of the four-byte bus width and 33MHz clock frequency gives PCI a 132Mbyte/sec theoretical maximum throughput, compared to EISA's 33Mbytes/sec.

Although PCI was largely developed by Intel, a company known for its 80x86 family of microprocessors, it was conceived as a processor-independent I/O bus. As an example of this processor independence, Apple Computer's newest PowerPC-based Macintoshes use the PCI bus.

LOADING UP

If you've shopped around for PCI-based systems, you've probably noticed that most of today's offerings have three or fewer PCI slots. This is problematic if you are planning to use a PCI-based system as a network file server; servers need lots of slots to handle multiple NICs and disk controllers.

According to Robert McNair, applications engineering manager for Intel's PCI components division (Santa Clara, CA), part of the reason for the relatively low PCI slot-count on today's machines is a bus-loading limitation. A second reason involves the large installed base of ISA and EISA adapter cards, which users will no doubt be reluctant to replace immediately. A system that also contains several EISA or ISA slots—in addition to PCI—can take advantage of this installed base.

As for the bus-loading issue, no matter what device is "driving" the bus (putting electrical signals onto the bus), it will only be capable of driving a limited number of devices. If there are too many electrical loads, signal voltages may not quite meet their required tolerances, resulting in signal errors. In extreme cases, circuits may be damaged if they are supplied with more current than they can handle.

PCI can drive a maximum of 10 loads, says McNair. In determining bus loading, the PCI chipset that drives the PCI bus itself counts as one load. Any PCI components (such as an embedded PCI graphics or SCSI controller, mounted on the motherboard) count as one load each.

PCI slots are counted as one and one-half to two loads each. Why not one load? If each slot could support only one load, it would create quite a hardship for the designers of PCI add-in boards. This is because you could only have a single PCI device on each card. By allowing cards to present a load equivalent to two PCI loads, the designers of the PCI specification gave board designers more freedom in the types of products they can design.

What these bus-loading rules mean is that a PCI bus will, in most cases, be limited to a maximum of four slots. (As an example, if the PCI chipset is taking one load and a PCI-based graphics controller is taking another, a system designer would have eight loads left to apportion among the slots, which equates to four slots).

System designers can get past the limitations outlined above by using PCI "bridges"—integrated circuits that serve as signal amplifiers and repeaters. Through the use of a bridge, a system designer can hang a second PCI bus off of another one. (But bear in mind that the PCI bridge itself presents a load to the primary bus; it usually counts as one load.)

Other types of bridges may connect either an EISA bus or an ISA bus to a PCI bus. The manner in which bridges are employed can have a significant impact on overall system performance. For example, consider Figure 1, which illustrates two different ways in which to bridge several buses together. In this figure, Approach A represents a cascaded configuration, in which PCI bus 2 is connected (via a PCI-to-PCI bridge) to PCI bus 1. The EISA bus, too, is cascaded from PCI bus 1. Because of this cascaded arrangement, all data flowing to or from any adapters plugged in to any of the PCI or EISA slots must cross PCI bus 1—putting an upper limit of 132Mbytes/sec on total I/O.

Approach B in Figure 1 shows a second arrangement, which I'll refer to as a peer arrangement because both PCI bus 1 and PCI bus 2 are bridged directly off the system bus. Neither one depends on the other to get or send data. In this case, total system I/O can go as high as 264Mbytes/sec—twice that of Approach A.

Figure 1 doesn't represent any specific computer system. Rather, it illustrates, in a general way, the difference between cascaded and peer-level PCI bus arrangements. Hewlett-Packard (HP, Palo Alto, CA) has begun using these two approaches in the company's NetServer line of network servers. Larry Shintaku, advanced development manager for HP's NetServer division (Santa Clara, CA), says that the company's entry-level NetServers use the cascaded-bus configuration (similar to Approach A, Figure 1)—as do most of the other PCI-based servers from other vendors. HP's high-end NetServers use the peer arrangement, for higher total I/O throughput capability.

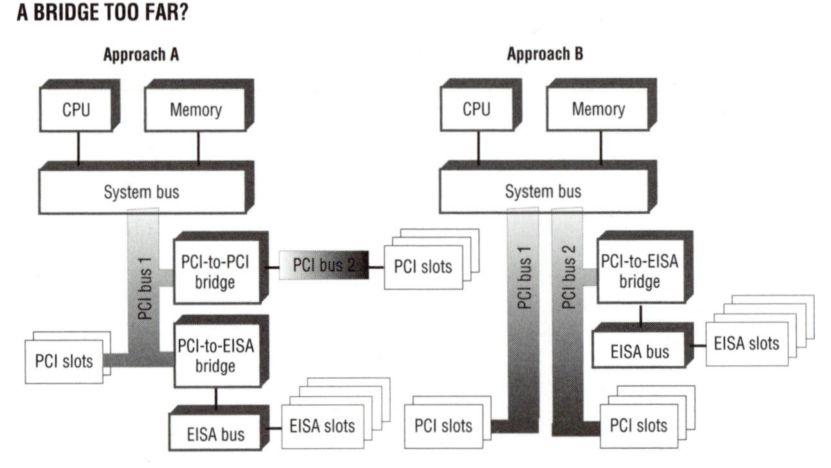

■ **Figure 1:** Bridges can be used to connect one PCI bus to another. Other bridges can link a PCI bus to an EISA or ISA bus. But how systems designers use these bridges can impact overall performance. In this example, Approach B offers higher total I/O throughout than Approach A, due to the peer-level arrangement of the two PCI buses in Approach B.

On the PCI slot-count issue, Shintaku takes much the same perspective as Intel's McNair, saying that bus-loading constraints are one reason that most of today's PCI systems have relatively few PCI slots, but that there's also a need to have enough ISA (or EISA) slots to accommodate older boards.

In terms of mechanical arrangements, PCI uses a multiplexed data and address bus. This means that the same electrical conductor paths are used to carry data and addresses. (It doesn't carry them both at the same time, however; it alternates as necessary, with one signal line indicating whether the information currently on the bus represents data or an address.)

By multiplexing the data and address information onto one set of conductors, PCI's designers managed to get a fairly low pin-count, with respect to EISA or other 32-bit buses. This reduces cost by allowing a smaller physical connector and takes up less "real estate" on motherboards.

YOU'RE IN MY SLOT

PCI's specifiers also came up with a clever way to conserve back-panel space on computers. Because PCI won't replace ISA or EISA overnight, most systems will carry two types of I/O slots, usually a mixture of PCI and EISA or PCI and ISA. There may not be enough physical slots on the back panel to accommodate as many I/O boards as a customer may want (network server applications, in particular, often demand lots of slots), therefore, PCI's designers came up with the concept of "shared" slots. A shared slot, for example, may (depending on its design) be used for either a PCI board/EISA board or a PCI board/ISA board.

Figure 2 shows the layout of a typical shared slot. It is a top view, looking down on the motherboard. EISA cards have their printed circuit boards on one side of the metal mounting bracket that is used to secure the board to the computer's back panel. With PCI, the circuit board is mounted on the other side of the bracket. Looking at the diagram, you can see that this allows two connectors (PCI and EISA, in this example) to be placed side-by-side in the space occupied by one physical slot.

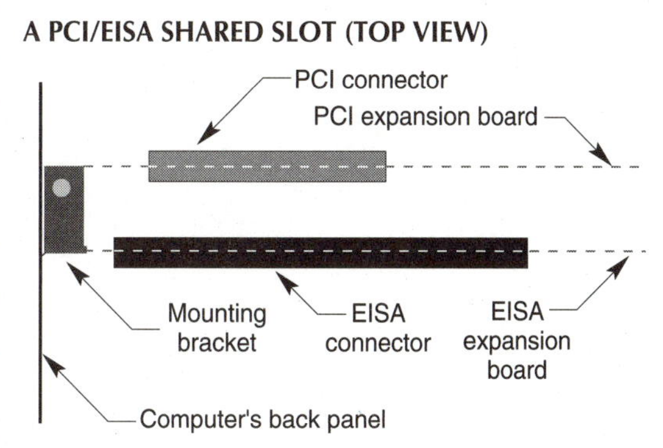

■ **Figure 2:** System designers can create "shared" I/O slots, which can be used for either of two types of expansion boards. The shared slot illustrated here will accept either a PCI or EISA card.

Shared slots don't increase the total number of slots—a five-slot system, for example, is still a five-slot system—but by sharing slots, a designer could come up with a system that has three PCI slots and three ISA slots. One of the slots would have to be a shared slot. This arrangement gives users the flexibility to use a combination of three PCI boards and two ISA, or two PCI boards and three ISA.

Intel's McNair notes that PCI's interrupt scheme is different from ISA's. PCI permits several devices to share a single interrupt, provided that all of the devices are capable of sharing interrupts. With the ISA bus, only one device is allowed to use a particular interrupt.

PCI can allow shared interrupts because it uses "level-triggered" interrupts, while the ISA bus uses "edge-triggered" interrupts (EISA can use either). "Level-triggered" means that the computer is sensitive to the voltage level that is on each interrupt line. For example, an interrupt line may be at a 5-volt level when there are no interrupts. When an interrupt occurs, it is signaled by the interrupt line going to 0 volts. (The voltages may vary or the logic inverted from this example, depending on your computer's design; the principle still holds true, however.)

With edge-triggered interrupts, the computer only responds to the transition between one state and the next. In other words, it only pays attention to the leading edge of the voltage waveform on the interrupt line.

When a PCI-based system receives an interrupt on one of the PCI interrupt lines (there are four of them), it knows that one—and possibly more—of the devices using that interrupt line needs attention. It then begins polling each device in turn, to find out which one sent the interrupt. Once the correct device is located, the system will immediately jump to the interrupt handler (also known as an interrupt service routine, or ISR) for that device.

When the ISR is complete, the system returns to the activity it was performing prior to the interruption. But, if the interrupt line previously mentioned is still indicating an interrupt (if the interrupt line is still being held low, in our example), the system will once again poll devices to find out which one is calling for an interrupt, then service that interrupt.

By using a combination of interrupt-driven and polling techniques, PCI can be much more flexible in its interrupt scheme than ISA. Although ISA supports more individual interrupts, the fact that each device must have its own interrupt line means you're more likely to run out of interrupts with ISA.

PCI's ability to have multiple devices sharing a single interrupt line is made possible by its use of level-triggered interrupts. If interrupts were edge-triggered, it is possible that some of the incoming interrupts might be lost. For example, if a second interrupt were to come in while the processor was still in the midst of processing a previous one, the processor would return from servicing the first interrupt but not be aware that a second interrupt had been asserted. With level-triggered interrupts, by contrast, any device assigned to that interrupt line can bring the line low to assert an interrupt. This line will continue to stay low until all pending interrupts have been serviced, ensuring that the processor doesn't miss any incoming interrupts.

PCI should offer enough I/O throughput to satisfy most of today's needs— especially if a dual, peer-level PCI arrangement is used. But PCI's developers have already made plans for expansion, by supporting a future 64-bit bus, and clock rates up to 66MHz. The compact pinout of PCI's multiplexed bus allows a second connector to be placed right behind the first, without taking up too much real estate on the motherboard. McNair says the 66MHz bus may not be needed for a long time—at least on PC-class machines. While high-powered engineering workstations may be able to take advantage of it, bottlenecks in other parts of PC systems may mean they can't effectively make use of the doubled clock rate.

This tutorial, number 87, written by Alan Frank, was originally published in the November 1995 issue of LAN Magazine/Network Magazine.

PCI and Future Bus Architectures

There was a time when network servers weren't expected to do much more than provide shared file-and-print services. Today, these same boxes are being asked to host applications (such as databases), act as firewalls, process intensive multimedia data, and support a multitude of peripheral devices.

And while network backbones have increased in speed to accommodate all of the requests going to and from servers and all the data that must pass through them, and as processor speeds and clock rates have increased, the servers themselves have had to endure a lot of finger-pointing as the source of traffic bottlenecks.

The I/O capacity and throughput of servers are the keys to smooth operation. But when the bus architecture has trouble shouldering the burden placed on I/O subsystems, the rest of the network notices. Often, even a fast CPU has to slow down to the bus speed when it wants to communicate with an adapter in the system.

A few years ago, Peripheral Component Interconnect (PCI) became widely adopted as the de facto bus for PC systems. Chances are, if you bought a system in the past few years, it came with multiple PCI slots.

But even though PCI has a higher transfer rate and clock speed than the previous EISA, ISA, and Micro Channel Architecture (MCA) buses, today's high-speed networks need something more.

There are three efforts to develop the next step beyond PCI: PCI-X, Future I/O, and Next Generation I/O.

In this tutorial, I'll first look at PCI and what it offers. Then, I'll examine the three proposals that hope to inherit PCI's place at the heart of I/O systems.

PCI OVERVIEW

Before the PCI bus architecture became the I/O system of choice, the EISA and ISA bus types were the most common.

For devices in EISA and ISA systems to communicate with the CPU, or host, they may be required to first go through an expansion bridge, the memory bus, the bus cache, and the CPU local bus. As you can imagine, this can lead to quite a bit of latency in processing I/O and data requests.

Although buffering, which stores signals if a bus is busy, solved some of the delay issues surrounding EISA and ISA, other problems remained—including what to do if more than one peripheral device needed to use the CPU local bus at the same time.

Because of EISA and ISA limitations, Intel and other vendors created the PCI specification, which is based on the idea of a local bus. This means that instead of peripheral devices having to jump through so many hoops to communicate with the CPU, each device can access the CPU local bus directly. The PCI architecture contains a bridge that serves as the connecting point between the PCI local bus and the CPU local bus and system memory bus. PCI devices are independent of the CPU, which means a CPU can be replaced or upgraded without any impact on the devices or the redesign of the bus.

Current PCI implementations are based on a 32-bit architecture and operate at 33MHz or 66MHz with data transfer rates of under 300Mbytes/sec. But as CPU power grows exponentially, in certain instances the current bus technology can't keep up with demands placed on the I/O system.

I/O systems in use today are mostly based on a shared bus architecture, meaning that memory is shared by the host and the devices. However, this approach does have its drawbacks. For example, if one peripheral device monopolizes the bus, then other devices trying to access the bus at the same time will see lower performance.

So, while most systems will continue to use PCI for communicating between the CPU and peripheral devices, in the next year or two PCI will coexist with a new breed of I/O solutions.

X MARKS THE SPOT

In the fall of 1998, Compaq Computer, Hewlett-Packard, and IBM proposed to extend the current PCI for cases where high-performance I/O is desperately needed. Dubbed PCI-X, this specification basically takes the existing PCI technology and moves it from 32 bits to 64 bits, and increases throughput and clock speed.

PCI-X operates at speeds up to 133MHz with transfer rates going higher than 1Gbyte/sec—a significant increase over the traditional PCI architecture. Additionally, PCI-X features more efficient data transfer and simpler electrical timing requirements, which is important as clock frequencies increase. PCI-X remains backward compatible with the current PCI bus.

The 64-bit nature of PCI-X's architecture brings it in line with other 64-bit technologies that are becoming common in the industry. Chip manufacturers such as Intel will start shipping 64-bit processors within a year, and the next versions of both NetWare and Windows NT (Windows 2000) will natively support 64-bit addressing. This means that when a system's bus and OS are both running at 64 bits, system performance naturally increases. However, most companies will have quite a task ahead to revamp drivers and other components to conform to the new addressing scheme.

The companies that developed the PCI-X proposal have submitted it to the PCI Special Interest Group (SIG), the body that oversees the PCI specification and its implementation.

Systems that support PCI-X could be on the market as early as the end of 1999. At first, expect to see PCI-X implemented at the server level, where higher levels of I/O are really needed. After that, this architecture could work its way down to workstations and even high-performance PCs within the next few years.

THE FUTURE'S SO BRIGHT

Interestingly, the same camp that's endorsing PCI-X—Compaq, HP, and IBM —is also throwing its considerable weight behind another post-PCI bus architecture.

What they are all working so hard on is Future I/O, which is based on a point-to-point, switched-fabric interconnect. This is an evolutionary leap ahead of the current PCI architecture, where the bus is shared by all devices. In the switched environment, a direct connection between each card and processor means that as new devices are connected, total throughput actually increases. The plan is to integrate the Future I/O protocol into peripheral devices so they can communicate with the switch in a more effective manner.

The initial throughput from Future I/O connections will be 1Gbyte/sec per link in either direction. Work on Future I/O, which is still in the specification stage, should result in one interconnect that can be used for communication among processors in a cluster, as well as for technologies that need the additional bandwidth, such as Fibre Channel, SCSI, and Gigabit Ethernet.

Because Future I/O aims to support a variety of connectivity types, three distance models are being designed. The first works for distances less than 10 meters (m) and involves ASIC-to-ASIC, board-to-board, and chassis-to-chassis connections. This model uses a parallel cable between chassis.

The second model supports distances of 10m to 300m and will be used for data-center server-to-server connections. This model will use fiber-optic or serial-copper cable with additional logic. The third model supports distances greater than 300m, which is made possible with additional buffering and logic.

The Future I/O Alliance plans to continue developing the specification during 1999 and hopes to have a final version by the end of the year. The group expects ratification of Future I/O in early 2000, with prototypes being demonstrated shortly thereafter. Products are expected to start shipping by early 2001.

THE NEXT GENERATION

Although Intel was instrumental in the creation of PCI, the company—along with partners Dell Computer, Hitachi, NEC, Siemens, and Sun Microsystems—has thrown its support behind a new bus architecture dubbed Next Generation I/O, or NGIO. Much like Future I/O, NGIO is based on a switched-fabric architecture rather than on the shared-bus model that is so common today.

Shared buses do not scale very well, which can limit the number of peripherals that can be supported by a system. Shared buses have other shortcomings, including some difficulty with peripheral and system configuration, difficulty in hot-swapping components, and distance limitations between peripherals and memory controllers. But point-to-point switched-fabric technology should address these drawbacks and provide a more efficient system.

In addition to the switched fabric, another innovation within NGIO is the use of a channel architecture—a more efficient I/O engine that processes requests from peripherals. Previously, this type of architecture was used only on mainframe systems. See the figure for an overview of NGIO's channel architecture.

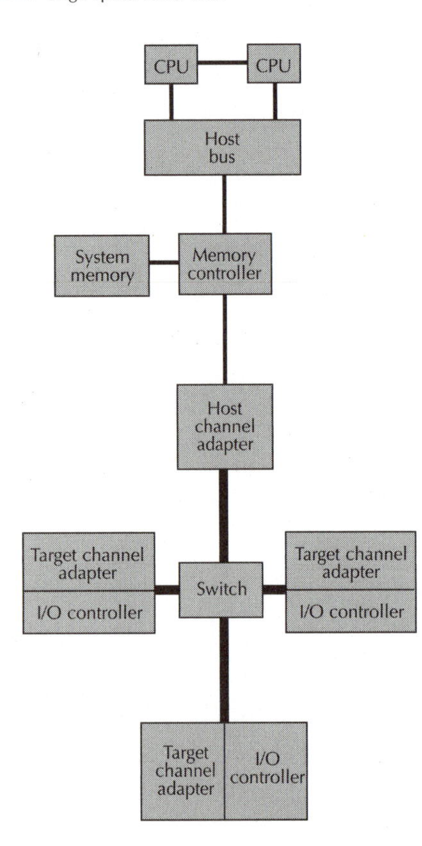

■ **Figure 1:** Generation I/O: By combining a point-to-point switched fabric with a channel architecture, Next Generation I/O offers a more scalable and efficient method of connecting hosts with peripheral devices.

NGIO's channel architecture includes a Host Channel Adapter (HCA), which is an interface to and from the memory controller of a host. It contains DMA engines and has a tight link to the host memory controller. NGIO also includes Target Channel Adapters (TCAs), which connect fabric links to the I/O controller. The I/O controllers can be SCSI, Fibre Channel, or Gigabit Ethernet, for example, which allows a variety of network and storage devices to be mixed within the I/O unit.

HCAs and TCAs can connect either to another channel adapter or to switches. The switches in an NGIO architecture let hosts and devices communicate with many other hosts and devices. Switches transmit information among fabric links, although the switches themselves remain transparent to end stations.

Each link in an NGIO system carries requests and responses in packet form. These packets consist of many cells, which is a bit like ATM formatting. These packets travel at speeds up to 2.5Gbits/sec. In fact, NGIO's Physical layer is very much like Fibre Channel's, which supports speeds of 1.25Gbits/sec or 2.5Gbits/sec.

The companies working on NGIO hope to release products that support the technology starting in 2000.

WHICH BUS TO TAKE?

All of these development efforts mean that within a year or two there could be a glut of I/O architectures on the market. Chances are good that not all of the technologies outlined here will come into widespread use, and industry analysts speculate that support from software vendors like Microsoft and Novell will be a key to success.

Currently, all three post-PCI architectures have support from big names in the hardware world, which is important since it is these vendors that will eventually develop servers, chipsets, boards, and other components that support one or more of the technologies currently on the table.

In the meantime, don't expect PCI to slip away quietly anytime soon. PCI has been the workhorse of the I/O world for several years, and even after new technologies are introduced to the market, you'll see systems that have PCI coexisting with other architectures.

RESOURCES

For more detailed information on the PCI bus specification, visit the PCI Special Interest Group's Web site at www.pcisig.com.

For an overview of PCI, see www.adaptec.com/ technology/whitepapers/ pcibus.html.

To see a detailed discussion of Intel's Next Generation I/O, see http://developer.intel.com/design/servers/future_server_io/ index.htm.

This tutorial, number 130, written by Anita Karvé, was originally published in the May 1999 issue of Network Magazine.

RAID

In 1987, UC Berkeley researchers David Patterson, Garth Gibson, and Randy Katz warned the world of an impending predicament. The speed of computer CPUs and performance of RAM were growing exponentially, but mechanical disk drives were improving only incrementally. As a result, they stated, "We need innovation to avoid an I/O crisis."

The authors famously proposed a solution in their paper, "A Case for Redundant Arrays of Inexpensive Disks (RAID)." They noted that PC disk drives were starting to match the speed of those supplied for mainframes and minicomputers—yet PC drives tended to be better in terms of cost per megabyte. Why? Because of standards such as SCSI, which enabled suppliers to embed functionality that had once required custom controllers.

Looking at what was available on the market, Patterson, Gibson, and Katz concluded that 75 PC disk drives could be lashed together to provide the capacity of a single mainframe drive—with lower power consumption, lower total cost, and 12 times the I/O bandwidth. The snag was Mean Time to Failure (MTTF), clearly much worse for an array of commodity drives than for a bulletproof, silver-plated mainframe unit.

Their paper therefore suggested the use of extra "check" disks, containing redundant information that could be used to recover data in the event of a disk failure. Once a failed disk was replaced, either by a human operator or by electronic switching, data would be reconstructed onto it automatically.

The rest is history, as just about anybody connected with networking is already aware. The I/O problem was widely recognized and, within a couple of years, Intel-based products like the Compaq Systempro (released in 1990) made RAID an expected ingredient in every midrange and high-end server.

So why discuss RAID now? Apart from the fact that we've never published a tutorial about it, the plummeting costs of both disk drives and the circuitry necessary to support disk arrays make RAID more relevant than ever before. It's now affordable for low-end servers and even standalone workstations.

When the first RAID-based products came out, the cost of controllers and capacious SCSI disk drives meant that servers could easily cost $35,000. Vendors were embarrassed by the "Inexpensive" in the RAID acronym and temporarily decided that the "I" stood for "Independent" instead.

Now, with 20Gbyte drives selling for under $200, it's time to be proud of RAID's low cost and leave the misnomer behind. The fact is, there's little that's independent about the drives in a RAID array.

By design, RAID technology hides the characteristics of individual drives from whatever operating system is run on top of it, presenting multiple disks as if they were a single, larger drive. It maps logical disk block addresses to their actual physical counterparts using a variety of algorithms. The differing organizations of data on disk are known as RAID levels, each with its own particular advantages and disadvantages.

RAID 0

RAID level 0 could more correctly be called "AID," because there's no redundancy about it. Data is merely divided into blocks, each one written sequentially to the next drive in the array. If there are four drives in the array, as shown in the figure , each logical I/O is broken into four physical operations.

The point of RAID 0 is performance. Theoretically, it can deliver n times the performance of a single drive, where n is the number of drives in the array. However, tuning the stripe size is important. If it is too large, many I/O operations will fit in a single stripe and take place on a single drive. If it is too small, each logical operation will be broken into too many physical operations, saturating the bus or controller to which the drives are attached.

Reviewers of RAID 0 products on workstations have commented that they offer little advantage with typical applications, such as word processors and spreadsheets. However, in cases where very large files must be opened or saved—on video servers, for example—they can be very beneficial.

RAID 1

RAID 1 is the simplest actual redundant array design, employing mirrored pairs of disk drives. As seen in the figure, it merely creates a duplicate of the contents of one disk drive onto another. While that fact makes RAID 1 easy to implement, it also makes it the most costly (100 percent redundancy) in terms of required disk overhead.

RAID 1's write performance is slower than that of a solo drive, since all data must be written twice. However, buffering on a controller usually hides this fact from the host computer.

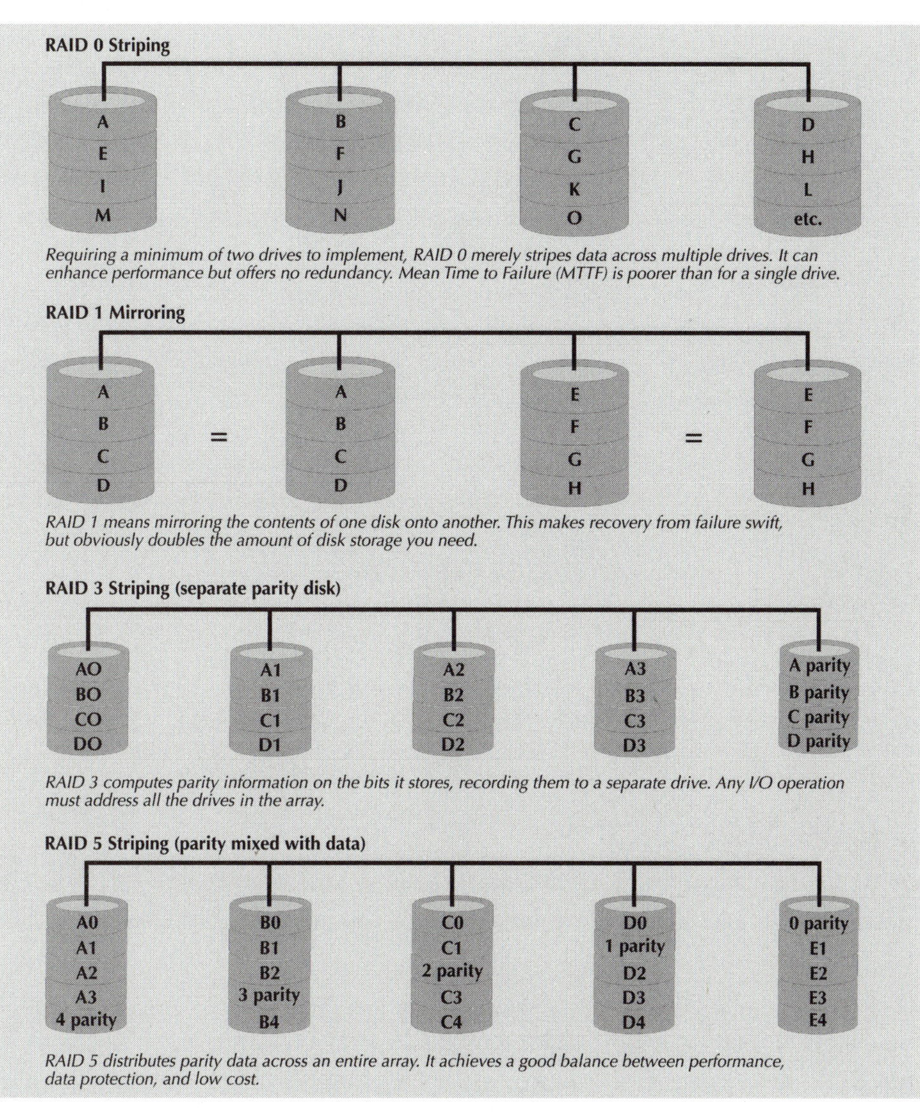
RAID 0 Striping
A
E
I
M
B
F
J
N
C
G
K
O
D
H
L
etc.
Requiring a minimum of two drives to implement, RAID 0 merely stripes data across multiple drives. It can enhance performance but offers no redundancy. Mean Time to Failure (MTTF) is poorer than for a single drive.
RAID 1 Mirroring
A
B
C
D
=
A
B
C
D
E
F
G
H
=
E
F
G
H
RAID 1 means mirroring the contents of one disk onto another. This makes recovery from failure swift, but obviously doubles the amount of disk storage you need.
RAID 3 Striping (separate parity disk)
AO
BO
CO
DO
A1
B1
C1
D1
A2
B2
C2
D2
A3
B3
C3
D3
A parity
B parity
C parity
D parity
RAID 3 computes parity information on the bits it stores, recording them to a separate drive. Any I/O operation must address all the drives in the array.
RAID 5 Striping (parity mixed with data)
A0
A1
A2
A3
4 parity
B0
B1
B2
3 parity
B4
C0
C1
2 parity
C3
C4
D0
1 parity
D2
D3
D4
0 parity
E1
E2
E3
E4
RAID 5 distributes parity data across an entire array. It achieves a good balance between performance, data protection, and low cost.

Striping, Mirroring, and Parity. RAID uses a variety of methods to organize data on a disk.

Reads can be faster, since it's always possible to retrieve data from whichever drive is available sooner.

RAID 2

RAID 2 is a bit-oriented scheme for striping data. Each bit of a data word is written to a separate disk drive, in sequence. Checksum information is then computed for each word and written to physically separate error-correction drives.

Unfortunately, I/O is slow, especially for small files, because each drive must be accessed for every operation. Controller design is relatively simple, high data-transfer rates are possible for large files, and disk overhead is typically 40 percent. However, while reliable, RAID 2 is seldom considered worth bothering with today.

RAID 3

RAID 3 introduces a more efficient way of storing data while still providing error correction. It still stripes data across drives bit by bit (or byte by byte). However, error-checking now takes place by storing parity information (computed via a mathematical function known as the Exclusive OR, or XOR) on a separate parity drive (see figure).

Given that parity values are simple to compute and write, RAID 3 arrays can perform swiftly. However, any I/O operation must address all drives simultaneously. This means that, while RAID 3 delivers high data-transfer rates, it is best suited to large files such as video streams.

RAID 4

RAID 4 modifies the RAID 3 concept by working with data in terms of blocks (as does RAID 0), rather than bits or bytes. This reduces processing overhead and can make for high aggregate data-transfer rates on reads. For writes, however, there is inevitable contention for the sole parity drive, making this RAID level relatively sluggish.

RAID 5

One of the most popular RAID levels, RAID 5 is again block-oriented and based on the storing of parity information. However, instead of placing parity data on a single drive, it distributes it across the entire array (see figure).

Because RAID 5 eliminates the parity-drive bottleneck, it enhances write performance. And due to the independence of all the drives in the array, read performance is tops among true RAID levels. Recovery following a disk failure is relatively slow, but reliable enough. All in all, RAID 5 achieves an excellent balance between performance, data protection, and low cost.

RAID 10 AND RAID 53

RAID 10 is also known as RAID 0+1 or 1+0 because it combines the elements of RAID 0 and RAID 1. It uses two sets of drives that mirror one another, as in RAID 1. Then, within these sets, data is striped across the drives (as in RAID 0) in order to speed access.

RAID 53, which should really be called RAID 30 using the above logic, combines RAID 0 and RAID 3. Again, it uses a striped array, as with RAID 0, but the segments of this are RAID 3 arrays. High data-transfer rates and high I/O rates for small requests are both offered—but at a price.

ENHANCING PERFORMANCE

RAID controllers can become a bottleneck, especially with high-speed interconnects like Fibre Channel, because of the calculations they must perform. For example, to perform a disk-write operation to a RAID 5 array, a Read-Modify-Writeback operation must be performed. First, old data must be read from both a data drive and a parity drive. Second, that data must be XORed. Third, new data must be written to the data drive. Fourth, new data must also be XORed with the parity data, and only then can the result finally be written to the parity drive.

One solution to this bottleneck has been to move the responsibility of calculating XOR data to the disk drives themselves. Seagate, IBM, and other vendors have released drives that can perform XOR calculations in parallel with other disks, without the aid of the RAID controller.

The industry is entering a period of rapid transition in I/O architectures. InfiniBand's 2001 products will couple I/O directly to host memory, offering transfer rates of up to 6Gbytes/sec, and RAID products will evolve to support such throughput. At the same time, the falling cost of controllers and drives will make RAID arrays ever more commonplace on the low end. Your next notebook computer may even offer you a choice of RAID levels.

RESOURCES

"A Case for Redundant Arrays of Inexpensive Disks (RAID)" by Patterson, Gibson, and Katz is archived at http://sunsite.berkeley.edu/Dienst/UI/2.0/Describe/ncstrl.ucb/CSD-87-391/.

Details on Storage Computer's proprietary RAID 7 can be found at www.raid7.com/wp_raid7afa.html.

The RAID Advisory Board Web site is located at www.raid-advisory.com.

A search engine designed specifically to find information about RAID and other storage-related topics can be found at www.searchstorage.com.

This tutorial, number 144, by Jonathan Angel, was originally published in the July 2000 issue of Network Magazine.

Storage Area Networks

As companies rely more and more on e-commerce, online transaction processing, and databases, the amount of information that needs to be managed and stored can intimidate even the most seasoned of network managers.

While servers do a good job of storing data, their capacity is limited, and they can become a bottleneck if too many users try to access the same information. Instead, most companies rely on peripheral storage devices such as tape libraries, RAID disks, and even optical storage systems. These storage devices are effective for backing up data online and storing large amounts of information.

But as server farms increase in size, and as companies rely more heavily on data-intensive applications such as multimedia, the traditional storage model isn't quite as useful. This is because access to these peripheral devices can be slow, and it might not always be possible for every user to easily and transparently access each storage device.

Recently, a number of vendors from all walks of the industry have been pushing a concept called the Storage Area Network (SAN). SANs provide more options for network storage, including much faster access than Network Attached Storage (NAS) and the flexibility to create separate networks to handle large volumes of data.

STORAGE SPACE

Before going into a detailed discussion of SANs, I'll examine how other methods of adding storage to the network work, and why the need arose to develop something beyond them.

The most basic way of getting storage devices on the network is to hang disk arrays or other storage devices off of servers, using an interface such as the Small Computer Systems Interface (SCSI) to make the connection. SCSI is a relatively high-speed interface that was developed more than 15 years ago.

This popular I/O interface connects not just storage-related devices such as tape and optical drives but also printers, scanners, and external drives.

SCSI has gone through a number of changes over the years, especially the speed the interface supports. Initially designed to handle speeds of 5Mbytes/sec, it supports a throughput rate of 160Mbytes/sec in its current iteration, Ultra3 SCSI. Within the Ultra3 specification is a subset implementation, Ultra160 SCSI, that's gaining popularity among SCSI vendors as well as server and workstation manufacturers.

While SCSI has been a workhorse over the years for connecting peripherals at a relatively fast speed, distance limitations have kept this particular interface from evolving very rapidly. The SCSI standards put a bus length limit of about 6 meters on devices. While this distance limitation doesn't really affect connecting storage devices directly to a server, it does severely restrict placing RAID and tape libraries at other points on the network.

This is where the concept of NAS comes in. NAS is straightforward in that disk arrays and other storage devices connect to the network through a traditional LAN interface such as Ethernet (or whatever the topology of choice might be). Storage devices would attach to network hubs much the same as servers and other network devices (see Figure 1).

NAS makes storage resources more readily available and helps alleviate the bottlenecks commonly associated with access to storage devices. However, NAS does have a few drawbacks.

First, network bandwidth places throughput limitations on the storage devices. Most NAS servers are placed on 10Mbit/sec or 100Mbit/sec Ethernet LANs, but even if a network is running at gigabit speeds, most NAS vendors today only offer interfaces up to Fast Ethernet.

Another downside to NAS is the lack of cohesion among storage devices. While disk arrays and tape drives are on the LAN, managing the devices can prove challenging since they are separate entities and are not logically tied together.

NAS has its place as a viable storage architecture, but larger enterprises need something a few steps beyond.

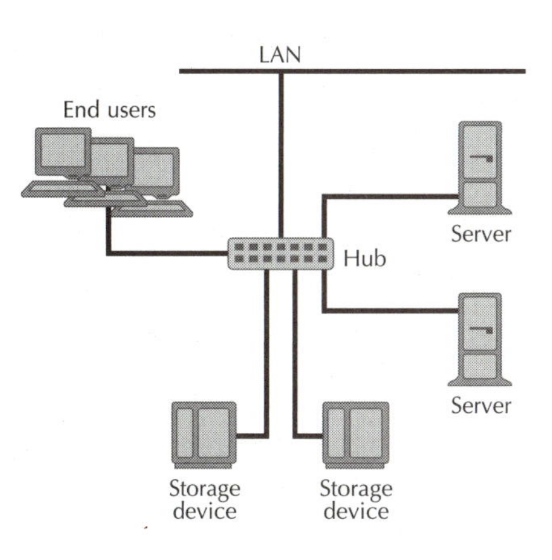

■ **Figure 1:** In a Network Attached Storage (NAS) scenario, storage devices such as RAID and tape drives are part of the LAN, making them accessible to any other network resource. Drawbacks of this model include bandwidth limitations, because NAS implementations are dependent on the underlying network topology.

SANS ARRIVE ON THE SCENE

SCSI-based storage and NAS-based configurations are both important ways of bringing storage to the network, but they are best utilized in situations where there is a relatively low volume of data traversing the links.

Large enterprises that want the ability to store and manage large amounts of information in a high-performance environment now have another option: the SAN.

In a SAN environment, storage devices such as DLTs and RAID arrays are connected to many kinds of servers via a high-speed interconnection, such as Fibre Channel. This setup allows for any-to-any communication among all devices on the SAN. It also provides alternative paths from server to storage device. In other words, if a particular server is slow or completely unavailable, another server on the SAN can provide access to the storage device.

A SAN also makes it possible to mirror data, making multiple copies available.

The high-speed interconnection that links servers and storage devices essentially creates a separate, external network that's connected to the LAN but acts as an independent network.

There are a number of advantages to SANs and the separate environments they create within a network. SANs allow for the addition of bandwidth without burdening the main LAN. SANs also make it easier to conduct online backups without users feeling the bandwidth pinch.

And, when more storage is needed, additional drives do not need to be connected to a specific server; rather, they can simply be added to the storage network and can be accessed from any point.

Another reason SANs are making big waves is that all the devices can be centrally managed. Instead of managing the network on a per-device basis, storage can be managed as a single entity, making it easier to deal with storage networks that could potentially consist of dozens or even hundreds of servers and devices.

FIBRE LOOPS

The interconnection of choice in today's SAN is Fibre Channel, which has been used as an alternative to SCSI in creating high-speed links among network devices.

Fibre Channel was developed by ANSI in the early 1990s, specifically as a means to transfer large amounts of data very fast. Fibre Channel is compatible with SCSI, IP, IEEE 802.2, ATM Adaptation Layer for computer data, and Link Encapsulation, and it can be used over copper cabling or fiber-optic cable.

Currently, Fibre Channel supports data rates of 133Mbytes/sec, 266Mbytes/sec, 532Mbytes/sec, and 1.0625Gbits/sec. A proposal to bump speeds to 4Gbits/sec is on the drawing board. The technology supports distances of up to 10 kilometers, which makes it a good choice for disaster recovery, as storage devices can be placed offsite.

SANs based on Fibre Channel may start out as a group of server systems and storage devices connected by Fibre Channel adapters to a network. As the storage network grows, hubs can be added, and as SANs grow further in size, Fibre Channel switches can be incorporated.

Fibre Channel supports several configurations, including point-to-point and switched topologies. In a SAN environment, the Fibre Channel Arbitrated Loop (FCAL) is used most often to create this external, high-speed storage network, due to its inherent ability to deliver any-to-any connectivity among storage devices and servers (see Figure 2).

An FCAL configuration consists of several components, including servers, storage devices, and a Fibre Channel switch or hub. Another component that might be found in an arbitrated loop is a Fibre Channel-to-SCSI bridge, which allows SCSI-based devices to connect into the Fibre Channel-based storage network. This not only preserves the usefulness of SCSI devices but also does it in such a way that several SCSI devices can connect to a server through a single I/O port on the server. This is accomplished through the use of a Fibre Channel Host Bus Adapter (HBA). The HBA is actually a Fibre Channel port. The Fibre Channel-to-SCSI bridge multiplexes several SCSI devices through one HBA.

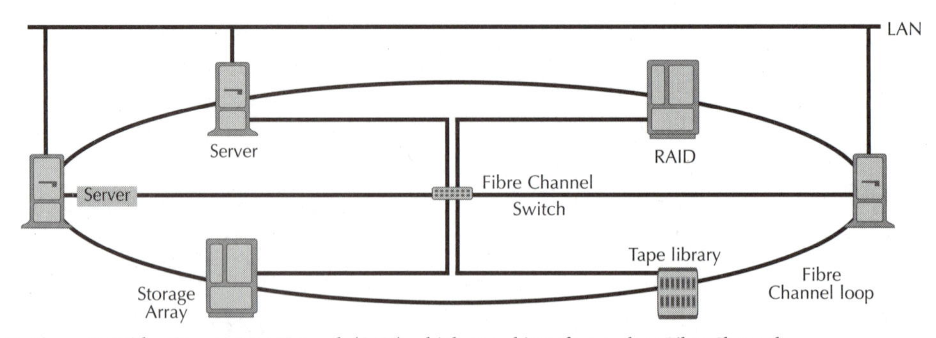

■ **Figure 2:** With a Storage Area Network (SAN), a high-speed interface such as Fibre Channel connects storage devices to servers and the rest of the network. A Fibre Channel switch or hub lets servers access any storage device on the loop.

The FCAL provides not only a high-speed interconnection among storage devices but also strong reliability. In fact, you can remove several devices from the loop without any interruption to the data flow. Also, packets sent over an FCAL are error-checked, and, if need be, packets can be re-sent if any are lost or corrupted.

Fibre Channel is really what has made SANs a reality, and future developments on the interface will likely bring more features, such as higher bandwidths.

ARCHIVING THE FUTURE

While a SAN architecture does appear to be the next step in the evolution of network storage, there are a few points that need to be more adequately addressed before SANs become more widely used.

One crucial piece to running a large SAN is software that administers and controls all devices on the network. While a SAN configuration inherently makes management easier than in the case of NAS, most companies will require a customized application to manage their SAN.

In a relatively small SAN implementation, customized software can be written to ensure communication among all devices. But as SANs grow, and as more vendors enter this space, simply writing management software will not be enough. Rather, there needs to be a standard way for components from different vendors to interact within the context of a SAN.

Vendors in the storage, and specifically the SAN, market have realized this shortcoming. Through vendor-neutral organizations and traditional standards bodies, these issues are being raised and dealt with.

SANs may require a bit more thought and planning than simply adding one storage device to one server, but as companies wrestle with reams and reams of information on their networks, this high-speed alternative that's always available should make wading through the information age a bit easier.

RESOURCES

For tutorials, white papers, and FAQs on Fibre Channel and how it relates to Storage Area Networks (SANs), visit the Fibre Channel Industry Association (FCIA) at www.fibrechannel.com. The FCIA was created in August 1999, when the Fibre Channel Association and Fibre Channel Community merged.

Strategic Research is a market research firm specializing in the storage market. An informative introduction to SANs can be found at www.sresearch.com/wp_9801.htm.

The Storage Networking Industry Alliance (SNIA) is an industry group of more than 75 companies working on storage networking standards for improved management. SNIA's Web site can be reached at www.snia.org.

A group of vendors has created an alliance focusing on the new 160Mbyte/sec version of SCSI. For more information, see www.ultra160-scsi.com.

This tutorial, number 136, written by Anita Karvé, was originally published in the November 1999 issue of Network Magazine

Network Wiring Hubs

It's no surprise that network administrators responded to the passage of the IEEE's 10BaseT Ethernet standard with a sigh of relief. The 10BaseT standard spells out the exact ways in which Ethernet data signals can be sent over high grade telephone wire, and that should make those professionals' daily lives a good deal more pleasant than heretofore possible.

Ostensibly, 10BaseT gives them a low-cost cabling alternative—unshielded twisted-pair wiring (UTP), better known as telephone wire—for transmitting Ethernet data signals. Millions of miles of telephone wire installed in buildings worldwide can now be safely used for Ethernet data transmission. The ability to use existing UTP and its resulting savings from not having to pull new cable may be a superficial benefit of 10BaseT. Not all UTP hidden behind walls can support Ethernet, so they often need new cables, anyway.

10BaseT's real benefits promise to overshadow the economics of not pulling new cabling. The true benefits of 10BaseT will be derived through increased control of Ethernet networks in general and network cabling plants specifically. The 10BaseT standard's most important feature might very well be that it gives Ethernet topological parity with IBM's Token Ring scheme, the industry's other popular networking technology.

The 10BaseT standard calls for connecting Ethernet workgroups in a star topology that focuses all network cabling in a single wiring concentrator, or hub. Token Ring's physical star topology, with its centralized wiring hub, has been one of the major reasons it has made substantial popularity gains against Ethernet in recent years.

AT THE LAN'S CENTER

Why would Ethernet managers want to concentrate the wiring at the center of Network Wiring Hubs 60 61 their networks? Token Ring's centralized architecture makes its LAN cabling systems, which are in reality electrical rings, easy to manage—and that's a distinct selling point to overworked network managers.

A description of a traditional Ethernet bus-style network explains the whys of centralized networks: in a standard Ethernet, the network cabling runs from one node (e.g., computer or printer) to another, in essence meandering from desk to desk in a continuous single strand.

This is a simplified bus network, which can also contain cable branches running off the main, or backbone, cable; in such branching cable systems, however, the problems described here are virtually identical to those affecting nonbranching buses.

Imagine the headache of troubleshooting this seemingly never ending stretch of cable on a bus: When workstations on a bus-based LAN experience network-related trouble, a technician may have to inspect the entire cable segment, which can be several kilometers long, before finding the problem. This can be time-consuming, frustrating, and costly.

Troubleshooting a hub-based network entails no such runarounds: Each workstation is attached directly to the hub via its own unshared cable. This means the technician troubleshooting a particular network node has to worry about only the cable segment running from the hub to that node, not the entire network wiring system. This can represent substantial time savings in diagnosing network failures.

A LOOK AT A HUB

Although each of the network hubs on the market are similar to each other, they're also dramatically different. All of the network hubs contain two basic components—a chassis and topology-specific modules—that allow creating easily reconfigurable networks. But each vendor has elected to work within its own design philosophy that makes for significant physical differences among those available.

A chassis is the hub's most visible component. It acts as the hub enclosure and serves as the interface between each of the individual modules. Individual vendor's hubs contain varying numbers of accessory slots, each of which accepts a single module.

The hubs usually contain an integral power supply and/or primary controller unit. Depending on a vendor's design philosophy, the controller unit may perform network management functions, serve as a repeater for all modules in the chassis, or act as the connection point to hubs located in other parts of a building or office of campus complex.

The modules serve as the link between the chassis and the network cabling and are thus cable-specific. That means they contain connectors that accept only certain types of connectors, each of which is associated with a specific type of cable. For example, a vendor may support IBM's Type 3 (IBM-certified UTP) Token Ring wiring and its associated connector with one module, thick and thin Ethernet coaxial cable with another, the UTP wire used in 10BaseT Ethernets with a third module, and ARCnet coax or fiber-optical cable with still other modules.

Each module usually contains connectors for multiple workstations. Placing two, four, or six connectors is common, depending of the intended use and type of cable to be attached to the module. For instance, thick coax connections require what is called a DIX connector, which is substantially larger that the RJ-45 used with UTP.

Some vendors offer modules that combine connector types, but only within a single type of network technology. For example, most hub manufacturers sell modules that support connections to both thick or thin Ethernet coax and 10BaseT UTP.

THEIR PHYSICAL DIFFERENCES

As is often the case in free enterprise, various network product vendors have applied their own proprietary sets of design rules to their hubs. For instance, some hubs are six to eight inches high, while others are 12 to 18 inches, and the physical and/or electrical junctions that connect the module to the chassis generally differ in each vendors' implementations.

The reasons for these differences range from the complex, such as needing to cram sufficient functionality onto a product, to the simple, such as merely ensuring that the hub is physically compatible with the vendor's existing products. These design differences mean, of course, that a module designed for one vendor's wiring concentrators generally will neither fit in nor operate with another vendor's hub.

Fortunately, these physical incompatibilities don't carry over to the hubs' abilities to communicate with each other. Although the modules themselves may be physically and/or electrically incompatible, the cable-specific jacks on them provide industry-standard connections.

For instance, the 10BaseT connectors (known as eight-pin modular connectors and sometimes called RJ-45 connectors, which are physically similar but electrically dissimilar) on one vendor's hubs are identical to those on another's. This is also true for the thin and thick coax, fiber-optic, Token Ring, and ARCnet connectors.

A WEALTH OF OTHER BENEFITS

The wiring hub's modular design allows vendors to integrate a wealth of other features and cabling options into them. As already noted, this architecture permits mixing the different cabling media—that is, fiber-optic, thick and thin cable, and UTP—into one wiring center. Without such hubs, expensive inter-network devices such as bridges and routers are required.

Just as importantly, the modular approach also allows mixing different media-access technologies, such as Token Ring and Ethernet or Ethernet and Fiber Distributed Data Interface (FDDI), in a single hub. This permits merging and managing the several different types of network technologies at a single point.

This capability is especially valuable to the growing number of large corporations where small clusters of Token Rings and Ethernets have sprung up independently. Again, it often means an organization does not have to purchase an internetworking device to connect and manage a variety of workgroups.

Vendors have also moved hubs out of the local workgroup and into so-called enterprise networks. Hubs can be daisy chained together—that is, connected one after another—to form large networks made up of multiple workgroups connected to separate-but-interlinked hubs.

On a more complex level, vendors are developing modules—usually in conjunction with nonhub vendors—that provide internetwork bridging and routing capabilities. These modules route and filter network traffic between the modules in hubs, whether the hubs are in a local or remote site. Filtering allows a network manager to improve internetwork performance and security by restricting the flow of specified types of data packets across internetwork borders.

BOON TO MANAGEMENT

Network hubs are ideal for centralizing network management capabilities. Numerous vendors have announced "intelligent" network hubs. Intelligence in this context refers to the hub's ability to accept management and configuration commands over the network cabling from a remote workstation.

Putting intelligence into a hub allows these devices to perform many functions that normally require an on-site technician's presence. For instance, network managers sitting at a centrally located network management station can not only turn remote hubs on or off, they can turn off individual modules and even individual ports within a module. They then can reroute traffic from the failed module or port to a working one.

This ability is vital in large, enterprise-wide networks, where a single manager may be in charge of a WAN spanning several cities or states. In such situations, it's not always economically feasible to maintain a technician on-site to troubleshoot occasional problems. The ability to reconfigure the hub from a remote site can thus mean the difference between keeping a network segment online or shutting it down until a technician can make the trip to repair the failed unit.

This tutorial, number 29, written by Jim Carr, was originally published in the December 1990 issue of LAN Magazine/Network Magazine.

Broadband Cable-based LANs

Networks can be classified as baseband and broadband. Baseband LANs, such as Ethernet, ARCnet, and Token Ring, are much more common in the office environment. Broadband LANs are popular where multiple services, such as closed circuit TV, data, and voice, are needed. Broadband is also popular in factory environments.

Broadband LANs work in much the same way that cable television works. Broadband LANs transmit multiple radio frequency signals on the same cable, usually coaxial but sometimes fiber. This ability to send many types of communication simultaneously over the same cable, including voice, video, and data, distinguishes broadband LANs from the far more common baseband LANs. To accomplish this feat, broadband LANs use a technique called frequency division multiplexing.

Frequency division multiplexing is used to put several channels on the same cable simultaneously. To understand frequency division multiplexing, think of the cable as a highway. A highway has a width, which determines how many lanes are possible; the cable has bandwidth, which also determines its capacity. Highway width is measured in feet; LAN bandwidth is measured in Hertz or cycles per second. It is the difference between a higher and a lower frequency— the greater the spread between the upper and lower frequency, the more information can be transported.

Each "lane" or channel on a cable uses a different set of frequencies. Just as cars travelling in different lanes of a highway do not collide, information occupying one frequency band or channel on a cable does not interfere with information on another band. Thus, frequency division multiplexing sends its different types of information at unique frequencies.

Frequency-division-multiplexed information is generally sent in analog, not digital, form. Digital signals are discrete—one or zero. Analog is continuous, like a wave. To reconcile this difference, digital computer data must be converted into analog form for transmission over a broadband cable. The conversion device is called a broadband modem. Once the broadband modem converts the data to analog form, it puts the data on the correct channel.

A TREE TOPOLOGY

Some common broadband LANs use a tree-and-branch topology. The root of the tree is the headend, or central retransmission facility. The trunk cable is attached to this root. Various branch cables are attached to the trunk cable. From there, user devices may be connected. Although most broadband networks use a single cable, some use a dual cable system, one for each direction to and from the headend. A dual cabling system has twice the bandwidth of a single cable system.

A headend is essential. All transmissions must pass through the headend. Broadband Cable LANs use a tree-and-branch topology. All transmissions travel up through the tree to the headend, where the signal's frequency is altered so it can travel to its destination. In this illustration, even though PC1 and PC2 are adjacent, their messages must travel through the headend because each device transmits on one frequency and receives on another. The headend is responsible for translating the device's transmit frequency to the receive frequency of another device. This frequency translation is called remodulation.

To illustrate how a broadband LAN works, let's follow a transmission, say a file transfer, from one PC to another PC. The PC sends the file to its broadband modem, where it is modulated into analog form. The modem sends data at one frequency, called the return frequency and receives at another called the forward frequency. These terms sound reversed because they are named from the headend's point of view.

Before getting to the headend, the signal probably passes through several splitters and couplers, where the signal loses strength. It also passes through amplifiers, where the signal gains strength. According to the principle of unity gain, the signal must arrive at the headend with the correct strength, usually at a level lower than the transmission strength.

At the headend, the packet is translated from the return frequency to the forward frequency, then sent back onto the cable to the receiving PC. The trip to the receiving PC is just like the trip to the headend, except in reverse. Once at the receiving PC, the receiving modem translates the file back to digital form so the PC can understand it.

Forward and return frequencies make a pair. The headend can handle more than one pair of frequencies, although it cannot mix and match them. One complex part of designing a broadband LAN is deciding which information will travel on which frequencies. In many designs, the bandwidth is divided into 6MHz slices, which is the same bandwidth a television channel uses. Each type of communication takes one or more slices. Thus, an Ungermann-Bass Net/One broadband network might run over two channels, a television security system runs over another channel, and voice communications take up other channels. In general, higher speed communication services, such as Ethernet and data services, take up larger chunks of bandwidth than slower services, such as security and voice communications.

UNITY GAIN

All components of a broadband system either amplify or weaken the signal strength. As radio frequency signals travel down a cable, they deteriorate or experience signal loss. Devices on the broadband network, such as splitters, couplers, power inserters, and equalizers, also cause loss. The amount of loss experienced by a particular signal depends on many factors, including the diameter of the cable, the components it encounters, the distance it has traveled, and its frequency.

Broadband LANs are equipped with devices called amplifiers to counteract this unavoidable loss. Amplifiers attach to the cable in certain places and regenerate signals to a level determined by the system design. Any distortion the signal has picked up is also amplified.

The goal in designing a broadband system is to reach unity gain. Unity gain means that the amount of loss caused by the components in the system is equal to the amount of gain caused by amplifiers in the system. Achieving unity gains is complicated since the amount of loss imposed by a component depends on the signal frequency. Thus, amplifiers and other components are placed strategically to keep signal conditions uniform as signals move from device to device.

Since different frequencies are affected differently by various components on the network, achieving and maintaining unity gain is an arduous task. Consequently, broadband LANs—far more so than baseband LANs—require trained professionals for design and installation. Also, broadband systems are best suited for large installations with large staffs of networking professionals, although smaller installations that require multiple service types can also benefit from broadband.

UPS AND DOWNS

Broadband shines in campus environments, especially if several different types of communication must travel to the same locations. Support for multiple services saves cable costs and provides the ability to add services as needed. For example, a company might start with a data network and later add voice, security, a building management system, environmental controls, process management, and closed-circuit television.

A centralized tree-and-branch wiring scheme makes more sense than stringing together multiple baseband LANs in different buildings. Another advantage is broadband's immunity to electrical noise. Broadband LANs typically use frequencies above most machine-generated electrical noise, which is in the low frequency range. Baseband LANs, except those using fiber, are not so fortunate.

The price for this flexibility is complexity. Designing, planning, and installing broadband LANs is extremely time consuming and difficult. Along with deciding where cables and devices go, a LAN planner must evaluate the radio frequency requirements. Hundreds, and possibly thousands, of calculations can be necessary for the design and installation of a broadband LAN cable plant.

Maintaining a broadband LAN is also more difficult than maintaining a baseband LAN. Over time, the radio frequency settings of the components will drift (literally going out of tune) and cause transmission problems. The components require periodic tuning, which is more tedious than difficult.

Most end-user companies are not prepared to install a broadband LAN unaided. Installations require the resources and expertise of a broadband design and installation company. Experienced network planners can guide an end-user company through the process of creating the network, deciding which services will be used, where different devices will be placed, and so on.

This tutorial, number 10, was originally published in the May 1989 issue of LAN Magazine/Network Magazine.

SECTION VII

Network Software

Network Applications

In its most generic sense, the term application applies to a task. For example, a widget maker needs to take orders and transmit these orders to a warehouse where the widgets are shipped to customers. The widget factory and warehouse must exchange inventory information. This whole procedure might be called an order entry and inventory control application. In this sense, other applications include list management, accounting, design, marketing, and sales—the tasks of any enterprise.

More specifically, the term application refers to the computer software used to get a job done. Thus, database management packages such as dBase IV, Paradox, and Oracle are called application software, as are other types of software such as WordPerfect, Co/Session, and 1-2-3. In this sense, application software is distinguished from system software, which is the software that makes computers and networks operate. Think of the application software running on top and taking advantage of the system software and hardware.

Finally, application refers to programs written to perform a specific task. For example, many users have written applications in the dBase language. These customized applications are written by and for end users, not by software vendors. This can get tricky, because some value-added resellers and system integrators write such customized programs to sell. The difference is they are not selling generic applications software as Microsoft, Borland and Lotus do. They, like the end-users themselves, are creating customized programs using the software of developers such as Borland and Microsoft.

Network operating system software, such as NetWare, LAN Manager, and VINES, provide some applications. This presents some complications and indicates some changing directions for the PC network industry.

Networks exist for applications. That is, users install networks to get a job done. Users can have computers, cable, interface cards, file servers, and protocols, but without applications software users can't do much but copy files from disk to disk. Network application software is what people use. The network is just the substrate upon which they use it.

APPLICATION TYPES

There are three types of applications—network-ignorant, network-aware, and network-intrinsic.

Network-ignorant applications are written for use on one computer by one person. These programs can run on a network in the sense that they may be stored on a file server and network users may run them at their workstations. Most of the time there are severe limitations on what these applications can do. Moreover, if two people try to use the program at the same time, data can be lost or corrupted.

For example, if two people try to work on the same 1-2-3 spreadsheet, the person making the last change to the spreadsheet will write over all the changes made by the user who first saved his work. The program has no way of keeping the users from destroying each other's work. It lacks concurrency control. On the other hand, 1-2-3 can be used safely by several people at the same time, as long as they are using different spreadsheets (and if they have a license to do so). But the standalone version of 1-2-3 does not provide functions to take advantage of the network.

Network-aware applications are a step above network-ignorant applications. Usually, they are network-ignorant programs modified to run on a network. These programs recognize they will be used by several users at a time. They have concurrency control features such as file and record locking to coordinate usage by multiple users. For example, when a Paradox user begins to modify an address in a mailing list database, other users who are also looking at the same database table are prevented from changing that particular address record. This is called record locking. When the change is complete, the change is displayed on the screen of every other user looking at the table.

Another network-aware feature is file locking. This is a less sophisticated and less used form of concurrency control. Instead of keeping users out of a particular record, they are kept out of the entire file altogether while another user has it open. Word processing programs are the primary users of this type of concurrency control.

Communications software and electronic mail are also network-aware applications. They use the network to extend the abilities of a PC and share network resources.

At the same time, even these network-aware applications use the network as little more than a peripheral sharing device. The file server holds the data and the program but does not do any processing. Users access the program as if it were local, but all the work is being done by their PC, including all concurrency control. This is changing.

Network-aware programs make up the vast majority of programs written for networks. They are a big improvement over network-ignorant applications and have gone a long way to spur the growth of networking. As they become more sophisticated, the distinction between network-aware and network-intrinsic is blurring.

NETWORK-INTRINSIC APPLICATIONS

Network-intrinsic applications actually share the processing power of several computers. Usually, although not always, this is done by dividing the application program into pieces. One piece is the server, which does data processing; the other piece is the client, which talks to the user. A database server is a good example of this application type. Its principles can be generalized for other network-intrinsic applications.

A database server is composed of front and back ends. The front end is responsible for formulating requests and displaying formatted data to the user. At the front end, users make queries, write reports, create new databases—all the tasks they do with any other database management program. The back end is responsible for managing and searching for data, concurrency control, and security. When a user asks for all the employees in the company database that make more than $50,000, this request is transmitted to the database server or back end. The database server then looks for all the employees making over $50,000 and sends these records to the front end.

In the network-aware method, one program, not two, runs in the user's machine. When the request for middle-income employees is made, the server downloads the entire file over the network to the user program. The user's PC then searches through the file to find the employees with the requested salaries. This takes up much more network bandwidth because the whole file is transferred, instead of just a few records. Other traffic includes concurrency control commands to lock various files and records as needed.

With a database server, concurrency control traffic is eliminated because the server takes care of it. Even more important, only the requested records are sent over the network. The result is a more efficient, safer, and better performing program and network. The two programs are working together to create one application—a true network-intrinsic application.

By implication, network-intrinsic applications have the ability to distribute data over the entire network. They can also distribute processes. This makes for distributed databases, compile servers, compute servers, multitasking communications servers, and many other applications in which programs cooperate across the network to get a job done.

Only a few network-intrinsic applications are available now. New network environments created by operating systems like OS/2 and network operating systems like LAN Manager will help their development by providing multitasking, more memory, faster processors, and programming interfaces that make writing network-intrinsic programs easier. It will take time.

UTILITIES AND APPLICATIONS

A category of applications we have not discussed is network utilities. Usually, utilities are programs written for network administration and management. One example is NetWare's SYSCON. Others include printer and disk management utilities.

Increasingly, utilities are included in network operating systems. For example, NetWare comes with numerous programs to administer and manage the network. But, even more striking than this development, is the way in which application software and system software are coming together.

This tutorial, number 9, was originally published in the April 1989 issue of LAN Magazine/Network Magazine.

Client-Server Computing

The term "client-server computing" has meant many things to many people. In the mainframe or minicomputer environment, it has been used to refer to the relationship between the host computer and its associated dumb terminals. In traditional local area networking terminology, it has also described the association between a personal computer acting as a "server" of data and applications files and the "client" PCs that request those files via a network operating system over LAN cabling.

In the newly emerging distributed network environments, however, client-server computing takes on a more specific definition: It refers to a relationship in which the server plays a more sophisticated role on the network, performing much of the processing formerly handled by its client PCs while still retaining its requester-server (i.e., data storage) responsibilities.

We'll focus on this definition of the client-server model here.

CLIENT-SERVER BASICS

As client-server computing systems have evolved, their creators have taken some parts from the centralized host world, and other parts from the decentralized PC environment. As such, client-server systems also combine benefits from both. From the host world, for instance, comes centralized data storage that can be centrally secured against unauthorized access. From the PC environment comes the standalone computing power needed to run powerful and easy-to-use applications and graphics packages unavailable with dumb terminals connected to a host.

In understanding client-server computing, it thus helps to understand how the traditional computing environments—that is, the host/terminal and file-server/PC—process data and application files: They do so in a totally one-sided fashion. In the former instance, all data manipulation takes place on the host; the terminals merely display the results of the mainframe's computation. Because the terminal is incapable of data manipulation, its functionality is thus limited; in particular, it offers almost no graphics capabilities.

In the standard file-server/PC relationship, the client PCs perform virtually all of the data processing. The server responds to data (and application) requests from the client, thus playing the role of an intelligent high-speed disk system by forwarding stored information to the appropriate client. This means that the server must first locate the requested files on its disk subsystem, then transfer them through its own memory and over the network cabling to the end-user PC.

When a user wants to access a database on the server, for example, the server first downloads the application software to memory in the user's PC. When the user queries the database for a particular data record, the server sends entire groups of data associated with it.

While this method does give users the data they need, it does so at the expense of efficiency: First, entire files rather than just needed data are sent across the network, unnecessarily using too much of the network bandwidth. And because the user's PC does the processing, each end user whose work requires heavy and frequent database access needs a high-performance computer with plenty of memory.

In the client-server computing environment, conversely, developers separate their applications into two components, a "front end" and a "back end," with the elements sharing the processing demands according to which is best suited for the task. This separation of responsibilities allows client-server systems to more efficiently use an organization's computing power and network bandwidth.

The *front end*, or client-based part of the application, provides the end-user interface—that is, the onscreen images the user follows while interacting with the application—as well as processing capabilities. As in the traditional network client-server model, the back end delivers server-based functions such as data lookup and retrieval.

In the client-server computing architecture, however, only the front end of the applications—not the entire application—is loaded into users' PCs when they start the program. Now, when a user's front-end application queries a database for a particular record, the back-end server-based software searches for the specific record and sends it—not entire masses of data—to the user.

This significantly reduces the volume of data moving across the network because entire databases are not continually being sent back and forth between server and client. This offers secondary benefits in that reduced traffic can also lower the risk of electrical or mechanical malfunctions compromising the integrity of data.

Another benefit of client-server computing: Because the server rather than the client handles much of the manipulation of data, it eliminates the need to give each employee who accesses the database a high-performance PC. Only the database server needs a large, fast hard disk, high-performance controller hardware, or multiple high-powered processor chips.

Although the most widely employed server in the client-server realm is the database server, a server in a distributed computing environment can be, among other things, an image or audio processor or an expert system.

In these situations, the database server, which usually resides on a dedicated computer, acts as an "engine" to drive the system. The applications in turn can request/update data in the database, using the database server's capabilities and benefits—for instance, improved security and centralized access—in the process.

A 'SEQUEL' IN THE WORKS

Because PC software applications seldom "talk" the same language and thus do not readily exchange data, software developers have had to agree on a common language for handling this data interchange. This language, called a Structured Query Language (or SQL, and pronounced sequel, as in another "Rocky" or "Star Wars" sequel), acts as the translator between applications. SQL is a high-level language that allows distributed databases to exchange information. Virtually all popular relational databases running on IBM PCs, Macintoshes, mini, and mainframe computers, support SQL.

Developed in the 1970s by IBM, SQL is an English-like query-type language that most database vendors have standardized on. SQL has become the standard for retrieving data from a relational database, and any database that supports SQL can theoretically exchange data with any other SQL-compatible database.

THE GOOD NEWS

The good news about SQL is that end users usually don't have to learn or even know about it. The application program should hide the SQL language commands (principally, these commands come in the form of verbs such as DELETE, SELECT, and UPDATE) from the user interface. A front-end e-mail application, for instance, could use SQL as a tool to locate a recipient's network address and determine the optimum way to "mail" a message. The database server operates in conjunction with a communications server to actually deliver the mail—all without the e-mail end user knowing about it.

The front-end component of a database server does not necessarily have to be a database application; this opens an amazing variety of possibilities. Any application that can make use of a database of information, including spreadsheets, point-of-sale software, and computer-aided drafting/engineering (CAD/CAE), can be incorporated into the database-server environment.

Take the example of a stock broker. It is not atypical for securities salespeople to have two or three computers and/or terminals in their office, each one connected to a different computer system. A networked PC might provide access to a server-based database of the broker's important customers, showing their holdings, priorities, etc. A terminal might provide up-to-the minute stock prices from an online timesharing host via a modem connection. And another PC might offer access to an order-entry system so the broker can process customer orders on the spot.

With a client-server system, the securities firm could deliver all those capabilities via a single client personal computer connected to a distributed database server. In this scenario, the database server acts as a transparent network manager, tracking down the requested services without questioning the end user for network address information.

When the broker's front-end application asks for stock quotes, the database server could automatically open a link through a communications gateway, retrieve information from the online host, and report back to the user's PC with only the specific data requested.

Similarly, the database server could also use the order-entry application's front-end instructions to provide an onscreen sales form. Or the database could display the customer's preferences—not to buy stock of companies with facilities in South Africa, for instance, to sell IBM stock when it reaches $110 a share, or to buy Microsoft when it drops below $90.

And the database server would provide all this information without the broker specifying where any of it is located.

This tutorial, number 23, was originally published in the June 1990 issue of LAN Magazine/Network Magazine.

Network Operating Systems

A network operating system (NOS) causes a collection of independent computers to act as one system. A network operating system is analogous to a desktop operating system like DOS or OS/2, except it operates over more than one computer. Like DOS, a network operating system works behind the scenes to provide services for users and application programs. But instead of controlling the pieces of a single computer, a network operating system controls the operation of the network system, including who uses it, when they can use it, what they have access to, and which network resources are available.

At a basic level, the NOS allows network users to share files and peripherals such as disks and printers. Most NOSs do much more. They provide data integrity and security by keeping people out of certain resources and files. They have administrative tools to add, change, and remove users, computers, and peripherals from the network. They have troubleshooting tools to tell network managers what is happening on the network. They have internetworking support to tie multiple networks together.

REDIRECTION

At the heart of the NOS is redirection. Redirection is taking something headed in one direction and making it go in a different direction. With redirection, an operating program does not know or care where its output is going.

You are probably familiar with DOS redirection. For example, the DOS command DIR > FILENAME will redirect a directory listing to a file instead of to the screen. The ">" tells DOS to give the results of the command to the entity on the right.

Network operating systems depend heavily on redirection, only in this case data is being redirected from one computer to another over the network cable, not over the PC's bus to local files or printers. Nevertheless, the operation is similar. If you type "COPY C: FILEA F:", FILEA will be copied from your local drive C: to the network drive F:. The NOS makes it appear to the COPY command that drive F: is local, when it really resides on another computer that is attached to the same network. The COPY command doesn't know or care that drive F: is across the network. It sends the file to DOS and the NOS reroutes the file across the network to drive F:.

Redirection can be done with printers and other peripherals. Thus, LPT1: or COM1: can be a network printer instead of a local printer and the NOS redirects files to these devices. With a NOS, users don't need to know about redirection; they just type the drive designator or print from their word processors as always.

SERVER SOFTWARE

The computer with drive F: must expect data, if the output from the user's PC can be redirected successfully. To do this, it must make its drive available to network users. This is part of the NOS's function at the server.

A NOS is made of a redirector and a server. Not all machines need to run the server software, because not all computers need to share their resources. But all network workstations must run redirector software because every client has to be able to put data onto the network.

With some NOSs, the computer running the server software cannot be used as a workstation. This is called a dedicated server. Novell's NetWare uses this kind of setup almost exclusively (although the low-end NetWare Lite can use nondedicated servers). With some other NOSs, all workstations on the network can also be servers. This a nondedicated server setup. This approach is used by Sitka and Artisoft, among others.

The two server approaches have advantages and disadvantages. Nondedicated servers allow for more flexibility, since users can make resources available on their computers as necessary. However, a nondedicated server approach requires that the users are willing to take some administrative responsibility for their computers and it necessitates that they be somewhat network-literate. Backing up the shared data, setting up security, and setting up access rights become more complicated and often become the responsibility of the user, not the administrator. Another drawback is that nondedicated servers often suffer some performance degradation when being used simultaneously as a workstation and as a server.

Dedicated servers have the opposite advantages and disadvantages. They are easier to administer since all data is in one place. They are faster because they don't have a local user to serve. On the other hand, it is harder to make resources available on an ad hoc basis, since setting up a server is more difficult and time-consuming. If a dedicated server fails, all users are forced to stop working because all resources are centralized. Your choice of dedicated or non-dedicated operation will depend on the work your network is doing.

FILE SERVICE

A file server's primary task is to make files available to users, although it also makes other resources available, including printers and plotters. File service allows users to share the files on a server. The server PC can make its whole disk, certain directories, or certain files available. The file server's hard disk becomes an extension of each user's PC.

The NOS can let the network administrator determine which users are allowed to use which files, for example, keeping the mail clerk out of the payroll file. Suppose a user wants to use a file residing on the file server's hard disk. Drive F: is set to correspond with the file server's hard disk. The actual process of setting up virtual drives has several names, including mapping, mounting, and publishing.

Now, suppose a user wants to run WordPerfect. At the F: prompt, the user types "WP" to load WordPerfect. WordPerfect is loaded from the server over the network, and into the user PC's memory. Meanwhile, other people can use WordPerfect from the file server (assuming there is a license for multiple users). WordPerfect makes sure no other user can get the document file being used by "locking" the file. With many applications, file locking allows other users to read the document but not edit it.

File service is an extension of the local PC. Applications work just as they would on a local PC. Some programs, however, have been designed to take advantage of the network, rather than just run on one. For example, some databases allow two users to edit the same table but not the same record and each user can see the other's changes.

The NOS provides much more than just file service; it provides security, administration, printer sharing, backup, and fault tolerance.

SERVER OPERATION

The server software makes a single-user computer into a multiuser machine. Instead of just one user, a server has many users. But we must qualify what we mean by "many users." A NOS allows many users to share the server's peripherals, printers, disks, and plotters, but it does not allow multiple users to share its processor. For now let's see how the file server allows users to share its peripherals.

In many cases, the file server is running the PC's native operating system (such as DOS or the Macintosh OS) as well as the NOS. When users' requests come in, the NOS receives and interprets them, then hands them to the operating system for execution. So if a request comes in to open a file, DOS opens the file and gives it to the NOS, which gives it to the user. If many users make requests at the same time, the NOS queues them and hands them to DOS one at a time.

High-performance NOSs, including Novell NetWare, Banyan VINES, and Microsoft LAN Manager, do not run DOS in the file server. DOS is replaced with a multitasking operating system, thereby gaining a performance advantage; however, they lose some compatibility and require dedicated file servers. In NetWare's case, it is a proprietary OS. VINES runs Unix; LAN Manager currently runs OS/2 but eventually will use Windows NT.

This tutorial, number 6, written by Aaron Brenner, was originally published in the January 1989 issue of LAN Magazine/Network Magazine.

Directories

Directories are the key to global networking; they're also a major thorn in network managers' sides.

With the advent of multiuser computing systems came the need to control who gets to do what on the system. The need for restraints and controls arose not only for reasons of basic privacy, but also because it's extremely risky to let inexperienced users have access to everything on the system, including critical system files.

On most mainframe and minicomputer operating systems, it's been standard practice for decades that before a new user can gain access to the system, an account must be created for that user. As part of creating the account, the system administrator must specify which file systems, file system directories, and individual files the user can access. This information about the account is then stored in the system's Access Control List (ACL) or an equivalent security database.

At the start of each computing session, the user logs in to the system using the established account name and password. The system then looks up the account in the ACL and checks the user-entered password against the previously stored password to see if they match. If the passwords don't match—or if no such account can be found in the ACL—the user's login attempt will be rejected, in which case, the user may try again, depending upon the established account constraints. If everything matches up, the user will be given access to the system—but only to those system resources previously defined in the ACL.

ENTER NETWORKING

As computer networking—another form of multiuser computing—developed and matured, developers of network OSs adopted essentially the same form of security architecture. However, one difference between network computing and the earlier mainframe style of computing is that with mainframes, the "system" (the computing system) is often one big computer (or a cluster of computers acting as one big one) that is time-shared among all users. All the users log in to the same system, and the system knows about each user. It's a cliché, but it bears repeating: Mainframe-style computing is a homogeneous, centralized model.

In the network computing model, which typically employs client-server architecture, the system is a complex collection of many different computers, some acting as clients, some as servers, and some as both (as "peers" in other words).

The decentralized form of today's networked computing systems is at once their greatest strength and their major weakness. The ability to add servers and divide up computing responsibilities makes networks inherently scalable and precludes the mainframe problem which is that even the most powerful computer is brought to its knees if you add enough users. But administering a collection of clients and servers can be a challenge, particularly if your enterprise encompasses hundreds or thousands of servers and tens of thousands of users. The administration problem is not due simply to the multiplicity of servers in a networked environment; most client-server applications have their own access control schemes, as well. For example, Oracle Server has its own security system, as do Lotus Notes and cc:Mail.

For users, it's a hassle to have to log in to each server they want to use. To relieve some of the pain, it's possible to "synchronize" passwords, so that users' single passwords are the same for all servers. This reduces the number of passwords users need to commit to memory. But it's still necessary for users to log in to all of the servers they need to access.

A PROMISE AND A THREAT

While directories present administrative problems, they are also playing an increasingly important role in client-server computing, particularly as the Internet brings all networks together—once you have video conferencing capability, you'll want the ability to reach someone. But even for the more prosaic, nonvideo-enhanced e-mail of today, there are times when you want to message someone but don't know that person's address. So, the strategy with directories must be to take advantage of their strengths, while finding ways to mitigate their drawbacks.

To illustrate the directory problem, consider Novell's NetWare (prior to version 4.0). It used what Novell called the NetWare "bindery" as the security database for NetWare servers. To give someone access to a NetWare server, a network administrator would have to create a user account and define what directories the user should have access to. This information would be stored in the bindery on that server. If there was more than one NetWare server and you wanted the user to have access to multiple servers, you would have to repeat the account-creation process on each server.

NETWORK USERS GANG UP

Network managers and administrators have been grappling with directories for almost a decade now, and the problem's been growing as networks have increased in scale. The Network Application Consortium (NAC) is one organization that's been working to change the directory services arena. Founded in 1990, NAC is composed of several dozen major users of networks. The list includes the Australian Bureau of Statistics, Carolina Power & Light, MCI, Nynex, Pacific Bell, Pacific Gas & Electric, the United States Marine Corps, and the University of Michigan. The NAC champions interoperability among diverse computing systems in general, and in the directory services arena, the organization encourages the use of the X.500 standard and the Lightweight Directory Access Protocol (LDAP) that derived from it. (LDAP was developed at the University of Michigan, so I'm not surprised to see the NAC championing it.)

The developers of network OSs have heard the cries from the user community and have taken steps to address the problem—but the issue is a big one, and it won't yield easily or quickly.

The term "Access Control List" implies a simple file (perhaps nothing more than a text file) that lists user account names, passwords, and permissions. For many systems, this is all an ACL is. But, driven primarily by the administrative burden created by having a multitude of noninteroperable ACLs, developers have come to envision a directory that would hold all account-related information for all users and resources on a network. This directory could handle user authentication when a user first logs in to the network. Whenever the user wishes to run an application, the application could consult the directory to see whether the user has the necessary permissions. Because of the amount of information it would need to contain and because of the need to organize it in some logical fashion, this directory would probably be closer to a database system than a simple text file.

SOME SOLUTIONS

Wouldn't life be much easier if there were only one directory and all the various network OSs and applications used it? The company that took the first steps in that direction was Banyan Systems. Banyan's StreetTalk directory is a centralized resource that all Banyan VINES servers in a given network consult for access permissions. StreetTalk is a directory that has an overarching, all-encompassing view of all the servers and users in a VINES network. Users log in once to StreetTalk and thereafter have access to all servers that the administrators permit them to access.

Banyan has expanded StreetTalk's charter by using it to manage other NOSs, as well. Banyan's Enterprise Network Services (ENS) can now manage NetWare, Windows NT, and Unix (Hewlett-Packard's HP-UX and IBM's AIX).

Not to be outdone, Novell developed its own Novell Directory Services (NDS), which debuted with NetWare 4.0. Patterned after the X.500 directory standard, NDS uses X.500-style nomenclature for users and resources.

X.500

Although network vendors are taking some steps toward consolidating directories, most of these initiatives are occurring within the vendor's own product family. But, if we're going to standardize on one directory, which will it be? Each vendor wants its own to be the standard. Vendor-independent standards are what's really needed.

Created in 1988, X.500 was intended to establish a standard for directories. One of the ISO protocols, X.500 is extremely ambitious and some would say over-engineered. X.500 can take some significant computing resources to implement, and for many years, there were debates about whether it *could* be implemented. Although X.500 directories haven't set the world on fire, the standard has been extremely influential in setting the general direction of directory development (witness Novell's use of X.500-style naming). Components of the standard include the Directory Access Protocol and Directory System Protocol specifications.

Seeing an opportunity to shed some of X.500's excess baggage, developers at the University of Michigan radically pared down X.500 and produced a "fat-free" version, which has now been standardized as the Lightweight Directory Access Protocol. To be precise, LDAP is a simplified version of the Directory Access Protocol.

As its name implies, LDAP provides a standard way to access and update directory information. In theory, any LDAP-compliant client (an LDAP-compliant Web browser, for example) should be able to access any LDAP-compliant directory, for purposes of adding, deleting, or modifying directory information.

In contrast to X.500, LDAP has proven wildly popular, and most major players in the world of networking have pledged support for the protocol. Some have even shipped such support. Netscape Communications' Directory Server was designed from the ground up as an LDAP-compliant directory. Netscape Navigator is an LDAP-compliant client. Novell recently shipped an NLM that makes NDS LDAP-compliant.

This doesn't make LDAP a silver bullet that solves the world's directory problems, but the standard should promote interoperability between various directories and clients, regardless of the vendor. This doesn't mean that things will necessarily converge on a single directory, however. There's now a multiplicity of directories, and that will continue, but LDAP will make it easier for them to interoperate.

While LDAP standardizes the directory access method, there's yet another problem that directory clients face when trying to access a directory. The client needs to know how to read the directory, which means it needs to understand the directory's schema—the basic organization of the directory. If all goes according to plan, the next release of LDAP (version 3) should include a method for clients to get the schema for any desired directory.

Also, the NAC is pushing the concept of a standard baseline schema, and work is under way to define the Lightweight Internet Person Schema (LIPS).

ARE WE IN SYNC?

If the world won't converge on a single directory, perhaps a different approach is needed. One possible approach is directory synchronization. Synchronization involves running some process that looks at two or more directories and gets them "in sync." This process works by adding any entries that are in one directory but not in the other, and replicating any other changes, such as a password change in one account to the corresponding account in the other directories.

NetVision's Synchronicity for NT synchronizes NDS and the Microsoft NT Domain Controller. It brings NT domains under the management of NDS and lets network administrators manage both their NT and NetWare servers from the NWAdmin program. NetVision, which is based in Orem, UT, has also announced Synchronicity for Lotus Notes. (For a review of Synchronicity for NT and Zoomit's VIA, see "Tying Directory Services without Knots," August 1997, page 102.)

Yet another approach is one jointly proposed by The Burton Group and the NAC. They've fostered the idea of a metadirectory—a "virtual directory" formed by combining all the information contained in a network's various directories. A metadirectory would allow an organization to manage all of its directories as a whole. The first implementation of such a metadirectory is VIA by Toronto-based Zoomit. [Editor's note: Microsoft acquired Zoomit in 1999.]

Clearly, we haven't reached directory nirvana, yet. But, as the saying goes, a journey of a thousand miles begins with a single step. We've taken quite a few steps (not all of them in the same direction), and we're at least starting to see the lay of the land. We've got quite a way to go, so I hope you're enjoying the sights.

This tutorial, number 109, by Alan Frank, was originally published in the September 1997 issue of Network Magazine.

NetWare Directory Services, Part One

One of the most eagerly awaited aspects of NetWare 4.x is its directory services scheme. It's also one of the most complex to describe. This discussion is necessarily a simplification of NetWare Directory Services (NDS). Since NDS comes from Novell, the NetWare 4.x documentation should be considered the definitive source on the subject. If you're installing NetWare 4.x, follow the 4.x documentation for specifics on NDS. For those wondering whether to make the jump to NetWare 4.x (and hence don't yet have the documentation), here's a brief overview of NDS.

BEYOND THE BINDERY

Versions of NetWare prior to 4.0 built and maintained, on each file server, a special database called the *bindery* to store information on each user, group, or other object the file server had to track. For example, when the supervisor of a NetWare 3.11 file server wants to give a new user access to the server, he or she uses Novell's SYSCON utility to create a new user, entering a new user name (the account name for that user) and telling SYSCON whether or not a password is required.

The fact that each NetWare file server, running NetWare versions prior to 4.0, maintains its own bindery can create a lot of administrative work in organizations with dozens or hundreds of file servers. If a supervisor wants to give a new user access to, say, six servers, he or she must log on to each server and follow the preceding steps. Wouldn't it be nice to be able to say simply: "Here's the account name for this user. Give her access to these six file servers"? What's needed is some sort of "global" directory, so users and other objects are known to the entire internetwork rather than to a specific file server. NetWare Directory Services has that orientation.

In the *NetWare 4.0 Concepts* manual, Novell defines NetWare Directory Services as a "*global, distributed, replicated* database that maintains information about, and provides access to, every resource on the network." The key words are global, distributed, and replicated. Global refers to the fact that entries in the NetWare Directory database are known to the entire network. Distributed means that portions of the NetWare Directory are replicated (a copy is kept) on various file servers. This setup ensures that, in the event of a file server crash, the NetWare Directory isn't lost. It also means that users won't be locked out of the network because one file server happens to be turned off or is otherwise inaccessible (if a wide-area link is down, for example).

While the bindery is a simple flat-file database, the newer NetWare Directory is hierarchical; it's logically organized in an inverted tree structure, with all key components branching out from the "root" at the top of the tree (see Figure 1). This tree structure also closely mimics the organizational charts of most companies, which permits network administrators to build a structure of account names that closely matches the company's organizational chart.

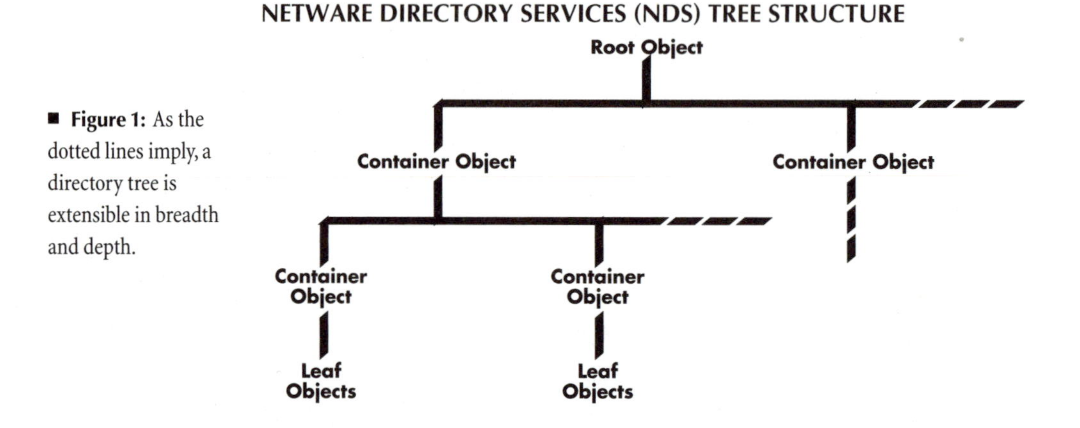

■ **Figure 1:** As the dotted lines imply, a directory tree is extensible in breadth and depth.

CATALOGING RESOURCES

Network resources such as users, groups, printers, print queues, and volumes, are cataloged in the Directory as objects. Objects can be either physical or logical. Some examples of physical objects are users and printers. Groups and print queues are logical objects.

Objects can also be classified in another way: as *container objects* or *leaf objects.* Container objects are so named because they contain one or more other objects. Leaf objects don't contain any other objects; they're at the ends of branches, hence the "leaf" designation. Some examples of leaf objects include users, NetWare servers, volumes, and print queues.

An object consists of categories of information, called *properties*. Some properties of a user object, for example, include login name, password restrictions, and group membership.

The container objects can be categorized into three types: the country object, the organization object, and the organizational unit. The *country object* is the highest-level container object (next to the root object) in the Directory (see Figure 2). The country object is optional, and it is not automatically created as part of the NetWare 4.x default server installation.

The *organization object* is one level below the country object (if country objects are used; otherwise it's directly below the root object). There must be at least one organization object in the directory—it's not optional. You would typically use the organization object to designate your company or organization (university or government agency, for example).

A level below the organization object is the *organizational unit.* It can be used to represent a division within your company or organization. There can be several levels of organizational units, so you can use them to designate departments or workgroups.

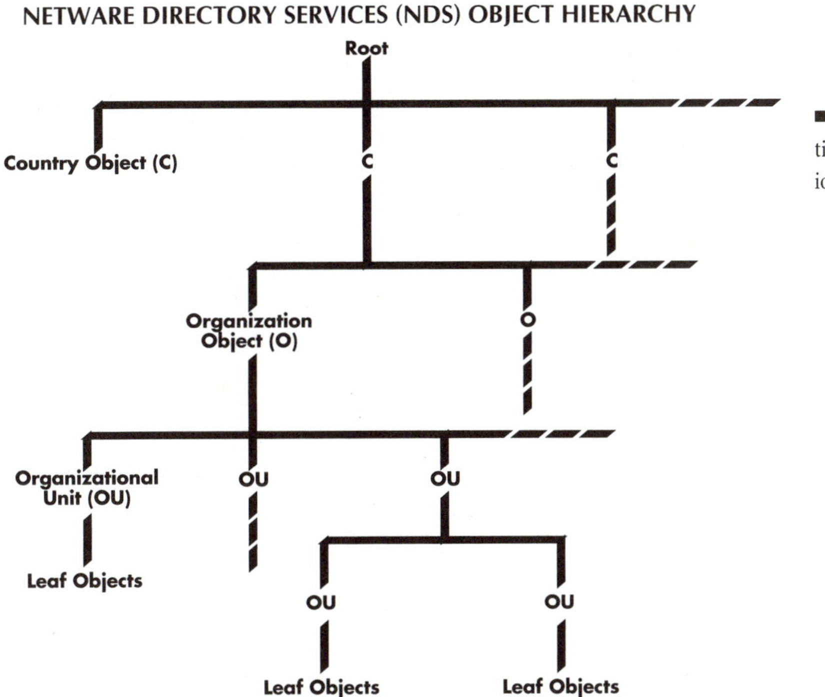

■ **Figure 2:** The relationship between various object types.

Note that there can be only one level of country objects (if you use country objects) and one level of organization objects. As mentioned earlier, there can be several levels of organizational unit objects.

Figure 3 shows part of a directory tree for a hypothetical organization, the Acme Auto Company. In this example, the country object is not used, so the organization object occupies the level just below the root object. Acme has operations in Germany and the United States, and the creators of the directory tree chose to use two organization objects: Acme_Germany and Acme_US. Acme_US has three divisions—Engineering, Sales, and Accounting—so organizational unit objects are used to represent them.

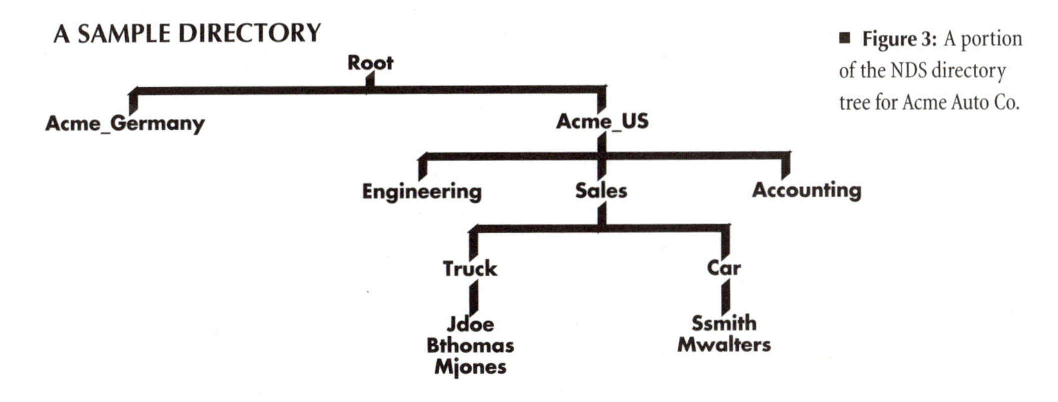

■ **Figure 3:** A portion of the NDS directory tree for Acme Auto Co.

Each organizational unit can have subgroups. For example, in the figure, Sales is divided into truck and car departments, so another level of organizational unit objects is used for these. Finally, we get to the users in these departments, who are represented by leaf objects. Each user's login name, which on many NetWare networks is made up of the user's first initial and last name, is listed.

An object's position within the directory tree is known as its context. In Figure 3, the context for user John Doe is TRUCK.SALES.ACME_US. Most leaf objects have a common name, and for user objects, the common name is the login name.

What Novell refers to as an object's complete name is formed by concatenating the object's common name with its context. In our example, John Doe's complete name is JDOE.TRUCK.SALES.ACME_US.

This tutorial, number 67, by Alan Frank, was originally published in the March 1994 issue of LAN Magazine/Network Magazine.

NetWare Directory Services, Part Two

In "NetWare Directory Services, Part One," the Tutorial began to explore Novell's NetWare Directory Services (NDS). That tutorial covered much of the basic NDS terminology and examined NDS' hierarchical structure, which resembles an inverted tree. This chapter explores the information in the NDS directory tree, which is distributed and replicated across the network.

The NDS directory is distributed for two major reasons: security and performance. Just as with the NetWare bindery, used in NetWare 3.x and earlier versions, the directory is the repository of all the information NetWare has about users, and it uses that information to control access to NetWare resources. If the directory were lost, no one, including the network supervisor, would be able to log in to the network. Thus, it's important to have more than one copy of the directory, just in case something happens to the original.

Performance, too, is improved by replicating the directory, particularly in wide area networks. If someone were to log in on a network segment that didn't hold the directory, it could take some time to authenticate the user. Having a copy of the directory reside in each LAN segment eliminates the performance issues.

BREAK IT UP

For large networks, which have a correspondingly large directory, it could be unwieldy to simply have replicas of the entire directory database stored in many different places. Instead,

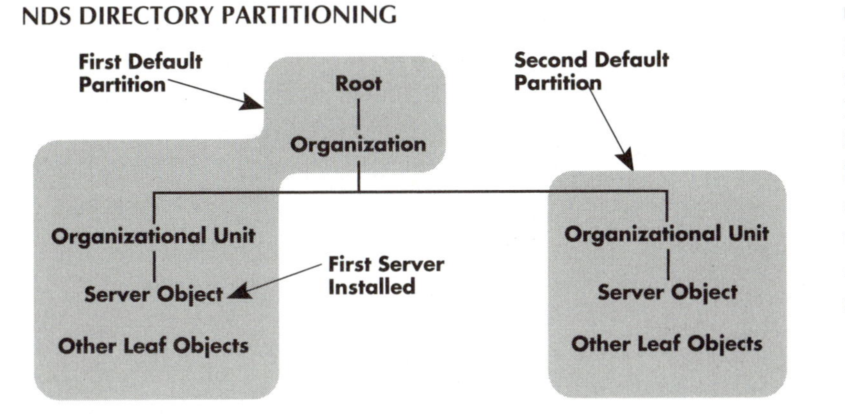

■ **Figure 1:** Here is an NDS directory tree with two partitions. The first NetWare 4.x server to be installed is the one on the left, so the installation creates the partition on the left. When a second server is installed later, a new partition is created.

NDS uses a distributed approach, in which the overall database is broken into portions known as *partitions.* (Don't confuse directory partitions with disk drive partitions, however.)

Directory partitioning lets NDS distribute portions of the NDS directory database among the various NetWare 4.x file servers in the internetwork. This arrangement means each file server isn't burdened with storing the entire directory database; a given file server might have only a single image of a directory partition, known as a *replica* of the partition.

Figure 1 is an example of how a directory tree might be partitioned, and how partitions are created by default, during the installation of NetWare 4.x file servers. In this figure, the partitions are the areas with a white background. Each partition consists of a container object and all objects contained within it.

IT'S AUTOMATIC

When the first NetWare 4.x file server is installed, the NetWare installation program will automatically create the first partition. The Root object is always included in the first partition.

Whenever an additional NetWare 4.x server is installed in a new container object, the installation program will, by default, create a new partition. This step is also illustrated in Figure 1.

As mentioned earlier, a partition replica is a copy, or image, of a partition. Only one *master replica* can exist for each partition, but you can also have other types of replicas, known as *read/write replicas* and *read-only replicas.* You can have as many of these replicas as you want.

You can view, but not modify, directory information in a read-only replica. Thus, NDS could consult a read-only replica to authenticate a user logging in to the network. Read/write replicas can be updated, so a network manager could add or delete objects within a partition using a read/write replica. Changes to a partition itself, however, can only be made by operating on the partition's master replica. A network supervisor would use the NetWare administrator graphical (Windows-based) utility or the text-based partition manager utility (for DOS-based workstations) to manage partitions and replicas.

The installation program places a master replica of the partition on the first server installed in a container object. If additional servers are installed in the same container object, the installation program places a read/write replica on those servers. By putting two or more servers into a container object, you automatically provide fault tolerance for that directory partition.

When you install a NetWare 4.x server, the installation program checks the network segments to which the server is connected. If it detects existing NDS directory trees, it accumulates the names of these trees. Then you have the option to make the new server part of whichever trees you select.

Once you've told the installation program which directory tree you want the server installed in, you then specify what the server's context, or location, in the directory structure, should be. You can specify a context that's within an existing directory partition or choose a new context. In the latter case, the installation program will create a new container object matching the new context, create a new partition, place the container object into the partition, create the new server object (the NetWare 4.x file server you're installing), and place the server object into the newly created container object. It will also create a master replica of the new partition, placing it on the new file server.

IS IT CONSISTENT?

To ensure that read/write replicas and read-only replicas are up to date with respect to their master replicas, servers holding NDS partition replicas frequently compare notes to check for inconsistencies. Any replica that is out of date gets updated from the master replica.

How does a server know if a replica is out of date, compared to its master? Apparently, replicas get time and date stamps, just as conventional files do, and any replica with a time stamp older than its corresponding master replica gets updated.

WHAT TIME IS IT?

It's therefore crucial that all NDS servers be synchronized with each other. In fact, this task is so important that Novell offers several different strategies you can pick from to keep all your servers in sync.

In the *single-reference time server* model, only one server in the entire directory tree is designated as the time server. If you elect to use this model, you designate all other servers as *secondary time servers.* Secondary time servers periodically synchronize themselves with the single-reference time server. When network clients log in, they will have their clocks synchronized to the "nearest" secondary server (or to the single-reference time server, if that happens to be the nearest server).

Servers can use either Novell's Service Advertising Protocol (SAP) to share time information, or servers can be custom-configured to contact specific servers for time updates. The simplest approach (and the default) is to use SAP. For large networks, where frequent SAP broadcasts would be a nuisance, Novell recommends the custom-configured method.

Without going into all the details related to time synchronization, assume that if yours is a small network with only a handful of servers, the single-reference time server with SAP communications should probably be your choice—that's the default. For large networks, particularly those with WAN connections, you'll want to carefully study the NetWare documentation to plot your time synchronization strategy.

DEJA VU?

Now for a short digression: Some years ago, a Macintosh networking product named DataClub came to market. DataClub was a "virtual server" that could substitute for an AppleShare server. It represented a clever blending of the peer-to-peer and client-server conceptual models, as data files resided on individual users' Macs throughout the network, yet the DataClub server appeared as an icon on the Macintosh desktop as if it were a single disk volume.

The tough technical problem that DataClub's designers had to tackle was how to build and maintain a distributed directory that could keep track of all those files. The solution: Each Mac that participated in the data "club" had a copy of the directory for the DataClub volume.

This solution presented a problem of its own, however, as any time you distribute a copy of something throughout a network you run the risk that some of the copies aren't in sync with the original. One way to make sure everything is in sync is to immediately update all copies the moment the original changes. But this approach could really slow file access to a crawl. Moreover, Macs that happen to be turned off during a directory update would miss the update and be left with old information.

IT'S A LITTLE FUZZY

The DataClub designers' solution was to design a system where individual Macs on the network would periodically exchange directory information to get copies of the directory in sync again. The designers referred to this technique as "fuzzy consistency." In other words, at any given moment, one or more copies of the directory might be inconsistent, but over time the system tended toward consistency. It was a clever and technically innovative solution to a messy problem.

It wasn't long before the company that developed DataClub was acquired by Novell. Novell still sells DataClub for the Macintosh. [Editor's note: as of 1994—not in 2003.]

There are some differences between DataClub's directory of files and the NDS directory of network resource objects. For one thing, each Mac running DataClub has an entire copy of the DataClub volume directory, whereas with NDS the total NDS directory tree is broken up into partitions, so each server need not maintain a copy of the entire directory tree. This approach is probably appropriate, as the DataClub method is best for small networks, while NDS can work with very large networks.

Still, the way that replicas of NDS partitions periodically get updated from master replicas is eerily similar to the way DataClub directory copies get updated. Is DataClub's "fuzzy consistency" one of the enabling technologies behind Novell's NetWare Directory Services? I wonder.

This tutorial, number 68, by Alan Frank, was originally published in the April 1994 issue of LAN Magazine/Network Magazine.

Printing

One of the most basic reasons to install a network is to share peripherals, including hard disks and printers. Sharing printers means only a few PCs need have printers attached, instead of each user having his own printer. Users send their documents, spreadsheets, and reports to the print server. Because the print server handles all the printing requirements for the group, the number of expensive printers can be reduced, and printing may be faster.

Once at the print server, the jobs are entered into a queue, where they wait for the printer to become available. Queues are just what they sound like— waiting lines for access to the printer. Jobs are normally serviced in a first-in-first-out order, although most print servers allow users to prioritize print jobs so they can be moved up or skipped to the top of the queue.

PRINTING IN ACTION

For a print server to handle multiple jobs at the same time, it must have a print spool. (Spool is reportedly an acronym for Simultaneous Peripheral Operation On Line.) A spool is hardware and software that controls a buffer. The buffer is memory that holds data. One or more print jobs may wait in the buffer while the printer is working on another job. Often, print jobs are spooled to the print server's hard disk.

When printing from a standalone application, the path from the PC to the printer is fairly direct. Using networked printers requires a more circuitous path. More opportunity for glitches exists, and they frequently occur. (See Figure 1.)

Like retrieving files from network drives, using networked printers requires redirection. Software in the user's computer captures the print job and sends it over the network. The application, a word processor for example, thinks it is printing to the local printer port, but the network client software redirects the output over the network.

Similar redirection takes place over the network, but the network operating system, not the PC operating system, is redirecting the output. The user's application tells the network

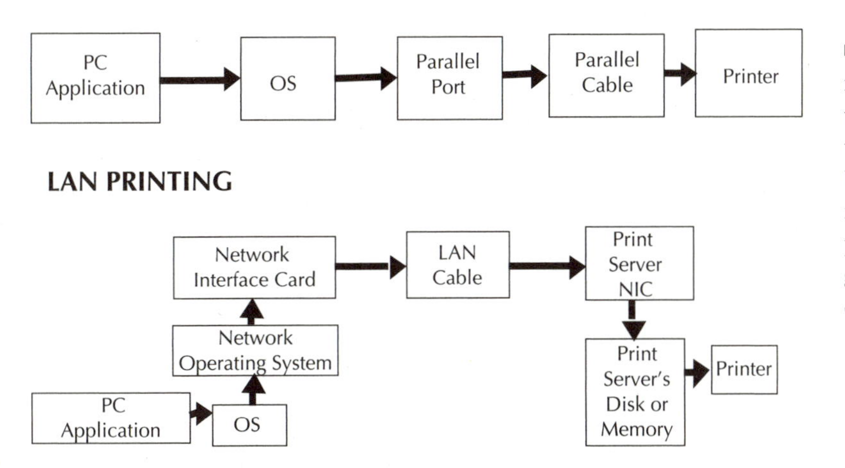

■ **Figure 1:** When printing from a standalone application, the path from the PC to the printer is fairly direct. Using networked printers requires a more circuitous path. More opportunity for glitches exists, and they frequently occur

operating system to redirect all output headed for the local LPT1 port over the network to, say, the print server's LPT3 port. The network operating system gets the job to the print server.

At the print server, the file is spooled to disk. Spooling may be done because the printer is busy. Spooling is necessary if the printer is not fast enough to take the whole file at once.

The print server's software handles the incoming job. If the printer is free, the document can be printed immediately. If the printer is busy, the document is spooled in a print file on the print server's hard disk. The file joins the queue, waiting its turn to be sent to the printer.

The print server's buffer feeds print jobs to the printer at the correct pace. Data waiting to be printed is stored in the buffer before it is sent. The larger the buffer, the faster the computer regains its full resources, since fewer disk accesses are necessary to feed the file to the printer.

In principal, redirection, spooling, and buffering and network printing should be easy; however, problems occur frequently. Although printer sharing was an original impetus for installing a network, networks still don't always share printers gracefully and easily.

PRINTING PROBLEMS

Problems crop up when trying to print on a network, including applications talking directly to the printer port, conflicting print spools, multiple print buffers, and multiple users. Some problems are inherent in having multiple people use the same printer.

A common problem is the conflict between the application and network. Some applications try to talk directly to the printer port and the network operating system does not have the opportunity to redirect the output. This problem occurs more often when printing to serial ports rather than parallel ports, because it is easier to handle a serial port than a parallel port. Therefore, more applications leave parallel operations to DOS. One solution is to configure the application to use the parallel port, where possible.

Another problem occurs when an application has its own print spool. The application will check the hardware directly to see when to send data, thereby causing problems for the network spool. The solution is to disable the application's spool.

Multiple print buffers cause problems. Print servers are not the only devices with print buffers. The application, PC, network, and printer may also have buffers. Usually, all these buffers will work together, but sometimes disruptions happen while one buffer waits for another. The solution is to use only one of these buffers. Performance will not be hurt significantly, and memory resources may be conserved.

A fourth network printing problem arises when multiple users share one printer. Each user sets the printer differently, different fonts, character widths, line spacing, etc. Often, the printer retains these settings. And while some applications, network operating systems, and print utilities allow users to send control codes to the printer, others don't. Where supported, these codes can reset a printer after every print job. This way, all users know the printer is starting with particular settings, no matter who has used it previously.

In some cases, the application's printer control codes may interfere with network operations or network printer control codes. For example, a printer control code from the application may be misinterpreted by the network operating system as a release printer code, leaving the user disconnected from the printer. Another symptom of such conflicts is garbled printing. Most likely, the application will have to be reconfigured not to use the printer control code commands, leaving the network operating system or print utility in charge.

The last set of network printing problems is inherent in the process. Multiple users have different requirements for paper, but most printers can only hold one type of paper at a time. The classic problem is a user prints his file, only to discover someone took out the letterhead he just put in the tray. Or that someone left green paper in the tray. One solution is to have multiple printers, each configured with a different type of paper. Some companies now manufacture printers with several paper trays and allow users to specify the paper tray. Electronic mail or some type of notification may be effective as well.

PRINT UTILITIES

Ironically, while an early reason to buy a network was to share peripherals, today network operating systems still do not share printers adeptly. All network operating systems support some kind of network printing, but functionality varies widely. Some network operating systems required the user to leave the application before setting up the print process. The user once had to explicitly tell the network operating system to look for a print job and then tell it when the job was over. Others required that all shared printers be attached to a file server. This, too, is inconvenient, especially when file servers should be locked in a closet for security purposes. Finally, many network operating systems will not allow users to send printer control codes to network printers.

Print utilities exist to solve these problems. Such utilities allow users to control the printer and print server from their PCs, without exiting their application. Users may manipulate the print queue, check which files are spooled, choose print codes, send jobs, reconfigure the printer, and perform other tasks. Print utilities are usually terminate-and-stay resident programs, so they can operate from within any application.

One feature of many printing utilities is the ability to automate printer configuration. For example, a printing utility will send printer control codes to reset the printer to default configuration after every print job. This way, every user knows how the printer is set before using it.

Another helpful feature is print notification. Users are notified when their print jobs have finished printing, a terrific convenience if the printer is out of earshot. Another useful feature is the ability to standardize line and form feeds so users don't have to do this manually. This reduces the chance users will go to the printer and change it, or leave it offline.

Many print utilities allow any PC on the network to act as a print server. When any PC can act as a print server, printers may be placed conveniently throughout the office. Such a print server can be dedicated to its purpose, or it may double as a workstation. While a nondedicated print server allows you to place printers easily and inexpensively around the network, it has some drawbacks. If many people need that printer, the owner of the print server/PC will be constantly interrupted when users pick up their print jobs and want to chat. Besides user distractions, the user may reboot or turn off the PC, thereby killing all print jobs in the queue. Some print servers are dedicated boxes, which sit on the network. These devices have a CPU and memory for spooling and buffering; however, without an on/off switch, a user cannot accidentally turn them off.

SMALL ADVICE

While printing on a network should be easy, it's not. Most problems arise because printing in general is not as easy as it should be. Pinouts, cables, and ports on the printers themselves create an endless stream of problems. Add the network, and printers become even more of a hassle. Fortunately, many network operating system vendors now recognize the need for easier network printing and have incorporated features into the NOS that once could only be obtained through a third-party utility.

This tutorial, number 12, by Aaron Brenner, was originally published in the July 1989 issue of LAN Magazine/Network Magazine.

OS/2 LANs, Part One

[Editor's note: Though OS/2 is not much of a factor in year-2003 networks, the technology that once made it unique is still important to grasp.]

OS/2, the operating system from IBM, is special for several reasons. The operating system has several features that improve networking significantly. These features are not necessarily unique—other operating systems such as Unix and VMS also have them—but OS/2 was the first PC operating system to include them. The most relevant features of OS/2 are multitasking and interprocess communications.

With these two features, OS/2 provides a much better platform upon which to build powerful multiuser, distributed applications—the kind that run on networks—than DOS. These network-intrinsic applications take advantage of the network to provide services and performance simply unavailable under DOS. For example, a distributed database, which allows a user to pull data from two different machines and combine it into one report, is a network-intrinsic application. It must run on a network. While it is conceivable to build such an application using DOS-based machines, the amount of work necessary is significantly greater and the performance is significantly lower.

OS/2 provides a better platform upon which to build these applications because of its Application Programming Interfaces (APIs), particularly for multitasking and interprocess communications. These APIs provide services to applications that other PC operating systems can't match. Crucially, they provide the basic mechanisms by which geographically separate programs can exchange commands and information. That is, OS/2 APIs have networking in mind. This becomes even more the case when OS/2 is combined with an OS/2-based network operating system such as Microsoft's LAN Manager or IBM's LAN Server.

VIRTUAL OS/2

Before delving into the intricacies of OS/2's multitasking and interprocess communication, we need to lay the foundations upon which they are built. The two most important pillars are OS/2's memory model and virtual device support.

OS/2 is a protected-mode operating system, which means it requires a microprocessor that uses protected mode, such as an 286, 386 or 486 chip, or one of IBM's new RISC-processor based PCs. Protected mode is best understood by comparing it to real mode. The 8086 and 8088 run in real mode. Under real mode, programs have direct access to a PC's hardware devices, including memory, disk drives, serial ports, keyboard, and screen. That means programs can bypass the operating system, in this case DOS, and do what they like with the hardware. For example, a word processing program may decide to send something to a printer port.

It can take over that port and DOS can't stop it, even if the program is going to interfere with another program that is already using the port.

Multitasking is impossible under real mode because there is no way to keep programs from interfering with each other as they use the PC's various components. The most acute form of this occurs when programs overwrite each other in memory. For example, a database program may use extra memory for sorting a database. In real mode, there is nothing to prevent another program, say, a spreadsheet, using this memory for a recalculation. When both programs try to put data in the same place, whichever program gets there first loses, since the second program writes over the data of the first. If the second program writes over some of the executing code of the first program, then not only is data corrupted, but the computer may hang since it can't continue with the first program.

Under protected mode, programs generally do not have direct access to the PC's hardware devices. The devices are protected from application programs, which is where the name "protected" comes from. But because they don't have direct access to devices, programs need some method to use them. That's where OS/2 comes in. OS/2 sits between the programs and the hardware, regulating access to the latter. OS/2's intermediary position between programs and hardware forms the basis of multitasking, which is the ability to run several programs at the same time.

How does OS/2 give programs access to hardware while keeping them from colliding with each other? The answer is twofold: virtual memory and virtual devices. Actually, virtual memory is a special case of a virtual device, but memory is so crucial to the workings of a computer it deserves special attention.

Virtual memory works as follows: Virtual memory introduces an additional step in a program's access to memory. Instead of simply asking for a physical memory address, as it does under DOS, programs ask for a virtual memory address. OS/2 then converts that virtual address into a physical address. This process is called mapping. In this way, OS/2 keeps track of where everything is, preventing one program from taking the memory of another by preventing it from directly accessing memory in the first place.

In addition, because OS/2 is handling their placement, program data and code need not always reside in memory. OS/2 may map a virtual address to a hard disk. It may decide it is better to put a piece of program code or data onto the hard disk rather than in memory. The program using the code or data still asks for the same virtual address. OS/2 just fills the request from another place. This process is called segment swapping.

OS/2 uses segment swapping when not enough RAM is available. OS/2 divides memory into 64KB chunks called segments. If there isn't room for a program to load completely, segments of program already in memory may be swapped to disk to make room for the new program. When the swapped-out data is needed, OS/2 swaps it back in and swaps another program out. The result is the amount of memory available is no longer equivalent to the amount of physical memory. Original versions of OS/2 use up to 48MB of memory, either on disk or in RAM, which is about 50 times the limit of DOS.

OS/2 handles devices similarly. Instead of directly accessing physical devices like parallel ports and video screens, programs are presented with virtual ports and screens. Applications write to or read from these virtual devices and OS/2 takes or sends the data. Each device has the appearance of its own keyboard, screen, and printer, but OS/2 is really just presenting each program with a virtual device, which it then maps to the real device as necessary.

OS/2's virtual memory and device management make it possible to run many programs at the same time. Key roadblocks of DOS are eliminated. Programs can use more memory, and they can no longer interfere with each other. But not only are virtual memory and device management the basis for multitasking, they are the basis for multiuser operation.

OS/2 MULTITASKING

No PC can actually do two things simultaneously, because a single processor can only perform one operation at a time. Processors can, however, do things so quickly the user thinks the processor is doing many things at once. And since they are doing things so quickly, processors often sit around and do nothing. Multitasking takes advantage of this idle time.

Suppose you are typing a letter using a word processing program like WordPerfect. The computer captures the letters you type and puts them on the screen. The time it takes to do this is a fraction of the time it takes for you to type the letter. Therefore, the computer is spending most of its time waiting for your next keystroke. But suppose it could do something else while waiting?

This is exactly what a PC running OS/2 does. The computer's idle time can be used by other tasks. While you are typing, the computer may download a file from another computer, recalculate a spreadsheet, sort a database, or draw a graph. To accomplish all these things at once, OS/2 divides the computer's time up into pieces called time slices. These last no longer than a fraction of a second. For one time slice, the computer works on the letter you are typing. During the next time slice, it stops working on your word processing and starts downloading a file that you requested. Then this task is suspended while the spreadsheet recalculation takes place. Thus it goes, around and around from one task to the next. Of course, the time slices are so small that you don't notice the machine is working on other things while you are typing.

The results are manifold. The ability to allocate a computer's time increases its efficiency, which in turn increases the user's productivity because time is not wasted moving among applications. Multitasking can also make multiuser operation possible. Because the operating system is automatically multitasking, less work is needed to make it do many things at once for many people. It is better able to handle many simultaneous requests from many users.

MULTIUSER OS/2

Multiuser operation does not, strictly speaking, require multitasking, however. A computer could work for many people one at a time. When one is finished working, the next could start, and so on. Obviously, this process is not what users expect of a computer system.

True multiuser operation requires the ability to accept input from numerous sources and redirect various output to multiple devices. Neither DOS nor OS/2 can do this alone. (DOS is single-user and single-tasking; OS/2 is multitasking but single-user; Unix is both multitasking and multiuser.) However, with the help of a network operating system, any of the three can provide services for multiple users. The difference is in the types of service they provide.

On a DOS-based PC network, the network operating system handles the multiple users for DOS. As the requests come in from the users, the network operating system feeds them to DOS one at a time. DOS then fulfills the requests. Then the network operating system ensures that the answers are returned to the users. There are limitations, however, to the type of requests DOS can fulfill. DOS can only do the simple input/output tasks for multiple people, such as reads and writes to a hard disk. It can't handle multiple programs in memory.

OS/2, on the other hand, allows users also to share processors, which is the basis for network-intrinsic applications. This is accomplished through multitasking, redirection, and interprocess communication.

OS/2 is multitasking. A way to assign different tasks to different users is needed to make it multiuser. This way is virtual devices and network redirection. Network redirection, as provided by a network operating system, tricks OS/2 into thinking it is mapping a virtual device to a local physical device, when in reality it is mapping that device to a remote physical device. For example, during one of its time slices, OS/2 can draw a graph for a remote user. The network operating system redirects the output from the OS/2 virtual screen over the network to the remote user's physical screen. In another time slice, OS/2 can accept keystrokes from another user, mapping them to a virtual keyboard. In this way, virtual devices and multitasking help OS/2 provide the basis for multiuser operation. But the network operating system is necessary to complete the multiuser picture.

The result is that OS/2 allows multiuser operation. It allows multiple users to share a processor whereas DOS only allows multiple users to share peripheral devices. This is extremely important for networking. It is OS/2's multitasking abilities and its support for virtual devices that make it such a good foundation for network-intrinsic applications.

This tutorial, number 14, by Aaron Brenner, was originally published in the September 1989 issue of LAN Magazine/Network Magazine.

OS/2 LANs, Part Two

OS/2 is not a multiuser operating system, but its multitasking ability provides the basis of support for multiuser client-server and distributed applications— the real network-intrinsic applications—especially once a tightly integrated network operating system is added. To make this transition, a mechanism by which the processes communicate is needed. This mechanism is an *interprocess communication (IPC).*

IPC OVERVIEW

There are many forms of interprocess communication. Microsoft Windows and Apple Macintosh users are familiar with one of the simpler types. Cutting and pasting data from one program into another using the clipboard is interprocess communication. OS/2 supports this type of data exchange, and much more, since it has more sophisticated interprocess communication capabilities.

OS/2 has a set of APIs that allow programmers to write applications that communicate while running concurrently on an OS/2 PC. More importantly, OS/2 APIs allow programmers to divide one program into multiple parts called processes or *threads.* These threads can communicate with each other. The advantage is increased speed and specialization. Different processes of a program may be tailored to fit particular tasks. Then, by running the processes simultaneously, instead of sequentially, each will complete its task more quickly.

There is also bigger advantage, OS/2's APIs allow one process to be shared by many others. This forms the basis of a client-server application. Imagine starting a program that sorts and stores raw data, like names, company names, addresses, and telephone numbers. We'll call this the database server application. Then start two more programs: a report writer and an address book. With OS/2 APIs, it is possible for the report writer and address book to get all their data from the database server at the same time. They don't have to handle sorting and storing the data; they just ask for and receive the data.

The advantages are tremendous. Because they don't have to handle data, the report writer and address book are much easier to program. Second, because each program has less to do, it can be more specialized. Third, because both use the database server, they can share information. For example, information added into the address book may be used in the report. Fourth, the programs can run concurrently, allowing a report to be run while the address book finds and displays addresses. Finally, data is not the only shared element. The address book might signal the database server to perform some task (other than a data transfer). This revolutionized what can be done on a PC.

Now imagine putting the address book on one machine, the report writer on another, and the database server on a third, and having them all work as if they resided in the same machine. Such a network-intrinsic application changes the face of computing.

OS/2'S IPC

An understanding of the tools of interprocess communication is essential. There are five mechanisms: shared memory, flags, semaphores, queues, and pipes.

Shared memory is a portion of memory that multiple processes can access. Process A puts data into the shared memory space. Then Process B gets it out. OS/2 ensures the two don't collide.

Flags are signals that processes give to each other. One process issues a flag and the others look out for that flag. When it is received, a particular action is taken.

Semaphores are similar to flags, except semaphores are used to exchange data, rather than for alerts. Semaphores might be used to coordinate the use of shared memory, for example.

Queues are a place for one process to put data in and another process to take data from. In OS/2, queues are not necessarily first-in, first-out. The receiving program may take data out of the queue in any order.

Pipes are like queues except they are first-in, first-out. One process opens the pipe and sends data through it to the receiving process.

In regard to the network, queues and pipes are the most important IPC. OS/2 queues and pipes are easier to use than many IPCs. A programmer can open, read, and write to pipes as if they were files, which are very familiar operations. OS/2 handles all the tasks behind setting up the pipes and queues. This is a tremendous boon for the programmer.

So far, all these IPCs work only within one OS/2 machine. With the addition of a network operating system, OS/2's IPC and multitasking features can be extended over a network.

NETWORK IPC

The key to extending OS/2's features over a network is an API for IPC. Many are available, but NetBIOS, APPC, and Named Pipes are the most important. None is exclusive to OS/2, but Named Pipes was developed for OS/2. The three have similar goals, but accomplish them differently.

NetBIOS is the basis of IBM's PC LAN Program and Microsoft's MS-Net. IBM originally developed APPC for communications among mainframes and midranges, but it has extended it to include PCs. Named Pipes was developed by Microsoft as an extension to OS/2's pipes. Because of the volume of programs in existence, more programmers are familiar with NetBIOS than APPC or Named Pipes.

The differences between the three IPC APIs are their sizes and ease of use. APPC is more sophisticated than Named Pipes and NetBIOS, but it requires much more memory. APPC and

NetBIOS are more difficult to write to than Named Pipes. Named Pipes contains much of the logic necessary to set up communication between two processes, which makes it easier to develop client-server and distributed applications. With so much network logic already embedded in the API, using Named Pipes can increase the application's performance since fewer messages need to be passed between processes.

By combining multitasking and interprocess communication, OS/2 provides a platform upon which to build multiuser, distributed network-intrinsic applications. Add a tightly coupled network operating system and OS/2 LANs can do things never done before on a PC.

This tutorial, number 15, by Aaron Brenner, was originally published in the October 1989 issue of LAN Magazine/Network Magazine.

Application Servers

One of the quiet revolutions changing the nature of computer networking is the growing use of special-purpose, or application servers. Whether they offer network users online access to facsimile capabilities, a mainframe gateway, or any of a variety of other sophisticated features, application servers add significant value to a network.

Unlike traditional network file servers, which act as intelligent, high-capacity data-storage and input/output systems, application servers perform specialized roles within a network environment. These may encompass managing printing queues, storing and routing electronic mail, or providing access to a database.

Application servers also reduce the demands placed on traditional file servers. Without having to handle special responsibilities, file servers can focus their processing resources on their primary jobs, network I/O and storing network data.

APPLICATION SERVER TYPES

[Editor's note: Even in 2003, the definition of "application server" has yet to converge around a single meaning. "Application servers" may be object-oriented middleware for distributed execution, or Web server/database combinations, or some other form of multiuser execution platform (such as Citrix MetaFrame or Microsoft Terminal Server,) as well as relatively simple single-purpose servers devoted to a particular application. This tutorial is about this last meaning of "application server."]

Like many areas in the continually evolving electronics industry, compiling a comprehensive list of application server types is probably impossible. Still, we can pinpoint at least nine distinct types of application servers: asynchronous communications servers, backup servers, database servers, e-mail servers, fax servers, image servers, optical disk servers, print servers, and directory servers.

Briefly, communications servers provide dial in and dial out access to networks; backup servers manage the archiving of a network's data-storage systems; database servers allow accessing data stored on a variety of computers; e-mail servers act as electronic post offices; and fax servers allow multiple network users to send and receive faxes online. Image-processing servers permit entering and maintaining digitally processed images, such as cancelled checks, into a database; optical disc servers grant access to the huge amounts of data stored on CD-ROMs, and print servers manage the printing process, including changing fonts and forms.

We'll describe the operation of four application servers—fax, print, management, and backup—in more detail shortly. Other application servers are described elsewhere in this book.

THEY WORK TOGETHER

Before continuing, it's important to clarify one point: With the exception of a traditional file server, application servers generally require the services of another server to operate properly. This means a network would contain several types of application servers, all working in tandem and relying on each other.

For example, a fax server requires access to a file server and the word-processing files contained thereon, before it can transmit data. And it might require using the resources of a print server for access to a laser printer. A database server requires access to data stored on a variety of other application servers and hosts, especially those where corporate databases are located.

Similarly, a key trend in the development of e-mail servers is to allow the e-mail "post office" machine to use the naming or directory service maintained by the network operating system, which is generally found on a file server. This simplifies managing the network because the administrator must keep track of and update only one naming service, not two.

TRANSPARENT OPERATION

One of the major strengths of application servers is their "transparent" operation to end users. That is, other than invoking an e-mail program, fax menu, or similar start-up program, most network users aren't even aware they're using an application server.

For instance, users of fax servers don't need to know they're accessing a device dedicated to sending and receiving faxes online. When they want to send a fax, they merely respond to onscreen prompts from the server software program by typing a destination fax machine's telephone number, the name of the file they want sent, the name of the recipient, and a short cover letter.

The fax server takes care of the rest automatically. It dials the phone number, transmits the fax via an internal modem, even redials the number should it get a busy signal. It also retrieves and stores incoming faxes without user intervention.

SERVING PRINTING NEEDS

Print servers were among the first of the special-purpose devices to appear on the network scene. Originally, they acted much like "traffic cops," starting and stopping print jobs or redirecting them to a specific printer, leaving functions such as font and forms management to the user.

Now, however, a print server can be configured so it lets one user print in Helvetica and a second in Times Roman without either user worrying about changing fonts. Print servers must be aware of which hard and soft fonts are available, which page description languages are installed, and the like. (Hard fonts are those provided by the printer itself; soft fonts are those downloaded from a PC application to the printer.)

Intelligent print servers know which printers are capable of certain tasks or have certain size paper trays, and direct output accordingly. Some advanced print servers can balance loads among multiple printers. Some also offer management and accounting features, enabling a network administrator to charge back users for the printers' services.

PUBLISHING THE FAX

A fax server can turn anyone with the correct destination phone numbers and the willingness to pay the associated line costs into an electronic publisher. In many ways, fax servers can replace both regular mail and electronic-mail systems. And they certainly reduce the time spent manually feeding sheets of paper into a standard fax machine.

A fax server is a hardware-software combination with connections to both the network—this can be any of the commonly used Ethernet, ARCnet, or Token Ring connectors—and the telephone system (via a standard RJ-11 modular jack). The fax server works in conjunction with a file server to let users send copies of their network-based electronic files—they can be word processing documents, spreadsheet forms, or graphics images—directly across the phone system to another fax machine.

In this application, users specify which files they want faxed, and the fax server accesses the correct file server to retrieve the needed files. It sends the fax out over the telephone system via its internal modem.

Obviously, this process eliminates several steps, including printing the original file on paper, then feeding it to a fax machine for transmission. It also holds more far-reaching ramifications, especially for organizations that need to fax out multiple copies of a single document.

With a fax server, the fax-sending process can be automated, much like a mail merge operation. This means, for example, that a company could use a fax server to "publish" dozens or even hundreds of personalized copies of a newsletter electronically, without requiring someone to manually perform the chore.

BACKING UP THE LAN

Several vendors have taken a similar approach to providing backup capabilities. Backup servers, as the name implies, are dedicated to providing centralized data archival (and recovery) facilities.

Legato Systems (Palo Alto, CA) has moved the backup-and-recovery process into the client-server environment with its NetWorker product. NetWorker is made up of client and server software, with the latter running on one or more backup servers. The client software, which runs on each PC on the network, determines which files to back up and routes them across the network to the right backup server. The server software manages the backup media—generally, tapes—and maintains an index of previous backups and their associated media volumes.

This system lets servers back up files on any network client. The online file backup history index also simplifies recovery operations—as users browse the index, they see historical views of the network file system at specified times. A lost file can thus be quickly identified and recovered.

ON THE DOWNSIDE

Naturally, there's a downside to using application servers. The most obvious, of course, is the cost of the additional hardware for each special-purpose server. And compatibility issues often create problems—that is, one application server may not operate on the same network with another. Placing numerous application servers on a network also adds complexity.

Because of these kinds of issues, application servers are more apt to be found in large corporate networks, where technical personnel are readily available, than in small, departmental workgroups. Still, many network users have discovered that the value added by application servers makes them well worth the expenses they entail.

This tutorial, number 34, by Jim Carr, was originally published in the May 1991 issue of LAN Magazine/Network Magazine.

Database Servers

Structured Query Language (SQL) is IBM's English-like database query language. It was developed to provide a relatively simple method for entering, retrieving, and changing data in an IBM database. Because IBM used it, SQL quickly became a standard other database vendors wanted to support. ANSI has standardized SQL as well. Using a common database language makes it easier to develop network database applications.

A database server is software that manages data in a database. It updates, deletes, adds, changes, and protects data, which is usually stored on a powerful computer with a large disk to which many users have access.

The database server software also protects data. Since network data can be accessed by many people, it must be protected from several potential problems. First, when two or more people work on the same database record at a time, they could overwrite each other's changes and cause data to be corrupted. Database server software must establish rules to regulate access to multiple simultaneous users. These rules are called concurrency control. Second, not every company employee needs access to every piece of data in the shared database. Therefore, a database server must also provide rules to regulate access, including passwords, access rights, and data encryption. These rules are called access control.

THE FRONT END

The database server does not contain the software needed for user interfaces, writing reports, generating applications, and other software associated with databases. The database server or back-end only manipulates data. Presenting, entering, and updating data is handled by the front-end. The front-end application usually runs on a different machine from the database server software.

Front-end software takes many forms because data can be manipulated and used differently. For example, Acme Widget Company has a database that contains customer information and inventory data. Different front-ends allow sales people to do promotional mailings to Acme's customer list. Outside distributors and salespeople on the road can dial into the database to learn which products are in stock. The service department can call up a customer's file when he calls and access all past ordering information. The accounting department can prevent customers from ordering more products if they are delinquent on their payments. Each group has a front-end tailored to their jobs, but they access the same database.

Different front-ends use a common access method—SQL—to communicate with the back-end. While the front and back ends communicate relatively easily, SQL is not a good language for nontechnical users. While programmers are not fazed by its syntax, salespeople, accountants, and technical support people do not want to type SQL commands.

In most applications, SQL will run "behind the scenes." Query-By-Example (QBE) is a simple front-end application. A QBE program asks users to provide an example of the type of data they want to get. A spreadsheet talks SQL to the back-end server, but the user continues to work with a familiar interface. If the user wishes, most front-end applications also permit users to execute SQL commands.

SERVER ADVANTAGES

A database server has several advantages over the traditional file server model of databases. In the file server model, there is only one database program. Although this program is stored on the file server, it executes in the workstation. The one piece of software performs both front- and back-end functions. When a user loads the database, the program file is downloaded to his or her workstation. When data is requested, the software gets the database file from the server, searches through the file, finds the requested records, and then downloads the information to the workstation and displays it on the screen. The whole database travels across the network from the file server to the workstation.

In a database server, the tasks are divided. The front-end requests the data, the request is translated into SQL, and it is sent over the network to the server. The database server executes the search on the machine where the data exists. Only the requested records are returned to the front-end using SQL and a network transport mechanism. This cuts down on network traffic and makes execution faster, especially when many people are using the same data.

Data is not the only traffic that is eliminated. In the file server model, concurrency control information must be handled by the workstation. With a database server, these commands do not have to be transmitted over the network, since they are handled locally. A database server provides better data integrity.

As mentioned, another advantage is that different front-ends can access the same back-end. This is not possible with the file server model.

Because a database server is intelligent, it can shorten the application development cycle and allow applications to execute faster. One method is stored procedures. Stored procedures are short programs that reside on the server, waiting to be executed. Any front-end can call the stored procedures. This saves programmers from including redundant code segments in every application. It also saves memory in the workstation. Triggers are similar to stored procedures; however, stored procedures are compiled code, and triggers are uncompiled code.

Despite the usage of SQL, no standard method exists by which all front-ends can talk to back-ends from different vendors. Each database server has its own access methods that the front-ends must use because of different types of SQL used as well as differences in transport protocols. The result is that applications written to work on one database server will not work on another. Although this is inconvenient for developers, it is a less serious problem for end users. When users choose a database management system, they choose a set of unique programming environments.

Database servers are an enormous improvement over other database management systems. They are faster, more powerful, more secure, more reliable, and easier to use. Their ability to accommodate a myriad of front-ends will make data sharing much easier. Their client-server design make them the next step in network databases.

This tutorial, number 16, was originally published in the November 1989 issue of LAN Magazine/Network Magazine.

E-Mail and SMTP

How E-mail Travels the Internet

Whether you're creating and sending your message from a proprietary e-mail package or an online service, your e-mail travels the same road that all Internet-based information—such as an ftp file transfer, a telnet session, or a Web page download—travels. That is, your e-mail traverses the Internet. The sender creates an e-mail message on an application. When the user sends the message, it is transmitted to the user's Internet mail server.

Figure 1 shows a simplified model of how e-mail travels from one user to another via the Internet. The sender creates an e-mail message on an application. The client system is known as a user agent, or UA. When the user sends the message, it is transmitted to the user's Internet mail server.

Once the message reaches the Internet mail server, it enters the Internet's message transfer system, or MTS. The MTS relies on other Internet mail servers to act as message transfer agents (MTAs), which relay the message towards the receiving UA. Once an MTA passes the message to the recipient's Internet mail server, the receiving UA can access the message.

THE FORMAT FOR E-MAIL

RFC 822 defines the standard format for e-mail messages, treating an e-mail message as having two parts: an envelope and its contents. According to RFC 822, the envelope contains information needed to transmit and deliver an e-mail message to its destination. The contents, obviously, are the message that the sender wants delivered to the recipient.

The envelope contains the e-mail address of the sender, the e-mail address of the receiver, and a delivery mode, which in our case states that the message is to be sent to a recipient's mailbox. We can divide the contents of the message into two parts, a header and a body. The header is a required part of the message format, and the sending UA automatically includes it at the top of the message; the user does not input this information. The receiving UA may

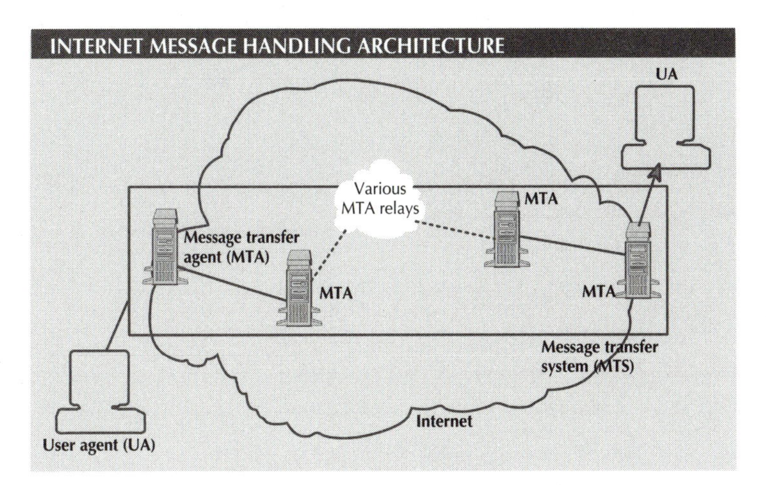

■ **Figure 1:** The Simple Mail Transfer Protocol (SMTP) allows e-mail-enabled client systems, or user agents (UAs), to send messages in ASCII text to other UAs. Internet mail servers act as message transfer agents (MTAs), relying on SMTP, conventional e-mail addresses, and the domain name service database to relay the e-mail message from server to server, until it reaches the mail server that services the recipient UA.

reformat the header information or delete it entirely to make the message easier for the recipient to read.

Take a look at Listing 1, which shows the e-mail message I sent to my corporate e-mail account from my CompuServe account.

LISTING 1—SAMPLE E-MAIL

```
Received: from arl-img-5.compuserve.com by mfi.com (SMTPLINK V2.11)
; Wed, 18 Dec 96 09:22:58 PST
Return-Path: <71154.2131@CompuServe.COM>
Received: by arl-img-5.compuserve.com (8.6.10/5.950515)
id MAA28405; Wed, 18 Dec 1996 12:24:54 -0500
Date: 18 Dec 96 12:04:12 EST
From: Lee Chae <71154.2131@CompuServe.COM>
To: Lee Chae <lchae@mfi.com>
Subject: Sample e-mail message
Message-ID: <961218170411_71154.2131_DHB86-1@CompuServe.COM>

  Here is the sample message you wanted to show in your Tutorial.
-Lee
```

In this sample message, the header information takes up the first 10 lines. An individual header consists of a field and a field value. For example, `To:` is a field, and `Lee Chae <lchae@mfi.com>` is its value. As you can see, the header contains detailed information about who sent the message, who is to receive the message, and how the message got from the sending point to the receiving point.

In this case, the header tells you that the message was sent from `Lee Chae` at the client system identified as `71154.2131@compuserve.com`. It was received by an MTA identified as arl-img-5.compuserve.com and sent to mail server mfi.com, which is the mail server that services my corporate e-mail system. The message is directed to my e-mail account, lchae@mfi.com.

In addition, the header displays the date of the message, the times at which the different MTAs received the message, and the unique ID of the message.

The body of the message contains the actual text the sender typed and is separated from the header by a "null" line. RFC 822 doesn't define the message body, as it can be anything the user enters, as long as it is ASCII text.

Aside from the message format, the other important standard is the e-mail address. You're probably familiar with the e-mail addressing system, but I'll go over it quickly, in case you aren't. Basically, a standard e-mail address usually follows the following form:

```
<mailbox ID>@<domain name>
```

The mailbox ID is the name of an individual mailbox on a local machine. In my case, my mailbox ID would be lchae. The domain name is the name of a valid domain registered in the Domain Name System (DNS), which is the distributed database that keeps track of the different host names, network names, and IP addresses used on the Internet. The DNS mail entries are the keys to Internet e-mail because they allow MTAs to find the machine specified in the recipient's e-mail address.

To finish with my example, the domain name for my e-mail address is mfi.com. Put it together and you get `lchae@mfi.com`.

SMTP: THE TRANSPORT

The transmission of an e-mail message through the Internet relies on the Simple Mail Transfer Protocol, which is defined in RFC 821. SMTP governs the way a UA establishes a connection with an MTA and the way it transmits its e-mail message. MTAs also use SMTP to relay the e-mail from MTA to MTA, until it reaches the appropriate MTA for delivery to the receiving UA.

The interactions that happen between two machines, whether a UA to an MTA or an MTA to another MTA, have similar processes and follow a basic call-and-response procedure. The main difference between a UA-to-MTA transaction and an MTA-to-MTA transaction is that with the latter, the sending MTA must locate a receiving MTA.

To do this, the sending MTA contacts the DNS to look up the domain name specified in the recipient e-mail address. The DNS may return the IP address of the domain name—in which case the sending MTA tries to establish a mail connection to the host at that domain—or the DNS may return a set of mail relaying records that contain the domain names of intermediate MTAs that can act as relays to the recipient. In this case, the sending MTA tries to establish a mail connection to the first host listed in the mail relaying record.

Now I'll use the case of two MTAs, a sending MTA and a receiving MTA, to illustrate the call-and-response mail transaction. First, as mentioned, the sending MTA chooses a receiving MTA, which may be the final destination of the message or an intermediate MTA that will relay the message to another MTA.

Next, the sending MTA requests a TCP connection to the receiving MTA. The receiving MTA responds with a server ID and a status report, which indicates whether or not it is available for the mail transaction. If it isn't, the transaction is over; the sending MTA can try again later or attempt another route. If the receiving MTA is free to handle a session, it will accept the TCP connection.

The sending MTA then sends a HELO command followed by its domain name information to the receiving MTA, which responds with a greeting. Next, the sending MTA sends a Mail From command that identifies the e-mail address from which the message originated, as well as a list of the MTAs that the message has passed through. This information is also known as a *return path*. If the receiving MTA can accept mail from that address, it responds with an OK reply.

The sending MTA then sends a Rcpt To command, which identifies the e-mail address of the recipient. If the receiving MTA can accept mail for that recipient (it may perform a DNS lookup to verify this) it responds with an OK reply. If not, it rejects that recipient. (An e-mail message may be addressed to more than one recipient, in which case this process is repeated for each recipient address.)

Once the receiving MTA identifies the recipient's address, the sending MTA sends the Data command. The receiving MTA accepts command by responding with OK. It then considers all succeeding lines of data to be the message text. Once the sending MTA gets an OK reply, it starts sending the message. The sending MTA signals the end of the message by transmitting a line that contains only a period (.).

When the receiving MTA receives the signal for the end of the message, it replies with an OK to signal its acceptance of the message. If for some reason, the receiving MTA can't process the message, it will signal the sending MTA with a failure code. After the message has been sent to the receiving MTA and the sending MTA gets an OK reply, the sending MTA can either start another message transfer or use the Quit command to end the session.

Once the receiving MTA accepts the message, it reverses its role and becomes a sending MTA, contacting the MTA next in line for the relay of the message. The process stops once the message reaches the Internet mail server that services the recipient specified in the Rcpt To e-mail address.

If at any point along the way an MTA can't deliver the e-mail—for whatever reason (for instance, if it can't identify the recipient address)—it generates an error report, also known as an undeliverable mail notification. The MTA uses MTAs identified in the return path to relay the error report back to the original sender.

SOME IMPORTANT FOOTNOTES

SMTP can handle only messages containing the 7-bit ASCII text defined in RFC 822. This means that alone, SMTP is ill fit to handle other types of data such as 8-bit binary data and other multimedia formats that more and more people are sending both within the body of e-mail messages and as attachments. However, as a solution to this limitation, the IETF developed the Multipurpose Internet Mail Extensions (MIME) protocol (RFC 1521), which packs multimedia data into a format that SMTP can handle.

In addition, you may have noticed that my description of the e-mail relay process stopped at the recipient's Internet mail server, or MTA. The reason for this is that a user agent can employ different methods to access, or retrieve, its e-mail from the MTA. For instance, the majority of companies rely on proprietary e-mail packages, such as cc:Mail, to handle e-mail operations on the local network. And individual users often use e-mail applications offered by online services such as CompuServe and America Online, which you can also consider proprietary e-mail programs. In both cases, the proprietary system acts as the last leg between the final MTA and the recipient UA.

However, you may want to keep an eye out for two important e-mail protocols: Post Office Protocol 3 (POP-3) and Internet Message Access Protocol 4 (IMAP-4). POP-3 is an older standard that defines a method for a client system, or UA, to access its e-mail. IMAP-4 is a new standard, rising in popularity, that offers a more robust UA. Both allow UAs direct access to MTAs for the retrieval of messages, although IMAP-4 gives you more options for handling and storing e-mail. You can find more information on POP-3 and IMAP-4 by looking up their RFCs, 1725 and 2060, respectively, on the Web. (There's a great Web site set up for IMAP at http://www.imap.org.) For more about IMAP-4, see "IMAP's New Territory."

This tutorial, number 103, by Lee Chae, was originally published in the March 1997 issue of LAN Magazine/Network Magazine.

The Basics of E-Mail Access

SMTP transports e-mail across the Internet; POP-3 and IMAP-4 deliver it to your desktop.

In the tutorial, "E-Mail and SMTP," I discussed how electronic mail travels across the Internet from your e-mail system to that of the recipient's. Simply put, e-mail relies on the Simple Mail Transfer Protocol (SMTP) to relay messages from mail server to mail server until they have reached the destination server. However, SMTP doesn't handle the transfer of mail from the mail server to the desktop.

In this tutorial, I'll discuss the last leg of the journey: how users access e-mail on their mail server from their client machines. There are a few different models for this operation and a few different protocols that can handle the task, including proprietary protocols found in commercial e-mail packages and two Internet standards-based protocols, Post Office Protocol3 (POP-3) and the Internet Message Access Protocol 4 (IMAP-4).

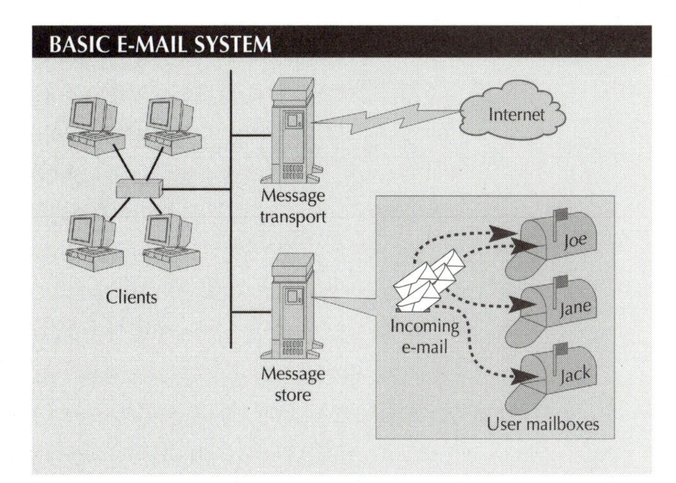

■ **Figure 1:** A basic local e-mail system consists of a message transport that transmits outgoing mail and delivers incoming mail to the message store, which stores mail in the appropriate user's mailbox. Users can then access their messages from their desktops using a client program that is running a message access protocol, such as POP-3, IMAP-4, or a proprietary protocol.

Companies usually employ proprietary e-mail packages to handle local e-mail operations, but standards-based systems are appealing. One, you can avoid the go-between, in this case a proprietary system, by having clients work directly with your mail server. Two, you can avoid having to install gateways in order to let your system communicate with the Internet. Finally, the enhanced features offered in IMAP-4 can potentially rival those found in proprietary packages, although IMAP-4 client packages are just beginning to enter the market in pretty primitive form.

WHAT'S IN STORE

As I mentioned, SMTP relays e-mail across a series of Internet mail servers until it reaches the designated recipient's mail server. It uses DNS tables and mail exchange records, also known as MX records, to locate the server. Once SMTP delivers the e-mail message to the final mail server, the message is placed in a message store, which is also called a post office. (If the recipient uses a proprietary e-mail system, such as cc:Mail, the e-mail is received by the product's Internet or SMTP gateway, which uses its own proprietary protocols to store the message in the post office for retrieval by the recipient.)

In its most basic form, the message store, located on a server, is usually some type of file system that holds delivered mail for access by users. You can think of the message store as one large directory with subdirectories dedicated to each user. These subdirectories are also known as mailboxes. A more advanced form of message store would allow users to create personal folders to store read and unread messages, and it would allow users to create archives for groups of messages. Other advanced features include the ability to perform keyword searches and to create a hierarchy of folders.

As a side note, the message store component of an e-mail system is usually located on a different machine than the message transport component, which handles the delivery of incoming messages and the transmission of outgoing ones. The reason for their separation is that the transport component usually handles a high volume of operations. So, if both systems were on the same machine, the traffic of inbound and outbound mail would experience a performance drag, as would users when they try to retrieve and manipulate mail in their mailboxes.

Figure 1 illustrates a basic architecture for a local e-mail system.

THE ACCESS MODELS

Once an e-mail message is delivered to the message store, it's ready for retrieval by the recipient. There are three basic models for message access: offline, online, and disconnected (see Figure 2).

The offline model is the most basic of the three. In the offline model, a client connects to the mail server and downloads all messages designated for that particular recipient. Once downloaded, the messages are erased from the server. Then, the user processes, manipulates, and stores the messages locally at the client machine. The advantages of the offline model are obvious. It requires a minimal amount of server connect time since the client accesses the server only periodically for mail downloads. Also, because the mail processing is performed at the client, the offline model won't eat up a lot of server resources. And, it requires less server storage space because mail is deleted from the server once it has been downloaded by a client.

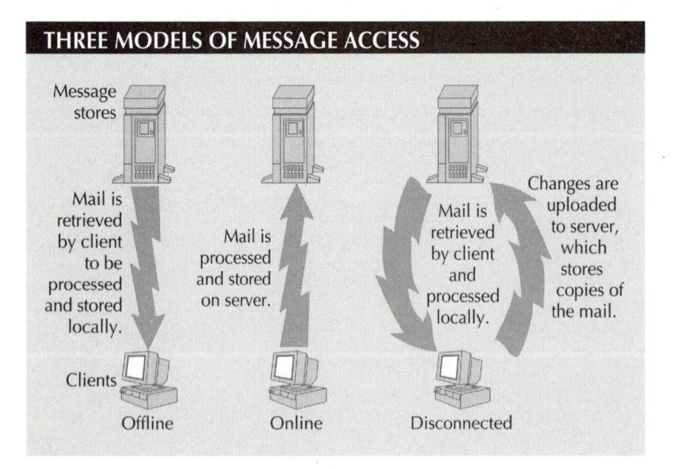
THREE MODELS OF MESSAGE ACCESS
Message stores
Mail is retrieved by client to be processed and stored locally.
Mail is processed and stored on server.
Mail is retrieved by client and processed locally.
Changes are uploaded to server, which stores copies of the mail.
Clients
Offline
Online
Disconnected

■ **Figure 2:** The message access transaction can be divided into three models: offline, online, and disconnected. The main difference between the three revolves around where the mail is actually read or processed and where the mail is stored.

Of course, the offline model does have drawbacks. For instance, because mail is downloaded to a particular machine, you must use that machine to access processed mail. So, if you downloaded e-mail to your desktop computer and then went on the road with your laptop, you wouldn't be able to read those same messages off the mail server from your laptop. And, because all processing and storage is performed at the client machine, the client must be endowed with enough resources to be up to the task.

With the online model, all e-mail processing and manipulation is performed at the server. In fact, all the messages remain on the server even after they've been read by users, although in some implementations, users can save e-mail to the local client.

As you might expect, the online model requires a constant connection to the server whenever users need to access and work with their e-mail. Consequently, this model's high connect time could mean that more users are sharing bandwidth at any given time. And, because the server is burdened with all message processing and storage, you'll need a more powerful server to handle an online e-mail system than you would with an offline system.

In addition, users can't work with their e-mail unless they are online, meaning they must be at a client that can access the mail server even to do the simplest task, such as reread an old message.

However, because all mail is stored and processed at the server, you can access your e-mail from any client that can connect to the server. For the same reason, this model requires fewer resources at the client machine. And, online-type systems usually offer enhanced features such as the ability to create a multitude of personal folders to organize e-mail, and the ability to create archives for storing messages.

The disconnected model combines elements of both the offline model and the online model. Using this system, a user connects to a mail server to retrieve and download e-mail. Then, the users can process the mail on the local client.

Once the user is finished with the messages, the user once again connects to the mail server and uploads any changes. With this model, the mail server acts as the main repository of the user's e-mail.

The disconnected model's strengths are that users can access e-mail from a variety of clients that have access to the mail server. Users can also process mail offline. This also translates to a shorter server connect time. However, the model does require a sufficient amount of resources on both the server and the client.

A MATTER OF PROTOCOLS

I mentioned earlier that proprietary e-mail packages currently dominate the business market. Most companies employ systems such as Lotus Notes, Microsoft Outlook, and Microsoft Exchange.

The majority of e-mail packages follow the online model of e-mail access. These products use their own protocols to retrieve messages from Internet mail servers and store them on local servers. Then, users can use the products' client programs to access and work with mail from message stores on the proprietary e-mail servers. The attractiveness of these proprietary packages lies in the added features they offer. For instance, they usually offer advanced mail manipulation such as folders, hierarchical folders, and archives. Some offer the ability to create bulletin board systems for general corporate communications. In addition, some offer message notification when a new e-mail arrives for a particular user. Also, they usually offer searching abilities, so users can find e-mails with particular headings, or look up the addresses of other users on the system.

The problems with these packages follow all the arguments you may have heard about relying on a proprietary solution: Mainly, you're locked into a closed system. For instance, if you want to be able to access your e-mail from a remote location, your proprietary package must support this capability. Or, let's say your SMTP gateway goes down. All outbound and inbound Internet e-mail will be blocked. If your e-mail server goes down, you won't be able to send or receive anything.

In contrast, Internet standards-based e-mail clients ought to be able to send e-mail as long as they have an Internet connection and know of an SMTP mail server that clients can use to send messages. In addition, as long as the client has an Internet connection, it ought to be able to reach your organization's Internet mail server to retrieve and process messages regardless of where the client is located or on what machine.

One of the first e-mail access protocols developed was the Post Office Protocol, the latest version of which is POP3. In essence, POP3 follows the offline model of message access, and it revolves around send-and-receive types of operation. However, there have been recent attempts to remodel POP3 so that it has some online capabilities, such as saved-message folders and status flags that display message states. But using POP3 in online mode usually requires the additional presence of some type of remote file system protocol.

IMAP-4 is the more advanced of the standards-based message access protocols. It follows the online model of message access, although it does support offline and disconnected modes. IMAP-4 offers an array of up-to-date features, including support for the creation and management of remote folders and folder hierarchies, message status flags, new mail notification, retrieval of individual MIME body parts, and server-based searches to minimize the amount of data that must be transferred over the connection.

POP3, IMAP-4 UPDATE

Currently, there are POP3 and IMAP-4 products available on the market, although the IMAP-4 offerings are limited. This status should change, as IMAP-4 is seen as the future of Internet-based e-mail. If you're just testing the waters, you should be able to find free implementations of both types of clients available on the Internet. Mac, PC, and Unix clients should all be represented. Again, the strengths of these protocols are that they are open and Internet-oriented. So, it won't take added steps to get your e-mail system to work with the Internet. In addition, users will be able to access mail from a variety of client platforms and from a variety of locations, as long as TCP/IP is up and running and users have access to the Internet.

This tutorial, number 106, by Lee Chae, was originally published in the June 1997 issue of Network Magazine.

E-Mail and MIME

Extending e-mail beyond simple text-based messaging.

In recent tutorials I've covered e-mail and how you can send and receive messages over the Internet. In this tutorial, I'd like to talk about MIME (Multipurpose Internet Mail Extensions) and how it allows you to use the Internet e-mail system to send more than just a simple, text-based message.

THE ASCII'S THE LIMIT

Initially, the Internet e-mail system was limited to simple text messages because SMTP, the protocol used to transport mail across the Internet, could carry only 7-bit ASCII text. In the United States, the main ASCII standard used for e-mail is US-ASCII. This version of ASCII offers only a basic set of characters—128 characters in all, each represented as a 7-bit binary number. This ASCII set was designed to cover the English alphabet, including both uppercase and lowercase letters, and the numbers 0 through 9, as well as some other characters.

However, our use of data, both at the workplace and at home, has moved beyond simple text. For example, with basic word processing programs, users can augment text with italics, boldface, bullets, and other types of enriched formatting. You can even embed graphics in documents. But because SMTP was developed to handle only basic text messages in a 7-bit format—as laid out by RFC 822, which defines the standard for Internet text messaging—these more sophisticated data formats can't be sent via e-mail.

The problem with US-ASCII lies with messages that contain information that doesn't fit into the 128 character set. For example, US-ASCII can't accommodate "rich text" characters such as an italicized or boldfaced ***a***. Even more, as the Internet is clearly an international forum, foreign language character sets aren't represented in US-ASCII.

And, the 7-bit format required by SMTP prevents users from sending other types of data via e-mail—for example, the 8-bit binary data found in many executable files and in files created by applications such as Microsoft Word.

This limitation looms large when you realize that the Internet e-mail system is perfectly situated to act as a data delivery vehicle. Business and personal e-mail accounts are now widespread, meaning users have a pipeline to each other from desktop to desktop. If e-mail could handle diverse data—for example, word processing documents and image files—users could share all sorts of data without having to ship disks or make any actual real-time network connections to copy or download files.

ENTER MIME

The developers of MIME found a clever way to work around the limitation: MIME packages different data types into a 7-bit ASCII format. That way, all e-mail, regardless of the data it contains, appears as standard e-mail messages to the Internet's SMTP servers. The beauty of the solution lies in the fact that SMTP didn't have to change to handle such data. In other words, the solution didn't require that all Internet mail servers be upgraded to a new version of SMTP (which would have been an extraordinary task considering all the Internet mail relays in existence). In effect, the transport system remained untouched.

Although the solution sounds simple, it took developers tremendous foresight to define the different data types MIME can handle. It also took a good amount of technical sleight of hand to standardize the ways in which these data types can be packaged as ASCII and still be readable in their original format once the packages have been unpacked. The solution is a work in progress; new data types, which the Internet Assigned Numbers Authority (IANA) must judge for inclusion in the MIME standard, are constantly emerging.

As defined in RFC 2045, MIME provides three main enhancements to standard e-mail. First, with MIME, e-mail can contain text that goes beyond basic US-ASCII, including various keystrokes such as different line and page breaks, foreign language characters, and enriched text. Second, users can attach different types of data to their e-mail, including such files as executables, spreadsheets, audio, and images. And third, users can create a single e-mail message that contains multiple parts, and each part can be in a different data format. For example, you could compose a single e-mail message that consists of a plain text message, an image file, and a binary-based document, such as a Word file.

A PACKAGE DEAL

RFC 2045 defines seven types of e-mail content that MIME can package and pass across the Internet: text, image, audio, video, application, message, and multipart. Each of these data types can come in a few different formats, or subtypes. (MIME also augments types and subtypes with certain parameters, which are specified in RFC 2045. However, these parameters contain too much detail to cover in this overview of MIME.)

Obviously, the text type of e-mail content supports messages carrying text. However, within the text type, MIME also supports the plain subtype, which is usually standard 7-bit

ASCII. MIME also supports the rich text subtype, which allows for some simple formatting features, such as page breaks.

The image type supports image files, and its subtypes include the GIF (Graphics Interchange Format) and JPEG (a compressed image format developed by the Joint Photographic Experts Group). In the words of RFC 2045, the video type supports, "time-varying picture images," and for now, MPEG (a compressed video format developed by the Motion Picture Experts Group) is its only subtype. The audio type supports audio data, and its only subtype is basic.

According to the RFC, there is no one ideal audio format in use today. So, the developers of MIME tried to define a subtype that would be the lowest common denominator. The basic subtype for audio signifies "single channel audio encoded using 8-bit ISDN mu-law" at a sampling rate of 8KHz.

The application type supports two types of data: data that's meant to be processed by an application, and data that doesn't fall into any of the other categories. For now, it supports the octet-stream subtype, which means the message can carry arbitrary binary data. Also, it supports the postscript subtype, meaning the message can be sent to print as a PostScript file.

A couple of additional notes on the application type: If a mail agent receives a message whose content subtype it doesn't recognize, by default it will attempt to pass the message on as an application type message with a subtype of octet-stream (or, application/octet-stream). Also, in the future you can expect the list of application subtypes to grow, as specific programs are accepted by the IANA for inclusion in MIME. For example, you could see an application/access or application/quark designation as these types of files are acknowledged by MIME.

The remaining two content types allow for special handling of an e-mail message. For instance, the message type allows an e-mail to contain an encapsulated message (the rfc822 subtype). The external-body subtype allows an e-mail to indicate an external location where the intended body of the message resides. That way, the user can choose whether or not to retrieve the message body. The message type also allows MIME to send a large e-mail message as several small ones (the subtype for this is partial). The receiving MIME-enabled mail agent can then open the smaller e-mail messages and reassemble them into the original long version.

Finally, the multipart type allows an e-mail message to contain more than one body of data. The mixed subtype allows users to mix different data formats into one e-mail message. The alternative subtype allows a message to contain different versions of the same data, each version in a different format. MIME mail agents can then select the version that works best with the local computing environment. The digest subtype allows users to send a collection of messages in one e-mail, such as the kind used with Internet mailing lists sent in digest form.

Finally, the parallel subtype allows mixed body parts, but the ordering of the body parts is not important. (For a quick summary of MIME content types and subtypes, refer to Table 1.)

TABLE 1-MIME CONTENT TYPES AND THEIR SUBTYPES

Content type	Subtype
Text	Plain text, rich text
Image	GIF, JPEG
Audio	Basic
Video	MPEG
Application	Octet-stream, PostScript
Message	RFC822, partial, external-body
Multipart	Mixed, alternative, digest, parallel

In addition, some companies and organizations are experimenting with their own content types. MIME accommodates this, allowing e-mail to carry these types of data, as long as the message has an x in its headers. The x tells the receiving MIME mail agent that the content type is experimental (more on MIME headers coming up). If the x isn't present, the receiving agent may try to convert the data according to some other convention, such as the application/octet-stream designation—an act that could turn the message into an unreadable mess. This is another fine example of the foresight that went into the development of MIME.

ROUND PEG, SQUARE HOLE

To package the different data formats into the 7bit ASCII format, MIME uses five different encoding schemes: 7bit, 8bit, binary, quoted-printable, and base64. But, as you'll soon see, only the quoted-printable and base64 schemes actually encode data. The 7bit, 8bit, and binary schemes merely indicate what format the data is in; they leave it to individual mail systems to select an encoding process based on the data format.

Because MIME-enabled mail agents perform the encoding automatically, you don't need a comprehensive understanding of how they work. So, I'll give just a brief review of what they can do and in which situations they are used.

The 7bit scheme tells mail agents that the message contents are in plain ASCII. For this reason, no encoding is necessary as all mail systems should support ASCII. The 8bit scheme indicates that the contents contain 8-bit characters. It's up to the mail agents to encode this information using their preferred means, if they have any. Because not all mail agents use the same encoding method for 8-bit characters, there's a good chance the 8-bit characters won't appear correctly when the e-mail is opened. For this reason, the 8bit scheme currently isn't a reliable encoding scheme. The binary scheme, because it is similar to the 8bit scheme, shares the same problem.

The quoted-printable scheme is used for text that contains a mixture of 7-bit and 8-bit characters. Essentially, it allows 7-bit characters to go unencoded and converts each 8-bit character into a set of three 7-bit characters. As a result, mail servers and mail agents see an e-mail containing only 7-bit characters.

The base64 scheme is used for data that isn't text, such as data that constitutes an executable file. It works by breaking the data down into sets of three octets, each set containing 24 bits. Then, it converts each set of 24 bits into a four-character sequence. (In other words, every six bits of data is represented by a character.) The characters used in the sequencing come from a set of 65 characters, all of which can be found in any version of ASCII. Because it's in ASCII, data encoded by base64 should be readable by any mail server or mail agent.

HEADS UP THINKING

Using MIME is simple. If your mail system supports MIME, it chooses the data type and encoding scheme—the packaging, as it were—for the user, depending on the contents of the user's e-mail and what file or files are attached to it. Once it does so, it adds MIME headers to the traditional SMTP headers found at the top of e-mail messages. (In the case of multipart MIME messages, headers will also appear in the body of the message.) These headers tell receiving mail agents that they've received a MIME message and indicate how the mail agents should handle the message.

The main headers are MIME-Version, Content-Type, and Content-Transfer-Encoding. The last two headers refer to the data type found in the message and the scheme used to encode the data, respectively. Some other headers are Content-Description, which lets you type in a description of the message (much like SMTP's Subject header), and Content-ID, which is akin to SMTP's Message-ID.

Listing 1 shows a sample MIME message. The e-mail message shown is a multipart message in encoded form. As you can see, the top-level headers explain that the message data is of the type multipart and its subtype is mixed (multipart/mixed). It also contains a boundary parameter. This tells the receiving mail agent where each part of the message begins and ends. MIME adds two hyphens (--) in front of the boundary value when it appears in the message. MIME identifies the end of the message with two hyphens, the boundary value, and two more hyphens.

LISTING 1—A MULTIPART, ENCODED MIME MESSAGE

From: joe_luthier@plucknplay.com

To: lchae@mfi.com

Subject: Info on Gibson guitar

MIME-Version: 1.0

Content-Type: multipart/mixed; boundary=17

--17

Content-Type: text/enriched; charset="us-ascii"

Content-Transfer-Encoding: 8bit

Content-Description: Greetings

As promised, I'm getting back to you about the Gibson Southern Jumbo guitar you were interested in. I've enclosed a spec sheet on the guitar, which is in Microsoft Word.

I <bold>guarantee</bold> that you'll <bold>love</bold> it!

--17

Content-Type: application/octet-stream

Content-Transfer-Encoding: base64

Content-Description: Spec sheet saved as MS Word file

<Encoded data for Word file would appear here.>

--17--

RESOURCES

To learn more about MIME, you can read the MIME-related RFCs (see "MIME RFCs"). Although these documents are aimed primarily at MIME developers and implementers, they are written in relatively plain language and should serve as a good education for those network professionals unfamiliar with the topic—and even for those experienced with MIME, but who have further interest in its details.

A Listing of MIME-Related RFCs

RFC-822: Standard for the Format of ARPA Internet Text Messages

RFC-2045: Part one: Format of Internet Message Bodies

RFC-2046: Part two: Media Types

RFC-2047: Part three: Message Header Extensions for Non-ASCII Text

RFC-2048: Part four: Registration Procedures

RFC-2049: Part five: Conformance Criteria and Examples

This tutorial, number 110, by Lee Chae, was originally published in the October 1997 issue of Network Magazine.

Mail-Enabled Applications

Token Ring, Ethernet, FDDI. It's all basically plumbing, and once it works, you give the network-access method about as much thought as you give to the copper pipes that bring hot and cold water when you turn on the tap. TCP/IP, IPX/SPX, and NetBEUI are the valves that control the flow of data. Copper pipes and flow-control valves interest only plumbers and architects. What the average person really needs is the end result of the plumbing system—the dishwasher, the showerhead, the lawn sprinkler system.

There's a new appliance on the horizon that is going to change how we view networking and networked applications, and that's mail-enabled applications. Mail-enabled applications deliver a new type of network appliance and new level of functionality. Mail will evolve from messaging to a complete communications medium.

Mail-enabled applications will deliver the infrastructure for e-mail management applications, workflow automation applications, document distribution, and forms processing. The groundwork is beginning to be laid, with mail engines from Novell, Lotus, Microsoft, and others.

Applications that take advantage of these mail engines are in their earliest stages. And from the looks of applications such as Beyond's BeyondMail for rules-based mail management and Reach Software's MailMan for workflow automation and eventually forms processing, the progress of mail-enabled applications seems quite promising.

TODAY'S E-MAIL IS MESSAGING

Right now, electronic mail is little more than an elaborate messaging scheme. Users on a network can send and receive messages, which are usually short and text-based, and sometimes have files attached. Users can ask for return receipts, send registered mail, forward messages to other users using the same e-mail program, and find out what time the recipient has read the mail. They can construct mailing lists from an address book or directory of e-mail users. With the right software, users on dissimilar computers, say a Macintosh user and an IBM PC user, can exchange messages with relative ease.

Despite the sophistication, e-mail systems remain fundamentally interpersonal communication. One person sends mail to another or to a group of people. The recipient or recipients read the e-mail, perhaps with a file attached, and respond to the message, usually by sending another e-mail message.

This isn't to disparage e-mail's power. E-mail can significantly increase an individual's productivity. It allows people to work offline, as it were, responding to messages on their time and their terms, rather than answering a phone call or having a face-to-face meeting. E-mail has been proven to increase workers' productivity, at least until the onslaught of messages becomes so great that any time savings is eroded by the time taken to manage the volume of messages.

MAIL BONDING

E-mail has the potential for becoming more than a method of interpersonal communication. It can become the platform upon which many other forms of communication occur and can serve as the skeleton for other applications. Mail-enabled applications are the merging of messaging with other application software.

The problem lies in the fact that most people do their work in one set of applications—WordPerfect, 1-2-3, and Paradox—and then want to communicate their ideas and share that work with their coworkers. However, messaging is not inherently part of a word processor, a spreadsheet, a database, or most other applications. Messaging is not even a part of most applications that could easily accommodate messaging, such as purchasing systems or contact management systems.

Most applications do not include mail or messaging capabilities for some very simple reasons. First, when these packages were conceived of and written, the concept of mail-enabling didn't exist or make sense. Many of these packages were written before networks became widespread.

Second, for those software publishers who do want to move toward mail-enabling, some roadblocks are in the way. There's no standard, de facto or otherwise. Each e-mail vendor implements a different scheme for mail-enabling, requiring applications vendors to choose among the options. This setup requires the application developers to choose correctly, or potentially lose business.

Also, many companies use more than one electronic mail system. For example, the mainframe terminal users may use PROFS, the VAX users may use All-In-One, and the PC LAN users may use any one of dozens of packages, but probably use Notes or Microsoft Outlook. To use a mail-enabled application, the interface must be able to work with all of the company's existing mail systems.

Any application that uses document routing or communication can become a mail-enabled application. Whether mail-enabled applications become widespread is a matter of how easily the problems can be solved, how easily existing applications can be adapted, and how easily new applications can be written. And that's a function of the available tools.

MAKING IT HAPPEN

Mail-enabled applications, although in their embryonic stages, have great potential. Take for instance, a consulting engineer whose client, a hospital, wants to purchase a boiler. Maybe the engineering firm has an old catalog lying around from one of their suppliers. Maybe an administrative assistant has to call the supplier's salesperson to send a new catalog to both them and their client. Either way, the engineers have to leaf through the catalog, looking for the specifications they need. Then they have to call the salesperson and ask for more detailed spec sheets on the various models and manufacturers.

Now consider the boiler manufacturer, who has to handle inquiry calls from customers across the country. The manufacturer is using a lot of resources in presale and postsale support calls. The company probably has an entire department set up to deal with simple calls, preventing their employees from resolving the more complex issues.

If customers could dial into the manufacturer's computer system, search the documents for answers to simple queries, and download the right specifications sheets and other documentation, then the sales support desk wouldn't have to spend its time manually sending out these packages, and they could spend their time on more meaningful calls. Once at the customer site or even at the manufacturer, the electronic documents could simply be forwarded from one person to another.

While it would seem that this scenario calls for an ordinary document management system, the key is document transport, which can be done via electronic messaging. Put the spec sheet—or any other type of file—in an envelope, and send it. But the forms and routing aspects that simple e-mail systems lack must be solved. The forms can be merely another user interface talking to a mail engine, and the routing can be a rules-based system directing messages to the right people.

Such mail-enabling is appropriate for any application that requires forms distribution. A purchasing system is another example, where one group of managers must approve a requisition if the purchase is below a certain dollar amount, and another set of managers must sign off if the purchase is over a certain amount.

If the requisitions were in the form of e-mail messages, and the system had the built-in intelligence to route the documents to the proper people, a mail-enabled application would be born. Ideally, the people wouldn't even have to use the same front-end mail applications as long as they subscribed to a standard back-end engine or a common interface.

POWER IN THE BACK END

Like databases, the new generation of mail applications follows the client-server architecture. Once considered a single application, mail has since been split into its many component parts: the core engine, the user interfaces, message storage, transport layer, gateways to other e-mail systems, and directory services.

As with databases, this client-server architecture allows each group of developers to concentrate on their strengths. For example, the engine manufacturers can concentrate on better message handling services, while the gateway manufacturers work to write faster and more functional gateways between dissimilar mail systems. Front-end application vendors, in the meantime, can offer e-mail interfaces and then build on top of them.

And like databases, no one mail engine has emerged supreme. Today, four groups are vying for the top spot: Novell, Microsoft, Lotus, and OSI. No clear winner has emerged. Message Handling System (MHS), which has been around since the mid-1980s, is the most mature. Originally developed by Action Technologies, MHS was sold to Novell. At one time, the majority of PC LAN mail applications used the MHS interface. Microsoft will outline its messaging strategy for its applications and network users. [Note: Microsoft ultimately announced its Messaging Applications Programming Interface (MAPI), which has been accepted by Lotus and other important messaging developers.] Lotus has announced the Open Messaging Interface (OMI), plus mail-enabled applications can be built using something as simple as the import/export function in its cc:Mail. [Note: OMI was the forerunner of Lotus's Vendor Independent Messaging (VIM) interface, which has failed to gather industry-wide support.] OSI's X.400 standard offers an internationally recognized method of store-and-forward messaging.

No standard communications language exists for mail-enabled applications in the same way that Structured Query Language has become the accepted method of querying relational databases, even though some differences exist among the implementations.

Instead of supporting each vendor's engine directly, developers could write to a common interface, which talks to the different engines. Developers would not have to write the core mail engine themselves nor would they have to worry about choosing the "right" engine. And developers would not have to spend their resources supporting different engines. Now, if only there were such an accepted, industry-standard interface.

[Note: the next three paragraphs are included for historical reference only—OMI became VIM, which is itself well along the way to the old standards' graveyard.] Lotus has proposed that OMI be that common layer and has placed the specification in the public domain. OMI is an Applications Programming Interface (API) that is published by Lotus, IBM, and Apple, and it offers a common interface for different vendors' mail engines, including directory, transport, and message storage functions. OMI will be available for a variety of platforms, including DOS, Windows, OS/2, Unix, and the Macintosh.

Applications make calls to the OMI interface to send and receive messages and for services such as looking up user names in the directory or storing messages in folders. OMI also provides developers the ability to access e-mail services in a modular fashion, such as directory, message storage, and transport. This way, developers and users can pick and choose services. For example, they may choose to use the network operating system's directory service instead of the mail systems'. This setup prevents developers and end users from being locked into using a particular service or vendor.

So far, Lotus has pledged to use OMI to mail-enable all its applications, including 1-2-3, AmiPro 2.0, and Freelance. IBM says it will include OMI technology in a future OS/2-based offering, and OfficeVision/2 will use OMI. Apple will support OMI in future versions of System 7. Still, OMI is a brand-new specification, and only time will tell if it is adopted. [Editor's note: Time has told—it wasn't.]

This tutorial, number 42, by Patricia Schnaidt, was originally published in the January 1992 issue of LAN Magazine/Network Magazine.

Workflow Applications

Information workers don't enter numbers or type words; they manipulate information, drawing conclusions and making projections. They forward that work or information to coworkers, who then make more projections and draw more conclusions. *Workflow computing* is designed to deal with the process of work, not just the end results.

Workflow computing is inherently a networked application. Although it could take place on a host-centric system, workflow applications take advantage of a network's distributed intelligence. Work can be distributed to the local sites, where workers can continue to use their familiar tools.

A workflow is a description of how work moves among workers and the operations required to process that information as it moves. The work process is decomposed into steps and dependencies. Workflows can be simple or complex. For example, on one end of the spectrum, an intelligent mail application can help prioritize and route a user's mail; on the other end, a workflow application can move a purchase requisition through the approval process.

According to the *Clarke-Burton Report* (The Burton Group, formerly known as Clarke-Burton, Salt Lake City), "The concept behind workflow automation is to extend the reach of computing beyond office automation, [to] bring the process itself—not just the work created by the process—into the realm of the computing system." Market research firm Forrester Research, located in Cambridge, MA, defines workflow computing as "computers and networks adding and extracting value from information as it moves through the organization."

Imagine the work and the process through which a company accomplishes its business as a river. Employees alter, detract, or add to this flow as it begins at the initial customer contact and ends with the product shipment or dispensed service.

Instead of wasting corporate time and money, while paperwork eddies in employees' in-boxes, whirls in their out-boxes, and becomes waylaid in the rapids that fall in between, work-flow software routes the work to the proper person. For example, a purchase order must be routed from the requester to one or more managers to the purchasing agent to the supplier and back to the requester. Different purchases require different levels of approval. Workflow software can automate these processes that are now decided manually.

Business cycles can also be shortened. Instead of managers tracking work through engineering, assembly, and marketing, for example, workflow software can monitor its progress, allowing people to concentrate on the specifics of their jobs. Managers and workers can know where a particular item resides in the process. Workflow software can help identify and eliminate bottlenecks.

Workflow software can reduce the difficulties of distance. Companies can more easily distribute their operations as their businesses dictate. Workers in far flung offices of the same corporation can communicate as easily as if they're local.

Workflow computing can also be externally focused. Companies' desire to communicate with its customers and suppliers will drive workflow computing. Companies will be able to establish tighter links to their suppliers. Customers can more easily communicate with their suppliers. Electronic Document Interchange (EDI) can become more than a primitive reality.

FOUR TYPES OF WORKFLOW

Intelligent routing is perhaps the most important quality of workflow software. Intelligent routing determines who is the next person in the process that must receive the work. This forwarding process can be accomplished in any number of ways, from elaborate process models to complex scripts to ad hoc arrangements.

Workflow computing software must provide facilities to track the flow of work throughout the system and to accommodate existing applications. The flow of work must be tracked and reported, so audit trails can be produced and managers can follow a work's progress. Also, you need to know who is accessing what information. Host-based, client-server, and personal productivity applications must be connected to the workflow system.

Workflow automation comes in four breeds: document flow, process automation, task automation, and workgroup tools. Companies will implement the different types based on their corporate practices and philosophies. For example, a more structured company is more likely to implement a process automation system, while an entrepreneurial company is likely to implement a workgroup or task automation system.

The early workflow systems take a document flow approach. Developed primarily by image-processing companies, these systems automate a paper-based process, such as loan applications or credit card billing. Very often, these systems completely simulate the paper-based process, from file cabinets to paper clips. In automating a purely paper-based process, many of these systems forfeit the benefits of altering the business practice to take advantage of a new technology. Automate an inefficient, manual process, and you will have an inefficient, automated process. These systems are also inflexible in that they create bit-mapped images of documents, which cannot be edited. Few provide automatic routing.

Process automation workflow requires a top-down approach. System designers and business managers analyze the process of work, then build the system based on that model. Once defined, the system is inflexible; it cannot be changed on the fly, forcing workers to adapt to a computerized process. AT&T's Rhapsody and NCR's Cooperation both implement this model.

Rather than automating the entire process, task automation takes the opposite tack and automates the individual tasks that make up the process. This is more of a bottom-up approach. Hewlett-Packard's NewWave Office is one such application, since it enables users to build scripts to accommodate tasks.

However, both process and task automation assume that a process is in place. If the work is accomplished without a formal process in place, neither type is helpful. In this case, workgroup or ad hoc tools are better. Workgroup tools give the power to the user, rather than to the top-level manager. Beyond's BeyondMail, which allows users to build in rules for routing e-mail, is one example. With workgroup tools, workers can define their workflows to their liking. Although they can customize their systems, workflow automation places the burden of system design and administration on the user. This approach is fine for power users, but tentative computer users may balk.

PARTS OR WHOLE?

No clear product strategy or vendor has emerged. Workflow products will come from the traditional office automation vendors, e-mail providers, vertical applications developers, and others.

Some vendors offer a complete system usually built around a shared database. Very often these workflow systems include bundled or encapsulated applications, such as e-mail, word processing, and print spooling. Essentially, these developers move the model of traditional office automation software into the world of distributed, networked intelligence. AT&T and NCR take this tack.

Other developers are moving toward a component approach. Instead of supplying the whole package from database to workflow, these vendors offer the individual pieces. They may offer one or more front-end pieces that work with other company's workflow database back ends. This pay-as-you-go approach enables corporations and network designers to pick and choose. While this approach is more flexible, it requires the user to act as an integrator, assembling the many pieces of the system.

Much of the current workflow fervor emanates from the electronic mail vendors. Mail-enabled applications are leading the way to workflow computing. The simplest of these provide intelligent mail routing that can be expanded to include routing documents and forms. These applications typically use an e-mail engine or back end of e-mail storage, routing, and directory services. Functionality is added to the e-mail front end until it becomes a full-featured workflow application.

Electronic mail vendors alone can't move the industry forward. Expect to see database vendors team up with e-mail vendors to provide expertise in delivering distributed databases. And to make intelligent routing truly useful, network operating system companies must improve their directory services so applications and users can locate and access networked resources.

SLOWING THE FLOW?

Workflow software is in its earliest evolutionary stages. Office automation software for the mainframe and minicomputer have existed for quite some time; PC groupware products have emerged, but neither offers enough functionality and flexibility. The next generation, exemplified by products such as Reach Software's MailMan or Beyond's BeyondMail will provide the basis for true workflow.

The movement to workflow computing won't happen without pain. Before such systems can become a reality, networks must be built that can mask the differences of incompatible computing systems. Vendors must provide open systems solutions, not proprietary platforms. The kinks of client-server software have to be ironed out.

Many of the challenges are not technical, but rather political or cultural. The impetus for workflow computing must come from a company's uppermost management, since it automates the very heart of the business. Only top management can ensure the steady progress and development of such a system. Also, employees may balk at such systems, which in the wrong hands can be used as a Big Brother monitoring system.

Nevertheless, the benefits of workflow computing can be great. Companies can eliminate steps that were necessary in paper-based systems but that have become wasteful in a world that conducts business online, thereby reaping the profits of just-in-time business.

This tutorial, number 43, by Patricia Schnaidt, was originally published in the February 1992 issue of LAN Magazine/Network Magazine.

Messaging-Enabled Applications

E-mail is a form of communication equal in utility to the telephone, fax, and paper mail, but e-mail is much more than sending memos or attaching files. Messaging is a transport for delivering information, and the next generation of messaging software will provide the underpinnings for applications that deliver new productivity gains to information workers.

Not only has this year's workgroup software been designed for a group of users, it is also suitable for the business-line applications, such as accounting, inventory, and materials planning. Scheduling, intelligent mail, forms, and workflow are a few of the applications built on the messaging infrastructure (see Figure).

A MESSAGING INFRASTRUCTURE

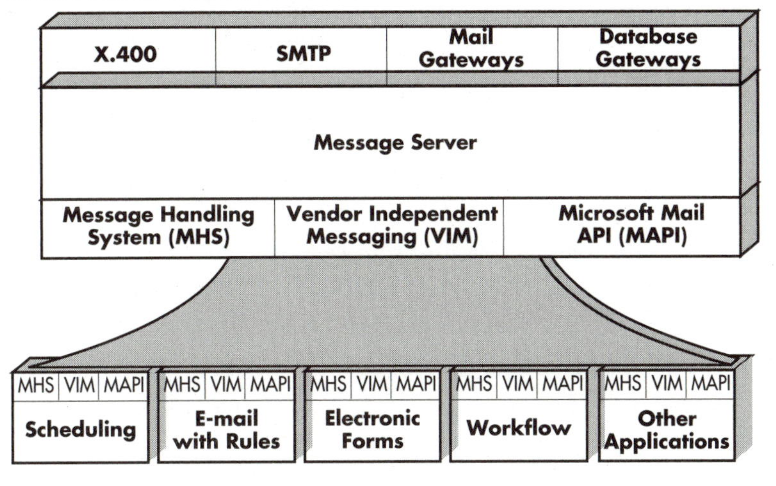

■ **The Message Backbone:** The next generation of workgroup software, which includes scheduling, intelligent mail, electronic forms, and workflow, will be built on a messaging infrastructure.

SCHEDULING

Scheduling and calendaring are typically the first applications to be added to a messaging backbone. Users can employ electronic calendars within their e-mail to set appointments for themselves and others. A scheduler vendor and an e-mail vendor may bundle their software, or a single vendor can write both applications. The more tightly integrated applications use APIs such as Novell's Message Handling Service (MHS), Microsoft's Mail API (MAPI), or Lotus' Vendor Independent Messaging (VIM).

INTELLIGENT MAIL

In an e-mail-reliant corporation, users can become inundated by the volume of incoming messages, many of which are noncritical or unimportant. Do you really care who has a used

Newton for sale? Or if the cafeteria is serving fish sticks on Friday? If e-mail stops enhancing productivity, people will stop using it.

One way for users to handle the e-mail barrage is to ask the software for help. Intelligent e-mail can make message-handling decisions for the user based on a set of rules, such as the message's urgency, the sender's identity, the subject matter, or keywords. Intelligent mail can prioritize, discard, or reroute the messages based on user-defined criteria.

Consider how easily users and administrators can configure and use the rules in any intelligent e-mail package. Users may have difficulty understanding how to configure the rules or how to change the rules for special circumstances, such as when they travel.

FORMS AND WORKFLOW

Electronic forms, or e-forms, also take advantage of a messaging transport. A company typically ventures into workflow with e-forms, because corporations are looking for ways to reduce the volume of paper and because the routing paths and interfaces for forms are straightforward. Many companies start with e-forms and workflow routing for expense reports or purchase order requisitions. Companies such as JetForm (Waltham, MA) have led the way, but WordPerfect has entered the market with its InForms.

If you implement e-forms, you will have to hurdle the security and authorization issue. You can't authorize an e-form by typing your name, because anyone could type your name, and it would look the same. Typing your name is the digital equivalent of marking a big "X" in the approval box. You must have a secure form of digital signature. Companies such as RSA Data Security (Redwood City, CA) are tackling this issue.

Electronic forms can also handle the results of database queries. With e-forms for data access, you can turn the query results into an e-mail message. For this setup to work, the e-mail front ends will have to understand SQL and the database APIs, or the database front ends must understand the mail APIs. BeyondMail 2.0 has this capability when used with Lotus Notes.

Workflow's promise lies not in simple tasks, such as routing travel itineraries or filing expense reports; the real utility resides in large-scale corporate applications. Workflow software can automate business processes so the software can routinely decide where to send documents.

For workflow to be effective for business-line applications, the system must be scalable and robust. It must be useable in a wide area network as well as with mobile employees. For these goals to be achieved, messaging and database systems must adopt some of each other's characteristics.

RIVAL ARCHITECTURES

As companies disperse their operations across the globe, they need a way to transport their corporate information. While "transport" implies a messaging infrastructure, business-line applications are primarily database oriented. For an application to be successfully distributed, it needs to adopt parts of a messaging architecture. Distributed applications need to allow access to information from a variety of locations, but users typically don't need quick response times. How updates and changes are distributed, or replicated, is critical to the success of an enterprise workflow system.

Whether you choose a database architecture, a messaging architecture, or a blend depends on your particular needs. For example, for an airline reservation system, a database architecture is better, because a fixed set of people is accessing a fixed set of information. Databases provide fast and robust access to structured data. Don't overlook that databases are a mature, robust, and secure technology with lots of add-on tools and experienced programmers and managers.

But a database architecture isn't always suitable. For example, for an information-publishing service, many different people will be downloading the information, which could be stored in files of any size and type, with graphics and video. Because databases are not flexible in handling unstructured data types and because many paths of communication exist, a messaging structure is better.

When you have a distributed system, you need an efficient way to update each site. A database, because it is built for real-time access, must propagate the changes as they occur or shortly thereafter. But people have a different set of expectations and needs for messaging. Because people don't need the messages instantaneously, the system stores the messages, then forwards them at a slightly later time, whether that interval is every half hour, hour, or day. The slower rate of information exchange makes distribution less costly in a wide area network.

The issue of message storage needs to be resolved. The client's capabilities are tightly coupled with the message storage's functionality. For example, the types of acceptable file formats, such as voice, video, and data, hinge on the message store's capabilities. Some messaging vendors say the message store belongs on the network, while others say it belongs on the client. For example, with Apple's Open Collaborative Environment (AOCE) and Microsoft's Windows 95, the message store and the file store are merged and reside on the client. With MAPI, you can do it either way. Banyan (Westboro, MA) says the message store should be server-based. Having the message store on the client makes users more mobile; having the message store on the server makes the system more robust.

IMPLEMENTATION ISSUES

The highly useful workgroup applications are not shrink-wrapped applications but rather require programming. Whether you rely on systems integrators or develop the experience in-house, customization is expensive. Your business-line applications must justify the development costs. Here's the conundrum: If messaging-enabled applications remain heavily dependent on integrators' expertise, the applications will be available only to the company's most highly leveraged workers. But for workflow to be useful, it has to be deployed across a corporation—it has to be as accessible to the receptionist as it is to the traders.

Lotus Notes is clearly the most successful workgroup software environment, and its base of third-party applications is expanding rapidly. Notes does require customization, but it is an extremely powerful tool.

Shrink-wrapped applications are a must. The first applications to be bundled will be the application suites owned by one manufacturer, such as Microsoft, Lotus, or WordPerfect, since including workflow in the package is considerably easier when one company owns all the applications. Next will come applications that adhere to industry-standard APIs, so users can mix and match.

THE VIRTUAL CORPORATION

The end game of workgroup software is that information sharing will become part of a company's workflow. Information will come from users' personal productivity applications as well as from corporate message servers, database servers, and information servers. If the industry can get information sharing to work, then users can build virtual corporations. Company-to-company communication stands to gain the most productivity from messaging.

Companies create individual departments to increase workers' productivity, but anyone who has worked in a large corporation knows that having departments isn't the most efficient way to work. But what if people in different locations and companies could work together for a project's duration, then form another team for the next project, all the while operating as if they worked in adjacent offices?

This tutorial, number 65, by Patricia Schnaidt, was originally published in the January 1994 issue of LAN Magazine/Network Magazine.

Threaded Conversation Databases

Groups that work together on a project or process often wish to record, share, track, and refer back to their work. Producing a manual (or a magazine), tracking a complex order or service transaction, and storing histories of problems and their solutions are just a few ways that strung-together chains or threads of conversations have proven their usefulness to organizations.

The basic feature list for a product that collaborators would find useful includes:

- a central storage location to maintain consistency;
- a structuring system that captures messages and responses in order and in the right context;
- remote access to the message store for users;
- control over privacy and access;
- the ability to move, reorder, delete, and duplicate messages;
- the ability to attach nontext documents, such as spreadsheet files, images, and audio clips, to messages and the ability to view or otherwise access them.

Two unrelated technologies—databases and e-mail—are tempting prospects for the job of recording, sharing, and retrieving collaborative work. Unfortunately, both candidates are not quite right for the job, though it's not immediately obvious why.

WHY E-MAIL CAN'T HANDLE IT

E-mail is a point-to-point technology; an e-mail post office is a transient storage facility, designed to move messages in and out rather than to keep them in one place. If only two people work on a project, an e-mail program with a reply function can serve as an adequate way to track the conversation. Each collaborator has a complete copy of both sides of the ongoing dialogue as the project moves forward. But there is not a single location for the data; responsibility for saving it, much less backing it up, can be clarified only by rules external to the e-mail program.

Now consider what happens with three collaborators, J, K, and L. First of all, there has to be an external rule that each collaborator will cc: the person they are not sending the message to: J messages K and cc:s L. Assume the best case—only one collaborator begins the process. Even if everyone responds to messages in a sort of token passing order, there will be at least one extra message with each "generation" of messages. Of course, in any actual situation, collaboration will not follow a strict synchronous turn-taking model, and two people will reply simultaneously to a message. Then the problem of manually incorporating extra messages into the two-way model thread begins to get complicated.

The number of superfluous messages in each generation increases with each additional collaborator. More importantly, the problems of synchronizing responses, manually reordering messages (which are, after all, simply sections of text with a message header), and navigating a long text document without markers and jumping-off points are subject to a sort of combinatorial explosion with additional collaborators. In a very short time, the collaborative message store, which is no more than an undifferentiated text file, will become unmanageable. Using e-mail to track a collaboration involving more than a handful of active participants requires a mass of external rules and a great deal of administrative effort.

CAN DATABASE FUNCTIONS HELP?

Databases have the advantage of storing data in a single place (or in a manageable collection of places). There are two familiar and well-understood database models that can be useful for storing textual data—field-oriented databases and freeform text databases. Field-oriented databases can have their components linked together logically, based on common fields (the relational model), or tied together with pointers (the hierarchical and network models). They are most useful for storing and retrieving sizable chunks of text when each of the chunks or messages is tied to a record with other, relatively short, fields, which can be used for indexing or sorting the database. (Sorting a large collection of messages by their titles or subjects, much less by their contents, would rarely be a useful way to present or browse them.) Typical field-oriented databases have limited functions for searching or otherwise manipulating the content of long text fields.

Free-form text databases usually make no attempt to break up a large body of text; they simply index all the nontrivial words in the text and expedite locating those words. Using Boolean operators to search a massive text can be very fruitful. For example, in a series of messages that track the steps of a LAN enhancement project, searching for "(Category or Type) and (3 or 4 or 5 or coaxial)" should find any messages discussing Ethernet cabling specifications. Such searches work well regardless of how skillfully the subject of the message was named and how well the message focused on a single subject.

Without field-oriented functionality, however, an indexed freeform text file is hard to navigate and only amenable to collaborating people with external rules and a substantial amount of external administration—for example, putting dates on messages and ordering them.

An additional tactic for categorizing pieces of text is assigning keywords—words that characterize the content of the text in some way and can be stored and searched separately. (Keywords can also be useful for keeping track of images and sounds that are not made up of text.) Though very large indexed text databases can be searched fast, the use of keywords can provide even faster performance.

Both database technologies can be useful for tracking threads of messages, but the messages need to be loaded into the database and presented to the collaborators. Establishing threads, notifying collaborators of updates, presenting maps of threads, and many other desirable functions would require custom programming in any of the standard database programs.

The combination of messaging and database functions has converged in several unrelated environments. The first widespread application of threaded conversations was for online bulletin board systems. Usenet newsgroups are threaded conversations. Forums or conferences on the commercial online services—CompuServe, Prodigy, and GEnie—also take this form. The most ambitious and elaborate commercially available product for tracking threaded conversations is Lotus Notes (Lotus Development, Cambridge, MA). Several other products, including Collabra Software's (Mountain View, CA) Collabra Share, Attachmate's (Bellevue, WA) OpenMind, and Trax Software's (Culver City, CA) TeamTalk have also been introduced to provide this function.

BASIC FUNCTIONS OF A THREADED-MESSAGE DATABASE

Displays. Entering messages must be straightforward, so a user who opens or logs on to the conversation database ought to be able to see a map of the conversation. A hierarchical display, similar to an outline, is a useful way to display sequences of comments and replies. Preset limits to the number of potential levels are probably undesirable. The ability to collapse some or all of the levels helps keep the overall structure clear.

Remote participation. Collaborators must be able to reach the message store over some kind of network. In addition, a gateway to regular e-mail products can let collaborators who don't have full access contribute.

Thread editing. People make mistakes. It ought to be easy to relocate a misplaced message or delete completely irrelevant ones. Actual messages don't always fit precisely in a threaded hierarchy. For instance, a single message may include two or more topics. It should be easy to duplicate a message and insert the copies at the places where they fit.

Scrolling. For large, ongoing threaded conversations, such as those on commercial online services, the day will arise when the space that the system has available runs out. For example, there may be a fixed number of message units allocated to a forum; when these slots are filled, each additional message bumps off an older one. This process is referred to as scrolling. A whole series of problems can ensue. It would be desirable to have configurable scrolling settings so that factors such as the message's position in a thread, the relative importance of a thread, or the author of a message could be factored into the scrolling behavior.

Once any message in a thread scrolls off, the meaningfulness of subsequent reply and commentary messages is apt to suffer, even if the title of the thread continues to be carried along.

THREAD DRIFT

A problem that fancy software features are not likely to solve soon has to do with the content of messages and the subject that is assigned to name a thread. Participants in online discussions sometimes refer to "thread drift" ironically, when they really mean that someone is drastically changing the subject.

Nevertheless, discussions can imperceptibly veer away from the original topic until the original name of the thread is a completely misleading guide to the content. More generally, there is nothing to stop participants in an unmoderated threaded conversation, or unskilled moderators, from using arbitrary subject names and wrecking the coherence of a conversation. Until the day that computers have the ability to interpret the meaning of sentences and paragraphs, which is nowhere in sight, external rules and human moderators/ censors/editors will be necessary to keep threaded conversations coherent to their users.

This tutorial, number 78, by Steve Steinke, was originally published in the February 1995 issue of LAN Magazine/Network Magazine.

Content Distribution and Internetwork Caching

At some point in the not-so-distant future, moviegoers will walk into their living rooms, plop down in front of a Web-enabled device, and watch a movie off the Internet. Or so it will seem. In fact, the movie will be delivered from a local server in viewers' homes, pushed there by a video server during the middle of the night.

Applications like this one are the promise of the new content distribution systems coming online. Combining elements of caching, content selection, and application development, these systems will enable profound, new, network-aware applications to make intelligent use of caches installed in the network. Little wonder that scads of vendors and providers are looking to jump into the space.

All this promise is a good reason to bone up on content distribution basics. Start off by understanding what people mean by caching, the precursor to content distribution. Then make sure you've got a good read on the benefits of caching; there are three reasons to go with caching, and not all are about boosting performance. Next, learn how caching works, and how content is kept up to snuff. Performance here is particularly important, and while many vendors might tout top-notch numbers, the cost of that performance may vary widely (see "The Price of Performance,"). Finally, find out why content distribution might be just the answer for certain applications.

SQUARE ONE

There are two basic types of network caches, though each performs the same function: copying popular content from an originating server so that it can be accessed more quickly by the user. Prepackaged cache appliances are the first type, bundling hardware and software together. Today, these appliances are installed in the carrier or enterprise network, but already vendors are talking about diminutive versions of these appliances for the customer premises, loaded with some 20Gbytes of memory.

The second type is caching software, which is likely to stay targeted at enterprises, as it requires users to purchase the necessary server hardware and install the software themselves.

The primary benefit of both types of caching is better performance. Moving content closer to the user reduces the number of router hops it has to traverse, so performance improves. Also, just cutting the distance between user and content speeds up the connection. Finally, a local cache server may be less loaded and better designed for performance than the originating content server. These factors cut page-load times by one-half or one-third of the time needed to retrieve the page from the original server.

While numbers like those might be reason enough to implement caching, there are other critical benefits. For starters, cache servers protect against the problem of instant popularity. Major events such as the release of the Starr Report or an Internet broadcast of a Victoria Secret fashion show cause huge traffic surges. Masses of users request a particular Web page, which swamps the server. Opening the network with faster optical or local loop infrastructure will likely only exacerbate this problem. However, replicating this content to other servers or caches distributes the traffic load around the Internet, alleviating the server bottleneck.

ISPs also win with caching and content delivery by reducing their reliance on costly Internet links. According to the Internet Research Group (www.irgintl.com), bandwidth at the Internet's core costs $800 per Mbit/sec per month, yet users at the edge pay only around $50 per month for 1.5Mbit/sec Digital Subscriber Line (DSL) links. By using caching, ISPs keep more traffic local, using less core bandwidth and narrowing this gap.

Finally, caches can be deployed locally to provide better oversubscription ratios. When carriers design their networks, they use mathematical models to balance the number of users vs. the amount of provisioned capacity to support those users. Rarely is there a one-to-one correlation between those metrics. This is particularly true of cable modem and DSL service providers, where the head-end links can easily be swamped if all users start downloading at once. Cache servers offload traffic, effectively enlarging the upstream pipe and preventing local saturation.

JUST THE BASICS

The key then is to shorten the distance between the user and the commonly requested content—not a particularly new science. PC design has long used fast memory caches to feed high-speed processors. Similarly, Web browsers use local-disk caching to improve local Web performance. This is why jumping back to a previous Web page gives the impression of much faster performance.

Network caches work on the same basic principles, but on a much grander scale. Unlike a private cache in the Web browser, however, the network caches select content based on the activity of hundreds and thousands of users.

Cache servers learn these patterns by intercepting user requests in one of two ways: transparent cache servers or proxy cache servers. Transparent cache servers sift through all traffic and hence require no modification to the end client. The key is that these cache servers see all of the traffic destined for the Internet. This means locating the cache server at a choke point in the network, like in front of a router connecting to the Internet. Alternatively, requests can be diverted to the server using a device like a layer-4 switch.

Then the cache server checks to see if the content is stored locally or not. If the content is already stored locally, it is delivered to the user. If not, the content is retrieved from the origin server, sent to the user, and possibly cached for future reference.

With proxy cache servers, network managers configure users' browsers to direct content requests directly to the cache. The proxy cache server then requests the content on behalf of the user. On the one hand, this lets network managers restrict the sites a user can visit. At the same time, though, this approach is obviously far more complicated, as it requires configuring each client. What's more, if the proxy cache fails, users cannot access the Web (see Figure 1).

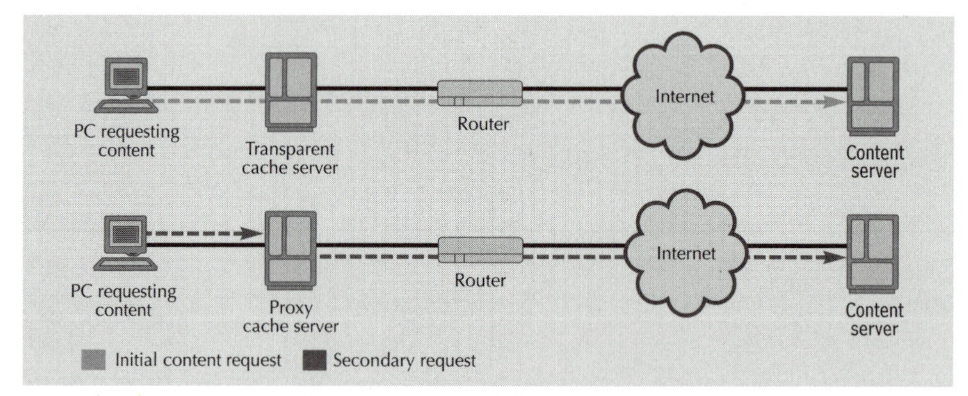

■ **Figure 1:** There are two types of cache servers on the market today. Transparent cache servers require no changes to the client. They sift through the Internet-bound traffic and either deliver stored content from cache or send the request on to the content server. Proxy caches require changing the client to receive all Internet requests. Content that is not stored in the cache is then retrieved for the user from the originating content server.

GUARANTEED FRESH

Simple enough, but how does a cache know when to discard content? Today's cache servers use either passive or active caching. With passive caching, caches check with the originating server about the freshness of the content, using the get if modified command in HTTP. Normally, a cache server uses the get command to request an object from a content sever. However, if the object is already stored, the cache server uses the get if modified command, which only pulls down the object if it has been modified since the last request. Then the cache server can compare the modification dates of the object from the server and the one in the cache, and serve up the newer one to the user.

The problem with passive caching is performance: Users have to wait for the cache server to check every request. Passive caching can be improved by checking freshness data, such as the time expiration date, in the object's header. When the object reaches a specified age, the cache server fetches fresh content.

Active caching improves caching performance by using heuristics to assess the life expectancy of an object. The server can calculate when an object entered the cache, how long it's been held, the origin IP address, or myriad other factors.

The big benefit here is that each request doesn't have to be checked. Instead, the cache server makes certain assumptions about the time to live for an object—say, two days, for example. During that interval, all requests for the object are serviced immediately from cache. Once those two days pass, the cache refreshes the object.

CONTENT DELIVERY

Content delivery is in effect another approach for keeping content fresh. While caching waits for customers to request information, content delivery lets companies proactively push the information into caches close to the user.

Here's how content delivery works. First, content providers designate content to be replicated to network caches. Here, the network manager either provides a list of elements to be replicated by the content delivery provider or uses tools that troll through the site, marking the elements to be replicated. The content provider then replicates those elements around the globe. Content consumers are then directed to the nearest server through DNS.

Geographic penetration of these caches is critical to the success of the content delivery system. The more caches deployed, the greater the likelihood that users will be able to retrieve a document or element from a cache located geographically nearby.

The sheer number of caches, however, isn't the only critical factor. A successful content delivery provider needs a high density of geographic coverage. With more caches in a given area, there's a better chance of finding a cache near the user. For example, Akamai claims to have 4,200 caches across 50 countries around the globe.

Many content delivery providers differentiate themselves by how they bring the content to the customer. Most replicate content across the Internet, but this can result in all sorts of performance problems. Some providers take a more costly approach, delivering their content via satellite into nodes comprised of satellite receivers and caches connected to the Internet. One start-up, Fireclick, preloads objects directly into a user's cache. Although this

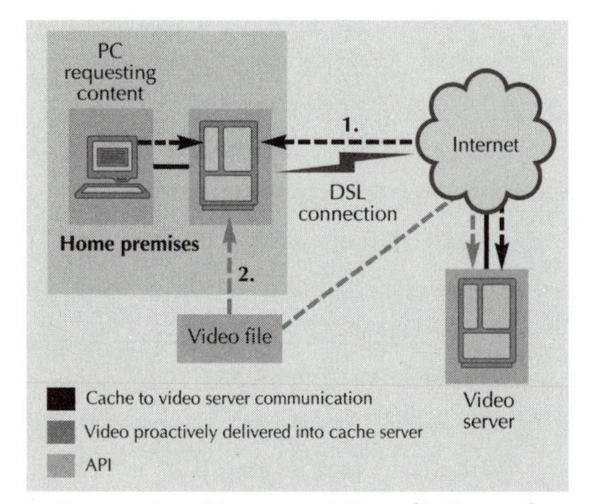

■ **Figure 2:** By writing custom APIs, applications such as video retrieval can be more intelligent, improving the user experience. First, the personal cache server shares the user's movie preferences with the video server (1). Then, during off-hours, the video server transfers movies likely to be watched by the user into the user's personal cache (2).

may well improve performance, users are likely to balk at the idea.

Many content distribution systems are also promulgating their own APIs to encourage developers to create network-aware software. Like the video server pushing a preselected movie into the user's cache (see Figure 2), or a main office distributing training videos into the caches of the field offices, these applications can be more intelligent about how they utilize network resources.

The result is a content distribution system that begins to look like an interactive and predictive television system. Users can request certain types of content through a Web browser and have it loaded into the cache. Even when users do not request the content, the different elements of the delivery network can work together to anticipate user needs. The cache server can monitor user preferences and send feedback to the content provider. The service can then issue the request to the content server to distribute the most appropriate material at the optimal part of the day.

THE PRICE OF PERFORMANCE

It should be pretty clear by now that keeping up with content requests is no mean feat. While the number of network caches available on the market has exploded, the price/performance ratio remains a critical issue.

Cache server vendors talk about content delivery performance in terms of the object ratio and byte hit ratios. The object ratio compares the average number of objects sent from cached data to the total number of objects served from cache. The byte hit ratio compares the average size of those objects served from cache against those objects delivered.

Beefing up cache performance to improve either of these factors is usually a matter of adding more memory and hardware. Your best bet is to emphasize memory, as it can be much faster than retrieving data from a disk. However, many vendors get great performance numbers by going a step further and clustering servers together. While this can boost performance, it can also dramatically increase the price tag.

Today, two protocols enable cache clustering: the Internet Caching Protocol (ICP) and the Caching Array Protocol (CARP).

With ICP , cache servers can pool their resources to detect if objects have been cached in one of the other servers. However, this approach requires additional instructions, which can affect performance. Cache server vendors can avoid the performance issues by using layer switches to link the servers, which in turn can cost hundreds of thousands of dollars each.

CARP solves the problem by running a hashing algorithm on the main network cache to determine the network caches in the array that should receive the URL request. The only snag is that the hardware can't forward network traffic as fast as a switch.

RESOURCES

There are a number of useful caching resources on the Internet: www.caching.com provides a good overview of how caching functions; www.web-caching.com also provides a great site for introducing caching and listing vendors that offer content distribution services, as well as offering numerous links to research, books, and other sites.

Speaking of books, consider *Web Proxy Servers* by Ari Luotonen, published by Prentice Hall. Duane Wessels will be releasing a book appropriately titled, *Web Caching*, published by O'Reilly and Associates in January 2001.

If implementations are what you're after, there are loads of vendors pushing caching and content delivery software. If open-source software is your beat, consider www.squid-cache.org, the site for Squid open-source caching software. It might not be the most scalable code out there, but it's free. For more info on Squid, ask the folks at IRCache (www.ircache.net); they've got Squid's original developer on staff, they run bakeoffs, and they do all sorts of workshops on caching.

This tutorial, number 148, by David Greenfield, was originally published in the November 2000 issue of Network Magazine.

Edge Side Includes

Caching and content delivery has been one of the few areas to experience strong customer acceptance in the last few years. Though growth has been down recently, feedback from early adopters is encouraging, and has validated the art of storing frequently requested static Web pages closer to the end user.

Static content—Web pages that don't change over a long period of time—is easily cached for quick delivery to end users via a content delivery network. But Web pages based on dynamic, frequently-updated content (such as sports scores) and personalized or customized content (such as localized sports scores) are in growing demand. These pages are usually created in a data center by a company's origin server, a database containing Active Server Pages (ASPs) or JavaServer Pages (JSPs), and a Web application server, which must work in real time, making proactive caching difficult.

This procedure entails a lot of processing at the origin server, which must format and deliver the data to the browser, hogging I/O, network, and memory capacity. The process becomes painful when repeated for every request, since the processing overhead expended to regenerate an entire page is very high. A bottleneck can occur as the origin server becomes overwhelmed, resulting in slow downloads or crashes, and ever-growing numbers of servers and load balancers must be deployed to right the balance.

THE PROPOSED SOLUTION

When in doubt, decentralize. Edge Side Includes (ESI) is a specification accepted by the World Wide Web Consortium (W3C, www.3w.org) that describes a means to push dynamic content from the origin server to multiple edge servers, closer to the end user. Offloading the origin server increases download speed and limits crashes because the origin server doesn't have to redesign Web pages every time a page element is updated—a process that doesn't scale to handle large numbers of simultaneous users.

ESI is the result of a collaborative initiative between Akamai (www.akamai. com), Digital Island (now Exodus, www. exodus.net), Mirror Image (www.mirror image.com), IBM, BEA Systems (www. bea.com), Oracle, and a handful of other server and Content Delivery Network (CDN) vendors. While Akamai is the ESI leader with over a hundred EdgeSuite (which includes ESI) customers, other CDN providers, such as Speedera (www. speedera.com) and Digital Island, have announced ESI-type services.

WHAT IT DOES

Essentially, the ESI specification is an Extensible Markup Language (XML)-derived language that tags and formats Web pages on origin servers, creating a template. The template is in turn composed of HTML fragments, which are specific pieces of text or customized content that can be treated separately from other Web page elements. A fragment can be a banner, a targeted ad, or a personalized welcome statement for example.

Most pages have fragments that are cacheable (static) and uncacheable (dynamic). However, to complicate matters, some fragments that are uncacheable may contain sub-fragments that are less dynamic and so can be cached. ESI tags indicate which fragments of a Web page can be pushed from the origin server to multiple servers at the CDN edge. The origin server then distributes the dynamic but cacheable content to multiple edge servers, where subsequent requests for that same content are sent until that edge-cached data is no longer valid.

ESI treats each fragment as a separate entity, each with its own cache and access profile. These parameters govern a fragment's Time to Live (TTL)—that is, the period of time in which the information contained, such as a sports score, is valid. While that particular fragment is valid, Web page requests can be served from the edge servers. When that fragment's TTL expires, the origin server is called upon to refresh the edge cache with the latest information.

Thus, the end user is much closer to the desired Web page information, and the usual caching benefits apply—even with supposedly uncacheable content. In broad terms, ESI seeks to separate the normally combined tasks of content assembly from content delivery into two separate processes that can be individually managed.

HOW IT WORKS

The ESI specification stands on four legs:

First, the language specification tags the template so that reverse proxies can understand ESI. ESI's XML-based heritage also enables interoperability.

Second, the Content Invalidation Specification (CIS) defines the rules used to invalidate data stored on the edge server, so that the origin server can send out new content to overwrite the outdated data.

Third, an Architecture Specification provides methods for HTTP intermediaries to control content.

Finally, Java ESI (JESI) Tag Library Specification provides the Java-written application-level interface that can communicate with ESI tags, as many dynamically assembled pages are Java-based.

These components are the building blocks for ESI's four essential functions:

1) Inclusion, which is the ability to fetch and include files that make up a Web page. Each file has its own configuration and control, TTL, and revalidation rules.

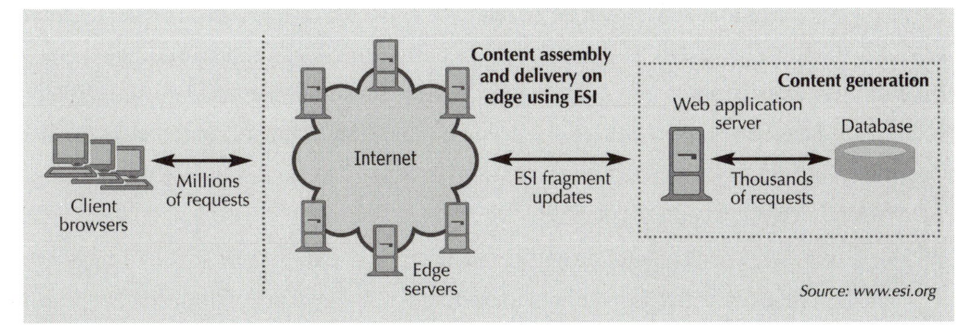

Pushing Content Delivery to the Edge. Here's a look at how Edge Side Includes (ESI) moves content from the database and Web application server in the data center core and pushes it towards the end user at the Content Delivery Network (CDN) edge.

2) Conditional inclusion, which provides conditional processing based on Boolean comparisons or other programmable variables, so that rules such as how a template is processed can be modified as needed.

3) Environmental variable support that allows a subset of standard Common Gateway Interface (CGI) variables, such as cookie information, to be used inside ESI statements or outside of ESI blocks. (CGI is a standard for external gateway programs to interface with information servers such as HTTP servers.)

4) Exception and error handling, which allows developers to specify where to send the browser if an origin site or document isn't available. Alternative pages or default behavior can be set for every fragment that forms a particular Web page. This same logic also supports what Akamai calls "an explicit exception-handling statement set:" If a major error occurs while processing an ESI-enabled Web page, the content returned to the end user can be specified in a failure-action configuration option associated with that ESI document.

As mentioned above, the CIS is central to ESI. It instructs syndication servers, content management systems, databases, custom scripts, application logic, and so on to send HTTP-based invalidation commands to the edge server and delivery network with instructions to overwrite the old metadata with fresh content from the origin server. (The origin server connects with the ESI intelligence over an optimized link.)

This setup has two advantages: First, Web site managers need only update content at the origin server, not at each edge server. The specification also ensures that old content is purged from the Web page in use in real-time, so that customers don't see outdated data (such as an out-of-stock item in an online purchase catalog) during the updating process, regardless of the number of edge servers that contain the old data. In other words, the revalidation process is distributed throughout the network. Administrators can integrate the CIS into their content management system using a database call, a script, or a number of other site administration methods.

This boils down to a simple way to enable Web content personalization at the network edge, whether on a private local server or a delivery network. More importantly, the caching rules involved in page assembly can be based on user agent and other header values, cookie values, the user's location or connection speed, or other programmable parameters. It's not clear, however, if these parameters can be customer-defined without a lot of vendor help.

Here are a few sundry points on ESI:

- Single Web pages don't have to be entirely marked up in ESI. Cacheable and uncacheable content can coexist on the page without impacting the performance of those fragments that are dynamically forwarded to the edge. It makes no difference to the proxy servers involved.
- ESI, as offered by Akamai, can run data compression on loads between the origin and edge

server as long as the requesting browser supports this compression. If it doesn't, the edge server can decompress the new content and send it to the browser uncompressed.

- Cached templates and fragments can be shared among multiple users so that many requests for popular, dynamic content can be fulfilled by using shared components delivered from the edge. For example, say a user requests a stock update for a particular company from a financial investments firm. The update is a dynamic, uncacheable fragment of a greater Web page that's cacheable. But once the fragment is updated, that revalidation is sent to each user's page that references the same fragment.

IT'S ALWAYS SOMETHING

Though ESI is promising, and most indications from early customer wins have been very positive, there's room for improvement.

ESI requires retagging the site with appropriate labels and commands that may require better tools that minimize the need to do this manually. Akamai already does this with its "Akamaize" tools that tag static content for the vendor's content delivery system.

"ESI demands retagging the site, and it's not as good for ASP sites," says Peter Firstbrook, a senior research analyst with the META Group (www.metagroup.com). "For most users, ESI might involve at least some site redesign."

Firstbrook also says that "[ESI] can't handle occasions when the page itself, not just the content of the page, is truly dynamically generated. It's very limited in terms of cache revocation, where vendors such as SpiderCache [www.spidercache.com] and Chutney [www.chutney.com] have lots more capability."

Firstbrook is referring to the standard way cached content gets served until its TTL expires—the only way old data gets dumped.

"But suppose that content was a sports score?" he says. "With such a dynamically changing bit of content, you either have to give it a very short Time to Live or not cache it. It could be valid for two seconds or two hours, depending on the game. So you want to be able to eliminate it from the cache if the score changes. Revocation is the method by which you can tell the cache to dump the score and get a new one from the origin site. This is the biggest problem with dynamic data."

A less dramatic point to consider is the need for buy-in from the developer community, which is tasked with learning and implementing the technology. Developers also want clarification on how ESI works with other Web markup languages, such as Server Side Includes (SSI, an ESI-based Speedera service similar to Edge Side Includes). These languages should be complimentary, but if several are processed by the same device, there must be a way to decide which markup language is paramount.

These hurdles are significant, but not deal-breakers for many businesses' e-needs. Look for further evolution in the ESI market, and increasing competition from other (non-Akamai) CDN providers as they ramp up their own ESI services this year.

RESOURCES

For an overview of Edge Side Includes (ESI), start with the main ESI site, www.esi.org.

For code samples, user guides, and testing tools, click on http://developer.akamai.com.

"Emerging ESI: Lower Costs, Better Performance," by Lori MacVittie, details step-by-step template markups and explains the programming commands involved. Go to www.networkcomputing.com/1301/1301ws1.html.

This tutorial, number 169, by Doug Allen, was originally published in the August 2002 issue of Network Magazine.

SECTION VIII

Internetworking

Internetworking

As local area networks become more and more prevalent and increasingly vital to the daily operation of an organization, the need to connect multiple LANs together has become as crucial as it once was to link individual PCs into a workgroup. More and more, it's likely that a worker linked into a firm's marketing department workgroup requires access to resources located on another LAN within the company—a database in the engineering network, for example.

This need has spawned one of the fastest growing areas of the LAN industry: The *internetworking* marketplace, composed principally of repeaters, bridges, routers, gateways, and, most recently, hybrid products called brouters and routing bridges. Internetworking products bring interconnectivity to workers linked into large, spread-out groups of LANs. They also play a major role in network management by allowing network administrators to segment, or divide, a single network into an assembly of multiple subnetworks. This subdivision can improve network performance—limiting the number of nodes on a network can reduce traffic over the workgroup wiring. It also facilitates security—internetworking allows restricting individuals to specified resources—and increases system reliability—when one workgroup goes down, it doesn't affect the entire network.

There are four primary types of internetworking products: repeaters, bridges, routers, and gateways (see Figures 1–4). (Beginning in 1994 or so, multiport bridges began to be marketed as switches, but switches usually provide the same fundamental functions as the devices traditionally known as bridges. In some cases, switches actually perform local routing functions as well.) Each internetwork product permits various levels of communication between individual networks; each also functions at a separate level within the OSI model.

REPEATING THE OBVIOUS

Repeaters offer the simplest form of interconnectivity. They merely regenerate, or repeat data streams (in reality, electrical signals) between cable segments. In their purest form, repeaters physically extend a network; repeaters operate at the Physical layer of the OSI model. Repeaters, for example, allow extending Ethernet network cable segments from 1,000 feet to more than 5,000 feet. In addition, they provide a level of fault tolerance by isolating networks electrically, so a problem on one cable segment does not affect other segments.

■ **Figure 1:** Repeaters operate at the lowest OSI layer. They regenerate electrical signals.

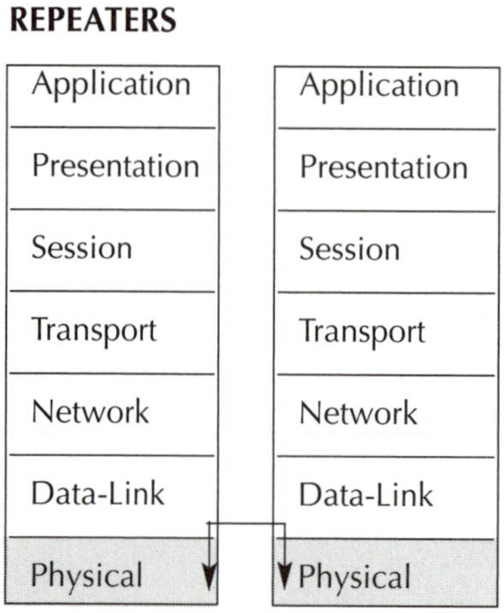

■ **Figure 2:** Bridges operate at the MAC sublayer and are capable of modest traffic control and network partitioning.

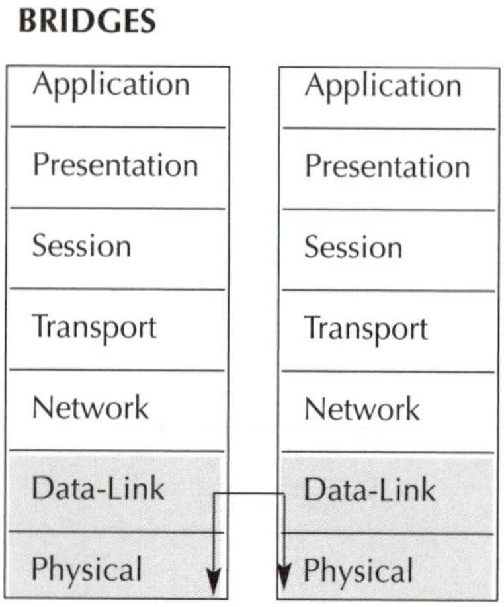

■ **Figure 3:** Routers operate at the network layer and are capable of stringent traffic control and network partitioning.

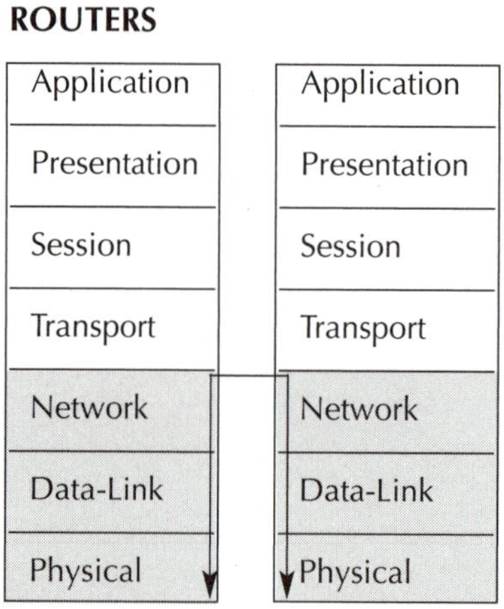

■ **Figure 4:** Gateways provide translations between two dissimilar computer systems, such as a PC LAN and an SNA network.

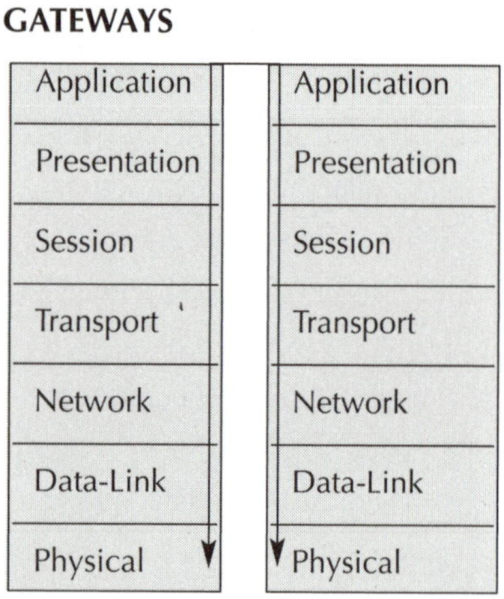

REPEATERS

Application	Application
Presentation	Presentation
Session	Session
Transport	Transport
Network	Network
Data-Link	Data-Link
Physical	Physical

BRIDGES

Application	Application
Presentation	Presentation
Session	Session
Transport	Transport
Network	Network
Data-Link	Data-Link
Physical	Physical

ROUTERS

Application	Application
Presentation	Presentation
Session	Session
Transport	Transport
Network	Network
Data-Link	Data-Link
Physical	Physical

GATEWAYS

Application	Application
Presentation	Presentation
Session	Session
Transport	Transport
Network	Network
Data-Link	Data-Link
Physical	Physical

Repeaters do not allow a network manager to isolate traffic; they regenerate every data frame or jam signal over all the networks they link. They do nothing to relieve the load on a network's bandwidth.

Bridges, on the other hand, isolate traffic to specific workgroups while still offering the ability to connect multiple LAN cable segments into a large logical network. Bridges operate one layer higher than repeaters in the OSI model; they operate at the MAC sublayer of the Data-link layer.

FILTERING TRAFFIC

Most bridges operate only between similar LAN technologies—between two Ethernets or two Token Rings, for example— but some do offer cross-technology capabilities. They regulate traffic by filtering data frames based on the destination address. When a frame's destination address is local, it is not forwarded by the bridge. When the destination address is remote—i.e., to a node on another workgroup—the bridge forwards it. Bridges automatically "learn" the addresses of the devices attached to their subnetwork.

More sophisticated bridges allow filtering traffic on a variety of factors, including frame size, source address, and type of protocol. Because filtering reduces network traffic, it can substantially increase overall network performance. Bridges operate independently of the upper-layer protocols which allows them to handle any transport protocol, such as the TCP/IP, IBM's SNA, and NetBIOS.

Bridges use custom filters to selectively reject or forward frames that match administrator-specified conditions, such as frame size, specific transport protocol (XNS, TCP/IP), or destination address. Custom filters can work on frames whether they're flowing into or out of a network; a filter can also forward only those frames that match user-defined criteria.

System administrators can use custom filters to help set up and manage administrative domains within a network; for example, a network manager could develop custom filters that isolate electronic mail domains. Custom filters can also restrict protocol-specific frames to certain preset domains. Similarly, filters could forward only specified types of frames.

Source-explicit forwarding (SEF) gives administrator-defined workstations exclusive frame-forwarding privileges on the internetwork. Designated stations can forward frames through a particular port on a routing bridge, while the frames of stations without SEF rights will be rejected. SEF thus permits a system administrator to limit access to normally secure or isolated network segments or resources.

These types of controls let network administrators manage their LANs better, permitting them to create secure domains and increase inter-workgroup efficiency.

Traditional bridges have offered transmission capabilities from only a single workgroup to another workgroup, but the move to centralized LAN management centers has prompted LAN manufacturers to market multiport bridges, now commonly called switches. Multiport bridges give network administrators the advantages of modular expansion and/or reconfiguration. By replacing one interface card with another—for example, adding an FDDI link to a modular multiport bridge—the administrator can keep up with an organization's changing network environment without completely rebuilding the network infrastructure.

THE ROUTER ROUTE

Routers operate at still another layer up—at the network layer in the OSI reference model. Routers connect logically separate networks operating under the same transport protocol (i.e., TCP/IP or SNA). Routers are thus protocol-dependent and must support the individual protocols being routed. A router allows multiple paths to exist in an enterprise-wide network, and is "intelligent" enough to determine the most efficient path to send a particular data frame through those multiple paths.

In a typical enterprise-wide network divided by routers, the separate networks are assigned unique numbers, and each independent network is managed separately. Routers automatically learn changes in a network's configuration, just as bridges do, within the limitation of the network protocol's ability to pass routing information between routing nodes. Routers are more complex than bridges, however, because the scope and scale of the internetwork are typically much greater than those of bridged environments.

Routers are particularly useful in organizations with multiple large networks connected to a single backbone. Because they have an inherently more difficult task, routers are generally slower than bridges. Newer routers, capable of routing packets at a LAN protocol's maximum bandwidth (with 10Mbit/sec Ethernet, about 15,000 frames per second), are erasing this limitation, however.

THE SPANNING TREE ALGORITHM

The spanning-tree algorithm allows physical loops to exist in a bridged Ethernet network. Loops, which are formed when there are multiple data paths between two segments of an Ethernet network, are particularly useful in mission-critical networks because they provide fault-tolerant redundancy and permit internetwork devices to find and use the most efficient routes between the other internetwork devices on that enterprise-wide LAN.

In a large multi-loop Ethernet, the spanning tree algorithm determines the most desirable path between segments and disables all other paths to eliminate redundant loops. (This path selection process is governed by options that can be selected by the system administrator.) Then, when the active path is unusable for any reason, spanning tree automatically reconfigures the network, activating the most desirable alternative path, until the original active loop is brought back online. Spanning tree permits connecting a corporate network to subsidiary networks via high-speed "active" lines; should either active line fail, a backup loop would be brought online automatically, thus ensuring continued communications.

Spanning Tree's ability to automatically sense trouble areas allows organizations to build large, reliable networks that are still easily managed from a central site; managing similar topologies created with routers alone requires a staff of competent network management personnel.

THE GATEWAY

Gateways act as translators between networks using incompatible transport protocols, such as between TCP/IP and SNA or between SNA and X.25. Gateways operate at the application layer of the OSI model.

One of the more common gateways is a communications gateway between a local area network and a mainframe or minicomputer; such a gateway generally places a special-purpose adapter card in a PC along with a standard network interface card. The resultant system serves as a shared gateway to the host for all the other PCs on the LAN. Such a gateway allows you to use a mainframe or mini as a network server, if desired.

The new internetworking products and features available combined with the old permit creating faster, more secure, and more cost-effective enterprise-wide networks—the kind now being demanded by multinational corporations.

This tutorial, number 21, was originally published in the April 1990 issue of LAN Magazine/Network Magazine.

Bridges

A data-link bridge is a device that connects two similar networks or divides one network into two. It takes frames from one network and puts them on the other, and vice versa. As it does this, it regenerates the signal strength of the frames, allowing data to travel further. In this sense, a data-link bridge incorporates the functionality of a repeater, which also regenerates frames to extend a LAN. But a bridge does more than a repeater. A bridge is more intelligent than a repeater. It can look at each frame and decide on which of the two networks it belongs. Repeaters simply forward every frame from one network to the other, without looking at them.

A bridge looks at each frame as it passes, checking the source and destination addresses. If a frame coming from Station 1 on LAN A is destined for Station 5 on LAN B, the bridge will pass the frame onto LAN B. If a frame coming from Station 1 on LAN A is destined for Station 3 on LAN A, the bridge will not forward it; that is, it will filter it.

Bridges know which frames belong where by looking at the source and destination addresses in the Medium Access Control (MAC) layer information carried in the frame. The MAC layer, which is part of the second layer of OSI Model, defines how frames get on the network without bumping into each other. It also contains information about where the frame came from and where it should go. Because bridges use this level of information, they have several advantages over other forms of interconnecting LANs.

WHY BRIDGE?

The most common reason to bridge is to improve network performance. Dividing one large network into two networks reduces the amount of traffic that flows over the entire LAN and therefore improves performance. Devices on both segments can still talk to each other via the bridge.

It is possible that a poorly placed bridge can reduce performance by creating a bottleneck. However, it doesn't take too much effort to discover the best place to put a bridge. For example, it doesn't make sense to split up 10 people whose workstations are physically close to each other if they frequently exchange information. A bridge between this workgroup and another workgroup, however, could improve performance dramatically. With the bridge, the two workgroups may still communicate transparently. Only communication between groups, not communication within groups, moves through the bridge.

Another reason to use a bridge is to change from one type of cable to another. For example, you may run twisted-pair cable in the offices and fiber-optic cable between buildings. Segments can be connected with a bridge, so "long distance" traffic can flow freely from one segment to another while local traffic stays local. Broadband and baseband Ethernet networks may be connected this way, too. A 16Mbps Token Ring backbone may use bridges to connect to several local 4Mbps Token Rings.

PROTOCOL IGNORANT

Because bridges operate at the MAC layer, they can interconnect LANs that use many different upper-layer protocols. Bridges are commonly referred to as protocol-independent. For example, the same bridge may connect networks running TCP/IP, DECnet, OSI, IPX, and XNS protocols. All these higher-layer protocols are encapsulated within the MAC layer. That is, the MAC layer is below the network layer where the upper-layer protocol information is kept.

A bridge will not allow a device speaking TCP/IP to talk to a device speaking IPX or OSI. That is a gateway's function. A gateway actually translates between protocols. A bridge simply passes frames back and forth, regardless of the protocols.

Many networks have more than one protocol running on them. For example, two groups of Sun workstation users may use TCP/IP most of the time and occasionally use OSI. A bridge between the groups will pass both TCP/IP and OSI frames. In fact, the bridge won't even know which protocol it is passing. But the two machines on either side of the bridge must use the same protocols for the message to make sense.

LEARNING AND FILTERING

A bridge is considered an intelligent device because it can make decisions based on situations it has already seen. To do this, a bridge refers to an address table. When a bridge is plugged in, it sends broadcast messages asking all the stations on the local segment of the network to respond. As the stations return the broadcast message, the bridge builds a table of local addresses. This process is called *learning.*

Once the bridge has built the local address table, it is ready to operate. When it receives a frame, it examines the source address. If the frame's address is local, the bridge ignores it. If the frame is addressed for another LAN, the bridge copies the frame onto the second LAN. Ignoring a frame is called *filtering*. Copying the frame is called *forwarding*.

The basic type of filtering is keeping local frames local, and sending remote frames to the other subnetwork. Another type of filtering is based on specific source and destination addresses. For example, a bridge might stop one station from sending frames outside of its local LAN. Or, a bridge might stop all "outside" frames destined for a particular station, thereby restricting the other stations with which it can communicate. Both types of filtering provide some control over internetwork traffic and can offer improved security.

Most Ethernet bridges can filter broadcast and multicast frames. Occasionally, a device will malfunction and continually send out broadcast frames, which are continuously copied around the network. A broadcast storm, as it is called, can bring network performance to zero. If a bridge can filter broadcast frames, a broadcast storm has less opportunity to brew.

Today, bridges are also able to filter according to the Network-layer protocol. This blurs the demarcation between bridges and routers. A router operates on the Network layer, and it uses a routing protocol to direct traffic around the network. A bridge that implements advanced filtering techniques is usually called a brouter. It filters by looking into the Network layer, but it does not use a routing protocol.

Other bridges are available that do true Network-layer routing. These routing/ bridges or bridging/routers are often used as the hub of an enterprise-wide network.

SOURCE ROUTING

Some Token Ring bridges, notably those from IBM, use a routing scheme called *source routing* to get frames from one network to another. The bridges we've talked about so far use transparent routing, which all Ethernet, and some Token Ring and FDDI, bridges use. With transparent routing, the frame does not know the route it will travel, nor do the bridges it passes over. Each bridge will forward a frame that is not local, until it finally reaches its destination LAN.

With source routing, the frame itself contains routing information. This information specifies the LANs and the bridges through which the frame will travel to get to its destination. The sending machine is responsible for putting this information into the MAC-layer header, which is the part of the frame that contains the source and destination addresses along with some other information about the frame.

For sending stations to know the route their frames will take, they must learn the layout of the entire network. This is done dynamically through a process called *route discovery*. During route discovery, frames are passed around the network. As they move from LAN to LAN, they are filled with information about the network. Each bridge puts three numbers into the frame: the numbers of the two LANs it connects and its bridge number. This information is then passed back to sending stations. Using this information, sending stations can then create a map of the network and appropriately route their frames.

Source routing is used primarily by IBM on its Token Ring LANs. Source routing does impose some overhead which might diminish network performance slightly. However, this is offset by the advantages of the routing scheme. Because the sending machine knows the route its frames will take, it can always choose the optimal path at the time of transmission, which is not possible with transparent routing. With transparent routing, the optimum path remains so until a bridge or a link fails. Also, source routing provides better management, since the path of a frame is immediately accessible from the frame itself. Finally, source routing bridges can be faster than transparent bridges, since they do not have to "look up" each frame to see if it must be bridged. The frame tells them immediately.

REMOTE AND LOCAL

So far we have discussed local bridges. Remote bridges connect two geographically separate LANs, mostly over a telecommunications link, such as a leased telephone line, a T1 link, a public data network, or microwave line. In remote bridging, the bridge is split into two devices. A bridge at one end puts frames destined for the other LAN out over the link. A bridge at the other end receives the frames and passes them to its local LAN. The process works in both directions.

Telecommunications links are not the only way to connect long distance LANs via bridge. Broadband networks and fiber-optic links can also bridge geographically distant networks. For example, bridges might be used to pass traffic over a fiber-optic backbone among the buildings of a university or business campus. Technically, this is not a remote connection, but the individual LANs may be several miles apart.

Either way, once bridges connect LANs over a longer distance, reliability and fault tolerance become more important. Bridges at both ends must take precautions against data corruption over the remote link.

SPANNING THE GLOBE

The first step in fault tolerance is redundant bridges. The IEEE 802.1D spanning tree algorithm allows redundant bridges to be configured on an Ethernet LAN. The stumbling block is that introducing parallel bridges creates a loop in the Ethernet topology, which is strictly forbidden by the rules of the Ethernet protocol. However, spanning tree manages those loops, so that frames don't circulate endlessly around the network. Without a backup bridge and spanning tree software, a failed bridge causes the network to be partitioned until the broken bridge is fixed.

According to the spanning tree algorithm, two bridges are set up, side by side. One is designated the primary bridge, and it is the only bridge to pass traffic. If the primary bridge fails, the traffic is automatically shunted to the backup bridge.

Spanning tree overcomes a major obstacle in bridging, but the backup bridge is idle as it waits for a failure. This is costly. If the backup link could carry traffic, the cost of the second link could be better justified. Here's where load balancing comes in. Using load balancing, traffic can be divided over the two remote parallel bridges. This provides much better performance, since not all traffic is going over one remote link. Since the spanning tree technology is still in place, if one bridge fails, the other can still carry all the traffic.

MANAGING BRIDGES

Since LAN configurations change constantly, it is crucial that bridges be easy to manage. A good bridge management package should allow bridges to be managed from a central location. A LAN manager should not have to be at the bridge but should be able to send instructions from a networked terminal or PC. Critical management functions include enabling and disabling bridges, changing security parameters, and changing the address filters and the protocol priorities dynamically. Many bridges allow the manager to download configuration information, thereby setting up the bridge to work as desired all at once.

Bridges should also provide information about what is happening on an internetwork. For example, a management package should report how much traffic is passing over the bridge, the type of traffic, how many errors occur and so on. With this information, the LAN manager can decide how to configure the network. It might be necessary to move the bridge and segment the network differently. Some bridges allow the LAN manager to set performance or error parameters. If these thresholds are exceeded, an alarm is sent to the manager's workstation.

Many bridges now support the Simple Network Management Protocol (SNMP). SNMP can be used to manage nearly any type of device, from a host computer to a multiport repeater. A bridge must implement the SNMP agent software, which sends information back to the SNMP management station. SNMP is most often used in TCP/IP networks.

Cooperation with an enterprise-wide management system is crucial. Such management systems include DEC's Enterprise Management Architecture, AT&T's Universal Network Management Architecture, IBM's NetView, HP OpenView, SunNet Manager, and the OSI Common Management Interface Protocol. These global management systems are essential to fill in the "big picture" of network management.

This tutorial, number 11, by Aaron Brenner, was originally published in the June 1989 issue of LAN Magazine/Network Magazine.

Ethernet Switching

If an Ethernet network begins to display symptoms of congestion, low throughput, slow response times, and high rates of collision, the reflexive response is to plan installation of pervasive higher-speed connections. However, it may not be necessary to make huge investments in Fiber Distributed Data Interface (FDDI) or Asynchronous Transfer Mode (ATM) technology.

Several approaches exist that can preserve much or all of the existing network's cabling and workstation interface card infrastructure while still greatly enhancing the throughput for users, even if demanding applications, such as multimedia production and video conferencing, are on a company's horizon. The most promising techniques, as well as the best return on investment, could well consist of installing the right mixture of Ethernet switches.

An Ethernet switch is, in principle, a multiport bridge. It concerns itself with the OSI layer two media access control (MAC) information in the frames it processes. It learns the MAC source addresses associated with each port as traffic appears, so little or no manual administration is required.

When a frame arrives at a port, the switch examines its MAC destination address. If the destination is local to the segment the frame is on, the switch filters the frame—that is, it ignores the frame and does not retransmit it. If the switch's address database associates the destination with another port, the frame is forwarded (or transmitted) on that port. If the frame's destination is unknown, the switch transmits it to every port except the incoming one.

Like bridges, switches segment traffic on one port from traffic on the others. Therefore, groups of heavy traffic producers, such as programmers, CAD jockeys, or multimedia producers, with their servers, can be isolated from more mundane users. The light-duty users will no longer be slowed by the traffic of heavy users, and the heavy users can be provided with as much as 10Mbps or more of dedicated throughput if needed.

Aside from having the marketing advantages of a hot new name, switch, rather than a tired old one, bridge, the current generation of switches also has some technical advantages over the bridges of yesteryear.

Switches often have a high-performance backplane that can support very high throughput—as high as the number of paths through the switch, multiplied by the throughput of each paths—from 60Mbps for a 12-port 10BaseT switch, to multiple Gbps for large switches with 100Mbps ports. Unlike traditional bridges, the optimal deployment of switches can result in aggregated total throughput; when everything works right, each additional switch can add to the total performance of the system.

Switches can also employ new tricks to overcome the historical drawbacks of bridges. Some switches read beyond the data link header in each frame and identify the network layer protocol. With this information, the switch can selectively filter specific protocols for security or performance purposes.

Other switches creatively deploy queuing buffers to combat the phenomenon known as blocking, in which particular paths experience a backlog or latency despite the ready availability of alternate paths. The advances in Application Specific Integrated Circuit (ASIC) technology that we have come to take for granted elsewhere have also bolstered the performance of switching devices.

SHARED 10MBIT/SEC PORTS

The earliest Ethernet switches, introduced by Kalpana (Sunnyvale, CA, now a subsidiary of Cisco Systems) in 1990, had sharable 10Mbps ports. The Kalpana switches use "cut-through" methods; unlike traditional bridges, which are classified as "store-and-forward" devices, the Kalpana switches begin forwarding a frame as soon as they read the destination address and look it up.

The cut-through technique reduces the latency, or delay, for a forwarding operation to about 40 microseconds compared to as much as 1,200 microseconds for a store-and-forward bridge (see Figure 1). Latency is a major factor for protocols such as NetWare's IPX (before burst mode became available), which have a lot of frame-by-frame acknowledgment, but is less of a factor for protocols such as IP, which can handle acknowledgment activity more efficiently. Cut-through switches look their best on traditional NetWare networks.

However, the cut-through method of forwarding can perpetuate and aggravate problems. Store-and-forward devices use the cyclic redundancy check (CRC) field to verify that frames are well-formed; they discard runts (short frames typically disrupted by a collision) and jabbers (overly long frames sometimes caused by a defective NIC). Cut-through switches pass these bad frames along. Jabbers often look like broadcast frames to a cut-through switch, and can be cascaded all over the network if nothing is done to eliminate them. Some switch producers let network managers choose between cut-through and store-and-forward operation.

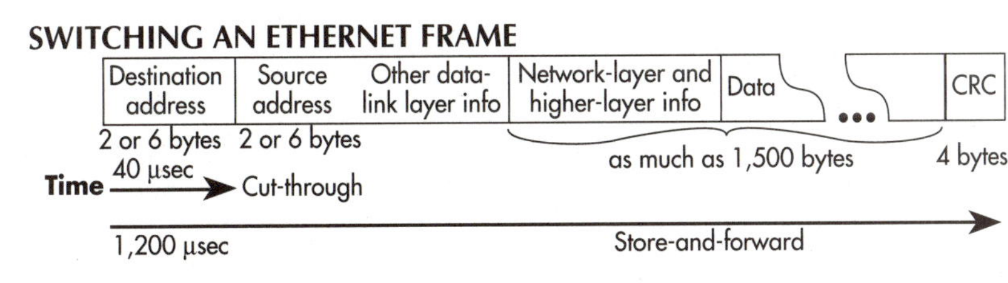

■ **Figure 1:** Cut-Through Vs. Store-and-Forward: A cut-through switch begins forwarding frames as soon as it reads the destination address. To perform the CRC data integrity function, a store-and-forward switch must read the entire frame before forwarding it.

Shared 10Mbps switches require no other changes to the network. Existing hubs, NICs, and cable can be used transparently.

Interesting variants of this class of switch are produced by Matrox (Dorval, Quebec) and XNet Technology (Milpitas, CA). These switches are cards that plug into the ISA or EISA bus of a PC and provide as many as 16 ports at a very low per-port cost in comparison with standalone switches.

DEDICATED 10MBIT/SEC PORTS

Workgroups with many high-bandwidth users may be interested in a class of Ethernet switches, pioneered by Grand Junction Networks (Fremont, CA), that segments a network, giving each node its own port (see Figure 2). Each user or device on one of these switches has a dedicated, collision-free 10Mbps connection to any other node. Because of the simplifications that result from unshared segments, these dedicated workgroup switches cost much less than switches that support multiple MAC addresses per port. Like switches with shared 10Mbps ports, dedicated port switches use existing NICs, cable, and hubs. However, none of the ports can be shared by multiple users.

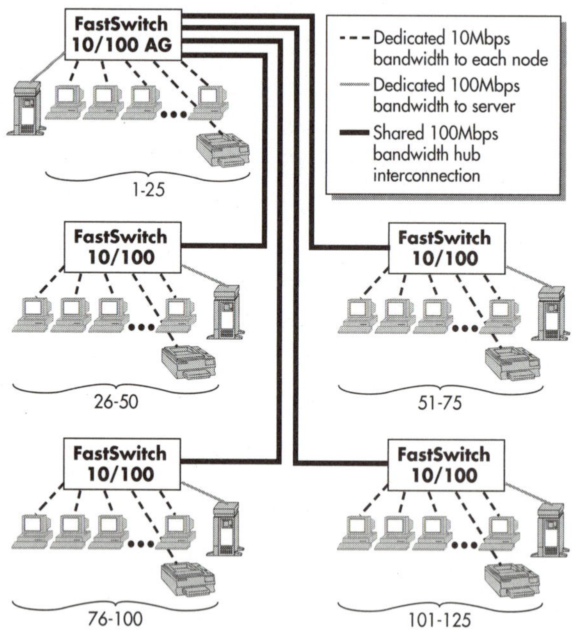

■ **Figure 2:** Scaling for Power Users: Each of the 125 nodes has a private Ethernet connection. Each server receives 100Mbps of bandwidth. The connection between switches is a shared 100Mbps link. Aggregate forwarding bandwidth is 850Mbps.

FULL-DUPLEX PORTS AND 10MBIT/SEC PORTS

A dedicated Ethernet connection doesn't need to listen for collisions, because no other traffic source is there to collide with. For this reason, such a connection can freely receive and transmit at the same time; in other words, it can operate in full duplex mode rather than the usual half-duplex. If incoming and outgoing traffic is perfectly balanced, the throughput on existing cables could be as high as 20Mbps.

At first glance, the advantage of full-duplex operation seems marginal because most workstation applications seem to receive much more traffic than they transmit; the "additional" channel will be idle much of the time. However, a lot of invisible acknowledgment and housekeeping traffic takes place on most networks, so, if the added cost of full-duplex cards and switches is not high, the purchase may be worthwhile for workstations.

In any case, server-to-switch traffic is likely to be better balanced between transmitting and receiving, and the added cost of a server NIC and one or two full-duplex ports on a switch can be more readily justified than the cost for upgrading a workstation to full duplex. Full-duplex workstation attachments could also be cost-effective for video conferencing, which normally produces symmetrical data traffic in both directions.

An additional option for server connections on NetWare networks is to connect more than one server NIC, either full- or half-duplex, to a switch. This solution requires a special NLM on the server that balances traffic between the NICs and overcomes NetWare's reluctance to permit the same network segment to have multiple server connections. Network Specialists (Lyndhurst, NJ) and Kalpana were the first companies to introduce these NetWare add-ons.

SWITCHES WITH HIGH-SPEED PORTS

If full-duplex or combined 10Mbps connections to a server (or a backbone) are still a bottleneck, the next approach is obtaining a switch with at least one 100Mbps port. FDDI is a mature, standardized, widely implemented technology. NetWorth (Irving, TX) and 3Com (Santa Clara, CA) are some of the leaders in providing combination FDDI/Ethernet switches.

For a high-speed server or backbone attachment, an alternative to FDDI is 100BaseT Fast Ethernet. Grand Junction has introduced a product with 24 dedicated 10Mbps ports and two 100BaseT ports. In theory, a 100BaseT switch could outperform a switch based on FDDI, because it would not have to perform translational bridging between FDDI and Ethernet.

With switches that include high speed ports, the NICs and cable that connect workstations and printers are unchanged. Only the connections to servers, routers, other switches, or workstations that require 100Mbps throughput need to be upgraded.

This tutorial, number 79, by Steve Steinke, was originally published in the March 1995 issue of LAN Magazine/Network Magazine.

Switching vs. Routing

It's not easy to tell a switch from a router these days.

There is a widespread misconception in networking circles that the difference between switching and routing is a simple binary opposition. In fact, it's very difficult to define the boundaries between the two functions. And needless to say, vendors call their product offerings by names that will result in the highest sales rather than by a careful study of where they fit in a logical taxonomy of devices. However, it's worth making an effort to clarify the distinction between switching and routing even if we're doomed to marketing-driven obfuscation in the long run.

As far as communication networks are concerned, the notion of switching begins with circuit switching. The telephone (or telegraph) company provides an electrical path that allows my instrument to connect to yours, perhaps with an operator plugging a connector into a jack. The telephone company first combined (or multiplexed) multiple calls on a single physical circuit using Frequency Division Multiplexing (FDM). You can think of these discrete calling paths as the first virtual circuits in the sense that there was no longer a one-to-one ratio between phone calls and wires (or insulators on the utility pole).

FRAME SWITCHING

However, FDM turned out to be insufficiently scalable for the demands of telephony. So, in the early 1960s the phone companies began digitizing voice signals and multiplexing them in the time domain using Time Division Multiplexing (TDM).

For example, with TDM a T1 line interleaves 24 phone calls among successive time slots within one frame, which consists of 193 bits. Bit 1 through bit 8 are dedicated to channel 1, bit 9 through bit 16 are dedicated to channel 2, and so on until bit 185 through bit 192 are dedicated to channel 24. The 193rd framing bit is used to synchronize the system. The interleaving process repeats 8,000 times each second. (Note that 193 bits per frame times 8,000 frames per second equals 1.544Mbits/sec, which is the throughput rate of a T1 line.)

Today, TDM phone call switching is circuit-oriented, though the devices that perform the switching function on digital circuits bear little resemblance to the mechanical switches that once selected paths made up of solitary electrical circuits.

One disadvantage to TDM—whether you're making a call with a TDM system or leasing a full-time digital line—is that your cost will be the same whether you fill every time slot with data or whether you transmit nothing (for our purposes, let's consider voice to be just another form of data). As we know, the data transmissions required for many applications are bursty, with intervals of high demand and no demand distributed almost randomly. Thus, TDM-based networks (or any circuit-oriented networks with hard resource allocations) are likely to be inefficient or otherwise not wholly suitable for data traffic.

Ultimately, dedicated circuits are a high-cost form of connectivity, and setting up a circuit via a switching system designed for voice communications results in long circuit-initiation times, as well as high cost.

PACKET SWITCHING

In the late 1960s the notion of packet switching was developed. The first commercial outgrowth of this technology was the X.25 network, which is still heavily relied on in much of the world, in spite of outbreaks of Internet fever in many countries. Packets on an X.25 network aren't slotted rigidly in the way that circuits in the TDM system are. Instead, the packets are created and transmitted as needed. Therefore, X.25 service can be priced by the packet or by the byte rather than by connection time or as a full-time circuit; you're using network capacity only when you're sending or receiving data.

Despite this freer form of multiplexing, X.25 networks are still connection-oriented, and a session between two nodes still requires a virtual circuit—the virtual circuit has just been unbundled from a fixed time slot.

Each packet in an X.25 network has a Logical Channel Number, or LCN. When a packet comes into a switch, the switch looks up the LCN to decide which port to send the packet out of. The path through the network of packet switches is defined in advance for Permanent Virtual Circuits (PVCs) and established on the fly for Switched Virtual Circuits (SVCs). With SVCs, call setup is required before data transmission can take place.

Incidentally, the protocol data units at Layer 3 are known as packets, while the Layer 2 protocol data units are called frames, at least when writers are precise. The X.25 protocols include Layer 3 functions, while frame relay, which is essentially X.25 with error correction and flow control removed, remains at Layer 2.

CONNECTIONLESS SWITCHING

LANs, which people began to develop in the 1970s, are a form of connectionless communication. When a Layer 2 bridge connects LANs, a form of frame switching takes place. When a Layer 3 router connects LANs, a form of packet switching takes place. IP and its precursors were the first wide area connectionless protocols used extensively. Each packet includes its source and destination addresses and moves independently through the network.

With these connectionless switching systems, for the first time there was no advance setup of the path needed. There was no advance agreement on the part of the intermediate stages committing to a specific level of service. There was no state maintained on the network's packet switches, which have no notion of circuits, paths, flows, or any other end-to-end connection.

There's an economic and a fault-tolerance factor you should keep in mind when considering the router-using flavor of packet switching. The economic issue revolves around the relative costs of computer cycles and communications circuits. If processing power is cheap, it's not unreasonable to figure out the routes every packet should take in order to make the best use of expensive circuits. If circuits are cheap, you may prefer to set up a mesh of connections using relatively dumb circuit switches, rather than dedicate a bunch of high-powered, special-purpose computers to routing packets.

The fault-tolerance issue (often expressed as the overheated claim that the Internet was designed to survive a nuclear war) is that connectionless networks are more resilient than connection-oriented ones. If a link or a router goes down, the overall system is designed to find routes around the problem. A broken connection-oriented network, designed to remember how to handle each circuit rather than to solve a routing problem for each packet, will likely need manual reconfiguration if it consists of PVCs, and if it consists of SVCs, it will at least need to perform a call setup.

So, even though routing is a form of packet switching, we can distinguish routing from other forms of switching by its connectionlessness. By "connectionless," we mean that any responsibility for maintaining the "state" of an end-to-end connection lies with the end nodes—the network itself doesn't track a circuit, connection, or flow. Why wouldn't we say that routers maintain state information when we know that they in fact maintain routing tables that govern the process? The short answer is that routing tables are applied indifferently to each packet; a switch uses some aspect of a connection (or flow) that the packet includes to select the path through the network.

When routers begin to maintain the state of circuits, connections, or flows (presumably in order to accomplish something worthwhile, such as increased performance), then in this more rarefied sense it's reasonable for them to call themselves switches. (To watch the lines blur even more, read about new strategies to switching and routing below in "New Twists.")

In a flat switched Ethernet network there is only one path between any two nodes. The forwarding tables in the switches might be considered to maintain state since they describe an end-to-end connection, though it's a degenerate case.

LAST WORD

So you can't distinguish a router from a switch by OSI layer. There are legitimate switching functions that can be performed at Layer 3 as well as Layer 2. It's tempting to conclude that switching is something that gets done in hardware, while routing gets done in software on a microprocessor. The grain of truth behind this idea is that performance is one of the most significant drivers behind the adoption of switching products, but the boundaries between software and hardware get thinner all the time, and they're no help with the logical distinction we're looking for.

In a switching network, you'll find the intermediate devices keeping track of—or remembering—qualities of the connection. In a pure routing network, the intermediate devices will be indifferent to anything but handing off packets to the next device, and they will not be distracted by any other information, upstream or downstream.

NEW TWISTS

Companies Take New Approaches to Switching and Routing

Cisco Systems has a new technique it calls tag switching. Ipsilon Networks, with a device that combines traditional IP-routing software with ATM hardware, has produced a technology it calls IP switching. Both of these new methods perhaps blur the distinction between switching (in the narrow sense) and routing.

However, the blurring of distinctions may be appropriate because it's desirable to add some of the beneficial characteristics of connection-oriented architectures to the familiar world of connectionless routers. In particular, switching can have a performance advantage over routing. Switching permits the assignment and guarantee of Quality of Service (QOS) characteristics, along with providing better abilities to control congestion. Also, switching enables precise path selection, which can be useful for diagnostic purposes.

In a tag-switching network, an ingress router will add a tag to a packet as soon as a packet enters the network. If the network consists of routers, the tag is an index into the routing tables of downstream routers that improves performance by simplifying the route lookup. But the tag can also be a Virtual Channel Identifier (VCI) for an ATM switch or a frame relay virtual circuit. If an edge router has identified a path via an ATM switch or frame relay virtual circuit as being the optimal path for a particular packet, it will assign the packet the appropriate VCI tag, which will allow the packet to use the path to "cut through" a switched network.

Cisco claims it will begin shipping tag-switching products by mid-1997, and it has submitted related protocols to the IETF, though the prospect for standardization before the end of 1998 is dim, especially because somewhat-competing proposals from IBM, Cascade, and 3Com are also in play. [Editor's note: These efforts ultimately resulted in the widely accepted MPLS standards.]

IBM, Cascade, and 3Com have endorsed Ipsilon's IP switching, though they all continue to deliver products based on other technologies that at least parallel, if not compete with, Ipsilon's. Ipsilon's switching relies on its Ipsilon Flow Management Protocol (IFMP, see Broadcast, *LAN Magazine*, June 1996, page 20). The specifics of the protocol have been published as RFC 1953 and RFC 1954, and RFC 1987 describes the related switch management protocol. Note, however, that informational RFCs such as these are not in any way an effort at standardization as it is normally understood. The Ipsilon protocols are proprietary, although they have been openly published and other vendors, including Digital Equipment, have built products with them.

To understand IFMP, you must first understand that network users have to define policies describing what sequences of IP packets constitute a flow. In most cases, flows will be identified by source, destination, and port matches—for instance, the ftp traffic between two nodes. Each qualifying IP flow is mapped to a virtual circuit in the ATM switch. If too many types of traffic are designated as flows, an ATM switch will get congested with too many virtual circuits; if too few are classified as flows, the network won't gain the maximum advantage from IP switching, because it will simply continue to route many flows.

The company claims that IP switching sets up a "soft state" across the network, as opposed to the "hard state" ATM switches establish. This is perhaps an alternate way of saying that a flow doesn't quite qualify as a connection, but it's still accurate to say that switching is taking place, not just routing.

Neither the Cisco approach nor the Ipsilon approach actually take full advantage of the capabilities of ATM, though they might claim this fact as a feature rather than a reservation. Tag switching has the advantage of supporting any Layer 3 protocol; Ipsilon initially supported only IP, though it has promised support for IPX. Tag switching is also more parsimonious with ATM virtual circuits than IP switching is—Cisco assigns a virtual channel to a route, and any traffic on that route will employ that virtual channel, while Ipsilon assigns a separate virtual channel to each flow, many of which might trace the same route.

This tutorial, number 105, by Steve Steinke, was originally published in the May 1997 issue of Network Magazine.

VLANs and Broadcast Domains

A technology plagued by confusion and lack of standards holds plenty of promise.

One primary difference between connection-oriented networks, such as ATM, and connectionless or shared medium networks, such as Ethernet and Token Ring, is the ease with which you can perform a broadcast on connectionless networks. On Ethernet and Token Ring networks, every frame is visible to all the end nodes on a segment or ring, even though normal network interface controller behavior is to ignore all the frames not intended for its own consumption. Thus, any frame flagged as a broadcast (which is typically done by making the destination field all 1s) can reach every destination on the segment or ring at once, with no special effort.

Broadcasts perform a number of behind-the-scenes, yet indispensable, network functions. When you turn on a NetWare/IntranetWare client machine, it broadcasts a Get Nearest Directory Server or Get Nearest Server packet to connect to the network and ultimately log in. Whenever an IP host or router needs to find the physical destination of an IP address, it generates an Address Resolution Protocol (ARP) broadcast packet, which asks something like,

"Will the node with IP address *x* please send me your MAC address, so I can address this packet to you?" (These addresses are stored in an ARP cache for a couple of hours or so—subsequent traffic may not require broadcasts.)

NetBIOS-based nodes maintain a list of addresses of other nodes they can reach; these lists are populated by periodic broadcasts that say, in effect, "Everybody tell me your name and address now." NetWare servers generally broadcast Service Advertising Protocol (SAP) packets that alert other nodes to the presence of file and print services. Routers advise each other of the availability of routes by broadcasting RIP (Routing Information Protocol), OSPF (Open Shortest Path First), or BGP (Border Gateway Protocol) packets. The AppleTalk Chooser constantly fires off broadcasts to find printers and servers. AppleTalk performs name resolution by broadcasting Name Binding Protocol (NBP) packets. In a recent sample I took on a subnet that has AppleTalk, NetWare, and IP traffic, but no NetBIOS or LAT traffic, the traffic at times consisted of more than 7 percent broadcasts.

Early LAN protocols—especially those like NetBIOS and AppleTalk, which were designed to be simple to set up and use, even when some connected machines were turned off part of the time, tended to make heavy use of broadcasts. With plenty of available throughput on the LAN, even relatively heavy broadcast traffic didn't pose a problem. TCP/IP, designed for wide area connectivity, where every bit/sec of throughput costs real money, is much more parsimonious with broadcasts than traditional LAN protocols are. (ARP broadcasts are limited to a particular subnet—they don't typically traverse a WAN link.) When people began to interconnect NetWare networks over wide area links, Novell's IPX was criticized for excessive broadcasts. Novell responded by delivering the NetWare Link Services Protocol (NLSP), which replaces RIP broadcasts with more efficient, less frequent NLSP broadcasts, and by introducing Novell Directory Services (NDS), which can eliminate much of the demand for SAP packets. In Windows NT networks, a Windows Internet Name Service (WINS) server can provide one of the functions performed by NDS: eliminating the need for name resolution broadcasts that request NetBIOS names over TCP/IP networks.

On a single-segment Ethernet network, the broadcast domain (all the nodes a broadcast frame can reach) is the same as the collision domain—all the nodes that must wait for each other to be silent before transmitting data. (A single Token Ring is also a broadcast domain.) However, multiple Ethernet segments connected via a switch or bridge form an extended broadcast domain made up of multiple collision domains. Bridges forward broadcasts out to every port, and they also forward frames with unknown destinations to every port. So traffic local to a collision domain is invisible to the rest of the broadcast domain, while broadcasts and frames with unknown destinations are flooded to every other node on the broadcast domain. Theoretically, thousands of nodes could be connected by bridging multiple collision domains. Just because you can do it doesn't make it a good idea, though.

There are two good reasons to avoid large, flat networks. The first is to provide the ability to administer the network effectively. In particular, it's often important to treat some network users differently than others. In terms of overall network performance, you may want to connect a group of users who all use a particular protocol, such as AppleTalk, to the rest of the network via a router rather than a bridge or switch. This approach keeps AppleTalk broadcasts from absorbing bandwidth over the entire network. As another example, a group of users with high-performance requirements—video conferencing users, for instance—might best be isolated from others. There may be security considerations that dictate special treatment for a specific set of nodes.

The second reason that large, completely flat networks become unworkable results from broadcast traffic itself. To a varying extent, depending on particular protocols, broadcast traffic can increase geometrically with additional nodes, while unicast traffic increases arithmetically. Segmenting networks with switches can extend indefinitely the bandwidth available for unicast messages, but at some point the broadcast traffic will swamp the whole network—a phenomenon that won't necessarily occur gradually. Instead, there may be some sort of cascading event where a high level of broadcasts generates even more broadcasts and there is a broadcast storm, which may require shutting down large sections of the network in order to regain control.

THE VLAN ARRIVES

As switches have become cheap and easy tools for segmenting flat networks, vendors have looked for ways for network managers to contain broadcasts without having to install routers on every segment. Virtual LANs, or VLANs, can best be thought of as logical broadcast domains, as opposed to the "physical" broadcast domains we have been discussing. Without VLANs, the traditional way to break up flat networks is to insert a router between the subnets you want to define. The IP addressing scheme supports subnetting in a way that many sites have managed to make work, though it would be hard to claim this method is flexible or easy to grasp. (NetWare networks are somewhat easier to configure as subnets because file servers with multiple network interfaces are routers by default.)

Beyond the protocols are the issues of router setup and configuration. Routers are difficult to administer, and it's easier than it ought to be to make catastrophic, or at least thoroughly frustrating, mistakes. Router ports are also substantially more expensive than switch ports with the same throughput. The basic idea of a VLAN is to allow an organization to customize broadcast domains according to its needs instead of accepting the limitations that come with Network layer protocols and with using routers to accomplish segmentation.

For instance, consider the network in the figure. Each switch has a 100Mbits/sec uplink to an IntranetWare server and multiple 10Mbits/sec ports connected to individual workstations,

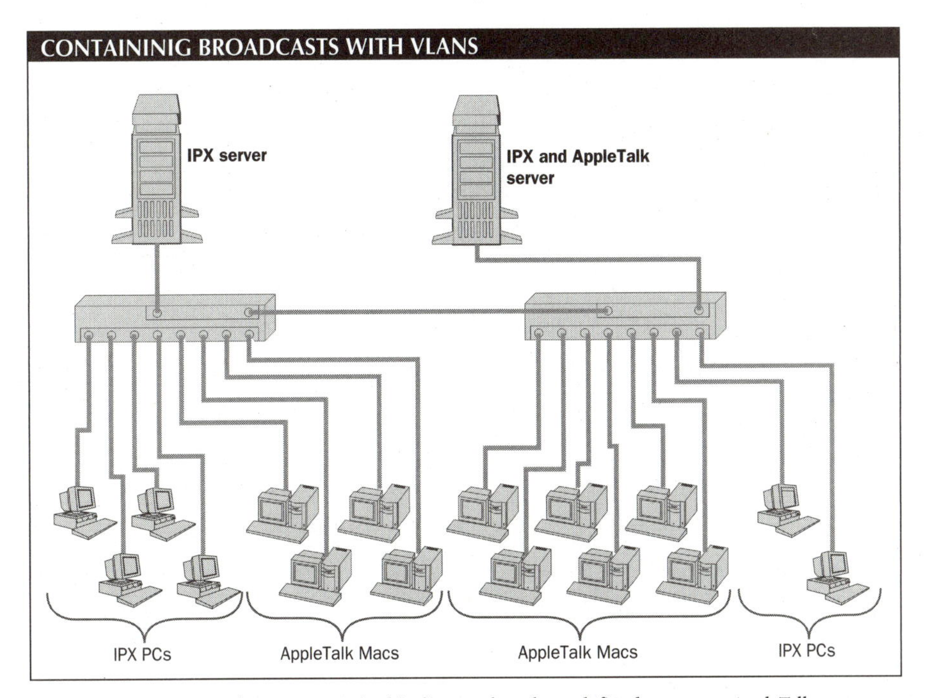

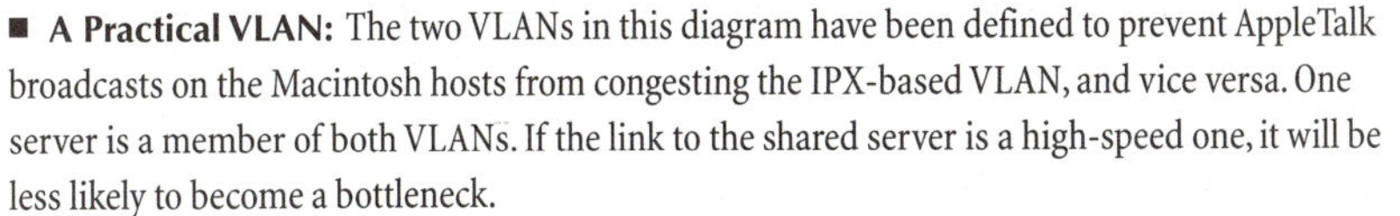
■ **A Practical VLAN:** The two VLANs in this diagram have been defined to prevent AppleTalk broadcasts on the Macintosh hosts from congesting the IPX-based VLAN, and vice versa. One server is a member of both VLANs. If the link to the shared server is a high-speed one, it will be less likely to become a bottleneck.

some of which are IPX and others of which are AppleTalk. The switches are connected with a 100Mbits/sec link. By default, this configuration constitutes a flat network: All the AppleTalk broadcasts are sent to the IPX nodes and all the IPX broadcasts are sent to the AppleTalk nodes. However, if all the AppleTalk nodes (along with the servers) are defined as VLAN 1 and all the IPX nodes (along with the servers) are defined as VLAN 2, then the broadcasts of each protocol will be segregated on their own VLAN.

DEFINING THE VLAN

VLANs can be segmented in several different ways, which is one of the reasons they can be confusing. The simplest way is to assign specific ports on a switch to VLANs. If an assigned switch port is connected to a shared segment—one with multiple nodes—all those nodes will be part of the VLAN, along with the nodes on other ports assigned to that VLAN. This VLAN segmentation is basically a matter of electronically isolating the ports of each VLAN. Within a given vendor's product line, port-based VLANs can often span multiple switches. However, interoperability across vendor boundaries awaits the adoption of the IEEE 802.1Q standard, which defines a tag field that identifies a frame as a member of a particular VLAN.

If VLANs are defined only by port, they typically can't overlap. This is unfortunate because permitting workstations or servers to be members of multiple VLANs is one of the most promising features of the technology.

The second way VLANs can be segmented is by MAC address. In this method, switches maintain tables of MAC addresses with their VLAN membership. With this form of membership, the nodes on a shared segment need not belong to the same VLAN. However, the fact that members of two VLANs reside in a single collision domain will diminish some of the traffic containment advantages of VLANs. Because MAC addresses are tightly associated with NICs, this Layer 2 VLAN definition has the advantage of portability. Wherever a workstation or laptop is plugged in across the switched network, the switches will recognize it as a member of the assigned VLAN.

VLANs can also be defined by their network or Layer 3 addresses. Many network managers, frustrated by the administrative overhead that comes with highly changeable networks configured with IP, could happily replace their existing subnet structure with VLANs. Vendors who support this form of VLAN definition generally offer tools that can automatically assign subnet members to particular VLANs. Once the node's VLAN membership is defined, the VLAN handles everything, even if the node is moved to a port connected to a different subnet or if the IP addresses need to be reassigned.

VLANS AND ATM

There is an interesting relationship between ATM LAN Emulation (LANE) and VLANs. Connection-oriented ATM doesn't have broadcasts—it has to invent them from the ground up. Because LANE, which makes it possible for traditional LANs to communicate over ATM, has no preconceptions or baggage with respect to broadcasts, it can readily include hooks to VLANs. Thus, the members of an ATM-based emulated LAN can readily be assigned to VLANs defined for Ethernet, Token Ring, and FDDI, or to their own VLANs. In many respects, understanding and setting up an ATM-based VLAN is more straightforward than is the case with traditional networks.

This tutorial, number 107, by Steve Steinke, was originally published in the July 1997 issue of LAN Magazine/Network Magazine.

Routing Protocols

Large internetworks spanning multiple workgroups, subnetworks, and host computers create special problems for system administrators. Like managers of local networks, they must deal with products from multiple vendors, get different network operating systems to communicate, and justify the costs of new services and applications.

But they also face the task of selecting, managing, and maintaining bridges, routers, and gateways—the internetworking devices that permit widely separated clusters of networks to communicate. These products offer important capabilities to help network managers solve particular problems, but each poses special implementation challenges, too.

This is especially true of *routers* and their associated routing protocols. Routers connect logically separate networks operating under the same transport protocol, such as the Transmission Control Protocol/Internet Protocol (TCP/IP).

Routers, which operate at the network (or third) layer of the Open Systems Interconnection (OSI) reference model, are *protocol-dependent devices.* That is, they must support each routing protocol on that LAN.

A WEALTH OF PROTOCOLS

A wide variety of routing protocols can be found on enterprise-wide networks today. Some are proprietary, or single-vendor, solutions developed specifically for use with a vendor's own products. Others are "open" in that they have been standardized by official sanctioning agencies.

Among the proprietary ones are the Internetwork Packet eXchange (IPX) protocol used by Novell's NetWare and the Interior Gateway Routing Protocol (IGRP) used by Cisco Systems. Open protocols include the Routing Information Protocol (RIP) for use with the TCP/IP suite and the OSI Intermediate System-to-Intermediate System (IS-IS) protocol. These were formulated or standardized by the Internet Activities Board (IAB) and the International Organization for Standardization (ISO), respectively. Since they are open, they can be used with multiple vendors' products in heterogeneous networks.

Other widely used routing protocols include the IETF's Open Shortest Path First (OSPF), DECnet Phase IV, the OSI's Connectionless Network Services (CLNS) protocol, and Apple Computer's Datagram Delivery Protocol (DDP).

ROUTING ALGORITHMS

The primary role of a router is to transmit similar types of data packets from one machine to another across wide area communications links such as T1 lines or Fiber Distributed Data Interface (FDDI) rings. Ideally, the router exchanges data by selecting the best path between the source and destination machines. It determines what is best via *routing algorithms,* which are complex sets of rules that take into account a variety of factors.

In operation, these algorithms' first task is to determine which of the paths on the internetwork will take a data packet to its destination. Because multiple paths often exist between any two routers, the algorithms are used to select the best paths. These decisions are based on a prescribed set of conditions, which might include the fastest set of transmission media or which network segment carries the least amount of traffic.

FLOODING THE NETWORK

Without microprocessors to perform the complex mathematical calculations required by routing algorithms, early routers were slow. The networks they ran on were equally low-powered, with little bandwidth and not complex. This meant that routers could be simple and operate without knowing much about where the other routers on the network were located.

These types of routers were isolated in that they did not exchange network routing information with other routers on the network. As a result, they forwarded data merely by flooding every path with packets. Data packets eventually reach their destination in this scheme, but flooding also risks creating routing loops, in which certain packets can travel around the network indefinitely.

Several measures can be taken to help flooding-type routers choose reasonable paths. One is called *backward-learning.* In this scheme, a router remembers the source addresses of all incoming packets and notes the physical interface it came in on. When it's time to forward a packet to that address, the router bases its decisions on this stored information.

Some routers avoid the entire issue of path-finding by relying either on a human or host computer to make these decisions. In the former case, the network manager provides each router with a block of *static* routing configuration information at start-up, including the information needed to make routing decisions.

In the host-router implementation, end hosts place information in every packet they place on the network. This information indicates every path and the immediate router the data must pass through to get to its destination. This is *source-routing.*

ADDING COMPLEXITY

More complex networks require dynamic routing solutions. In large wide area networks with multiple links between networks, routers perform more efficiently when they understand how the network is linked together. An *integrated router* does this by exchanging information about the network's topology with other integrated routers. As a result of this exchange, integrated routers create *routing tables* that show the best paths between the various links on the internet.

Algorithms for integrated routers must be able to quickly determine the network topology. This process, called *convergence,* must take place rapidly, otherwise routers with obsolete or incorrect data about the network can send data into dead-end networks or across unnecessary links.

DISTRIBUTED ROUTING

Routers can be arrayed in centralized or distributed configurations. Central routers with intensive CPU resources and lots of memory control how data packets are moved around. These relatively expensive central routers receive topology information from remote, less-powerful "slaves," then build routing tables and pass them along to the remote routers as needed.

The newer, increasingly sophisticated routers with their own CPU and memory resources available make centralized routing obsolete. In the *distributed routing* environments now prevalent, all routers on the internetwork can calculate routing algorithms quickly and efficiently.

The two main distributed-routing algorithms are a *distance-vector algorithm,* or the *Bellman-Ford algorithm,* and a *shortest path first,* or *Dijkstra* algorithm. Both are in wide use, in proprietary and standard protocols. For instance, RIP (used with TCP/IP, Xerox Network Systems [XNS], and IPX), DECnet Phase IV, and Cisco's IGRP implement distance-vector algorithms. OSPF, used in TCP/IP, is a shortest path routing algorithm.

Distance-vector routers create a network map by communicating in a periodic and progressive sequence with each other. This information exchange helps them determine the scope of their network in a series of router hops that reveals more information about the network.

Here's how the algorithm works: When the network is started, each router knows about only the networks it is connected to directly. Each router then advertises information about its immediate connections to the routers it is directly connected to. By incorporating the updates it receives from routers nearest to it, each router learns about networks connected to its neighbors, two router hops away. Additional updates expand each router's knowledge hop-by-hop.

Shortest Path First (SPF) routers update each other and learn the topology of an internetwork by periodically flooding the network with *link-state* information. Link-state data includes the cost and identification of only those networks directly connected to each router.

SPF routers send link-state data to all routers on a LAN. This allows OSPF routers to perform two important subsequent steps. First, they use the link-state data to build a complete table of router and network connections. Then each router calculates the optimal path from itself to each link. In this repetitive process, each router checks the potential pathways between the links on the network, eliminating the most indirect path in favor of the shortest. That path, with any routers connected to it, is put on an active list. This process goes on until all possible shortest links are active.

TCP/IP ROUTING

The TCP/IP protocol suite offers a slightly confusing routing picture, partly due to its nomenclature, and partly due to the structure of the internetwork that spawned TCP/IP.

The Internet community uses the word "gateway" to describe a device that the rest of the internetwork marketplace calls a router. In the Internet, the devices serving as gateways do so within two interconnected environments: One, they link the subnetworks within individual universities and research institutions into a large network, and, two, they act as the physical links between individual universities and research institutions.

Also confusing is the Internet's use of the terms *interior gateways* and *exterior gateways* to describe certain gateways, or what are normally called routers. These gateways (hereafter called routers) perform specific duties within the hierarchical Internet architecture, with the duties matching the services required by this structure.

The TCP/IP routing scheme used in the Internet relies on interior gateways, or routers, (noted with "I" in the illustration), to move data packets within an autonomous system, such as the network at a university campus. Exterior gateways, or routers, (noted with "E"), pass data packets between these autonomous systems.

INTERNET ROUTING ARCHITECTURE

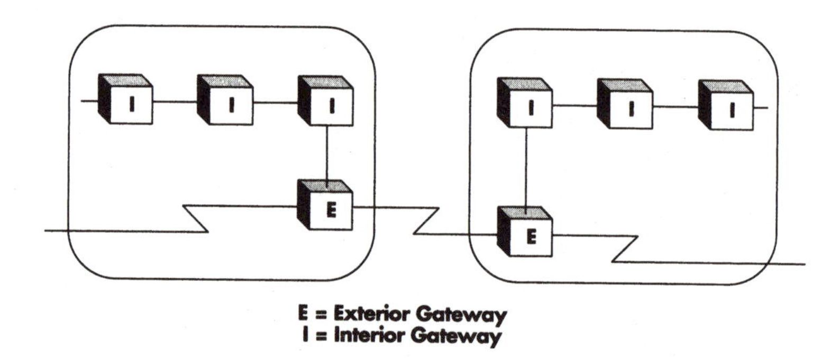

■ **Figure 1:** The TCP/IP routing scheme used in the Internet relies on interior gateways, or routers, (noted with "I" in the illustration), to move data packets within an autonomous system, such as the network at a university campus. Exterior gateways, or routers, (noted with "E"), pass data packets between these autonomous systems.

An interior router moves information *within* an *autonomous system.* An autonomous system is a group of networks under the control and authority of a single entity. An example is the networked computing resources at a single university. Among the Interior Gateway Protocols (IGPs) are RIP and OSPF.

An exterior router moves data from one autonomous system to another—that is, from one university's internet to another's. The TCP/IP Exterior Gateway Protocol (EGP) is an example of this type of protocol.

This tutorial, number 35, by Jim Carr, was originally published in the June 1991 issue of LAN Magazine/Network Magazine.

Configuring A Router

Learn the ins and outs of the smartest internetworking devices.

While switches and bridges can often configure themselves automatically—at least to the point where traffic will flow and some of the objectives of these devices will be fulfilled—the plug-and-play, full-service router doesn't exist. The subject of router configuration is enormous, encompassing the specific behavior of hundreds of hardware interfaces, protocols, and parameters, as well as many vendor-specific methods and applications. However, there are several tasks common to most router applications that are worth trying to understand better.

Whether the router provides a simple ISP connection from a two-node network over an analog line or serves as a multi-gigabit collapsed backbone connected to numerous ATM and FDDI segments, there are several fundamental configuration tasks you need to know about. You must configure the physical interfaces to LANs and WANs. On each interface, you need to enable support for needed traffic protocols. For each interface-protocol combination, you must either define routing tables or configure support for an automatic routing table update protocol. To implement traffic policies, including security and priority setting, you need to define filters for each interface-protocol combination.

PHYSICAL INTERFACES

With the exception of special purpose "one-armed" routers, routers have at least two physical interfaces. While wide-area connections require trickier and more esoteric information for configuration, local-area connections are not always trivial to set up. For example, Ethernet interfaces that can handle either 10Mbit/sec or 100Mbit/sec links may need to be explicitly configured to auto-negotiate the speed. In addition, the congestion control some 100BaseT vendors provide won't work on auto-negotiated links. FDDI networks include powerful fault tolerant features, but they must be configured properly on the router interface.

Setting up WAN interfaces can be a challenge if your experience is limited to LANs. First, there's the Physical layer, a line whose other end you generally don't control (your two-wire or four-wire cable most likely gets terminated in a telephone company facility somewhere). You're probably at the mercy of the telephone service provider for information about whether the line is working properly.

If the link is a dial-up analog line, you will need to set up modems, perhaps by creating a script that issues AT commands to the modem and that sends along a user name and password to authenticate a session. ISDN modems can gracefully and rapidly dial their destinations, but there are a multitude of decisions to make with respect to the two BRI channels. Do you want the second channel to be active all the time, or only when the first line is saturated, or never? If the second channel is up and a call comes in, should one of the lines unbond itself and answer the incoming call?

Whether it's a dial-up or an always-live connection, the link will need to be configured for one of the framing protocols—PPP, frame relay, LAP-B (Link Access Protocol-Balanced, for X.25), HDLC (High-level Data-Link Control), or another Layer 2 protocol. Often, the necessary information will come from the telecommunications service provider. For example, you will need to tell the interface which Data Link Connection Identifier (DLCI) to use any time a link is made up of a frame relay permanent virtual circuit. Encryption and data compression may need to be configured at Layer 2 also. Frame types are often closely linked to a particular interface and a specific service, so many configuration choices can be handled by the manufacturer's default values and by the information your service provider is obligated to supply.

KNOW YOUR PROTOCOLS

Configuring one or more Network-layer protocols on any given interface is the task that first comes to mind when you think of configuring routers. Before we discuss network protocols, it's important to recall that most modern routers are also bridges. If you have multiple protocols running on your network—and who doesn't these days—you have to decide what to do with each one when it gets to the router.

Some packets, such as LAT and NetBIOS, can't be routed, because they don't have Network-layer source and destination addresses. Therefore, if you want the users connected to two interfaces to be able to run NetBIOS-based peer-to-peer networking, for example, you will have to enable bridging between the two interfaces. Then, unless you explicitly filter particular kinds of traffic, every protocol that isn't routed from one interface to the other will be bridged. For example, if you route only IP and IPX and turn bridging on for the sake of NetBIOS, you will also be bridging AppleTalk broadcasts.

Other packet types, such as AppleTalk and old versions of IPX, can be routed; yet, even when they are routed, they might still clog a wide-area link with more overhead than better-behaved protocols would. If you bridge these chatty protocols over a dial-up line, whether analog or ISDN, you better hope there isn't a meter going "kachingg" every minute or two, because those lines will be up around the clock—unless you're a genius at filtering and spoofing.

The most common network protocol you're likely to want to route is IP. Typically, each router interface will be assigned an IP address that is part of the same subnet as the other nodes on the segment. (For more details on IP addresses and subnetting, see the Tutorial "IP Addresses and Subnet Masks," page 27, October 1995, by Steve Steinke.) The interface will also need to have a subnet mask specified. In most cases, you will also specify a default gateway address and a DNS server address. If you are setting up Internet access, the values for these addresses will be provided by your ISP.

If there are Novell servers in your network, you most certainly have IPX traffic. By default, any NetWare or IntranetWare server with multiple interfaces will route traffic between them. Whether you are configuring a single-purpose router or a server that functions as a router, the essential configuration step is to assign a unique network number to each interface on the IPX internetwork. IPX has one significant advantage over IP in that all you need for a new subnet is an interface and a network number. You don't have to be concerned with scarce, carefully reserved 32-bit addresses, address classes, and subnet masks as you do with IP.

One peculiarity of IPX is that there are four possible Ethernet frame types on Novell networks: 802.2, 802.3, Ethernet II, and SNAP (Subnet Access Protocol). If there is a mixture of these frame types on a network, there can be odd effects. For example, some routers can only route one of these frame types on an interface at a time, so if bridging is disabled, devices configured for the nonrouted frame types may be invisible.

SUPPLYING ROUTES

Aside from accepting incoming traffic and forwarding outgoing traffic on its interfaces, a router must keep track of how to move packets a step closer to their ultimate destination. The mechanism for storing this information is the routing table. (If bridging is enabled, bridging tables must be present. Bridging tables associate specific MAC addresses—actual physical nodes—with specific interfaces, while routing tables are concerned only with associating network addresses with specific interfaces.)

There must be a separate routing table for each enabled network protocol: one for IP, one for IPX, one for SNA, one for AppleTalk, one for VINES IP, one for DECnet, and so forth. The good news is that there are routing information protocols that can automatically populate and update routing tables. The bad news is that there may be several of these routing information protocols for each network protocol, and you might have to tweak each of them separately.

Sometimes, especially with point-to-point links, there's no need for the traffic and configuration overhead of dynamic routing tables. In these cases, you can define static routes that will accomplish your goals. There may also be times when you choose to employ a combination of static and dynamic routes.

For traditional IP networks, the Routing Information Protocol (RIP) is widely used. There are two versions of RIP, with valuable extra features in version 2. You may have to decide whether to listen for RIP broadcasts, send RIP broadcasts, or both. There are features of RIP designed to stamp out unproductive routing loops—split horizons and poison reverse—which you might choose to implement. Some RIP implementations can be set for triggered updates, where updates are governed by route changes rather than by a fixed schedule.

The OSPF (Open Shortest Path First) protocol for dynamic-routing table updates has become increasingly important in recent years because it uses less network throughput capacity than RIP for updates. Additionally, it converges on a stable configuration after changes occur more quickly than RIP.

OSPF, RIP, and Cisco's proprietary IGRP (Interior Gateway Routing Protocol) and EIGRP (Enhanced IGRP) are all designed primarily for use within a particular Internet domain or Autonomous System (AS)—that is, an area with common administration and a specific routing strategy. Other routing information protocols are better suited to networks with inter-AS traffic, such as the Internet backbone. These protocols include EGP (Exterior Gateway Protocol) and various versions of BGP (Border Gateway Protocol).

Special-purpose applications might also require separate routing tables or parameter settings on existing routing information protocols. Multicasting, for example, can be accomplished only if routers are configured to handle it. Networks that implement the Resource Reservation Protocol, RSVP, might select tortuous routes when not all the routers in a domain support it.

RIP, IGRP, and EIGRP have also been implemented for IPX-based networks. The IPX link state protocol, analogous to OSPF in the IP world, is NLSP (NetWare Link Services Protocol.) NLSP can greatly reduce the fat of frequent RIP broadcasts, which have traditionally consumed scarce WAN bandwidth on large Novell networks.

POLICY AND SECURITY

With properly configured interfaces, network protocols, and routing information protocols or static routes, the traffic will go through routers. In fact, traffic that you'd prefer not to have routed may go through. Routers can be configured with filters to help secure the network and to keep it focused on your organization's goals.

Filters can be configured on the German model—everything not explicitly permitted is prohibited—or the Italian model—everything not explicitly prohibited is permitted. Thus, you may elect to filter all traffic except that which comes from specific source addresses. Or you may want to forward every protocol except DECnet and VINES IP. Any characteristic that is transmitted in a packet can be used for filtering: source or destination addresses, protocol types, port IDs, application-specific indicators, and even particular data. This is the case because filtering is implemented primarily by defining a bit mask and comparing the masked value in each packet to some other value.

Numerous filters, with various dependencies and Boolean relationships to each other, may have to be defined for each protocol on each interface. Fortunately, most vendors have many useful filters predefined, and some have tools for defining, testing, and implementing them efficiently and safely.

Configuring routers is not a job for amateurs, or even for careless professionals: Ask the folks at Network Solutions and other ISPs who have browned out chunks of the Internet in recent months with mistyped command lines. The spirit of this tutorial is to alert readers to the scope of the task, and, perhaps, to help organize and make more manageable the prospect of learning to master this job.

This tutorial, number 111, by Steve Steinke, was originally published in the November 1997 issue of Network Magazine.

Token Ring Internetworking

While most of the internetworking frenzy has been directed at internetworking Ethernet LANs, connecting Token Ring LANs encompasses a separate set of issues. Unlike Ethernet, bridging predominates in Token Ring environments. Source routing, which is used in Token Ring, is a more sophisticated form of internetworking than transparent bridging, and this has staved off the onslaught of routing. Yet, as MIS builds increasingly complex networks, routing will become a more critical element in Token Ring internetworks. When internetworking Token Rings, you have three options: source-routing bridging, routing, and source routing transparent.

SOURCE ROUTING

The simplest way to interconnect Token Ring networks is to bridge them using source routing. In transparent bridging, which is used for Ethernet LANs, the internetworking takes place at the Medium Access Control (MAC) sublayer of the OSI Data-link layer. In source routing, the internetworking occurs at the Logical Link Control (LLC) sublayer, a half-layer "higher" than the MAC sub-layer. This higher level of operation, combined with a richer frame format, gives source routing a higher level of functionality than transparent bridging.

Unlike transparent bridging, in source routing, the sending and receiving devices help determine the route the frame should traverse through the internetwork. The route is discovered through broadcast frames sent between the source and destination frames.

Source-routing bridge hardware is simpler than routers, making them easier to build and less expensive. Source routing also has advantages over transparent bridging. Because the same path is used for the duration of the session, the process is efficient, except for the setup time to determine the route. Source routing optimally uses parallel and redundant paths, so links are not left idle. It also allows IBM controller timing to be properly set for extended LANs.

Source routing has its disadvantages. The route discovery process requires more overhead than transparent bridging. If many devices are simultaneously performing route discovery, the network slows down considerably. The bridge selects the best path at the time of the session's commencement; if the link is heavily loaded, the two devices are stuck with a possibly inefficient path for the session's duration. Transparent bridges choose a new path per frame. Source-routing functionality must be engineered into the bridges, making them more expensive and complex than transparent bridges.

Two types of route discovery frames are used: All-Routes Broadcast (ARB) and Single-Route Broadcast (SRB). An ARB frame traverses all possible routes between end stations, while an SRB takes only one route between end stations. To determine the best path between sending and receiving stations, the sending station transmits an SRB.

When the SRB reaches its destination, that station issues an ARB. Multiple copies of the ARB travel over all possible routes back to the originating station. Note that an ARB can never travel the same ring twice nor exceed the hop count limit—source routing uses the spanning tree structure, which can be configured so only one copy of the frame reaches each ring.

As the frame passes through a bridge, the bridge inserts its own bridge number and the numbers of its attached LAN segments into the frame's Routing Information Field. Each Token Ring segment and bridge are assigned identifying numbers, and the combination of LAN and bridge numbers is unique for each bridge. The bridge also indicates the maximum frame size, which will be used to negotiate the frame size for 4Mbps and 16Mbps Token Ring.

The originating station receives an ARB frame back for each possible path from the sender to the destination as well as the maximum frame size. The originating station selects the most efficient path based on the first frame to return, the number of hops between the source and destination, and the frame size. For the transmission session's duration, that route is used.

When a frame already contains a route, the frame's Routing Information bit is set to notify the bridge that routing information is available. The source-routing bridge examines this bit to filter frames. If the bit is not set, the bridge ignores it. If the frame is not a broadcast (and therefore is data) and the bit is set, then the bridge checks to see if its own ring and bridge numbers are contained. If not, the bridge ignores it. If a match occurs, the frame is forwarded over the link to the next LAN. The bridge sets bits to indicate that it has been copied, which prevents frames from circulating endlessly in the network.

The majority of Token Ring networks use source routing, but it is not appropriate in some cases. For example, if you need to connect Token Ring and Ethernet networks, bridging won't be sufficient.

NETWORK-LAYER ROUTING

You can also interconnect multiple Token Ring networks with routers. Whereas bridges operate at the Data-link layer, routers operate at the Network layer, one level higher. Routers deliver a higher level of functionality than bridges, but they incur greater overhead. Routers must be built specifically to handle Network-layer protocols, such as TCP/IP or IPX/SPX. Many also include bridging functionality for protocols that can't be routed.

As MIS departments consider consolidating SNA and PC LAN networks into one "super-network," many are looking to routing. Companies with Token Ring LANs also look to routing to provide the same level of security, flow control, and path control that Ethernet LANs and routers provide.

In a routed network, routers communicate with each other, learn where the end-station devices reside, and manage the traffic flow. They require a specific protocol to do so. For example, most TCP/IP routers use the Routing Information Protocol (RIP), but some use Open Shortest Path First (OSPF). Routers can detect congestion on a particular link and send network traffic over a less congested link.

If a particular link fails, the router reconfigures the network around the failure. This convergence presents a strong advantage over bridging. Source routing determines its paths statically; if a link fails, the session must be restarted. With routing, if this convergence happens quickly enough, the user won't notice the difference. If the network does not converge quickly enough, the session is dropped and the user loses the connection.

One problem with routers in Token Ring LANs is most IBM networks don't include routable protocols. SNA/SDLC and NetBIOS are not routable protocols. They both lack Network layers and thus can't be handled by routers.

One solution is to use a router that encapsulates Token Ring traffic into TCP/IP packets. This method adds the overhead of encapsulating your traffic into TCP/IP at the transmission side and de-encapsulating it at the receiving end.

If you encapsulate SNA traffic, you may run into timing issues. In an SNA network, the session will be dropped if the controller does not respond within the preset time. Some MIS shops reconfigure their VTAM tables to allow for a slower response time. Some manufacturers add controller-like functionality into their routers to fool the front-end processor into thinking that the router is a cluster controller. "Poll-spoofing" is a tricky feat to accomplish. Test this solution carefully before you implement it in a production network.

Another solution is to add TCP/IP to your network. If TCP/IP runs on all hosts and LAN servers, TCP/IP routers will manage the traffic. This solution makes the most sense if you are largely a TCP/IP shop or if your MIS department is willing to invest in learning TCP/IP technology.

A third solution is to implement a nonstandard routing method (available from different manufacturers) that gives good response time and can quickly reconfigure a network after a failure.

SOURCE ROUTING TRANSPARENT

Source routing transparent is a truce between the Token Ring source-routing camp and the Ethernet transparent-bridging camp. It provides a way for Token Ring and Ethernet LANs to interoperate.

Source routing cannot operate in a transparently bridged LAN, since the frames lack routing information and the bridge has no way of knowing that the frame should be forwarded. You can use transparent bridges in a source-routing environment; however, the routing information is ignored, and the advantages and overhead of source routing are wasted.

Frame size is a problem when you have both 4Mbps and 16Mbps rings, since transparent bridges can't indicate the correct frame sizes, while source-routing bridges can. In many instances, duplicate transparent and source-routing bridges must be used to carry the two types of traffic, which clearly is inefficient.

The upcoming IEEE Source Routing Transparent (SRT) specification addresses the coexistence and interoperability of Ethernet, Token Ring, and FDDI. A source routing transparent bridge allows both source routing and transparent data to be passed. An SRT bridge will source route frames with embedded routing information and transparently bridge those that lack this information.

Source routing transparent implements two logical paths for the two frame types. All frames are filtered to determine if the routing information bit is set. If it isn't, the frame takes the transparent path. If it's set, then it takes the source-routing path.

Source routing transparent defines three frames that can be used for route discovery: All-Route Explorer (ARE), Specifically Routed Frame (SRF), and Transparent Spanning Frame (TSF). The ARE frame is equivalent to source routing's ARB frame; it traverses all routes between end stations. The SRF is issued in response to the ARE. The TSF lacks routing information but performs the SRB's function.

These new route discovery methods are not supported by all IBM source routing software, so changes in some end-station source-routing software will have to occur.

The SRT standard allows the end station to discover the route, which source routing does not permit. The destination station performs this function in three instances: when it becomes the source, when it gets routes from the ARE sent by the source, or when the destination picks up the route from the data frames that were sent from the source. The latter allows the destination to discover the route without creating overhead.

In SRT, end stations have more robust options for choosing the best route for a particular session. For example, they can reserve more than one route to be used as backup if the primary route fails during transmission. Route selection criteria include the first frame returned, the lowest number of hops, the first route returned with the largest frame size, or any combination of the above. Once the route is determined, the path is inserted in the data frame, and the bridge decides whether or not to forward it based on the routing information.

SRT does not enable a transparently bridged LAN to talk to a source-routed LAN; it cannot translate among different frame formats. SRT allows the mixing of source-routed and transparently bridged networks on the same internetwork, alleviating the need to purchase duplicate hardware.

This tutorial, number 48, by Patricia Schnaidt, was originally published in the July 1992 issue of LAN Magazine/Network Magazine.

Mainframe Gateways

IT is experiencing a revolution. Data is no longer locked up in the glass-walled, air conditioned, computer room. Information is a key part of a company's strategy, and as such, must be easily accessible. The network can be used to bring that corporate data to the desktops of the users. Once the data appears on the user's PC, he or she can manipulate it by changing variables, trying different views, and plotting graphs. Or the LAN-to-mainframe link may be part of a mission-critical application such as an airline reservation system. Or perhaps users simply need to share electronic mail with mainframe users.

Getting this information requires a physical link between the PC and the host. A mainframe gateway is a combination of hardware and software that allows PCs, PS/2s or Macintoshes on a LAN to communicate with a mainframe. We will limit our discussion to gateways to IBM SNA mainframes, such as the 30xx, 43xx, and 9370.

GATEWAY TYPES

The hierarchy in the mainframe world is straightforward. The mainframe is the central repository of data and applications. Access to this information is granted through a system that runs on the mainframe. This system is usually Virtual Telecommunications Access Method (VTAM). VTAM contains information about every device on the SNA network. Terminals are wired to cluster controllers. The terminal family is referred to as 3270.

Introducing PCs into the equation allows more sophisticated applications to be put into place. It also disrupts the hierarchy, since a PC can act as both a terminal and a cluster controller. Most importantly, the PC brings about drastic changes in the politics in IT.

The most basic micro-to-mainframe link is the coax adapter. The two de facto standard coax adapters are from IBM and DCA (now Attachmate). The PC also runs terminal emulation software so it can act as a 3270 terminal. But the coax adapter is not a LAN-to-mainframe connection. Each PC must have an adapter and coax cable. If it is remote, it needs its own synchronous modem. With a network, however, fewer of these components are needed, and their cost can be distributed among a number of users.

A DFT coax gateway is basically a LAN version of the coax card. Unlike the other types of gateways, a DFT gateway allows a PC to emulate a terminal. Distributed Function Terminal (DFT) is a terminal mode that has largely replaced the older Control Unit Terminal (CUT) mode. DFT assumes the device talking to the mainframe is intelligent (as opposed to a terminal). In CUT mode, the cluster controller is responsible for displaying the data, whereas in DFT mode, many of the controller's functions are distributed to the intelligent device.

Like a DFT terminal itself, a DFT coax gateway may have up to five concurrent sessions. This means that at most five people can communicate with the mainframe at the same time. One way to increase the number of concurrent sessions is to install multiple coax cards in the gateway PC. Another method is to buy a coax gateway that uses multiplexing to support up to 40 sessions on a single card.

A DFT gateway is cost efficient if you already have a cluster controller in place. They are also relatively easy to install. However, they have few sessions to offer users.

Because most users are located at a different location than the mainframe, the SDLC gateway is widely used. A Synchronous Data Link Control (SDLC) gateway connects geographically remote users to a mainframe using modems and telephone lines. One PC on the LAN is designated the gateway and it emulates the 3174/3274 controller. This PC contains the SDLC board, gateway software, and a synchronous modem.

Typically each SDLC gateway supports up to 32 workstations with a total of 128 simultaneous sessions. (The 128 limit was traditionally a limitation of NetBIOS.) As for modems, 9600 bps and below is fine for interactive work, but file transfers and remote applications will require 19.2 Kbps and above. An AT with a standard gateway adapter will support speeds up to 19.2 Kbps. For speeds of 56 Kbps, an intelligent gateway adapter will be required.

Keep in mind that the modem bandwidth and the adapter card's CPU must be divided among the users. Although 128 simultaneous sessions is the theoretical limit, the practical limits are set by the type of applications the users run.

The *Token Ring Interface Coupler* or TIC is the newest method of connection. Made by IBM, the TIC is basically a Token Ring interface adapter for cluster controllers, midranges, front-end processors, and mainframes. It can be a local or a remote connection.

Depending on the TIC card and Token Ring, transmission is at 4Mbps or 16Mbps; at either speed, it is significantly faster than coax or SDLC methods. The TIC is the most logical method if users on an existing Token Ring need access to a local mainframe. A TIC is also essential for cooperative processing applications built on APPC.

Each workstation on the Token Ring can be addressed directly, or one PC may be set up as the gateway. Although direct addressing provides performance benefits, it increases management efforts. Each workstation must be defined in VTAM as a Type 2 Physical Unit, thereby adding overhead at the mainframe level and increasing the amount of work for the people who maintain the system. From the CPU's and systems programmer's point of view, it is more efficient to designate one PC as the gateway. The mainframe polls the gateway PC, and the gateway is responsible for polling each workstation.

A fourth gateway type is a channel-attached gateway. This is the highest performance method, and is most often used when other mainframes, mini-computers, and engineering workstations need to communicate with the mainframe. Because a coax gateway and an SDLC gateway communicate through a cluster controller or front end processor, they are limited by the controller to a raw throughput of 56 Kbps. A channel-attached gateway is connected directly to a mainframe's I/O channel, and while speed varies according to manufacturer, you can get bandwidth of about 20Mbps.

SESSION ISSUES

Most gateways are PC-based. The gateway PC can run either DOS, Windows, or OS/2. The workstations communicate with the gateway via a communications protocol, most often NetBIOS. Some gateways tailored for the NetWare environment use IPX. The OS/2 gateways generally use Named Pipes.

The most basic function of a mainframe gateway is distributing the mainframe's sessions to the PC users. Gateways allocate these sessions or Logical Units (LUs) statically or dynamically. A PC or terminal explicitly defined in VTAM is said to have static or dedicated LUs. So a particular user, say John in Accounting, always gets a particular session, say LU #2. In dynamic pooling, sessions are grouped together. So one pool could be printer sessions, another, terminal sessions. Pooled sessions are distributed on a first-come, first-served basis.

Users with high priority or who deal with sensitive information should have dedicated sessions. Dedicated LUs simplify management, since the mainframe personnel always know who has access to which session. The disadvantage is that VTAM tables become very large. And every time a user is added, the tables must be updated. Dynamic allocation insures flexibility and eliminates waste. A mainframe gateway should support both methods.

On the whole, mainframe gateways offer very little in the way of management capabilities. Functionality can range from the very simple—reporting which user has which LU—to the very complex—reporting information back to IBM's NetView. This is one area where improvement is needed.

WORKSTATION SOFTWARE

The gateway PC takes care of the basic communication and session handling with the mainframe. But each workstation must run software as well. The first key item is terminal emulation. The PC running DOS, OS/2, Windows or the Macintosh must be able to act as a 3270 terminal. Of the 3270 family, the 3278 and 3279 color displays are the most widely emulated. The 3178 and 3278 monochrome displays are also popular. If you need to display the full screen formats of the larger displays, each PC may need to be equipped with a VGA or better graphics adapter.

In addition to terminal emulation, a gateway may also need to provide printer emulation. This will permit PC users to print mainframe data to their local and LAN printers. Most gateways emulate the 3287 printer; some emulate the 3286.

File transfer is as essential as terminal emulation. Most support IBM's host-based file transfer program called Send/Receive or IND$FILE. Most have a proprietary file transfer that is generally a great deal speedier than IBM's. Some increase the speed of Send/Receive by using larger buffers. On the other hand, that requires more memory on the PC side. The gateway should also allow users to transfer files in background mode. With background file transfer, a user can set up a file transfer and switch to an interactive terminal session while the transfer takes place.

Most gateways allow users to have multiple, simultaneous sessions on the mainframe. Some gateways limit this to a specific number, say two terminal sessions and one printer session per user. Others allow one user to hog every session the gateway has available. Most people would find it difficult to keep track of more than three simultaneous sessions (one for calculation, one for interactive work, and one for printing). It is very useful if the software supports windowing, so the various sessions can be overlaid on a single screen.

The PC keyboard must be remapped to emulate the 3270 terminal keyboard. The difference between the <ENTER> and the <RETURN> keys on the 3270 keyboard must be reconciled. The PF keys must be added to the PC keyboard. These problems are compounded by the differences in the XT and AT keyboards. Often, the workstation software allows the user to pop up a map of the keyboard. Some allow the user to remap the keys on the fly. Another handy feature is to allow users to build macros to automate common functions, such as logging in or retrieving mail.

If the users need to access a mainframe application that makes use of SG2 or APA graphics, special software is required. SG3, the older graphics method, has been replaced by vector graphics (which is also called All Points Addressable). In APA graphics, the mainframe sends only the vector points, and the terminal draws the image on the screen. The 3179-G and 3279-SG3 displays support graphics.

By the time an operating system, a network operating system, and gateway software is loaded into a PC's memory, there's precious little RAM left. Generally, the amount of available RAM and the functionality is a tradeoff. The high end functions, such as graphics or APPC, tend to require a great deal of RAM. If you need these high-end features, consider a gateway that supports expanded or extended memory or one that is able to page to disk. Or consider one that supports OS/2 workstations.

This brings us to a hotkey to DOS. While many gateways offer the user the ability to hotkey to DOS, in reality, there is often little RAM left to do more than a directory listing. Also, when you switch to DOS, often times the file transfer or other process will stop because DOS is not a multitasking operating system. Some vendors have developed a workaround. But if you want true multitasking, you will need OS/2, 32-bit Windows, or UNIX.

PROGRAMMING INTERFACES

Terminal emulation, file transfer, and print services are basic, but essential functions. As the PC becomes a partner with the mainframe, rather than a souped-up terminal, more sophisticated applications are possible. Application programming interfaces (APIs) are the enabling mechanism. APIs allow programmers to move part of an application down from the mainframe and onto the PC. So for example, a loan application can be entered and verified on the PC, then sent to the mainframe. Mostly, APIs are used for far simpler tasks. Users can automate simple tasks, such as downloading files.

Nearly all gateways have some type of programming interface, whether proprietary to that manufacturer or compatible with IBM's. HLLAPI (High Level Language Applications Programming Interface) and 3270-PC API are two of the best-known IBM APIs. They allow you to present information in a way different from the traditional mainframe. For example, mouse support is enabled through HLLAPI. EEHLLAPI or Entry Emulator HLLAPI is a subset of HLLAPI. These APIs are popular because they do not require any changes to be made on the host side.

Enhanced Connectivity Facilities/Server Requester Programming Interface (ECF/SRPI) is a newer API for cooperative applications. It is easier to implement than Advanced Program-to-Program Communications (or APPC). APPC is used in applications where the PC or midrange is considered the peer of the mainframe. This is a radical departure from the traditional terminal-midrange-mainframe hierarchy. APPC is primarily used in high-end transaction processing applications, such as done in airline reservations or banks.

This tutorial, number 22, was originally published in the May 1990 issue of LAN Magazine/Network Magazine.

SNA and LU6.2 Connectivity

When IBM introduced Systems Network Architecture (SNA), its strategic wide area networking product family, in 1974, mainframes dominated computer environments. Because they were ruled by mainframes, the computer networks in use were hierarchical, or tree-structured, with the mainframe at the top of the inverted tree. This arrangement required lower-level systems on the network to communicate with each other through the mainframe rather than directly with each other, which was a tremendous waste of host-based resources.

The popularity of personal computers in the 1980s changed the structure of corporate computing, however. No longer were computing resources located in or controlled from a centralized location; they were now distributed throughout an organization, in dozens (even hundreds) of PCs as well as the mainframe.

The growth of PC-based networks further strengthened this distributed environment. With hundreds of PCs connected to networks—all needing to communicate with each other—it became intolerable for most organizations to rely solely on the traditional SNA-type network. No longer could they allocate host resources to managing communication among the dozens of computers on the network.

In 1982, IBM responded by adding a protocol called Logical Unit (LU) 6.2 to SNA. LU6.2 makes all computers peers on an SNA network, including hosts. With LU6.2, the mainframe no longer plays dictator to its "slave" counterparts on the network. LU6.2 brought other changes to SNA as well. Rather than forcing devices connected to SNA to act as dumb terminals incapable of handling processing, LU6.2 permits cooperative processing. Each processor on the network can do what it does best rather than rely on the mainframe. It also allows the dynamic allocation and tuning of SNA networks, substantially reducing the need for operator intervention.

SNA'S STRUCTURE

SNA, with IBM's market presence behind it, has been a de facto data processing and networking standard for many years. IBM has developed and installed sufficient hardware and software to make SNA the world's most widely installed network topology, with more than 22,000 sites.

As a protocol suite, SNA offers functionality similar (and is an alternative) to the Transmission Control Protocol/Internet Protocol (TCP/IP) and Open Systems Interconnection (OSI) protocols. Like these protocols, SNA offers a layered approach to communications. From the top down, SNA's seven layers are: end-user, network-addressable unit services, data flow control, transmission flow control, path control, data link control, and physical layers. Together, they provide services that are more or less synonymous with those of OSI and TCP/IP. However, the layers' functions do not completely correspond from one protocol to another.

These layers handle the following tasks. The physical layer (layer 1) moves data between computers on a network. The data-link layer (layer 2) uses the Synchronous Data Link Control (SDLC) protocol to pass data across the physical interface. The path control layer (layer 3) manages routing and traffic control, packing data together to increase throughput. The transmission control layer (layer 4) initiates, manages, and concludes transport connections or sessions, controlling the data-flow rate between layers 3 and 5. The data-flow control layer (layer 5) determines which LU can transmit next and helps manage error recovery. The NAU services layer (layer 6) handles presentation services to layer 7. The end-user layer (layer 7) provides the user interface.

In practice, SNA and LU6.2 use only layers 4 through 6. In this tutorial, we'll describe the functions and features these layers provide.

DEFINITIONS OF PHYSICAL AND LOGICAL UNITS

Physical Units		**Logical Units**	
No.	**What it defines**	**No.**	**What it defines**
PU1	Terminals	LU0	Application presentation services
PU2	Controllers	LU1	3287 emulation, network & remote job
PU2.1	Enhances	LU2.0	Entry
PU3	Not defined	LU2 3278/9	Terminal emulation
PU4	Front-end processors	LU3	3270 Printer connections
PU5	Mainframe hosts	LU6.2	Peer-to-peer communication

LU, PU FUNCTIONS

Before continuing, let's define logical unit, and its hardware-specific counterpart, the physical unit. Logical units (or LUs) and physical units (PUs) are the two primary functional entities in an SNA network. Both LUs and PUs are also referred to as network-addressable units (NAUs). (For a breakdown on LUs and PUs, see "Definitions of Physical and Logical Units.")

In SNA parlance, PUs are physical systems or nodes connected to each other via cabling. PUs are also known as Node Types, or NTs. IBM has defined five node types, including hosts (PU5) and terminals (PU2), and left a sixth available for future definition. For the sake of simplicity, we'll refer to node types/physical units only with the PU designation—such as PU2.0—although it's just as accurate to label PU2.0 as NT2.0.

Logical units are electronic entities connected by sessions. Just as there are five PUs, there are five LUs, including LU6.2, that make SNA network resources, including disk files and processing cycles, available for application software. A node on an SNA network can be a physical unit (such as hardware) as well as a logical unit (such as a logical session connection).

In essence, LU6.2 acts as an interface, or protocol boundary, between SNA and an end user's application. Closely associated with LU6.2 is PU (or node type) 2.1, which is an enhancement to PU2.0 for cluster controllers. PU2.0, also known as 3270 terminal emulation, is the protocol most often used to connect personal computers into an SNA network. PU2.0 also lets other devices on an SNA network access a mainframe by emulating a cluster controller.

PU2.0 has some limitations, however. Most importantly, it allows devices to access only a host, not other peer nodes. And it requires one or more System Service Control Points (SSCP), which start and stop sessions. PU2.1 remedies these shortcomings.

PU2.1 includes all the features of PU2.0, with major upgrades. It allows SNA devices to connect to peer nodes without mainframe assistance, and it permits running multiple sessions simultaneously. And while PU2.0 permits only peripheral node connections, PU2.1 provides peer-to-peer connectivity. PU2.1 also requires no central control point to manage session services.

LU AND PU TYPES

There are five LUs and an identical number of PUs. We've described one of each already. The table provides a brief description of the full set of physical and logical units.

Of the logical and physical units listed in the table, LU3 is outdated. LU6.2, which can be considered a subset of SNA, consists of a base set of features plus 41 options. IBM implements LU6.2 across its entire product line, although the specifics vary from product to product. In its most popular form, users know LU6.2 as Advanced Program-to-Program Communication (APPC).

Users purchase APPC in the form of a developer's toolkit, which allows them to create transaction programs capable of using entities called APPC verbs. These APPC verbs are used within the transaction program to get LU6.2/APPC to perform functions the program needs carried out. Although APPC and LU6.2 are synonymous, their verb names differ slightly. With a few exceptions—for example, APPC/PC does not support some of the more obscure features of LU6.2—APPC functionality is virtually identical to that of LU6.2.

LU6.2/APPC functions primarily as a resource allocator and controller. This means that LU6.2/APPC ensures that programs have access to network resources when they need them, and that network resources are not corrupted, such as when two users attempt to make simultaneous updates to the same file.

IBM has made APPC its preferred LU6.2 implementation in a variety of its systems, including its PCs, System/38/36, AS/400, and 9370 mainframe. IBM has said that APPC is its strategic product for distributed processing in the minicomputer and PC environments. IBM sees APPC as one way of taking PC-to-mainframe market share away from terminal emulation applications.

ON THE PC SIDE

APPC/PC is IBM's offering at the low end of the LU6.2 market. Two versions are available. Version 1.1 is designed for IBM's Token Ring and PC LAN Program networks and standalone micros. The OS/2 Extended Edition APPC is part of the Communications Manager for OS/2.

While backed by IBM, the memory-resident APPC/PC program has met resistance from end users. It consumes 164 KB of RAM, which is too much for most PC users to spare. With only 640 KB of DOS-addressable memory, many users don't have enough RAM to run DOS, APPC/PC, and applications software on a network. This is not an issue for end users running OS/2, which can access up to 16MB of memory; however, only a small percentage of users in the PC environment are using OS/2.

Because of the memory limitation, third-party applications developers have been slow to release software that takes advantage of APPC/PC. If more users migrate to OS/2, developers are likely to release more applications written to APPC/PC.

Another of IBM's LU6.2 implementations is Advanced Peer-to-Peer Networking (APPN). APPN adds network management capabilities to LU6.2 peer-to-peer communications services. APPN also includes a routing capability that can create new routes between nodes on a network dynamically.

This tutorial, number 31, by Jim Carr, was originally published in the February 1991 issue of LAN Magazine/Network Magazine.

Wireless WANs

How many times have you rushed through the airport, notebook smacking against your leg, trying to find a pay phone with a phone jack so you can dial in for messages between flights? For the most part, airport pay phones don't have jacks, and you can't use your wired modem.

And how many times have you checked into a hotel room, only to discover that you can't plug your notebook's modem into a jack in the room? The jack and wires are fused, or the jacks are nonstandard. "DataPorts" on hotel phones are few and far between, though they are very welcome.

You're trying to mate a mobile technology—notebooks—with a stationary concept—dial-up. With your notebook, you can compute anywhere, but the need to communicate with your office or with your customers ties you back down to land.

Then you eyeball your portable cellular phone, sitting there in your briefcase, right next to your landlocked notebook. You're not the only one. Wireless technologies will be able to provide communication for laptops, notebooks, Personal Digital Assistants (PDAs), and even less intelligent devices, such as soda machines.

Talk of wireless networking has accompanied the arrival of notebooks and the talk of PDAs. To be successful, PDAs will have to be wireless. Notebooks are likely to exist in a quasi-wired world; they could go either way. Most of the loudly discussed wireless technologies are for in-building, or local, usage.

Companies such as Motorola, NCR, and Windata are building wireless network products that are designed to be used locally. Even the companies coming out with wireless PCMCIA cards, such as Xircom and Proxim, are using spread-spectrum technology, which is designed for local transmission.

But as the universe of roaming computer users grows, they'll roam farther. And they'll need universal connectivity—wireless connectivity.

UNWIRING THE WAN

Five types of wireless technologies may serve: packet radio, packet cellular, circuit cellular, satellite, and paging. Packet radio, packet cellular, and circuit cellular are the most important.

- **Packet radio.** Mobile users can use the existing packet radio networks, such as RAM Mobile Data and ARDIS, to connect their computing devices to their networks.
- **Packet cellular.** Data communication can also take place over the US cellular network.
- **Circuit cellular.** Data can be carried over cellular voice channels today, in a manner similar to the way data is carried over analog lines by conventional modems. Modem vendors, including AT&T, PowerTek, and Vital, will play a big role in this market.
- **Satellite.** Low earth orbit satellites will carry voice and packet data. Motorola is building its Iridium network to provide worldwide voice coverage. Satellite bandwidth will be too small to serve the needs of most LAN users.
- **Paging.** The same networks that serve beepers may apply, despite low bandwidth.

FOUR FACTORS

Which technology or technologies take hold depends on their coverage, cost, performance, and adapter.

Service coverage must be nationwide and penetrate buildings. Many current wireless LANs fail in that they are local—and often can't cross buildings or even floors.

For nomadic computing to take off, communication coverage must be ubiquitous. Both packet cellular and circuit cellular will offer greater geographical coverage than the packet radio networks, because a huge cellular infrastructure already exists to handle voice calls. Advantage in for packet and circuit cellular.

What is the market price of convenience? Dialing out from hotel rooms and racing from pay phone to pay phone in airports will be cheaper than using wireless communication. According to Forrester Research (Cambridge, MA), circuit cellular costs $.50 per minute, while packet radio and packet cellular charge $.04 per packet and up. Advantage in for packet radio and packet cellular. Cost and performance are a delicate balance. Although packet technologies are lower cost, they cannot elegantly handle fax. Packet cellular has been demonstrated at 19.2Kbps, which is a far cry from the speed LAN users are accustomed to. Fax service, because of the nature of the transmission, requires a circuit connection. For those on the road, faxing is as important as dialing back into the company network. Ad in for circuit cellular.

The adapter is critical. The different technologies' adapters vary in size, weight, cost, and power consumption. The new PCMCIA cards will help reduce the footprint of these adapters from the size of a brick to the size of a credit card.

Initial PCMCIA cards will use spread-spectrum technology (for LAN usage), as well as circuit cellular (for WAN usage). The adapter should be a fraction of the notebook or PDA's cost—anywhere from 10 percent to 25 percent. The adapter must also be simple to insert and remove. Make the adapter too big or difficult to install, and no one will want to transport it or figure out how to connect it.

Power consumption is key. The adapters draw power from the notebooks, placing yet another demand on the notebook's already short-lived battery.

Wireless vendors are acutely aware of the power crunch and are working to manage the power consumption. A packet cellular adapter can be combined with a cellular phone and use the same radio and antenna. The circuit cellular modem cards, as mentioned, will be implemented as PCMCIA cards.

A LOOK AT PACKET CELLULAR

Forrester predicts that by 1997, packet cellular will have an installed base of 1.4 million users in the US, which will be more than half the wireless installed base.

CelluPlaN, in particular, is projected to be the leader. It is a consortium composed of IBM and these nine cellular carriers: Ameritech Mobile Communications, Bell Atlantic Mobile Systems, Contel Cellular, GTE Mobile Communications, McCaw Cellular Communications, Nynex Mobile Communications, PacTel Cellular, Southwestern Bell Mobile Systems, and US West

NewVector Group. Together, they propose a cellular packet technology called Cellular Digital Packet Technology (CDPD).

Packet cellular has a number of advantages. Because the technology relies on the existing cellular voice network (with some modifications), current cellular users will be familiar with it: they'll see cellular packet as an extension of their existing voice service. They'll feel comfortable buying and using it.

Packet cellular used the existing cellular infrastructure, with some modifications. The initial investment in CelluPlaN was $400 million added to the $9 billion cellular partners can use to subsidize their data service investment off their substantial voice revenues.

Also key to packet cellular's future success is its use of existing cellular radios and antennae. The cellular companies have the infrastructure in place—they can simply add a few more components. Most of the carrier and end-user components will cost less because they're already manufactured in such high volume.

ARDIS and RAM, the packet radio networks currently in use, cannot match packet cellular on a number of points. ARDIS and RAM have fewer users, smaller geographic coverage, lower bandwidth (from a LAN point of view), and require specialized equipment.

THE TECH BEHIND THE CURTAIN

The technology behind CelluPlaN is CDPD, which allows data transmissions to be overlaid onto the existing analog voice networks. With CDPD, data is transmitted in short bursts and only when the system detects an idle time between voice calls. Data can be sent without needing a dedicated channel, so data transmission will not impact the system's ability to carry its bread-and-butter—voice.

CDPD sends packets (just like on LANs) across one or more channels as the idle time occurs. Even during peak times, channels are idle between the time when a call is terminated or handed off and the time when the channel is reassigned to a new call.

A technique called channel hopping is used to select the channel that is least likely to be used for a voice call. This technology exists in today's cellular systems to hand calls off to new cells.

Cellular poses a few unique problems for data. When talking over a cellular phone, you can (impatiently) wait out the clicks, blackouts, and detect that the control and user-identity information, and user-authentication procedures are also in place. When a user connects, the system assigns a temporary ID for addressing. With nationwide cellular handing calls from cell to cell, it will be harder to physically eavesdrop—you would have to be sure you hit the right cell control site. The advent of digital cellular will make it harder still. But data encryption seems like the sensible thing to do.

CDPD proponents are publishing their specifications, which will allow multiple vendors to implement the technology. The proponents, except for IBM, are service providers and therefore benefit from other companies manufacturing adapters, modems, and other equipment. The CDPD organizers are also working with the Cellular Telecommunications Industry Association and the Telecommunications Industry Association to make it a standard.

BEYOND THE USUAL APPLICATIONS

The target users of wireless WANs are no surprise: people in sales, service, or those who travel frequently. By giving salespeople a "nomadic office," they're forced to spend more time with customers and less time at their desk. With networks, they can get more immediate access to information, such as inventory and pricing. Service providers can also use wireless WANs. Typical users may be rental car agencies, waitresses, baggage handlers, and loading dock workers. Traveling managers can stay in touch with their offices via e-mail and wireless connectivity.

But some applications may be unique. Consider for example, the soda vending machine that calls a central computer when it runs low on change or out of Mr. Pibb. Gas and electric meters could be imbued with intelligence so they notify the utility company how much gas or electricity the customer has used, and meter readers no longer knock on doors. Parking meters could do the same thing, I suppose.

This tutorial, number 52, by Patricia Schnaidt, was originally published in the December 1992 issue of LAN Magazine/Network Magazine.

Dial-Up Internetworking

Enterprise network occupies a large slot in the industry's lexicon. The term conjures visions of an intercontinental network connecting hundreds of sites and tens of thousands of computers. But enterprise networks of this scale exist in only a small number of companies.

The vast majority of networked companies don't have a multitude of offices flung across the globe, populated by workers who need constant, heavy-duty access to each other's information. Most companies only need remote access on an occasional basis and make relatively minimal demands of many applications.

Look at companies' telephone line purchase patterns. Most phone lines used for data purposes are digital leased lines; to a lesser extent, companies buy fractional T1 and full T1. 56/64Kbps leased lines fill most of the need.

Cost is one reason for the plenitude of 56Kbps lines. Most companies outside the Fortune 1000 can't justify the high price of a T1, unless they piggyback data on their existing private

T1 network used for voice traffic. Per month, a 56Kbps line costs in the hundreds of dollars; monthly costs for a T1 fall in the thousands. [This pricing is on the high side in 2003.]

ON A SMALLER SCALE

The term *enterprise* can and should be applied on a smaller scale. An enterprise network connects every location in your company—even if there are only a few zone office locations. For these smaller, but no less important, networks, internetworking solutions—outside of the usual recommendations of high-speed multiprotocol routers pumping data down multiple T1 pipes—apply. Smaller sites need the same level of connectivity but not the same level of horsepower. If your site is one of these smaller networks, dial-up internetworking is likely to play into your plans.

Most companies don't build private networks for voice, but many build private data networks. Still, for most companies, dial-up lines are the most convenient way to communicate. The public switched telephone network is never more than a phone jack away. It's convenient, it's cheap, and it's reliable.

For data, the drawback has historically been speed. For dial-up, 2,400bps and 9,600bps modems may be the norm, but they provide insufficient speed for more than a single user running an application that requires minimal bandwidth.

Advances in technology have made dial-up a more viable alternative. The majority of the telephone infrastructure is fiber optic, not copper, which reduces the number of transmission errors and translates into higher throughput. With VLSI integration of chips and the use of digital signal processors, modems can operate at higher speeds, at a lower cost. Add some compression software, and a 14.4Kbps modem can achieve data throughput of 20Kbps to 57Kbps. These aren't exactly Ethernet speeds, but neither are they at the 2,400bps modem level.

The bottom line is always the biggest motivator; cost is a reason for using dial-up lines over leased lines. With dial-up lines, you pay for what you use, and only what you use.

With leased lines, you pay a flat rate, no matter how much or how little you use them. For occasional usage, dial-up lines make firm financial sense. According to the calculations of market research firm Infonetics (San Jose, CA), shown in Figure 1, if a company needed one hour per day of connectivity between San Francisco and Pittsburgh, the cost of a leased 56Kbps line would be $1,759 per month, but with dial-up networking, the same connect time would cost $327. That's a savings of $17,000 per year. Think about how many hours of usage you would actually need. An hour a day is enough time to exchange 60MB of information, assuming the connection operates at 25Kbps, according to Infonetics.

The flip side of the raw cost consideration is the throughput cost. Although the cost of using POTS (Plain Old Telephone Service) is cheaper, the time costs are less apparent but are of increased importance if the throughput becomes a problem. After all, there is also a cost factor in the time it takes to transfer files or to run an application.

DIAL-UP VS. LEASED LINE COSTS

Distance	Dial-Up Service	Leased 56KBps Line	No. of Hours to Break Even Price
10 miles	$12 per month + $.60 per hour	$137	208
10 miles	$12 per month + $9.60 per hour	$732	75
10 miles	$12 per month + $14.40 per hour	$1,224	84
10 miles	$12 per month + $15.00 per hour	$1,759	116
10 miles	$12 per month + $15.00 per hou	$1,931	128

Source: Infonetics, San Jose, CA

■ **Figure 1:** Dial-up internetworking can be more cost effective than leased 56Kpbs lines for users who require a part-time connection between two sites.

SPONTANEOUS INTERNETWORKS

You could benefit from dial-up networking if your company has any one of the following scenarios:

- *You want to connect multiple remote offices that have relatively low traffic demands.* Most offices don't need constant, heavy access either back to headquarters or to their branches. Users transfer some files and log into e-mail. The high cost of leased lines has typically prohibited offices with low bandwidth requirements from internetworking, but with the lower costs of dial-up internetworking (and advances in speed), such connectivity can be cost-effective. Small offices can benefit from the increased communication that, until recently, companies with big budgets have been enjoying. For example, a headquarters may allow access to its networks from various branch offices. Another scenario is to let one company access a particular portion of another company's network, say an inventory database or online information service. Many companies have already done so on an informal basis, quite successfully, using standard telephone lines.
- *You want to telecommute.* As notebooks proliferate and as workers become increasingly mobile, the need to access the home network becomes paramount. Dial-up internetworking provides one solution. Asynchronous communication servers and cellular communication are others. Anyone who has used remote communications software to dial in to a network knows the pitfalls. Remote communications software over a modem can be slow and doesn't always work well. And you can't run any significant applications; at best, you can download or upload a small file (hopefully a compressed one). Cellular holds some promise, but until digital cellular is implemented, the bandwidth offered on the voice net-

work is pitifully low. With dial-up internetworking, the remote user participates as a full-fledged node on the network, entitled to all of his or her access rights and applications. With communications software, you are merely an emulation of your former self.

- Although with dial-up, the speeds can be slow and you wouldn't be eager to restructure a database while remote, you can easily check your e-mail and run similar applications. (Hint: Keep your applications software local. You might also want someone at headquarters to administrate your electronic mailbox for you; it can be dreadfully slow if done remotely.) With higher speed lines such as Switched 56 or ISDN, you could run a CPU-intensive application, such as a database or CAD program.
- *You want to manage networks.* Remote communications packages are the tried-and-true answer to remote network management, yet these applications provide relatively limited control and functionality. With dial-up internetworking and the capability to run as a full network node, you can do anything remotely that you can do locally. Systems integrators and in-house administrators can take advantage of such products to provide support remotely for their customers.
- *You want to back up leased lines.* Leased lines are typically quite reliable but nevertheless subject to outages. Dial-up routers can be in place waiting to provide a backup to a leased-line service. When you need the bandwidth, it's there. In this way, you can build a hybrid of a public and private network.

HOW TO DO IT

To engage in dial-up internetworking, you don't need much equipment: a dial-up router on both ends of the link and a telephone line. Dial-up routers are new, with products currently or shortly available from Centrum Communications (San Jose), CMC (Santa Barbara, CA), DCA/ICC (Alpharetta, GA), NEC (Sunnyvale, CA), and Telebit (Sunnyvale).

In essence, a dial-up router isn't much different from a "regular" router, except that it handles dialing up the remote site for the user. When the router detects traffic coming in, it checks whether the data is destined for a remote site. If so, the dial-up router uses its built-in modem to dial that site's telephone number, establishes the connection, and transfers the data.

After a certain period of inactivity, the router breaks down the connection. Dial-up routers will most likely end up in locations where the users lack computer expertise, so find a product that requires minimal human intervention (and, of course, supports remote management).

As with ordinary routers, support for multiple protocols, such as NetWare IPX/SPX, TCP/IP, and Apple AFP, is critical. The router must be able to accommodate the different network users who dial in. Also look for standards-compliance. A TCP/IP router that uses Point-to-Point Protocol (PPP) will be able to communicate with other TCP/IP routers using that protocol; a router implementing a vendor's proprietary WAN protocol will be able to communicate only with its own kind.

Look for routers that are designed to survive in a WAN environment. With a dial-up, you pay for every packet sent over the wire; operating in that environment requires you to be packet-wise. NetWare routers should implement Burst Mode and filter Service Advertising Protocols to reduce the sheer numbers of unnecessary packets.

Some dial-up routers locally acknowledge a NetWare server's keepalive packets, a process that also reduces overhead. "Spoofing" the server with local acknowledgements can be tricky business, and it has to be carefully implemented. Whatever the network protocol, compression is key, both for the packet header and for the actual data portion.

Security is of paramount importance. Dial-up lines are inherently less secure than leased lines; you don't know where the line physically goes and who can access it in the process. Passwords are the bare minimum. Call-back, where the user calls in and the router calls back to a predetermined number, is useful. Tighter security, including better authentication schemes, is needed.

This tutorial has focused on low-speed dial-up lines, but other options exist. Switched 56 is a dial-up, circuit-switched service that provides 56Kbps of bandwidth. Available only in the United States, the phone companies have priced Switched 56 rather aggressively. With the impending deployment of National IDSN, IDSN may indeed become a real service. A basic rate ISDN interface can deliver 64Kbps or 128Kbps of bandwidth for the cost of one or two Switched 56 lines

This tutorial, number 53, by Patricia Schnaidt, was originally published in the January 1993 issue of LAN Magazine/Network Magazine.

SECTION IX

Network and Systems Management

Network Management

Although the physical location of the personal computers on a local area network seldom changes, networks are still dynamic entities. That is, the logical makeup of any network fluctuates from moment to moment.

For example, the number of data and application files in use or stored away, the amount of available disk storage space, the number of users logged in to the network, and the volume of traffic passing through the network cabling all change continually. Moreover, a network offers users a distributed-processing environment, with some processing performed by a centrally located server, some done at users' workstations, adding even more activity to the network.

Keeping this conglomeration of network hardware, software, cables, and the people using them working efficiently comes under the ambiguous term of network management. It's ambiguous in that managing a network can range from the simple to the complex, from a moment's quick fix of plugging in a misplaced network cable to a day-long search for an obscure disk problem.

Network management can be as simple as creating a boot diskette for a new user and making sure that user has proper access to network resources. (Although in truth these jobs may not be all that simple in some widely distributed networks.)

Or managing a network can include daily disk-maintenance duties—backing up network files or defragmenting disk directories. Or it may mean troubleshooting the network, trying to discover why some users are experiencing slow network response. Or it may include reconfiguring a remote internetwork device to improve overall system performance.

In short, network management incorporates an almost unlimited list of duties—basically, doing whatever it takes to keep the network running smoothing and efficiently, with minimal or no downtime.

This job has grown even more difficult as networks have become larger and more complex. The evolution from small workgroups of often identical PCs to large internetworks made up of dissimilar machines—IBM PS/2s, Macintoshes, PC clones, printers, communications gateways, and bridges and routers—has brought more power to the desktop while adding immense complexity to the network manager's job.

Fortunately, vendors are developing more and better tools—some software-based, others complete systems that provide onscreen maps of network resources—to help in the endless task of managing a network.

FROM SIMPLE TO COMPLEX

Network management tools, whether they are as application-specific as a performance monitor or as comprehensive as IBM's mainframe-based NetView, help bring some order to the potentially chaotic network management environment. They give network managers information and capabilities they can use in the battle to keep their networks running trouble-free.

Whether they are intended to merely find cable breaks or to pinpoint the cause of a network slowdown, network management tools are vital to the network manager's day-to-day life. They can help ensure uptime and network reliability, maintain predetermined performance levels, manage network resources optimally, plan for expansion, maintain company security, track network use, and provide a basis for charging customers for network time.

For example, knowing how many network users regularly access a laser printer—and how long they have to wait for their printed material to appear— can help a company decide when it's time to add a second printer. Knowing which workstations generate the heaviest traffic lets a network administrator predict possible bottlenecks—bottlenecks that can be avoided by adding internetwork devices such as bridges or routers.

FIVE FUNCTIONAL AREAS

At a basic level, network management requirements generally fall into five functional areas: configuration management, fault management, security management, performance management, and accounting management.

Configuration management applications deal with installing, initializing, booting, modifying, and tracking the configuration parameters or options of network hardware and software.

Fault management tools provide an audit trail, or historical overview, of a network's error and alarm characteristics. These types of tools show a network manager the number, types, times, and locations of network errors. These errors might be dropped packets and retransmissions (on an Ethernet) or lost tokens (on a Token Ring).

Security management tools allow the network manager to restrict access to various resources, from the applications and files to the entire network itself; these generally offer password-protection schemes that give users different levels of access to different resources. For instance, a user in marketing could be allowed to view, or read a data file in accounting but not be permitted to change or write to it.

Security management is also important in managing the network itself—for instance, only certain individuals (such as network administrators) should be permitted to change configuration settings on a server or other key network devices.

Performance management tools produce real-time and historical statistical information about the network's operation: how many packets are being transmitted at any given moment, the number of users logged into a specific server, and utilization of internetwork lines. As already noted, this type of information can help network administrators pinpoint areas or network segments that pose potential problems.

Performance management tools generally allow polling individual network devices for component-specific information. A communications server might provide information on throughput for each serial port, while a file server might report the number of users logged in, what applications they are using, and the number of active files. This information can then be studied to determine which gateways, servers, or routers are being used heavily and may need added capabilities in the future.

Accounting management applications help their users allocate the costs of various network resources—a public data network gateway, access to a mainframe session, or printer time—to those using them. These applications provide information about session start up/stop, user logins and resource use, and audit trail data. Companies can then use this information to bill departments internally or customers for computer and/or network time.

BUILT-IN NOS MANAGEMENT

Most network operating systems (NOSs) provide some level of network management capabilities; in particular, almost all the leading NOSs offer password-protection schemes that limit users' access to network resources. Novell, for instance, implements its NetWare management scheme through user profiles, which define not only the user's access rights, but the users' classifications (supervisor, workgroup manager, console operator, or user), which also determine the resources they can access.

In this scheme, a supervisor has access rights that allow reconfiguring and upgrading the entire system. The workgroup manager, available with NetWare 3.X, controls only the resources of a single user or user group. This concept allows a supervisor to distribute some of the responsibility for maintaining the network to others around a large network.

A user with console operator access rights can run NetWare's FCONSOLE utility, which allows monitoring and controlling a variety of network performance criteria, such as print queues. The user can access only those resources allowed by the supervisor (or workgroup manager with NetWare 3.X). Although users can access the NetWare management utilities, their rights to actually perform management functions are severely limited.

Although other NOSs' access schemes may differ in specific features from NetWare's, they all offer similar resource-restriction capabilities that give the network managers control over their networks.

PROGRAMMABLE MANAGERS

Many other network product vendors also offer specific network management products that address more-detailed needs. These include Sun Microsystems's SunNet Manager [now known as Solstice], Hewlett-Packard's OpenView, IBM's NetView for AIX, and Cabletron's Spectrum.

Both Sun and Hewlett-Packard designed their network management applications to work with other vendors' "agent" applications that add specific functionality to a system. For example, various agents can perform monitoring and controlling capabilities on gateways and routers.

Other products, however, deliver only partial solutions. These devices include protocol analyzers, which provide configuration and performance data but no accounting management capabilities.

MANAGEMENT STANDARDS

As networks have grown larger and become increasingly heterogeneous in nature, so has the need for industry-standard network management protocols (and products) that operate across a wide range of vendor offerings. The first of these protocols, the Simple Network Management Protocol (SNMP), was developed by the Internet Activities Board in 1988. SNMP generally relies on the User Datagram Protocol/Internet Protocol (UDP/IP) as the underlying mechanism for transferring data between different types of systems and networks, though IPX and AppleTalk have been employed successfully by some products.

Briefly, SNMP is a protocol that defines the communication between a network management station and a device or process to be managed. SNMP's three-layer architecture (network management stations, agents, and a common set of protocols that binds them together) operates with a management information base (MIB) and a structure of management information (SMI). The MIB and SMI are network management concepts that allow defining each network element so these elements can be monitored and controlled by the management stations.

Though widely accepted, SNMP has several limitations. For one, it is considered by some to be too simplistic for managing the large, global-style networks evolving today, and its manager-to-agent architecture leaves it incapable of managing true enterprise-wide networks, which can require manager-to-manager systems as well. Because products based on it are widely available—hundreds of vendors make compatible products—SNMP remains the network management protocol of choice for most PC-based network managers.

HOST-BASED SYSTEMS

Two mainframe-based network management systems with wide industry support are IBM's NetView and AT&T's Unified Network Management Architecture (UNMA.)

Although proprietary in nature, these products enjoy broad end-user support because of their associated vendors' large installed bases of computers. With the protocols already available, many users incorporate their primary vendor's network management products into their networks as a matter of course. IBM's NetView permits nonIBM networks to access the NetView host via its NetView/PC and LAN Network Manager gateway products. IBM also supports SNMP in many of its products.

This tutorial, number 26, by Jim Carr, was originally published in the September 1990 issue of LAN Magazine/Network Magazine.

Simple Network Management Protocol—SNMP

The most widely adopted standard provides a way for devices and consoles to communicate.

As a multitude of forward-looking academics, vendors, and standards bodies recognized in the late 1980s, network management is a big job in many different ways. In particular, management is too big a job for any one vendor to handle. Manageability requires instrumentation in many devices from numerous vendors; it requires accessibility or transportability across the network, as well as on dedicated out-of-band channels; and it requires programming interfaces for a multitude of general purpose, as well as specialized, applications. It's all too evident that multivendor network management is impossible without a standard.

In developing such a standard, the ISO bet on a horse named CMIP (common management information protocol), while the Internet world bet on the Simple Network Management Protocol, or SNMP. And you know what happens when bloated, unwieldy OSI standards compete with (relatively) sleek Internet standards—Internet standards get implemented in products and OSI standards collect dust on shelves. Since 1990, when RFC 1157—the Internet Engineering Task Force document that defined SNMP—became an Internet standard, practically every widely deployed network management system has used SNMP.

SNMP management has three components. The first component is the Structure of Management Information (SMI), which is a sort of toolkit for creating a management information base, or MIB. The SMI identifies the permissible data types and spells out the rules for naming and identifying MIB components. It defines the structure of the SNMP naming mechanism. The MIB, which represents the second component of SNMP, is a layout or schema for information relevant to managing networks and their component devices. The third component, which the Simple Network Management Protocol itself defines, is the different possible payloads or Protocol Data Units (PDUs) that form legitimate management messages.

SNMP is most often implemented over IP. In fact, UDP port numbers 161 and 162 identify SNMP agents and SNMP managers, respectively. But nothing in the standard prevents SNMP messages from being delivered with TCP, HTTP, or nonInternet protocols, such as IPX or AppleTalk's datagram delivery protocol.

THE SMI

The SMI specifies the allowable data types in the MIB, and it spells out the ways data can be represented. Also, it defines a hierarchical naming structure that ensures unique, unambiguous names for managed objects, which are the components of a MIB. Compared with the objects that general purpose programmers use to build applications, SNMP-managed objects are very simple and stripped down. MIB objects typically have six or so attributes. For instance, an object usually has a name, such as ifInErrorsortcpAttemptFails; an object identifier in dotted decimal form, such as 1.3.6.1.2.1.2.2.1.14; a syntax field, which selects one of several possible data types such as Integer, IPAddress, or Counter; an access field, which selects among "not-accessible," "read-only," "read-write," and "write-only"; a status field consisting of either "mandatory," "optional," "deprecated," or "obsolete"; and a text description of the object.

MIB objects are static; they're compiled from a text-like description language to a binary form that agents and managing processes can load. While the standard MIBs—MIB-2 is the current generic TCP/IP MIB standard, and specific MIBs have been adopted for bridges, printers, and other entities—provide a useful common denominator for management application developers, most vendors find it necessary to define their own proprietary objects to take advantage of their products' capabilities.

The SMI is two layers of abstraction away from the sort of management data that IT staff and users care about. It sets the rules for defining MIB objects, which are one layer of abstraction away from management data. All these abstract rules and reserved words make it possible to have machine-readable specifications that remain comprehensible to humans. The SMI enables a vendor to write an SMI-compliant management object definition (perhaps one called PacketsContainingTheWord-Spam), run the text through a standard MIB compiler to create executable code, and install the code in existing agents (for instance, agents with the right hardware and software to count instances of the word "Spam") and in management consoles that could then begin generating reports and charts about such occurrences.

MIB

The MIB is a hierarchical name space, with the registration of its nodes administered by the Internet Assigned Numbers Authority (IANA). According to the hierarchy in the figure, the object ID for every relevant MIB object must begin with either 1.3.6.1.2.1 (which represents the mib-2 node) or 1.3.6.1.4.1 (the enterprises node). The term MIB is also used to refer to specific collections of objects used for particular purposes. Thus, MIB objects under the mib-2 node include the RMON (remote monitoring) MIB objects and other generic MIB objects, while those under the enterprises node include all proprietary MIB objects. It has been estimated that there are 10 times as many proprietary MIB objects as there are generic ones.

Each MIB object has a value associated with it according to the syntax part of its specification—for example, the number of packets that have come in since the last system reset, the number of clock ticks since the last reset, or the administrative state of a router. When a MIB object is instantiated on a device or other entity, the value associated with the object is sometimes called a MIB variable. SNMP agents store MIB variables and send them to SNMP managing processes (or consoles) when they are asked to. The value of a MIB variable is the basic unit of useful management information.

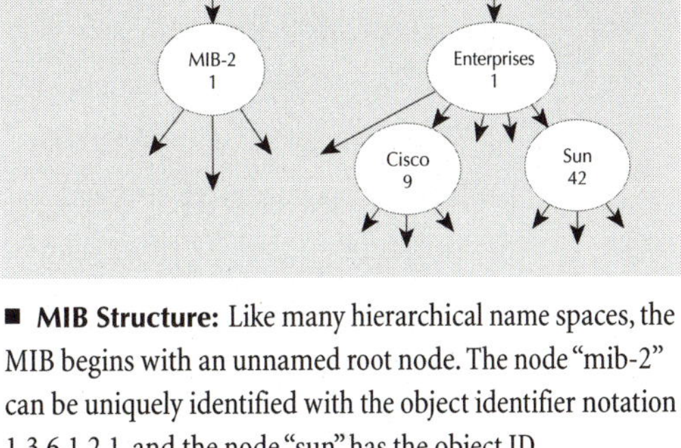

■ **MIB Structure:** Like many hierarchical name spaces, the MIB begins with an unnamed root node. The node "mib-2" can be uniquely identified with the object identifier notation 1.3.6.1.2.1, and the node "sun" has the object ID 1.3.6.1.4.1.4.2. Administrative responsibility follows the tree structure. Thus Sun can assign the nodes beneath "sun" though it must register the names with the IANA.

MIBs are often divided into groups of related objects. MIB-2 has 10 groups, including "system," "interfaces," "ip," "tcp," "udp," and "snmp." The RMON MIB for Ethernet segments has nine groups, including "statistics," "history," "alarm," "event," and "capture."

While RMON is an extension of the MIB, traditional SNMP agents are not capable of capturing most RMON data. Special RMON probe devices or built-in probe functions that can see an entire segment are necessary to collect and forward RMON information. An RMON probe not only captures more data and processes it more than ordinary device agents, it can reduce traffic by storing intermediate results locally and forwarding them to an application on demand.

AGENTS AND MANAGING PROCESSES

SNMP managed entities—hubs, routers, database managers, toasters—must be outfitted with an agent that implements all the MIB objects relevant to them. The managing process must also be aware in advance of all the MIB objects implemented on entities it is expected to manage. Management consoles, such as HP OpenView, that communicate with many kinds of devices must allocate incremental resources for each different type of entity—it's no wonder these platforms are resource intensive.

SNMP proper describes the packet layout (or protocol data unit) for messages between management agents and managing processes. In the first version of SNMP, there are only five possible messages. The GetRequest message with one or more object instances asks for the values of those object instances. The GetResponse message returns the object instances with their values. The GetNext-Request message provides an efficient method for the managing process to search tables of values. The SetRequest message writes a value to the agent, providing SNMP with the ability to configure and control entities remotely. (It's probably not necessary to mention that managed objects with an access attribute of "read-only" can't be Set.) SetRequests are acknowledged with GetResponse messages. Finally, Traps are messages initiated by the SNMP agent and sent to the managing process. They are not acknowledged.

SNMP has a crude and insecure mechanism for access control and authentication in the form of a "password" called a community name or community string. Among other problems, the community name is generally visible in unencrypted form in each SNMP packet—short work for any hacker with a protocol analyzer. With such severe potential threats, many network managers have been reluctant to implement the Set or write parts of SNMP, using it only to monitor devices and collect statistics. Of course, SNMP is a much less valuable technology as a result, and the potential for creating self-healing networks and other such marvels is cut off.

The people who developed SNMP were fully aware of the security problem and began working on it even before the first version of SNMP was standardized. In 1995, however, the standardization process for SNMP-2 broke down with competing approaches to a secure SNMP solution. One aspect of the problem was the fact that a secure SNMP-2 standard could not interoperate with the original SNMP because the original is irretrievably insecure.

SNMP-2 has other enhancements aside from security—it provides improved support for systems and applications management, as well as other extensions; it enables manager-to-manager communication, and therefore, a more distributed management model; and it makes retrieving tabular data more efficient with a new message type.

However, it seems clear that 1997 will pass without any vendor implementations, much less standard implementations, of secure SNMP. A potential savior is the possibility of adapting HTTP to transport management data. There are few criticisms of the SMI or the MIBs that structure management information. HTTP has at least two reasonably secure implementations, though the protocol itself has other problems. Under this plan, management agents would become a kind of specialized Web server and managing processes would exist as browsers. A series of valuable discussions on this subject is available from the Simple-Times electronic newsletter at www.simple-times.org. In fact, the Simple-Times electronic newsletter is a terrific source for any information having to do with SNMP.

This tutorial, number 102, by Steve Steinke, was originally published in the February 1997 issue of LAN Magazine/Network Magazine.

Policy-Based Networking

If the most common utopian dream of a hard-pressed network manager is the self-healing network, the second most common is the policy-driven network. It goes something like this: Upper management establishes policies—a set of principles concerning the availability of computing resources to employees and departments; the objectives and resources that will be made available for safeguarding the organization's data and computing assets; and factors that are sufficiently important to create exceptions to these principles. These policies are then entered into the distributed system, after which they are cascaded out to each desktop, server operating system, application, directory service, database, router, and VLAN.

This (fantasy-based) procedure automatically builds a call-response system, which informs users of any policy-based constraints or limitations that might be obstructing their access to certain resources. The system also points departments and users in the right direction if they need to change something: "Press 1 if you have forgotten your password. Press 2 if your department needs video conferencing at the desktop every day. Press 3 if you need encrypted e-mail."

No doubt, this pleasant dream will remain forever unattainable. But standards organizations and vendors are making efforts to help automate some of the steps of policy implementation for networked systems.

POLICIES, PRACTICES, AND PROCEDURES

Policies are more general and abstract than practices, which in turn are more general than procedures. For example, policies might be, "Every employee will have access to e-mail and calendaring applications." Or, "The organization has legal and moral obligations to maintain the confidentiality of customers' personal information with respect to anyone outside the organization or anyone within the organization without a specific need."

Sample practices might be, "New employees will be assigned e-mail and calendaring accounts on their first work day. They will be scheduled for training within 30 days." Or, "Only members of the Customer_Confidential group will be permitted to look up confidential customer data. Physical security devices will be employed to assure better confidentiality than a password-only system provides."

Sample procedures might include such statements as, "Employee mail accounts are named according to the following system: Firstname_Lastname. Duplicate first and last names are resolved by having the employees with the least seniority insert their middle name in this way: Firstname_Middlename_Lastname. Names not resolved by this method will be referred to the Chief Directory Officer for resolution. It is the user's responsibility to archive messages. The IT organization will maintain backups of current messages." Here's another example of a procedure: "Members of Customer_Confidential will receive read-access rights to tables X, Y, and Z in the Customer database. They will use a card key and a six-digit PIN to log in. All the data in tables X, Y, and Z will be encrypted using Triple DES."

Policies are generally expressed in ordinary business language; they don't address implementation methods. Practices address implementation methods, but not at the level of specific products, data structures, and keystrokes; these specifics are covered in procedures. The advantage of having three specification levels is that changing particular products or vendors should not generally require changing practices. Additionally, practices that do a better job of carrying out policies can be adopted without high-level policy debates. Wise organizations encourage broad participation when formulating policies. Widespread understanding of why policies are the way they are make the derivation of practices from policies smoother and less politicized than would be the case if policies were mandated by a small group of executives or IT staff. Moving from practices to procedures is a technical (and budgetary) matter that should be relatively free of organizational politics.

Different organizations have various levels of commitment to developing and maintaining documents related to policies, practices, and procedures. Large, distributed enterprises with numerous resources will typically have a greater need than small organizations for documenting uniform policies, practices, and procedures. For educational institutions, spelling out user responsibilities and prescribing consequences for abusive activities might have a high priority. Financial institutions will certainly have strong incentives to explicitly document the steps they take to secure data and prevent tampering. To minimize bureaucracy, every organization needs to determine which aspects of its business need to have policies, practices, and procedures spelled out in detail.

COMMON POLICY TARGET AREAS

The most common subjects for policies include distributing resources, defining security requirements, and setting performance expectations. Resource distribution might include defining the accounts and access permissions users are entitled to; spelling out the kind of desktop configurations various classes of users can expect; defining the role of mobile-computing and work-at-home resources; and establishing the applications the organization makes available and users' obligations with respect to those applications.

Security policy is a huge area that includes many topics not directly related to distributed systems policies. For example, organizations need procedures for physically securing the premises, as well as for teaching users how to resist "social engineering." Nevertheless, there are many important areas of overlap between security policy and distributed systems policy. Backup and restore procedures are one example. Disaster recovery plans are another. Other related topics include the usual network security items, such as data integrity, confidentiality, continuity of service, and nonrepudiation, which are implemented via such tools as authentication, access control, virus protection, encryption, a public key infrastructure, and intrusion detection.

Performance policy is often a matter of setting priorities among users, groups, and applications. Should a resource become scarce, the most important operations should be the last to suffer. The biggest constraint in most environments is wide area throughput, and many products and proposed standards that refer to policies emphasize the particular practices and procedures that prioritize WAN traffic. Routers and switches are often equipped with multiple queues; traffic is steered to the shortest queues based on addresses, port numbers, and other characteristics, providing a crude sort of Quality of Service (QoS) hierarchy. The RSVP Working Group is working to define the Common Open Policy Services (COPS) protocol, which will make it easier to prioritize traffic on IP networks that define resource reservations. Of course, one of ATM's strengths is its rich set of built-in mechanisms for defining QoS.

Policy fulfillment and Service Level Agreements (SLAs) also come together in performance and availability. Many performance and availability goals can be more aptly described as policy statements than as SLAs, because an organization's internal "customers" are generally not paying for IT services and thus have no incentive to accept particular service levels or to negotiate them realistically.

POLICIES AND DIRECTORIES

In order for practices and procedures to be implemented, a great deal of information must be captured and maintained, as well as easily accessed. If directory proliferation wasn't a serious problem before installing a policy-based network, putting a set of policies, practices, and procedures in place would itself cause a directory overload. In addition to applications, NOSs, and host gateways—each of which requires its own passwords and access control mechanisms—you have to keep track of desktop configuration standards and compliance; IP addresses, DHCP leases, and DNS data; firewall settings; public keys; digital signatures; remote access authentication and accounting; and traffic priorities—all the things policies affect. Without interoperable directories, each of these policy areas could result in yet another directory. In addition, as Internet commerce becomes more and more feasible, the need to tie organizational roles to purchasing authority and to other restricted activities will require new directory information and access to a public key infrastructure—potentially even more new directory fodder.

Fortunately, vendors and standards bodies are addressing the directory-proliferation problem by developing extensible, replicable, communicative directory services, thereby eliminating the need for new directories for every new application. Directories must be extensible because no matter how complete their features might be, not every aspect of applications, services, security measures, and corporate roles can be anticipated. Replication allows organizations to copy relevant portions of a directory to a database close to the users and applications that need the information. Directories that can communicate with other directories don't have to be comprehensive and universal, so the problems of synchronizing multiple repositories are minimized.

The NetWare Directory Service, NDS, has been extensible and replicable since its introduction, and version 5 fully supports Lightweight Directory Access Protocol 3 (LDAP-3), the first truly workable version of that directory access protocol. Microsoft's Active Directory, due out with Windows NT 5.0, will for the first time support LDAP and an extensible schema. Metadirectories, such as Zoomit's VIA (the company was acquired by Microsoft in 1999), provide an alternative means of binding together disparate directories.

NOS directories have long been the most important databases for storing server account and basic security information, including password authentication, server access control mechanisms, and even local data encryption mechanisms. Extended versions of these directories are natural repositories for the policy data of the future. Novell extended the role of NDS for managing desktop configurations and application software with its ZEN Works products. NetWare 5 also enables NDS to perform many public key infrastructure functions. The NetWare 5 directory also lets NDS be a DNS server and a DHCP server. The Microsoft Active Directory is also tightly integrated with DNS and DHCP, as well as PKI elements.

In 1996, Microsoft and Cisco Systems announced an initiative called Directory Enabled Networks (DEN), which aims to define directory schema extensions for networks, especially in the areas of performance and traffic prioritization. This work will help minimize the need for multiple directories and reduce the risks of duplicating data. After incorporating input from several hundred companies, Microsoft and Cisco turned the DEN project over to the Desktop Management Task Force (DMTF), which folded the DEN schema into its Common Information Model (CIM). (CIM is the DMTF's candidate specification for managed objects independent of protocols, OSs, programming languages, and object models.)

Not all the information relevant to traffic control and policy-based transactions belongs in a directory. Dynamic, fast-changing data about networking devices, for example, probably won't work well in a repository designed for relatively long-lived data and whose replicas converge on a single data model over a period of minutes or hours rather than in real time. This need for supplementary data storage facilities is an additional incentive for using metadirectories that can tie multiple directory services together.

Several other vendors have begun to move beyond the dedicated directories for each application or device. Integrated security vendors, including Check Point Software (Redwood City, CA) and Internet Devices (Sunnyvale, CA), have announced intentions to support LDAP in their products, and have taken steps to design user interfaces that provide a broad view of security and traffic management policies. 3Com has announced an ambitious framework called Policy-Powered Networks. It includes Policy Server, an LDAP-compliant directory that holds the data and procedures for determining specific policy actions; Policy Manager, an application that lets users define rules; compliant switches, routers, and NICs (called enforcers), which carry out the actions specified by the Policy Server; other directories, which the Policy Server can query via LDAP; and compliant applications that can request specific services or classes of service.

Developing usable, consensual policies and practices will always require hard work. Over the next few years, vendor efforts and the development of standards will make implementation simpler and less error prone than it has been.

This tutorial, number 121, by Steve Steinke, was originally published in the August 1998 issue of Network Magazine.

Service Level Agreements

Service Level Agreement (SLA) details the responsibilities of an IT services provider (such as an ISP or ASP), the rights of the service provider's users, and the penalties assessed when the service provider violates any element of the SLA. An SLA also identifies and defines the service offering itself, plus the supported products, evaluation criteria, and QoS that customers should expect.

Enterprises that use outsourced services, including those from ASPs and telecommunications vendors, rely on SLAs to guarantee specific levels of functionality, network bandwidth, and uptime. In fact, research firm IDC (www.idc.com) notes that nearly 97 percent of the large companies (those with 2,500 or more employees) it queried required an SLA for network availability in the next 12 months. And why not? SLAs are the key to ensuring consistent QoS, performance, and uptime in business-critical computing environments.

SLAs first came into prominence in 1998, when the Frame Relay Forum released its "Service Level Definitions Implementation Agreement," or FRF13. The document's guidelines defined acceptable parameters for several key characteristics of frame relay service, such as frame transfer delay, frame delivery ratio, data delivery ratio, and service availability (or uptime).

As important as the guidelines themselves were—outlining the specific performance and availability metrics that users of frame relay services could expect—FRF13 also served a more important purpose: It indicated clearly that frame relay providers finally felt confident that they could meet standards they themselves had authored, thus assuring customers of their ability to deliver credible service.

Providers of various other outsourcing services have followed suit, and the SLA industry has burgeoned. In fact, IDC predicts that the market for managed- and hosted-service SLAs will grow from $278 million in 1999 to $849 million by 2004.

Generally, SLAs complement other contractual agreements that cover a variety of details, including corrective actions, penalties and incentives, dispute-resolution procedures, nonconformance, acceptable service violations, reporting policies, and rules for terminating a contract.

These contracts generally fit under what some analysts call Service Level Management (SLM), which provides managing and service contract capabilities. Currently, SLM tools on the market, such as FireHunter from Agilent (www.agilent.com) and Eccord Enterprise from Eccord Systems (www.eccordsystems. com), provide flexible ways to define SLAs and monitor their effectiveness.

AN SLA'S COMPONENTS

The ASP Industry Consortium (www.allaboutasp.org) has identified four areas that require detailed SLAs: the network itself, any hosting services supplied by the vendor, the applications it hosts and manages, and "customer care," which includes help desk services.

Each area contains its own set of elements, metrics, typical industry ranges, and criteria for calculating these metrics. For instance, the network SLA would include details on bandwidth, performance, and QoS. SLAs should also detail the nature and types of tools required for users and service providers to monitor and manage them.

NETWORK SLAS

The elements of a network SLA should cover the characteristics of the network itself, connection characteristics, and network security. The network SLA identifies the IP performance levels that a service provider guarantees in the course of delivering application services to its customer. While some enterprises accept a "best effort" delivery standard via the public Internet, others demand that their providers offer service over private IP networks that allow specific guarantees for application availability on a customer-by-customer basis.

A network SLA should define the type of network infrastructure that the service provider will deliver. Understanding the nature of a network's physical components helps providers set customer expectations on the performance levels they'll receive. The network SLA also spells out network availability, measured in percent of uptime, and throughput, measured in bits per second.

While 100 percent uptime might be every enterprise's goal, 99.5 percent to 99.9 percent are more realistic averages. One key element of a network SLA is specifying penalties for downtime during critical business hours versus overall downtime: For instance, downtime at 2 a.m. may not disrupt the typical enterprise's business, but it could be unsatisfactory in an e-commerce environment.

When specifying through-put, a network's capacity is detailed in the capacity of the backbone connections within the network's core. Typically these run from 56Kbits/sec (a dial-up connection) up to 10Gbits/sec (also known as OC-192).

Another key part of guaranteeing network service is the connection SLA, which spells out acceptable data losses and data latency (or data delays), plus bandwidth provisioning. In the past, this was alluded to as the basic bandwidth—for example, 56Kbits/sec, T-3, OC-3, and so on—for which customers are billed.

According to the ASP Industry Consortium, few service providers actually detail provisions for data loss (which results from dropped packets in saturated IP networks) in their SLAs; those that do will often guarantee 99 percent packet-delivery rates. While real-time applications such as Voice over IP (VoIP) or interactive media couldn't operate effectively at such a loss rate, packet loss in the 5 percent range is acceptable for typical Web browsing.

Data latency, as with data loss, is critical in VoIP and multimedia environments where delays must not impact end-user performance; real-time interactive applications require response times of 100 milliseconds (ms) or less. The ASP Industry Consortium notes that Web browsing, on the other hand, remains viable at 250ms. In practice, U.S. and European network providers often guarantee a round-trip delay of 85ms between the routers in their core networks.

Another key part of guaranteeing network services is the security SLA, which defines the applications, data, and services that are protected while in transit over the service provider's network. Unlike the hard-and-fast metrics of a network SLA, a security SLA is determined by customer requirements and is more subjective.

Issues specific to this SLA include the level of encryption, such as DES or TripleDES; the point in the network, such as the access point, where data is encrypted; use of public or private encryption keys; and whether certain applications require encryption at all. A network security SLA must also take into con-sideration how the encryption/decryption process impacts network performance and QoS. Finally, it should include severe penalties for security breaches.

HOSTING SLAS

Hosting SLAs ensure the availability of server-based resources, rather than guarantee server performance levels. As such, hosting SLAs should cover three critical areas: server availability, administration of servers, and data backup and the handling of storage media.

A server availability SLA, measured in percentage of uptime, should guarantee a minimum of 99.0 percent uptime, based on a rolling 30-day period. Although a hosting SLA's objective should be to deliver always-on (100 percent) server availability, 99.5 percent to 99.9 percent uptime is more realistic.

The server-administration component of a hosting SLA details the management responsibilities of a hosting service. Specifically, these spell out the acceptable response times for restoring failed servers, as well as define metrics for performing data backups.

For example, a hosting SLA should mandate that a host provider respond to a restoration request for a failed server within a set period of time (such as one or two hours); it should also guarantee that the server will be returned to service within another specified period (such as 12 to 24 hours).

In addition, it should outline the percentage of scheduled data backups that will actually be conducted; ASP Industry Consortium guidelines for this indicate that 99 percent of planned backups should be completed. The data backup SLA should also specify frequency of backups—a full nightly backup is typical—and require that the hosting service protect backup tapes by storing them offsite for a predetermined time period (such as 30 or 60 days).

On a higher level, a data backup SLA might also require the hosting service to create and regularly test an overall disaster-recovery plan. This could include contingencies for "hot site" functionality, which would give a customer access to temporary computing facilities when the customer's own site is unavailable due to a catastrophic event, such as a hurricane or an earthquake.

THE APPLICATION SLA

Applications generally utilize a variety of OSI Transport-layer services. For instance, Web-based applications rely on TCP/IP, which provides specific services such as file-transfer functions. This interaction can impact application SLAs in several ways.

Most importantly, unlike traditional network and server SLAs, where lower-layer (layers 2, 3, and 4 in the OSI model) cell/packet/frame metrics are easy to define, application SLAs are impacted not only by the transaction-processing execution of the application itself but also by the delays introduced by lower-level protocol error-handling procedures.

Thus, application SLAs require the institution of application-specific metrics—that is, definitions of performance levels that relate to application utilization. For example, an application SLA should define the percent of user interactions, such as downloads or data requests, to be executed without failure.

It should also define the acceptable time lapse between a user's request for data and the moment the updated data screen appears, as well as an acceptable bit-per-second rate for data transfer in a transaction session. The time-lapse guideline works in conjunction with the execution guideline, ensuring, for example, that while a download is deemed successful even if it takes several hours, it would still violate the SLA for taking so long.

IN THE PENALTY BOX

As noted, effective SLAs levy penalties against service providers who violate the terms of their contracts. Generally, these come as credits against future service.

In addition to refunds for lost time or poor performance, penalties for SLA violations should also consider the impact of a violation on an enterprise's business. It is, for instance, unreasonable for a service provider to offer a refund just for a specified amount of time lost when an enterprise suffers financial loss because of an SLA violation.

Lisa Erickson-Harris, a senior analyst with Enterprise Management Associates (www.ema.com), says she has heard of cases in which a service provider actually issued a check for a violation. That, however, is rare, so enterprises should ensure that their SLAs provide penalties that cause their service providers significant "pain."

SLA MANAGEMENT TOOLS

Tere´ Bracco, an analyst in research firm Current Analysis' (www.currentanalysis. com) enterprise infrastructure group, warns it is "buyer beware" when negotiating and managing SLAs. Consequently, she recommends that enterprises require service providers to offer them comprehensive tools for monitoring and managing their SLAs.

The best SLA tools let enterprise IT managers look at network performance the same way the service provider does. SLA management tools such as ViewGate Networks' (www.viewgate.com) Inteligo, typically deployed by the service provider but used by the enterprise via a Web interface, allow IT personnel to monitor network performance and manage their network SLAs on the fly, just like the service provider itself.

SLA tools should also monitor and manage an SLA's metrics in real time, rather than providing mere historical views of past performance.

RESOURCES

The ASP Industry Consortium Web site (www.allaboutasp.org) contains a wealth of information on Service Level Agreements (SLAs), including an excerpt from a 75-page white paper on SLAs developed by the group.

The Information Technology Association of America (ITAA) Web site (www.itaa.org) also offers a variety of useful information about SLAs. This includes presentations such as "Managing Technology to Deliver SLAs" and an ITAA SLA Library available only to members.

You can find templates for sample SLAs at www.nextslm.org, a Web site developed and maintained by several networking vendors.

The book *Foundations of Service Level Management*, (Sams; ISBN: 0672317435) by Rick Sturm, Wayne Morris, and Mary Jander, provides recommendations for Service Level Management (SLM) strategies and SLAs.

This tutorial, number 155, by Jim Carr, was originally published in the June 2001 issue of Network Magazine.

Protocol Analyzers

Today there is a wide variety of troubleshooting tools for analyzing network problems. Cable testers, designed simply to register common electrical faults, may include sophisticated time domain reflectometry and digital signal processing to precisely localize fault conditions and rapidly examine performance over a broad bandwidth. Handheld troubleshooting tools may combine some cable-testing capabilities with NIC and hub testing, as well as with higher-layer protocol tests such as Ping and Traceroute. SNMP consoles, once found only on management platforms such as OpenView and SunNet Manager, are now commonly supplied with individual devices. They are also attached to Web servers, and used for network drawing and documentation tools, as well as built into handheld diagnostic devices.

As a result, many of the tasks that once required a protocol analyzer can be resolved with other tools. Nevertheless, most of the hardest and subtlest problems still require protocol analysis for complete resolution.

ANALYZE THIS

In its simplest form, a protocol analyzer captures all the traffic on a medium, parses it according to the rules of any network protocols that are present, and displays the results. Ethernet, the most widespread shared-network medium, is the most common interface for protocol analyzers, but any other Physical-layer or Data-link-layer medium that can redirect signals to a probe will have protocol analysis tools. Of course, on shared media and high-speed media, the analysis software will have to work harder than it would on point-to-point lines or low-throughput networks (see figure).

Most shared-medium network technologies—Ethernet, Fast Ethernet, Token Ring, and

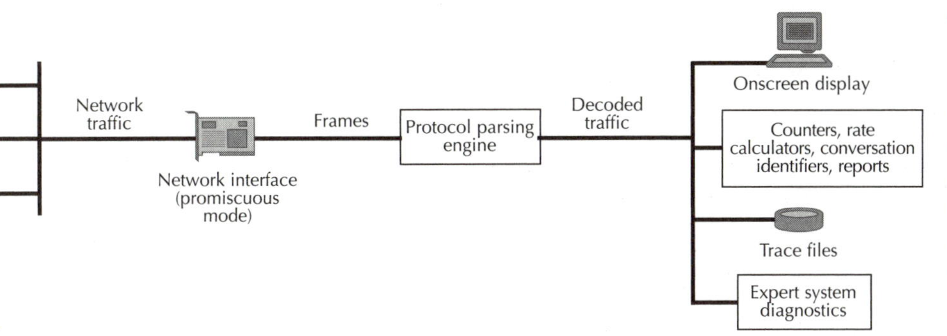

■ **Figure 1:** Protocol analyzers capture network traffic, parse each frame to identify the protocol that defines it, and pass the decoded traffic along for display or for further analysis.

FDDI NICs—support a "promiscuous mode" operation, where all the traffic on a segment or ring can be processed. Ordinarily, a NIC processes only frames with its own MAC address as the destination, as well as broadcast frames, which are intended for everyone by definition. If the NIC is switched into promiscuous mode, it can also capture unicast and multicast traffic sent to other nodes.

Most current protocol analyzers run on ordinary portable PCs, though some of the high-end models use proprietary OSs and hardware for detecting rare error conditions that ordinary NICs are blind to. (Some NICs that operate in promiscuous mode suppress bad frames, which is undesirable for protocol analysis if your network produces such things.)

The analyzer software takes the frames received by the NIC and typically writes them into a big capture buffer in RAM. Ordinary commercial CPUs can keep up with a saturated 10BaseT network, but not with 100BaseT. One option for dealing with high-speed networks is to apply filters before capture, eliminating particular protocols or particular addresses that seem extraneous to the problem. Alternatively, many analyzers now have dedicated hardware that can capture full-duplex, 100BaseT traffic—normally a package of RAM with as direct a connection as possible to the network, and an interface to send captured data to the analyzer software.

From a captured frame it's easy to display the source and destination MAC addresses. A Type identifier field indicates what higher-level protocol defines the payload of the frame. Applying the rules for IP, IPX, AppleTalk, SNA, DECnet, and other layer-3 protocols lets the software display the network addresses for the source and destination nodes, and branch out and unpack the higher-layer protocols. Once the Application layer is reached, the decoding job is finished and the captured traffic is ready for display.

KEY FEATURES

One essential element the analyzer adds to the display is a time stamp for each packet, which may be absolute or relative to some other time. Packet order and the duration of delays between particular packets are often crucial diagnostic information. Some analyzers have the valuable feature of supporting multiple views for the different protocol stack layers, reducing the quantity of data and simplifying the display.

Filtering the captured traffic at this point is also a valuable technique for focusing on the most relevant data. It's also quite useful for the analyzer software to convert hexadecimal data fields to ASCII text, at least when the payload is e-mail, text files, or HTTP data. Yet another very useful feature lets the user substitute names—either DNS or NetBIOS or NetWare device names, or arbitrary mappings of addresses to meaningful names—for numeric addresses in the capture display.

Traditionally, protocol analyzers stopped at this point and left it up to the often-flabbergasted user to figure out what was going on. Users needed to know in detail how the steps of a login procedure were carried out and what the meaning of various time-outs and error conditions were. They needed to understand how RIP and OSPF operated to define the transit paths through the network. They needed to understand how long SNA would be willing to wait for an acknowledgment before timing out and disconnecting.

While the definition of protocols is usually well-documented, the way they behave on a live network and the way they interact with one another is not generally well-documented, and often known only to engineers and programmers who have spent large portions of their careers examining protocol analyzer trace files.

The protocol analyzer makers responded to the shortage of experience in this area with software-based expert systems that could identify multi-packet patterns and suggest error conditions that might have caused the symptoms. Hewlett-Packard's Expert Advisor and Network Associates' Expert Sniffer were two of the earliest implementations of expert systems technology for protocol analysis, but most other vendors do much the same thing nowadays.

Filtering is another key feature in protocol analysis, as I mentioned earlier. Filtering before capturing runs the risk of not capturing the key problem indicators, but is sometimes unavoidable if the traffic level is too high for the analyzer to capture in real time. It's better to ignore rationally chosen traffic streams than to ignore random frames whenever the software can't keep up. Filtering after capture is an essential part of simplifying the display, and has no disadvantages because the data remains in the capture buffer or trace file, even if you filter too aggressively and fail to display the key data on the first attempt.

Aside from filtering by protocol and addresses, most analyzers permit you to define filters based on bit patterns anywhere within the frame. For example, you could easily search for packets containing specific text or with custom combinations of multiple layer parameters.

Capture buffers can be written to disk as trace files. The Network Associates Sniffer format is the most commonly used format, employed by practically every vendor. Trace files that can't be diagnosed at one site or by a particular model of analyzer can be sent elsewhere for deeper investigation.

DOUBLE DUTY

Most modern protocol analyzers also perform network-monitoring functions. As long as they're vacuuming up all the traffic on a segment, it doesn't take much work to count total frames, collisions, error frames, and broadcasts; to keep track of which conversations generate the most traffic; or to display the rates at which these events take place. These statistics can be presented as graphs or speedometer gauges, or captured to files in order to generate reference baselines for future comparison.

If you're familiar with what an RMON system does, you'll probably recognize the similarity among these network-monitoring functions. RMON probes are generally distinct from RMON consoles with SNMP as the protocol that allows them to communicate, while protocol analyzers were not originally designed with this standardized distribution of functions. However, many protocol analyzers today can accept frames captured by an RMON probe and decode them as if they had captured them directly. This is a valuable capability, as the rise of switched networks threatens some of the utility of the protocol analyzer as a diagnostic tool.

Because protocol analyzers depend on promiscuous-mode NICs (or on some form of signal duplication or "mirroring"), their reach is bounded by the edge of the Ethernet collision domain (or the presence of a bridge or router on other topologies).

In the past, when dozens or even hundreds of users shared a single network segment, the protocol analyzer could capture all the problems and clues from a single location. Today, as users are divided into smaller subnets, or even into microsegments with a single switch port per user, the scope of the protocol analyzer has shrunk. On a switched Ethernet network with full-duplex ports, the protocol analyzer can't even capture traffic without a special capture port on the switch.

In some cases, switches come with RMON probes built into each segment. Unfortunately, these implementations are usually stripped-down versions of RMON that don't include the packet capture function that a protocol analyzer would require. Sometimes switches support a backplane connection that could give the protocol analyzer access to all of the traffic on the switch, but more often port-mirroring functions only support the redirection of one port at a time to the protocol analyzer.

One interesting extension of protocol analysis is the relatively new field of intrusion detection. The same sorts of "expert analysis" and pattern recognition that are employed to identify the signatures of misconfigured routers and broadcast storms can also be used to lock on to suspicious patterns of failed logins and inappropriate file browsing. The front end of the protocol analyzer is much the same as the front end of the intrusion detector; the primary difference lies in the patterns they are trained to detect.

GLORY DAYS

In some respects, the heyday of the protocol analyzer has passed. Ten or even five years ago, network protocols were not implemented as sure-footedly as they are today. Many early networking mistakes have been solved and left behind, despite the fact that a network manager's day is still a full one. A modern switched network can practically be a plug-and-play environment.

Nevertheless, when a problem comes along that the cable testers, the handheld troubleshooting tools, and the SNMP management platform can't pin down, the only alternative is to fire up the old protocol analyzer and start capturing and decoding packets.

This tutorial, number 131, by Steve Steinke, was first published in the June 1999 issue of Network Magazine.

Troubleshooting Ethernet Problems

The most common medium for local networks, 10BaseT-compliant Ethernet, has proven itself to be reliable, forgiving, and relatively trouble-free. As 100BaseT becomes more prevalent, you can expect a few additional problems, but nothing resembling the days of coaxial-cable-based Ethernet. Nevertheless, it's important to be prepared to cope with any difficulties that arise, and this tutorial gathers together information about the tools you can use to solve Ethernet problems.

Most 10BaseT and 100BaseT links have three hardware components: a NIC in a client or server computer; a port in a hub, switch, router, or other device; and a cable connecting the first two components.

The parts of the network interface that can cause trouble include the driver software that talks to the computer's OS, the configuration settings in the OS, and perhaps some configuration software or hardware-based jumpers or switches that configure the interface card.

The port at the other end of the cable from the computer is quite straightforward on a hub, though on a switch or a router there could be complex configuration issues, such as VLAN definitions or policy settings.

The cable will often consist of a patch cord with eight-pin modular plugs (often called RJ-45 connectors) on each end; a wall plate with an eight-pin modular jack; a section of cable running inside the walls or ceilings to a punchdown block in a wiring closet; perhaps another jumper cable connecting the punchdown block to a patch panel; and finally a patch cord that runs to the hub, switch, or router port (see figure). The good news about the cable segment is that once it is correctly installed, it will be reliable indefinitely—as long as no one reconfigures it incorrectly.

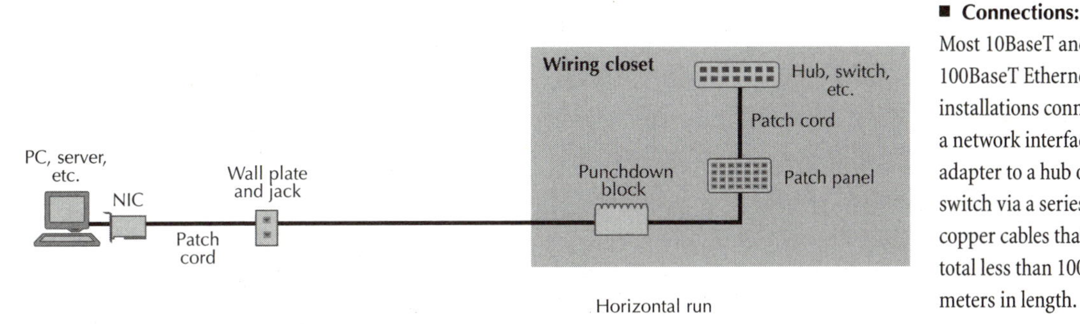

■ **Connections:** Most 10BaseT and 100BaseT Ethernet installations connect a network interface adapter to a hub or switch via a series of copper cables that total less than 100 meters in length.

EVERYDAY TROUBLESHOOTING

The first thing to check when there is an apparent Ethernet link problem is the link light. Unfortunately, 10Base2 and 10Base5 (coax-based Ethernet) don't typically have link indicators. But practically every NIC you'll encounter nowadays has a link light that lights up whenever the interface receives link pulses from the other end. (10BaseT interfaces generate link pulses about 60 times each second when there is no data on the cable.) Practically every hub, switch, or router also has a link LED that stays on steadily when it receives link pulses. (For the rest of this tutorial, I'll save space by calling the devices at the other end of the cable from the computer "hubs." Unless it's spelled out otherwise, when I refer to hubs, you can substitute "switch" or "router" and so on.)

Thus, if the link lights at both ends of a questionable link are lit (and if they go off when you unplug the cable), you can be confident that the cable is wired and connected correctly, that the NIC and the hub are powered up and capable of receiving data, and that the most likely source of the problem is the configuration of the computer or the network interface.

If only one link light is on, the cable segment is almost certainly bad; the interface at the lit-up end is successfully receiving link pulses, while the interface at the other end is sending them but not receiving them. If possible, plug the computer into another cable that supports a working system. If the computer works on the other cable, you know the computer is configured properly. If a computer known to be working properly shows the one-link-light-out symptom on the original cable, you have added reason to suspect the cable segment.

At this point, a cable tester will quickly narrow down the problem to a particular length of cable or to a connector along the path. If you don't have cable testing equipment, your best bet is to simply substitute in known good cable lengths until the link lights come on at both ends. In this situation and many others, having a laptop computer known to be configured correctly for Ethernet connectivity, along with a good patch cord, will allow you to rapidly substitute the computer (the most easily misconfigured component) and the computer patch cord (the most exposed and vulnerable part of the cable link)—thus speeding confirmed diagnoses for the most common problems.

If both link lights are off, don't blame the cable right away. Hubs often have other LEDs besides the link indicators, such as power and traffic indicators. If the traffic indicator shows data, then at least some ports of the hub are working correctly. Hubs are designed to automatically "partition" or isolate misbehaving ports, so a malfunctioning port might cause the link lights to be dark.

To check this, plug the cable into a different port on the hub and see if the link lights come on. (If the device is a router or a VLAN-configured switch, it may be necessary to reconfigure the port before you connect the cable.) If a known good hub port doesn't light up the link lights at both ends, you should connect a known good computer, such as your laptop, to the computer end of the cable segment. If the lights stay out at this point, then something along the cable path is definitely the culprit.

Remember that 10BaseT cables connecting one hub to another must "cross over." (The cable must also cross over when you connect one computer to another to create a minimal two-computer network.) That is, pins one and two on one end must be connected to pins three and six, respectively, on the other end, and vice versa. Most hubs have one or more crossover ports with a switch, so you can use a standard cable and let the switch do the crossing over. Really sophisticated hubs detect the need for crossing over and do the job automatically, so you don't have to worry about whether your cable is crossed over or not.

The other indicator light you'll commonly find on hubs is the "collisions" light. A flicker of the collisions light every few seconds is to be expected on a normal network, but anything resembling a constantly lit collision indicator likely indicates serious problems. The most common reasons for excessive collisions are violations of the 10BaseT rules: no links longer than 100 meters and no more than five repeated segments, with no more than three of the segments populated and no more than four repeaters (hubs) between any two nodes.

Ethernet normally becomes aware of collisions in the first 64 bytes of the frame, and networks with too many repeaters or overly long segments can experience collisions after the 64th byte-time. Since the shortest allowable frame is 64 bytes, whole frames could be accepted by nodes on the network, despite their being corrupted by a collision, if late collisions occur. (For 100BaseT networks, there can be at most two repeaters linked by a five-meter cable, and late collisions occur after 512 bytes. Frames are always padded to a minimum length of 512 bytes for this reason.)

ADVANCED TOOLS

With built-in LEDs, a network-configured laptop, and substitute cables, you can consistently weed out bad configurations, as well as shorted, open, or miswired cables and connectors. A laptop or other correctly configured computer will generally be able to log on to a NetWare or Windows NT network or ping an IP address if the link lights come on. (Windows

95/98 clients generally have ping.exe in the Windows directory, and ping programs are also standard with Mac OS and Unix implementations.)

If you can't log on to a server when others can, or if you can't ping a destination that responds to other pings, then the cable comes under suspicion again, even if the link lights are on. Particularly at 100Mbits/sec, Category 5 cable can be sufficiently out of compliance to prevent a link from working, without obvious electrical or mechanical faults. Inadequate connectors, electromagnetic interference, or unsound wiring techniques are three of the most common ways the cable link might fail while still letting the link lights come on.

A cable tester that can measure near-end crosstalk and attenuation across the frequency band from 0Hz to 100MHz is the only way to be sure of a good cable link for 100BaseT. (10BaseT is more forgiving, but not infinitely so.) If you have a lot of coaxial cable-based Ethernet to troubleshoot, cable testers that estimate the distance of problems along the cable using time domain reflectometry can save a lot of trial-and-error operations. The big names in cable testing are Fluke (www.fluke.com), Microtest (www.microtest.com), Wavetek Wandel Goltermann (www.wavetek.com), Datacom Textron (www.datacomtextron.com), and Hewlett-Packard/Agilent Technologies. Of course, a cable plant certified by qualified cable installers is likely to hold up for many years and keep cable problems to a minimum.

One useful gizmo is a 10BaseT/100BaseT indicator. Psiber Data Systems (www.psiber.com) is one manufacturer of what the company calls "link testers." Cards that can run at either 10Mbits/sec or 100Mbits/sec are widespread, and while they are generally designed to autonegotiate the data rate based on what they're connected to, the autonegotiation protocols were not fully baked and interoperable when early products shipped. This means an interface can get stuck at 100Mbits/sec and freak out the poor 10Mbit/sec-only interface at the other end. The 10BaseT/100BaseT indicator is inexpensive and about the size of an electric toothbrush, with an eight-pin modular plug on the end and LEDs that indicate whether the connection is 10Mbits/sec or 100Mbits/sec, and whether it's full duplex or half duplex.

The next step up from a cable tester is a handheld troubleshooting tool. (Fluke has the broadest line of these.) These devices measure broadcast traffic, network utilization, and errors. In most cases they will list all the nodes on the network and indicate who the top senders and receivers of traffic are. These statistics will often quickly narrow down the possible causes of a problem, and then the troubleshooting tool can test hub, NIC, and cable performance individually. For example, overly long frames, known as "jabbers," almost always indicate a defective network interface or corrupted driver software. Also, if utilization consistently remains at levels higher than 60 percent, while collision rates remain low (let's say less than 5 percent), the likely problem is overall congestion. The solution is to segment the network, perhaps with a switch that gives the most congested nodes a dedicated path.

The most comprehensive handheld troubleshooting tools analyze traffic at the Network layer. These tools distinguish IP, IPX, AppleTalk, and other protocols and present statistics about the top talkers for each. Some of these devices even act as SNMP consoles, collecting, for example, switch and router port statistics and capturing RMON data. Many of the problems these devices can identify, such as duplicate IP addresses and addresses for nonexistent subnets, are not Ethernet problems, so I'll say no more about them here.

The ultimate Ethernet troubleshooting tool is the protocol analyzer (see "Lesson 131: Protocol Analyzers," June 1999, page 26). At the high end of the market, there are specially designed network interfaces on some protocol analyzers that can isolate and identify Physical-layer problems that might not be picked up by an ordinary NIC in a laptop computer running protocol analysis software. For 100BaseT networks, most protocol analyzers have add-on buffers that permit full wire-speed captures—comprehensive decoding software can't keep up with continuous levels of 100Mbits/sec, much less full-duplex traffic. Of course, like the handheld tools, protocol analyzers can look at all seven layers, not just Ethernet's Data-link and Physical layers. Cable testing functions, generally built into handheld testers, are not an option with protocol analyzers.

The main drawback of a protocol analyzer as an Ethernet troubleshooting tool, however, is the training and sophistication required for a user to become effective. Nevertheless, the most complex and thorny fault conditions will continue to require protocol analysis technology for the indefinite future.

This tutorial, number 135, by Steve Steinke, was first published in the October 1999 issue of Network Magazine.

Managing the Desktop

A "black box" means you don't have to concern yourself with a device's inner workings. Don't worry yourself with details, just let the product engineers take care of what's going on inside. The problem: Network managers have many black boxes attached to their networks, but users expect them to know what's inside. Except you can't tell your user, who's complaining that he can't print and the network doesn't work, that the manufacturers of these black boxes—workstations, printers, and software—assumed that the network managers didn't need to be concerned with the magic inside.

Network management has largely been focused on managing network devices, not the devices attached to the network. Administrators savvy to the Simple Network Management Protocol (SNMP) can tinker with the internals of routers, bridges, and hubs, deftly setting device parameters and diagnosing problems. But networks have far more workstations, adapter cards, and printers than internetworking devices. And those are mystical entities, mostly without SNMP agents.

The Desktop Management Interface (DMI) could help bring those dark corners of the network into broad daylight. Products, including servers, desktop computers, and adapter cards, implementing DMI for management will begin to ship this fall. The Desktop Management Task Force (DMTF, Hillsboro, OR) has finalized its DMI specification, and in July 1994 it provided a DMI software development kit for DOS, Windows, and OS/2. Product makers can use this kit to write DMI agents for their products, which ideally will make them easier to manage.

DESKTOP MANAGEMENT INTERFACE ARCHITECTURE

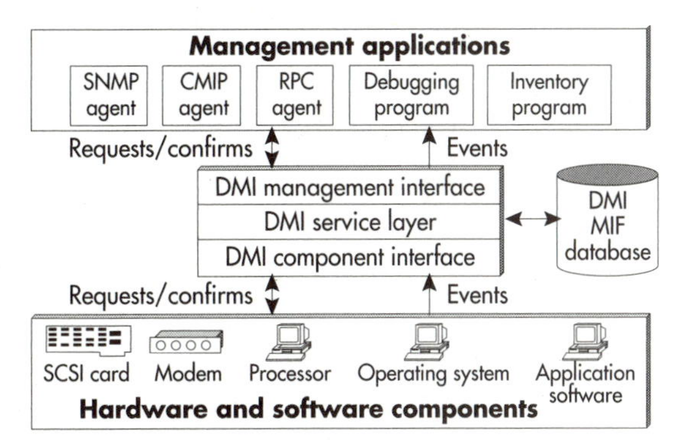

■ **Figure 1:** A management application can access and configure the managed components via the DMI Management Interface (MI). It can also respond to events occurring in its managed components. The local agent provides the services for managing the individual components. The Management Information Format (MIF) file describes the component's manageable aspects.

Five product types can benefit from DMI: PC platforms, such as desktops and servers; hardware and software components, such as operating systems, application software, video cards, fax modems, and network adapters; network and local management applications; peripherals, such as printers and mass storage devices; and management consoles.

For every DMI-managed device, the network manager will be able to determine what the device is, who made it, where it's installed, and other relevant information. With those capabilities alone, taking inventory of network devices might be less of a nightmare than it currently is. Network managers will be able to glean this information from their desks, using management software, rather than having to physically go to the site and unscrew the device cover to look inside. Beyond that, manufacturers can use DMI to configure and diagnose their products. Now that's a real boon to productivity.

HOW DMI WORKS

Implementing DMI-based management involves the management application, the local agent, which runs on the desktop computer, and the hardware and software components being managed. These parts utilize the three elements of DMI: the Management Interface (MI), the Service Layer, and the Component Interface (CI). Figure 1 shows the DMI architecture and its relation to devices.

The management application can retrieve a list of managed components, access the specific components, and configure the components via the DMI Management Interface. The management application can also respond to events that occur in its managed components.

Each DMI-managed desktop has a local agent, which provides the services for managing the individual components. The local agent presents a common interface to management applications. It is also responsible for coordinating requests from management agents with actions to be executed by the components. The local agent is the embodiment of the DMI Service Layer. The Service Layer must be tuned to a specific operating system, and the first implementations are for DOS, Windows, and OS/2. Implementations for Unix and Windows NT are on the drawing board.

Because the Service Layer runs directly on the desktop computer, its memory footprint is crucial. Efficiently running on existing systems, including low-end PCs, is essential for the success of DMI. One reason SNMP is not widely used on desktop computers and their internal components is because of heavy memory and processing requirements. For the DOS Service Layer, the DMTF targets the TSR to be 14KB and loadable into high memory. In Windows, the Service Layer will be a Dynamic-Linked Library (DLL). For memory-constrained systems, the local agent can also operate as a network proxy, just as in SNMP.

The local agent communicates with the managed components via the DMI Component Interface. The Component Interface gets and sets the component devices' attributes. Component categories include PCs and servers, network adapters, printers, operating system software, application software, and modems.

But the DMI isn't only about the network manager initiating actions. DMI-managed components may initiate events when some undesirable action has occurred, such as discovering a virus or exceeding a preset limit for disk space. The component notifies the Service Layer, which in turn communicates the event to the management application. From there, the network manager can be notified or an action may be taken automatically, depending on the management application itself.

The DMI relies on a Management Information Format (MIF) file to describe the manageable aspects of a particular component. The MIF, which is an ASCII file, guarantees a basic level of information, called the standard component ID group, that even existing equipment can provide. In the standard component ID group, the product name, version, serial number, and the time and date of last installation are mandatory. The ID number is assigned on the basis of a device's order of installation relative to the other system components.

Each group also contains a component name, ID, and class. A management application can use the class identifier to find groups of a certain type, such as LAN adapters.

Each component typically has one or more groups in the MIF, and each group contains one or more attributes that describe the component. The product manufacturer chooses to include either DMTF-specified groups or private groups. For example, a fax modem manufacturer might write MIFs for the standard fax group, the standard modem group, and a private MIF for its own product.

DEVICES GET MIFFED

With the release of the DMTF developer's kit, three classes of products will be manageable: desktop and server systems, adapter cards, and printers. Future classes include software, mass storage, and servers. Let's get an understanding of what parameters the DMI lets you see and control.

The systems group accesses information relative to a PC motherboard. With a DMI-instrumented PC, the network manager will be able to find out the system name, location, primary user name, user phone number, system uptime, system date and time, bus architecture, and slot count. The system's MIF also defines another group of attributes that relate to plug-in cards, such as video cards. For those, the network manager will be able to find out the model, part number, revision level, warranty start date, warranty revision, warranty duration, and support phone number.

For the adapter card MIF, the network manager can discern the driver type (Open Data-Link Interface, Network Driver Interface Specification, NetWare 3.x, NetWare 4.x, LANtastic, and so on), topology type (Ethernet, Token Ring, LocalTalk, T1, and so on), connector type (RJ-45, BNC, DB-9, for example), bus width, version and size, board number, permanent and current network addresses, data rate, buffer memory size, total number of bytes and packets both transmitted and received, and the total number of errors transmitted and received.

For printers, the network manager will be able to determine whether the printer is a part of the IEEE 1284 or Network Printer Alliance (NPA) group. For NPA printers, administrators can discover how much memory the printer has, whether it's duplex, and its speed, for example.

WHAT ABOUT SNMP?

If you're thinking the DMI sounds remarkably like SNMP, then you're on track. The difference is that DMI was designed to run on desktop systems, which typically don't have the extra horsepower to run resource-intensive SNMP code. The DMI MIF is sufficiently akin to the SNMP MIB that if the management-application vendor so desires, MIFs can be mapped to MIBs. Standard SNMP network management platforms, such as Sun's SunNet Manager, will be able to communicate with DMI applications. The DMI can also be mapped to use the IEEE 802.1B LAN Management protocol, which is Common Management Information Protocol (CMIP) over Logical Link Control (LLC) or CMOL.

BEHIND THE DMI CURTAIN

The DMTF was started by Intel in 1992, and critics viewed it as an Intel-centric specification. The DMTF now operates independently of Intel and is a nonprofit organization. It has three levels of membership, depending on the level of involvement your company prefers, from defining the specifications to implementing them.

Few people would be surprised to find out that Intel plans to incorporate DMI into all of its networking products, and in fact, its 10/100Mbps Ethernet cards, the EtherExpress Pro/100 includes DMI agents. Intel also talks of incorporating DMI directly into its CPUs.

DMI also has some big-league support. For example, Microsoft will implement DMI in Windows 95—at least in some form, according to Chris Thomas, the DMTF evangelist. The LAN Adapter Working Group includes 3Com, IBM, Intel, Hewlett-Packard, IBM/Pennant, Lexmark, QMS, Tektronix, and Xerox. The PC Platform Working Group includes AST, Compaq, Dell, and HP.

WHAT DOES IT ALL MEAN?

To say networks are heterogeneous is nearly a cliché, but as many different vendors' hardware and software products sit inside desktop computers as make up the physical network. Peering into desktop devices is largely a manual labor. But DMI provides a vendor-independent, software-based means of managing desktop pods of heterogeneity.

DMI applications will debut this fall, first for PCs, LAN adapter cards, and printers, but eventually DMI will worm its way into operating systems, applications software, and SNMP management platforms, giving network managers a way to see and control most things that reside on the network.

This tutorial, number 71, by Patricia Schnaidt, was originally published in the July 1994 issue of LAN Magazine/Network Magazine.

Asset Management, Part One

Controlling costs is a fundamental business practice, but many companies fall short in one area: managing IT assets.

According to a 1997 Sentry Technology Group (Westboro, MA) survey of 1,600 companies, approximately 10 percent of total IT purchases can be described as inefficient spending. In other words, on average, 10 percent of a company's IT spending goes to unnecessary or underused equipment and software. The survey projects that in 1997 alone, companies could have wasted a total of $66 billion through bad IT spending practices.

One way to combat this problem is asset management, the careful tracking and analysis of

IT equipment and software through their life cycle within an organization. By instituting a solid asset management system and adhering to good asset management practices, a company can realize significant cost savings.

Asset management addresses two primary issues: How a company spends money on its IT infrastructure, and how it ensures it is receiving the full value of its investment. For this reason, asset management is not confined to the IS department, but reaches across departments and disciplines within an organization. For example, it touches upper management, which validates purchase requests; accounting, which tracks expenditures; and individual departments, which request resources to meet their IT needs.

In this tutorial (part one of two on asset management) I'll introduce you to the topic of asset management by uncovering where IT assets exist in your company and exploring the various forms they can take. I'll also point out the most common ways companies waste money and resources by mismanaging these IT assets.

DO YOU KNOW WHERE YOUR ASSETS ARE?

Understanding exactly what IT assets are is the first step of any asset management strategy. Can you list what assets your organization owns? While such devices as computers, servers, and routers are obvious assets, many others are overlooked, leading to redundant spending and mismanagement.

Perhaps the single biggest investment is tied to your computers—both client workstations and servers. Of course, each computer you own is an asset you need to be aware of, but you also need to account for the other assets linked to each computer. For example, computer memory, including RAM and hard disks, is an asset, as is the computer's processor. In addition, any peripheral device attached to or installed in the computer, such as a CD-ROM drive, modem, or NIC, is an asset.

Aside from a computer's hardware, the software running on each machine is another asset you need to account for. This includes the operating system software, the applications you've bought and installed on the machine, upgrades, any in-house developed applications running on the machine, and so on.

Your network infrastructure also contains a multitude of assets, including printers, hubs, routers, switches, and cabling. Depending on your network environment, the list can go on and on.

While physical assets are fairly obvious to understand and identify, other assets are not. For example, software licenses, hardware and software warranties, maintenance and support contracts, and outsourcing deals all represent money your organization has spent to build and run its IT infrastructure.

To this list you can add any IT equipment your organization leases, the data lines for your

WAN and remote access services, and any Internet service you purchase. Because you've invested money in these items, you ought to consider them IT assets.

HOW COMPANIES MISHANDLE ASSETS

Companies mismanage their IT assets in several ways. One of the most common has to do with how companies value and use their computers. For instance, each computer in a company represents a dollar value, but in many cases this value isn't necessarily the mere purchase price.

When a company first purchases a set of computers, each computer costs a fixed amount. However, if you follow the computer through its life cycle in an organization, you see that the amount of the investment usually changes. For example, after the IS department receives a purchased set of computers, computer support usually configures the machines with the appropriate settings to work in the company's network environment. IS also installs software and any other peripheral devices users need. So, as you can see, the computers have already increased in value.

In addition, not all computers receive the same setup. For example, the ones headed for the accounting department receive the appropriate financial software and database front ends, while the remaining systems are outfitted with general applications such as an e-mail client, a word processing program, a browser, and so on. Obviously, each set of computers now has a different value, depending on what software is used.

The value of these computers may change again when they receive new peripherals, such as a faster modem, or additional RAM and software upgrades.

If your company neglects to track the assets that have been added to each computer, it may underestimate the computer's real value and not receive a full return on its investment. For example, once one of these machines reaches the end of its life cycle, you should redeploy associated software licenses to workstations that are still in use. Also, you should reclaim usable hard disks and RAM, as well as other peripheral devices such as NICs and internal modems. By doing so, your company won't have to spend more money purchasing these items for computers that need them.

Without a good assessment management system in place, you might not be able to detect these "hidden" assets in an old computer. As a result, you may end up wasting money by purchasing redundant IT assets such as RAM or software licenses.

SOFTWARE SNAFUS

Along with PCs, software licenses represent another major IT expenditure, as well as another common area of mismanagement.

As previously mentioned, licenses are sometimes buried within older computers, which often descend down the corporate food chain to other users. For example, a computer purchased two years ago for an employee might have an associated database license attached to it. After a couple of years, the computer might be passed down to an administrative assistant. If the assistant doesn't need database access, that license is going unused and is a wasted asset.

Overbuying is another form of license mismanagement. For example, if your organization buys a 100-user, concurrent license for an application, and only 50 users use it at any given time, you've just paid for twice as many licenses as you actually need.

In addition to overbuying licenses, companies often overspend for them. Typically, this problem is linked to how an organization procures software for various departments, and how those departments communicate with one another.

Often, departments within an organization make IT decisions according to their own needs. For example, the personnel department, having determined that Access is the best database for its needs, decides to switch to Access for its 100 employees and puts in a request to buy the software and the appropriate licenses. Meanwhile, another department in the company is growing frustrated with its antiquated database package and decides that within the next six months it will also purchase a new system for its 50 users.

If the company doesn't have centralized procurement, it may end up investing in 100 Access licenses for the personnel department, then buying another set of 50 licenses within a few months. If it had a proper procedure for managing its software needs, the company could get a better deal from the software supplier by purchasing the licenses in a larger volume.

Ultimately, a company can optimize its buying power through volume purchasing. But, by not knowing how and when its departments purchase software, the company may let this opportunity to save money slip through the cracks.

When dealing with vendors, knowing what IT assets you already have and which ones you actually use gives you some leverage. To take the previous example, let's say your organization purchased 150 licenses for Access. When an Access upgrade is released, the vendor may call you and say it's time to upgrade those 150 licenses. If you don't know how your company uses Access, you may just stay the course and purchase those licenses.

However, if you've kept track of your organization's Access usage, you may realize that not all 150 licenses are being used. Maybe your company is regularly using only 100 of those licenses. With this information, you can make an accurate purchasing decision and avoid purchasing 50 unneeded licenses.

Along with overbuying software and overspending for licenses, companies sometimes buy software they don't need. For example, a company may purchase a software suite for its employees, who use only three of the applications. In this case, the company is paying for additional applications it doesn't need. It may be cheaper for the company to buy only those applications that are used.

In general, it's a software vendor's duty to sell you software. The more you know about the software your company already owns and how it uses it, the more immune you'll be to persuasion. Of course, this also applies to hardware vendors.

SIGNING ON THE DOTTED LINE

Aside from assets such as computers and software, your company probably has assets in the form of maintenance and support contracts, warranties, and leases. Like their physical counterparts, these assets are often mismanaged, leading to a poorer bottom line for your organization.

In general, companies need to keep track of the maintenance and support contracts and warranties they enter into with outside vendors. If a new computer has problems and is still under warranty, investing the in-house time and labor to fix that computer is a waste of money. Without a good system for tracking warranties and tying them to individual properties, it's hard to know whether an individual computer is still under warranty.

An even worse scenario is when a company actually pays a vendor for support or maintenance on IT equipment and then puts in-house staff to work on it. In this case, the company is wasting not only the money it has spent on vendor support, but also the money tied to its own IS support person.

Another example of a contract pitfall is the company that has lost track of which support contracts apply to which IT properties. If the company no longer uses the software or equipment under contract, it is paying for support it doesn't need.

Equipment leasing also contains some hidden money traps. As mentioned, your organization ought to keep track of the life cycle of its IT assets and how these assets change during that cycle. In some cases, you may find you've added investments to these assets. For example, say you add memory to several machines you've leased. If you don't keep track of these additions, you won't know that the computers contain the memory—which is rightfully yours and could be used productively elsewhere in the company—when you return them to the leasing vendor.

Overall, your organization needs to keep track of the various vendors with whom it does business. This practice can uncover other forms of asset mismanagement. For instance, you may discover you're paying multiple vendors to provide similar services. By consolidating vendors, you can reduce this redundant spending. Also, by offering more of your business to fewer vendors, you may be able to leverage your buying power to cut better deals.

COMING SOON ...

So far, we've discussed where IT assets exist in your company and the ways in which companies often mismanage them. In an upcoming tutorial, "Asset Management, Part Two," I'll give you some guidelines about how to develop an asset management plan and system for your organization. Along the way, you'll learn how asset management can benefit your organization, both operationally and financially. In addition, I'll cover the types of asset management solutions currently on the market and some practices that will help ensure that your company is spending its IT dollars wisely.

This tutorial, number 118, by Lee Chae, was originally published in the May 1998 issue of Network Magazine.

Asset Management, Part Two

In "Asset Management, Part One" (May 1998) I introduced the topic of IT assets and pointed out some common ways organizations mismanage them. I also explained how companies lose money by purchasing items they already own and underusing equipment they've already purchased. Both practices shrink the value of an organization's IT dollar.

This month I'll offer you a quick guide for putting together an IT asset management system, and I'll also take a quick look at the asset management market and offer you some tips for managing IT assets (see "Best Practices for Asset Management").

BUILDING THE BASIC SYSTEM

Putting together an asset management system requires three steps. First, you need to create a database to house information about your organization's IT equipment. Second, you need to set up a process in which you inventory and track IT equipment and enter this data into the asset database. Third, you need to analyze this data so that it becomes a resource for evaluating the status of your IT equipment, informs your decision-making about how best to use the equipment, and helps you purchase the appropriate types and amount of new equipment.

To build a database that serves as an information repository for your organization's IT equipment and software, you need to identify each piece of equipment that enters your organization with an entry that contains such descriptions as the vendor name, product name, model number, and unique identification. You also might want to append each entry with additional information, such as the item's purchasing price, the date it came in, and other specific information. For instance, for a computer you might want to note its BIOS, any important enhancements it contains (such as RAM), and so on. For software, you might want to note the version number and whether it's Y2K-compliant.

Once you have your database's parameters worked out, you need to populate it by inventorying your equipment. The best place to start is in your receiving room (or whichever department handles new and incoming equipment at your organization). Besides noting what equipment is on its way, you also need to invest some time in marking or tagging the equipment when it arrives. Some organizations use bar code labels, which can save money in data-entry labor.

For equipment such as hubs, tagging is easy. But for other types of equipment, the process is more complicated and requires more attention. For instance, when you tag a new PC, you also need to tag any additional assets it houses, such as hard disks and extended memory. By marking these components and noting the system in which they reside, you can retrieve and reallocate them when you retire the system. The same is true for modems, network interface cards, and, of course, keyboards and monitors.

Tagging also helps you track equipment's movement within your organization. According to several studies, organizations tend to underestimate the amount of equipment they actually own. (For instance, according to Hewlett-Packard, one of its customers estimated that it owned 700 PCs. After conducting an actual count, the company discovered it had 1,200 PCs.) As a result, organizations often buy more equipment than they need.

Especially in large organizations, it's easy for equipment to fall through the cracks. Think about a computer's life cycle. Typically, it changes hands, locations, and functions numerous times. By tracking the PC, you know where it is and who is using it. Conversely, you know whether it's being underused or misused.

Inventorying incoming equipment and putting that information into a database is not a one-time deal. Rather, you need to establish a process by which all incoming equipment is entered into the database and then tracked throughout its life cycle in the organization.

After you've established your inventory process and entered your new equipment's information, you need to turn your attention to those assets you already own. The reason for starting with new equipment is that machines have life cycles within organizations—you don't want to waste time marking equipment that's on its way out the door. So, when inventorying equipment already in use, be sure to work your way backward from recent purchases to older gear.

EXTENDING YOUR SYSTEM

Knowing what equipment your organization owns is valuable information, but you can extend the value of this knowledge by keeping track of other data points with each item entry. For example, you could include information about where equipment is being used, who is using it, and how. Doing so helps your organization get the most out of its assets. Say you have a set of PCs, each with 128Mbytes of RAM, being used by data-entry staff. You can perhaps make better use of the RAM by reallocating some of it to other PCs in your organization that need upgrades (while leaving the data-entry staff's systems fully functional).

The more information you keep on each item, the more analysis you can perform on your IT assets. For instance, for software you might want to keep track of who has a license for a particular application. Then, by metering the application's usage, you know how many copies are being used. By analyzing this information, you can make more accurate purchasing decisions when investing in upgrades. For instance, if many licenses are not being used, you could scale down your order and allocate the remainder of your budget to other areas.

Overall, it's wise to keep track of how much money you've spent on a product during its lifetime. Say you buy a set of PCs that, over time, require additional investments in service and upgrades. If you keep track of how much money you spend on each PC over its lifetime, you know better what your total cost of ownership is for each one. With this information, you know which PCs offer the best price and performance, which brands and models cost more than you thought, which ones require the most repairs, and so on. When it's time to invest in a new set of PCs, you can make better purchasing decisions, as well as more accurate forecasts for your IT budget.

At its most advanced employment, an asset database also contains information about IT-related leases, contracts, and warranties. Keeping track of this data will give you a clearer picture of the contracts and services you have invested in, and how they are being used by your company.

USING THE DATA

The primary benefit of an asset management system—and the main justification for investing the time and money to build one—is the information it provides about how your organization acquires and uses IT equipment. But there are other benefits. For example, if you tie your asset system to a help desk system, you can increase productivity, financial savings, and end-user satisfaction.

Say an end user calls the help desk with a PC problem. The support person can tap the asset database to find out what PC that user has and what software he or she is running. Because the user won't have to spend time describing his or her environment to the support person, the call will be shorter, and the user can go back to being productive sooner.

As a side note, if the problem and resolution are tracked with the piece of equipment or software causing the problem, help desk operators can check the problem list associated with that item and troubleshoot a new call more quickly. Also, you know what types of equipment cause the most problems, and you can use this information when evaluating the next round of purchases.

A good asset management system can also help when your organization is planning to migrate to another software environment or hardware platform. By checking what equipment and software is in use and where they are located, you can plan deployment better. For instance, you'll know which machines aren't ready for the new system and what upgrades you need to perform. By having this information ahead of time, you can minimize user downtime and frustration when you cut over to the new system.

PRODUCTS ON THE MARKET

You can divide the asset management market into two rough categories: low-end building blocks and high-end multicomponent packages. However, no offering provides a total solution. Your final system will be a combination of product, policy, and hands-on effort.

In the building-block category, you'll find products such as inventory applications and tracking and metering programs. These tools help you build and populate your asset database. However, for the most part they are limited to querying workstations and servers for information on the equipment they house and software they run. They won't necessarily keep tabs on a switch or a hub.

The high-end packages typically provide you with a full-blown asset database. Usually, the database lets you track physical inventory, lease and contract information, and financial information on your assets. Many of these packages let you tie the asset system to a help desk component and, in some cases, to your organization's financial system. Some developers in this area are Peregrine Systems, Bendata, Tangram Enterprise Solutions, and Janus Technologies.

THE BOTTOM LINE

In "Asset Management, Part One" I explained several ways you can use asset data to benefit your organization. In the end, however, the main reason for implementing an asset management system and program is to gain control of IT costs and determine how your organization can spend its IT dollars more wisely. The bottom line is that every dollar your organization spends should help the organization reach its business goals.

BEST PRACTICES FOR ASSET MANAGEMENT

To manage and reduce your organization's total cost of IT ownership, you can buy all the asset management technology you want, but without the appropriate policies and practices in place, you'll never reach your goal. Here are three asset management practices you should adopt in your organization.

1. **Consistent acquisition.** Standardizing on a finite set of hardware and software lets you minimize the number of variables you introduce into your IT environment, making it easier to manage, support, and evaluate your equipment.
2. **Consistent configuration.** Make sure you configure your equipment and software consistently. Similarly, consistent acquisition reduces the number of variables in your operating environment and makes management and support easier and less costly.

3. **Consistent management.** As problems appear in your IT environment, prioritize them and place those causing the biggest financial hits at the top of the list. That way, when you resolve these problems, you can realize more immediate financial returns (and look like a star in the process). Then, move to the smaller problems. Make sure you apply these solutions to similar setups within the organization.

RESOURCES

Here are a few sites on the Web that contain white papers and other documents on IT asset management. In addition to visiting these sites, you can also go to your favorite search engine and search on the keyword phrase "IT asset management."

www.assettracking.com A site devoted to asset management and Y2K issues. It contains white papers on both topics, as well as links to other sites, articles, and resources on the Web.

www.metagroup.com/publicat.nsf/web+pages/asm A META Group article that contains several asset management case studies.

www.tallysys.com/tally/pubs/ A vendor site that contains a series of white papers on desktop asset management that covers issues such as PC inventory, Y2K issues, and software license management.

www.apsylog.com/justifying.html A vendor page that lists seven ways to cost-justify asset management.

www.assetpro.ca Click on the "Asset Management" link in the navigation bar when you get to this vendor site. You'll find a definition of asset management as well as a breakdown of the different processes it involves.

This tutorial, number 120, by Lee Chae, was originally published in the July 1998 issue of Network Magazine.

Server Performance Benchmarks

"IQ tests measure whatever IQ tests measure." That's the common wisdom in applied psychology, but it's just as true when applied to computer benchmarks.

A benchmark suite is designed to measure and compare the relative performance of computer systems. In some cases, a benchmark may be extremely specific: Back in the mid-1980s, as technical editor of Portable Computing, one of my tests measured the relative CPU performance of different DOS-based laptops by launching WordPerfect for DOS, loading a very large document, and timing how long the system took to replace every letter e with the letters xyz. Crude, yes, but that special-purpose benchmark did its job. The benchmarks I'm going to talk about here are more general, and quite a bit more sophisticated.

WHAT BENCHMARKS DO

A well-crafted benchmark is designed to meet one of two goals: to measure the performance of an entire system based on a well-defined set of tasks, or to attempt to isolate the performance characteristics of a particular subsystem. You can use the benchmark results to compare systems or, if you know the performance characteristics that you require for a certain project, you can use benchmarks to "size" the hardware and/or software appropriately.

In the laptop performance test example, I used a well-defined set of tasks—using a certain word processor, a certain test document, and a certain word processor operation—to crudely test a single aspect of laptop performance. Had I been concerned with measuring the performance of word processing software, I would have tested many word processing packages on one laptop computer.

There are many better-designed benchmarks in general use. Many are administered by independent organizations, such as the Standard Performance Evaluation Corp. (SPEC) and the Transaction Processing Performance Council (TPC). Others are created and promoted by vendors, such as Intel's iComp Index, which measures relative performance of the company's various x86 processors.

Several magazines have also written benchmark suites, such as ZD Labs' WinBench suite (www.zdnet.com/zdbop/), whose results are cited by many Ziff-Davis magazines, including *PC Magazine, PC Computing,* and *Computer Shopper*. Another respected test is Khornerstone, which is used by our sister publication *UNIX Review's Performance Computing* (www.performancecomputing.com) for its Tested Mettle server reviews.

A challenge when designing benchmarks or deciding how to interpret them is knowing what you wish to benchmark. Of course, a benchmark measures performance, but which performance do you mean? Some tests measure a system's raw CPU throughput, attempting to minimize the impact on relatively slow I/O systems or network throughput; such tests would be appropriate for a CAD workstation, but not for a file/print server. Other tests stress a storage subsystem by moving large files across a SCSI bus, or by attempting to flood the network; neither test would be greatly affected by raw CPU performance.

The best bets are so-called real-world benchmarks, which attempt to measure how a system would respond in a realistic situation; my old laptop test tried to measure real-world results. No benchmark can truly be real-world, but if used judiciously, the proper benchmark may provide valuable information for comparing systems, as well as assessing what size system may be required to run a certain application.

Benchmark results may only be valid on the exact configurations tested, and only if the primary application you use that server for has characteristics (such as I/O centric or CPU intensive) similar to the benchmark software (see "Rules to Bench By"). Also, note that an application's performance characteristics may change as the workload increases; a transaction server might offer a flat response time when handling 100, 500, or 1,000 messages per hour, but it might choke at 2,500 messages per hour.

RULES TO BENCH BY

When reading benchmark results, particularly tables or charts that compare various systems' performance, be certain you are comparing apples to apples.

It is imperative that you know which benchmark was run, and which version. Benchmark results are generally not comparable between tests or different test versions.

Know the exact system configuration used for the benchmark, including processor model and speed, the amount of L2 cache, the amount of RAM, and which peripherals were in use. In many cases, the operating system and OS version will be important.

UP TO SPEC

Perhaps the most widely known benchmarks have historically come from SPEC (www.spec.org), a nonprofit corporation based in Manassas, VA. SPEC is an umbrella organization that covers three groups, each with its own benchmarks: the Open Systems Group (OSG), which produces server and processor benchmarks for Unix and Windows NT; the Graphics Performance Characterization (GPC) group, which tests the graphics performance of OpenGL and X Windows systems; and the High-Performance Group (HPG), which tests numeric computing systems such as engineering workstations.

A number of system tests are designed by SPEC OSG and are licensed to testing organizations (typically vendors) who run those tests. The testers may publish the results, and they may also send the results back to SPEC, which reviews the results and publishes them on its Web site. As only tests published on the SPEC Web site have been checked by the organization, they (and we) recommend that you look at the SPEC site as the official results repository.

SPEC OSG's flagship benchmark suite, CPU95, isn't particularly valuable for measuring the performance of a network server; its components measure a system's CPU performance, not the OS or I/O functions where many servers find their bottlenecks. OSG's newer SPECweb96 benchmark, however, is well-respected as a test of Web servers.

SPECweb96 runs on any server that supports HTTP/1.0, and it measures the basic Get performance of static pages using random searches. The higher the benchmark score, the better. As of mid-summer 1998, the three highest SPECweb96 results were all Unix systems:

NCR's four-CPU model 4400 (score: 7,800), Silicon Graphics' eight-CPU Origin 2000 (score: 7,214), and IBM's 12-processor RS/6000 Model S70 RS64-2 (score: 7,013).

I found it interesting to note that Hewlett-Packard submitted two results for its uniprocessor NetServer LH 3/400; it scored 2,131 when running Netscape Enterprise Server 3.5 for NetWare, and the score was 1,342 using Internet Information Server 4.0 for Windows NT 4.0.

Another test to watch is SPECsfs97, which measures system file-server performance for Network File System (NFS)-based servers, which are typically Unix-based. SPECsfsS97 tests CPU, mass storage, and network components. Unfortunately, only four vendors (Digital Equipment, IBM, Network Appliance, and Sun Microsystems) have submitted results to SPEC, and that's not really enough to help you make decisions.

TPC-C AND TPC-D

No new product preview with a server manufacturer would be complete without the vendor mentioning either their record-breaking TPC-C or TPC-D results. Using those results, manufacturers hope to prove they're offering either the most powerful transaction processing systems ever built (at least until next week) or the best value—or both.

The TPC (www.tpc.org) is a nonprofit consortium based in San Jose, CA; its 42 current members are primarily systems manufacturers (like the Acer Group and IBM) or database software developers (like Oracle and Sybase). Because the members are often arch competitors, the TPC is widely perceived as being unbiased. But because the members are corporations, there's no bones that the purpose of TPC is primarily to help vendors sell hardware or software, as well as to provide timely access to competitive test data.

Unlike SPEC, TPC releases its benchmark software freely into the public domain. Like SPEC, vendors run their own tests, and may submit the results back to the TPC for publication. The two main tests are designed to measure real-world performance of servers and software.

The simpler TPC-C test measures online transaction processing—in particular, the speed of entering new-order transactions into a nine-table database while the system is simultaneously executing payment, order-status, delivery, and stock-level queries. The results, presented in a unit called tpmC (transactions per minute for TPC-C), represent the number of new-order transactions performed without allowing response time to drop below five seconds; the higher the number, the better. Based on the total retail price of the tested system (including hardware and software), a price/performance value known as $tpm*C* can be derived. For a list of the current top 10 TPC-C machines in both performance and value categories, see Table 1 and Table 2. [Editor's note: This data was current in early 1998.]

The more complex TPC-D test is designed to test decision-support systems. Instead of updating databases with new orders, TPC-D submits complex queries against nine read-only database tables. The sizes of the databases can scale from 1Gbyte to 3Tbytes, so make sure

The Top 10 $tpmC Machines

Rank	Machine	Processors	tpmC	$tpmC
1	Compaq ProLiant 5500 6/200	Four 200MHz Pentium Pro	11,649	$27
2	Unisys Aquanta QS/2 Server	Four 400MHz Pentium II Xeon	17,700	$27
3	Compaq ProLiant 7000 6/400	Four 400MHz Pentium II Xeon	18,127	$27
4	Acer Altos 19000 Pro4	Four 200MHz Pentium Pro	11,082	$28
5	Dell PowerEdge 6000	Four 200MHz Pentium Pro	10,984	$30
6	Compaq ProLiant 3000	Two 333MHz Pentium II	8,228	$31
7	Compaq Server 7105	Four 200MHz Pentium Pro	11,359	$33
8	NEC Express 5800 HX4100	Four 200MHz Pentium Pro	12,106	$33
9	Unisys Aquanta HS/6 Server	Four 200MHz Pentium Pro	13,729	$33
10	NEC Express 5800 HX6100	Six 200MHz Pentium Pro	14,144	$33

■ **Table 1:** These figures are effective as of July 1998. All of these systems were running Windows NT 4.0 and Microsoft's SQL Server.

■ **Table 2:** This table shows the top 10TCP-C clusters and machines as of July 1998.

The Top 10 TPC-C Clusters and Machines

Rank	Machine	Processors	Software	tpmC	$tpmC
1	Compaq AlphaServer 8400 5/625	500MHz Alpha 21164	Digital Unix 4.0D and Oracle8	102,542	$140
2	IBM RS/6000 SP Model 309	233MHz PowerPC 604e	AIX 4.2.1 and Oracle8	57,054	$148
3	HP 9000 V2250	160MHz PA-RISC 7300LC	HP-UX 11 and Sybase11.5	52,118	$82
4	Sun Enterprise 6000	167MHz UltraSparc	Solaris 2.6 and Oracle8	51,822	$135
5	HP 9000 V2200	160MHz PA-RISC 7300	HP-UX 11 and Sybase 11.5	39,469	$95
6	Fujitsu GranPower 7000 Model 800	250MHz UltraSparc	UXP/DS 20 and Oracle8	34,117	Y(yen) 57,883
7	Sun Ultra Enterprise 6000	167MHz UltraSparc	Solaris 2.6 and Oracle8	31,147	$109
8	Compaq AlphaServer 8400 5/350	266MHz Alpha 21064A	Digital Unix 4.0A and Oracle7	30,390	$305
9	Tandem ServerNet Cluster	300MHz Pentium II	Windows NT Server 4 and Oracle8	27,383	$72
10	HP 9000 K580	160MHz PA-RISC 7300LC	HP-UX 11 and Sybase 11.5	25,363	$77

you're comparing similar database scale results. The test produces three metrics: the power metric, based on a geometric mean of response time on all of the test's SQL queries (called QppD@Size); a throughput metric measured in queries per hour (called QphD@Size); and a price/performance metric (total hardware/software price divided by the QphD@Size metric). The power metric measures the scalability of a system, while the throughput measures its response time. Because of its complexity, vendors must have TPC-D ratings approved by independent auditors before the TPC will publish the results.

MEASURING IT UP

Before you use any benchmarks, keep in mind that the vendor runs the benchmark and, in particular, chooses the platform, picks the best configuration, runs the tests, and has the option *not* to report numbers that don't reflect well on the brand. Furthermore, the tests by their very nature are somewhat simplistic and won't reflect the actual application mix you're installing onto any server.

So, would I choose a single benchmark to compare servers or to help scale a server to my needs? No. No single number can do a complex server justice, any more than a new car's zero-to-60 measurement reflects the true capability of an automobile. That said, I would recommend perusing the TPC-C benchmark results next time you're shopping for a transaction server, and the SPECweb96 results to similarly benchmark Web servers.

RESOURCES

Links to numerous benchmarking organizations can be found at www.nullstone.com/htmls/benchmark/benchmark.htm.

Intel's latest iComp Index results can be found at www.intel.com/procs/perf/index.htm. Additional information on benchmarks that Intel recommends can be found at www.intel.com/procs/perf/PentiumII/index.htm.

SPEC's home base is at www.spec.org, and the TPC can be found at www.tpc.org.

An out-of-date, but still interesting, FAQ on benchmarks was written by Dave Sill; the latest version I can find, dated March 16, 1996, is at http://sacam.oren.ortn.edu/~dave/benchmark-faq.html.

This tutorial, number 122, by Alan Zeichick, was originally published in the September 1998 issue of Network Magazine.

SECTION X

Network Security and Backup Systems

Security

Prevention is the key when it comes to network security. Identifying and stopping intrusion—in all its forms—is what security is all about. But identifying a potential intrusion is not always obvious, or likely. The usual security suspects—Soviet spies, CIA agents, and industrial espionage—make great headlines, but they don't pose real risks to the average company. However, just because you're not building the next secret weapon doesn't mean that you're not at risk from security breaches. Far more often, security risks come from acts committed out of human error, greed, malcontent, or machine error.

Physical theft, electronic tampering, and unauthorized access are just three of the more obvious threats to network equipment and data. Physical theft includes people stealing computers, taking floppies with data, and tapping into the cable to siphon off information. Electronic tampering covers computer viruses and other malicious reprogramming. Unauthorized access, the most common threat to security, usually occurs when people see information they shouldn't.

There are literally hundreds of approaches that can be taken to deal with these threats. Just as there are many forms of home security—from a lock on the door to a 24-hour guard—there are many forms of network security. And as the type of home security you use depends on your neighborhood, valuables, insurance, and the amount of money you have, the type and amount of prevention your network needs depends upon the importance of the company's data, the expense of computer equipment, the likelihood of intrusion, and the amount of money you can afford to spend.

NETWORKING IS A RISKY BUSINESS

Networks seriously increase access to your information, and with access comes the responsibility of restriction and control. In addition to the usual sources of security breaches—people taping passwords to their monitors and using scanners to electronically eavesdrop—networks invite a whole host of other vulnerabilities. It's easy enough to drop another workstation or server on the network or add another application. Add the ability to dial into the network system, and you pose an even greater risk.

There is no simple formula for calculating your security needs. The amount of security depends upon the threat you perceive. In some cases, the need for security is clear: banks, airlines, credit card companies, the Department of Defense, and insurance companies. In other cases, the risks may be less obvious. Allowing any worker to examine the payroll file makes for disgruntled employees. Your personal calendar indicates when you are out of town. The following are some of the more common risks to network security.

- Your network can be a danger to itself. Being made of mechanical components, a network can do itself damage when disk heads crash, servers fail, and power supplies blow. Tape and disk platters get old and go bad. Bugs, such as in an out-of-control operating system process or one with a faulty memory mapping, destroy data. Monitor mechanical equipment for wear. For critical components, keep spares onsite or, if warranted, online.
- Your network is physically vulnerable. Thieves and other intruders can physically break into your building, wiring closet, or server room and steal or vandalize equipment and data. When a file is erased, very often it physically remains on disk or tape—only the entry to the directory structure is removed. Sensitive documents may be printed out and left lying around the office, waiting for prying eyes or thieving hands.

Your first line of defense is the simplest: Use locks, guards, and alarms to protect against these physical vulnerabilities. Lock servers in a room and lock wiring closets, permitting access to only those with a key. Sensitive data must be completely wiped off the media when deleted. Shred all sensitive printouts. Bolt expensive equipment to the floor or to a desk. A slew of products exist to prevent intruders from physically taking equipment. Most involve locking equipment with metal bars, in steel cabinets, or with large chains. Others sound loud alarms to deter the thief. These products can help to keep your equipment from being physically stolen (it also makes them difficult to move from one station to another). If your security needs are extreme, you might employ biometric devices. Biometric devices use a physical aspect of people, such as their fingerprints, to verify their identity.

The next step is to secure the cable. Copper cable gives off electromagnetic radiation, which can be picked up with listening devices, with or without tapping into the cable. One solution is to switch to fiber-optic cable, which does not emit electromagnetic signals and is more difficult to tap without detection.

Diskless PCs are a popular security measure. A diskless PC lacks floppy and fixed drives. Users must boot the computers off the file server. With no drives, no way to remove data physically exists. However, be aware that diskless PCs with serial and parallel ports and expansion slots are insecure. A user can insert a removable disk into an expansion slot and remove data. Or the user can attach a printer.

Another step is to physically limit access to data sources. Use the keyboard lock on PCs and file servers. Lock file servers in closets or computer rooms, thus preventing direct access and forcing intruders to circumvent network security. Rooms with doors and locks are good places for printers and other output devices since printed data may be as sensitive as electronic data.

- Viruses are potentially one of the most dangerous and costly types of intrusion. Although they are relatively rare to a well-kept network, the penalties inflicted by a virus can be severe. Your network is vulnerable at any point it contacts the outside world, from floppy drives to bridges to modem servers. At these external contacts, your network's messages can be intercepted or misrouted. Workers take notebooks on the road and may come into contact with a virus-infected computer. Users may take work home, where their home computers are infected. Demonstration programs, bulletin boards, and even shrink-wrapped software may have viruses.

  Protecting your network against a computer virus is much the same as protecting it from unauthorized access. If intruders can't access the network, they can't unleash a virus. However, many viruses are introduced by unwitting authorized users. Any new software should be suspected of having viruses. Although programs from bulletin boards may sometimes be infected, several software companies have shipped shrink-wrapped software that was infected with a virus. While specialized programs can look out for viruses and limit the havoc they wreak, no program can prevent a virus. It can only deal with the symptoms.

- Intentional threats are also potentially damaging. Employees and outsiders pose intentional threats. Outsiders—terrorists, criminals, industrial spies, and crackers—pose the more newsworthy threats, but insiders have the decided advantage of being familiar with the network. Disgruntled employees may try to steal information, but they may also seek revenge by discrediting an employee or sabotaging a project. Employees may sell proprietary information or illegally transfer funds. Employees and outsiders may team up to penetrate the system's security and gain access to sensitive information.

- Workstation file systems present a threat to the network. DOS is easy to circumvent. Intruders can use the many available programs to get at a hard disk and remove data, even if security programs are at work. For this reason, high security installations may want to use a different operating system, one with a different file system. Unix has sophisticated file security, and additional programs are available for even more protection.

- Your network radiates electromagnetic signals. With an inexpensive scanner, experienced electronic eavesdroppers can listen in on your network traffic and decode it. Shielded cable, such as coax and shielded twisted pair, radiates less energy than unshielded cable, such as telephone wire. Fiber-optic cable radiates no electromagnetic energy at all—since it uses light instead of electrical signals to transmit—and it's relatively easy to detect taps into a fiber cable, since these decrease the light level of the cable. If your installation demands maximum security, Tempest-certified equipment shields electromagnetic emissions.

- By far the most common network intrusion is unauthorized access to data, which can take many forms. The first line of defense against unauthorized access should be the workstation interface. Login passwords are a must. Nearly all network operating systems will not give workstation users access to network resources without the correct password. To make passwords more effective, the administrator should assign them and change them at random intervals. Don't let users post their passwords on their monitors or desk blotters. Use mnemonic passwords to help users remember.

Software is available to blank a user's screen or lock the keyboard after a certain definable period of inactivity. Other software will automatically log a user out of the network. In either case, a password is required to renew activity. This prevents the casual snooper, but not a determined one.

A more secure method to stop unauthorized access is an add-in card for each workstation. This card forces the workstation to boot up from a particular drive every time. It can also enforce some kind of user validation, like a password. If the card is removed, the workstation is automatically disabled.

- Your network administrators present yet another risk. If you give them free rein over the applications and data, you're exposing your network to unnecessary risks. Your network administrators manage the network, not the data on it. Administrators should not have access to payroll information, for example. Similarly, don't fall victim to the fallacy that the department heads should have complete access to the network and its information just because they are in charge.
- Finally, your network is subject to the whims of nature. Earthquakes, fires, floods, lightning, and power outages can wreak havoc on your servers and other network devices. While the effects of lightning and power outages can be minimized by using uninterruptible power supplies, you'll need to store backups of important data (and perhaps even equipment) offsite to deal with large-scale disasters.

THREE FORMS OF DATA SECURITY

Information security entails making sure the right people have access to the right information, that the information is correct, and that the system is available. These aspects are referred to as confidentiality, integrity, and availability.

Information stored on a network often needs to be confidential, and a secure network does not allow anyone access to confidential information unless they are authorized. The network should require users to prove their identities by providing something they know, such as a password, or by providing something they possess, such as a card key. Most network operating systems and many applications packages use passwords.

In government circles, this aspect of security hinges on secrecy; access to information is granted according to security clearance. In commercial circles, this aspect of security comes more from confidentiality, where only users who need to know the private information have access.

Guarding access to information is one aspect of security; the security system must also guarantee the information itself is accurate, referred to as data integrity. In providing data integrity, for example, a network ensures that a $14,000 bank account balance isn't really supposed to be $14 million. The system must verify the origin of data and when it was sent and received. Network operating systems grant users access to files and directories on a read, write, create, open, and delete basis. Word processors lock files so more than one user cannot modify the same file at the same time. Databases use record locking to provide a finer granularity of access control.

The third aspect of security is network availability. Although not commonly thought of as part of security, a secure network must also ensure that users can access its information. The network must continue to work, and when a failure occurs, the network devices must recover quickly.

SOLVING SECURITY PROBLEMS

Whatever type of security you implement, diligent watchfulness is important to its success. To help, network operating systems include audit trails that track all network activity, including which workstation has tried to log in to a file server three times unsuccessfully or which files have been changed when they should not have been altered.

Some audit trails can sound alarms when certain events take place. For example, the system manager may want to know when certain files are open, or when unusual traffic takes place. Audit trails will also keep a running log of all that takes place, so the network manager may be able to detect a pattern of intrusion.

Protecting against internal threats requires you to control access to files and applications on a need-to-know basis. Only grant access if users present valid reasons to access the application or data. Use the network operating system's security features to restrict access. Keep audit trails of who accesses what files and when. Enforce the use of passwords.

Such access privileges may be assigned by file, by user or a combination of both. For example, users with a certain security level may read and write to certain files. Those with lower security levels might be restricted to reading these files.

The network manager should create a profile of access privileges for each user. This profile, which is executed when the user logs on, restricts the user to authorized data and devices. Profiles may also be set up for data and devices, limiting their access to only authorized users. Profiles make managing security easier since they provide a consistent method of assigning and maintaining network privileges.

Once a user has workstation and network access, other security barriers can be put in place. Most network operating systems have many levels of access control that limit what resources are available, which data can be accessed, and what operations can be performed. These include restricting who can read and write to certain files, directories, applications, servers, and printers.

To reduce the risk, limit connections to the outside world. When you must make connections, use call-back modems, encryption, and virus-detection software. With call-back modems, users must dial into the system, verify their identity, then the modem calls the user back at a predetermined telephone number to establish the connection. Encryption scrambles data into an unreadable format so even if the packets are intercepted, the message remains nonsensical. Upon receipt of the message, only the people who know the private code, or key, can unscramble the data. Virus-detection software will identify many viruses and disable them if possible.

Biometric devices are a rather drastic security measure. Biometric devices use a person's physical characteristics to verify an identity. The verifying physical characteristic varies. Some use fingerprints, others use voice recognition, others scan a person's retina. Biometric devices are quite costly and are for highly secure environments.

ENCRYPTION

Passwords, locks, access privileges, and even biometric devices do not always deter the determined intruder. A common tool like a protocol analyzer can be hooked up to the network and the intruder can watch all data, including passwords, pass by. Data encryption is the answer.

With encryption, data is scrambled before transmission, making it unreadable as it passes over the wire, even if it is intercepted. To scramble or encrypt this data, its bits must be transformed according to an algorithm. The data is transmitted, and at the receiving end, a system of keys is used to decode the bits into intelligible information. Keys are necessary for encoding and decoding.

Encryption usually requires extra hardware because of the processing power required. Hardware-based encryption schemes are more difficult to crack than software-based methods.

A common data encryption standard specified by the U.S. government is Data Encryption Standard (DES).

DES defines how the data should be encrypted and the specifications for an electronic key. It uses one 64-bit key for encryption and decryption. This can cause problems because the key must be in the hands of the sender and receiver. The only way to get it from place to place is to transmit it. Transmitting the key introduces a security threat. The Public Key System, with matched public and private keys, is a solution.

Encryption may be done before data is stored or transmitted. Some networks only encrypt data when it is sent, which makes wire tapping more difficult but does not keep intruders from taking data from a disk. Other networks also encrypt data on the hard disk. Data is encrypted as it is written and decrypted as it is read from the disk. Having encryption working in both places keeps network data much more secure. Encrypting passwords, as NetWare 386 does, is sometimes sufficient to deter the casual data thief.

To further enhance encryption's effectiveness, keys should be changed at random intervals. This prevents intruders from discovering either the key or the time the key is changed. Alternative keys should be available, too, in case the original set is compromised.

The best network encryption schemes hide much of the encryption hassle from end users by taking care of key management and encryption automatically.

DEVELOP A SECURITY PLAN

Make a planned attack to secure your network. Once your network has been hit by a virus or a data thief, it's too late to start thinking—you should already be acting. Start the planning process by naming a security administrator, who may or may not be the same person as the network administrator. The security administrator works with the network administrators and department heads to develop a security plan.

You must evaluate the dangers to your network. You need to examine its vulnerabilities, the points at which it is susceptible to attack. Then you must identify the threats, or possible dangers to the system, such as a person, an object, or a natural disaster. Vulnerabilities take several forms, including physical, natural, mechanical, communications, and human.

Unintentional, intentional, and natural threats exist in your network, but the majority are unintentional. Users and system administrators commit errors—they delete the wrong file, they disable access to a directory, they corrupt a data file, they never change their passwords, or they write them on their desk blotters. To counter unintentional errors, train your users and administrators about the network and its applications. Keep regular backups of the applications and data, for after a virus infection or data loss, restoring the damaged or lost files may be your only choice.

Reinforce the need to not write their passwords next to their computer or give them to anyone else. They should use passwords that are fairly difficult to guess. For example, users' passwords should not be their first names or spouses' names. Passwords that include numbers are much more difficult to guess. Users shouldn't type their passwords while someone is watching. Users and administrators should change their passwords frequently. Administrators shouldn't use supervisor logons as their "usual" logons.

Don't over-secure the network. Security procedures generally limit freedom to access the network, so implement them carefully. If you restrict access to certain directories, users may not be able to cut and paste freely from one document to another. Users will balk at elaborate security procedures that interfere with their jobs. They will find ways to circumvent the network security procedures, such as storing data on their local hard drives, not on the server, where it would be protected and backed up. Carefully balance the need for security with the security procedure.

For any security plan to work, the employees must take it seriously. The most effective action you can take is to educate your users and administrators on why your security plan is important. When people understand why controls are necessary, they are more likely to cooperate. Make it clear to prospective and current employees that everyone is expected to cooperate. Establish clear consequences for failure to cooperate. Be specific about policies and procedures. Write them down and give everyone a copy. Make sure each individual knows what to do. Don't overdo it. Insofar as possible, make it easy to cooperate. Enlisting the support of employees is probably the single most cost-effective security precaution a company can take.

Finally, natural threats, such as power failures, earthquakes, and other such disasters are a rare but real part of life. Develop a disaster recovery plan to deal with natural disasters and follow it. Archive important data and emergency backup hardware offsite in a secure facility. Keep enough in the archive for you to get your business up and running (and relatively current) should your primary facility get flattened—it could happen!

This tutorial, number 44, by Patricia Schnaidt, was originally published in the March 1992 issue of LAN Magazine/Network Magazine.

IP Security

Security extensions to IP bring authentication and privacy to the Internet.

IP is the underlying technology of networks used by the government, academia, and the corporate world, as well as the public Internet, to send myriad types of packets all round the world. Yet, despite the strength of IP as a WAN protocol, security has never been its strong suit.

This layer 3 protocol oversees packet forwarding through different types of networks. But, IP's ubiquity and wide acceptance as the basis for both public and private networks has led to great concern over the issue of security.

Several types of attacks have been known to take place over IP networks. One is called IP spoofing, in which an intruder tries to gain access by changing a packet's IP address to make it appear that the packet came from somewhere else. Other attacks include eavesdropping on an IP transmission. This can be done by using a protocol analyzer to record network traffic. Another type of IP attack involves taking over a session and masquerading as one of the parties involved in the communication.

The IP security problem will probably only get worse as companies rely on the protocol more and more for remote communications. Also, Virtual Private Networks (VPNs), which allow companies to create a private connection over the public Internet, require strong safeguards. Although VPNs do exist to a certain extent today, most industry watchers will tell you that for this concept to really fly, authentication, encryption, and other security measures need to be in place. The hope is that an IETF standard called IP Security (IPSec), an extension to IP, will be the catalyst for private and secure communications over the Internet.

IPSec is actually a suite of protocols being developed by the IETF. The suite includes the Authentication Header (AH), which addresses authentication for IP traffic, and the Encapsulating Security Payload (ESP), which defines encryption for IP data.

The Authentication Header ensures that the packet has not been altered or tampered with during transmission. It can be used in combination with the ESP (if you need privacy as well as verifying authenticity), or it can simply be used to verify the authenticity of a regular IP packet. The AH also allows the receiver to verify the identity of the sender.

AUTHENTICATION HEADER

A traditional IP packet consists of an IP header and a payload, which can consist of a TCP or UDP header and data. If the AH is used, it immediately follows the IP header.

The format of an Authentication Header is shown in Figure 1. The first field in the AH is the next header field; this is an 8-bit field that tells which higher-level protocol (such as UDP, TCP, or ESP) follows the AH. The payload length is an 8-bit value that indicates the length of the authentication data field in 32-bit words. The reserved area is a 16-bit field that's not currently in use; this field has been set aside for future use, and therefore is always set to zero.

The Security Parameters Index (SPI) and the sequence number fields come next. SPI is a 32-bit number that tells the packet recipient which security protocols the sender is using. This information includes which algorithms and keys are being applied by the sending device.

The sequence number tells how many packets with the same parameters have been sent. This number acts as a counter and is incremented each time a packet with the same SPI is bound for the same address. The sequence number also guards against a potential attack where a packet is copied and then sent out to confuse the sender and receiver.

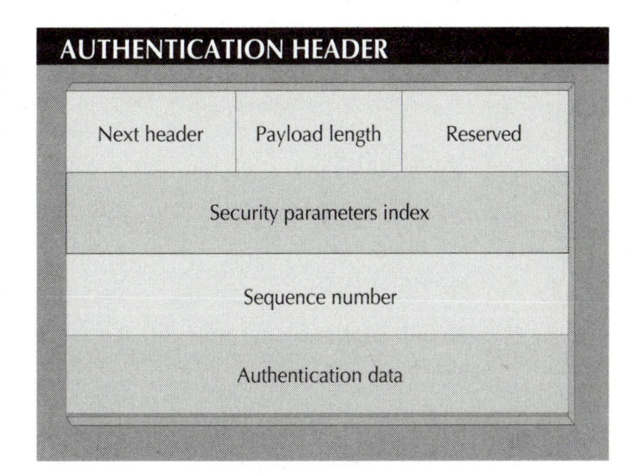

■ **Figure 1:** The Authentication Header (AH) resides between the IP header and the Encapsulating Security Payload (ESP), or other higher-level protocols, such as UDP or TCP. The chief goal of the AH is to let the sending and receiving parties know that the data came from where it says it did, and that it was not changed during transit.

At the end of the AH is the authentication data, which is a digital signature for the packet. To authenticate users, the AH can use either RSA Data Security's Message Digest 5 algorithm or the U.S. government's Secure Hash Algorithm. The IETF is also looking into other authentication algorithms, such as hashed message authentication code.

ENCAPSULATING SECURITY PAYLOAD

ESP is the protocol that handles encryption of IP data at the packet level. It uses symmetric, or secret key, cryptographic algorithms like Data Encryption Standard (DES), and triple DES to encrypt the payload. The default method is 56-bit DES. The only point of concern about which algorithm is used is the current regulations of individual countries. Many governments put a cap on the strength of encryption exported outside their borders.

As shown in Figure 2, the ESP includes several parts, the first of which is the control header that contains the SPI and the sequence number field. The SPI and sequence number serve the same purpose as in the AH. The SPI indicates which security algorithms and keys were used for a particular connection, and the sequence number keeps track of the order in which packets are transmitted.

The SPI and sequence number are not encrypted, but they are authenticated. The next few parts of the ESP are encrypted during network transmission.

The payload data can be of any size (subject to the normal limits of IP) because it's the actual data being carried by the packet. Along with the payload data, the ESP also contains 0 bytes to 255 bytes of padding, which ensures the data will be of the correct length for particular types of encryption algorithms. This area of the ESP also includes the pad length, which tells how much padding is in the payload, and the next header field, which gives information about the data and the protocol used.

The last piece is the optional authentication data. This field contains a digital signature that has been applied to everything in the ESP except the authentication data itself.

To decide whether ESP or AH is best, network managers or security officers need to ask whether they only need authentication or if they need both authentication and encryption. Because AH doesn't provide encryption capabilities, if a scenario requires both features, ideally ESP makes better sense since it does offer both authentication and encryption.

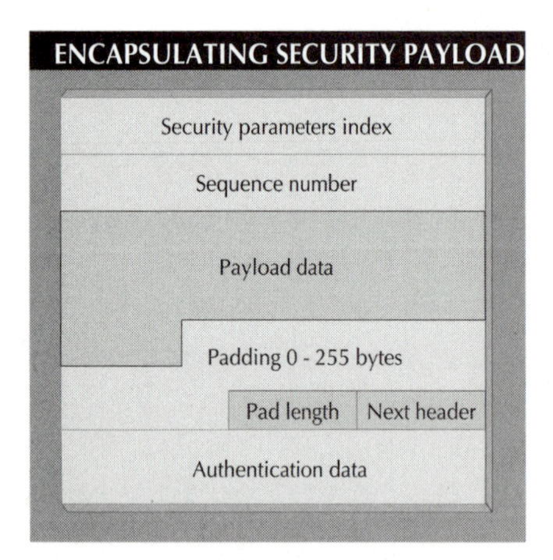

■ **Figure 2:** The Encapsulating Security Payload (ESP) protocol follows a standard IP header (or the Authentication Header, or AH, if (used) and provides encryption and authentication of the packet.

KEY MANAGEMENT

Together, the IPSec ESP and AH protocols provide privacy, integrity, and authentication of IP packets, but they are not the complete package. To round out the standard, the IETF also includes a protocol that provides several services, including negotiating which protocols, algorithms, and keys will be used in a communication; verifying the identity of the other party; and managing and exchanging keys.

The Internet Security Association and Key Management Protocol (ISAKMP)/Oakley key exchange protocol automatically handles exchange of secret symmetric keys between sender and receiver. The protocol integrates ISAKMP with the Oakley method of key exchange.

ISAKMP is based on the Diffie-Hellman model of key generation, in which the two parties share information beforehand to ensure the identity of the other party. Under Diffie-Hellman, two parties generate their own public values, which they send to the other party. The two parties communicate through UDP messages. Each party then takes the public key they received and combines it with a private key. The result should be the same for both parties, but no one else can generate the same value.

Although ISAKMP is an automated method, it does allow for the level of trust in keys to be controlled, much like administrators currently control network passwords and how often they are changed. With ISAKMP, the SPI (that 32-bit number that contains security protocol information for a packet) can be reformatted at specified intervals.

ISAKMP/Oakley supports three methods of key information exchange: main mode, aggressive mode, and quick mode. Main mode establishes what's known as the first phase of ISAKMP SA.

SA, or Security Association, is the method of keeping track of all the details of keys and algorithms in an IPSec-compliant session. The SA includes a wide range of information, including the AH authentication algorithm and keys, the ESP encryption algorithm and keys, how often keys are to be changed, how communications are authenticated, and information about the SA's lifespan.

The ISAKMP main mode sets up a mechanism that's used for future communications. This mode is where agreement on authentication, algorithms, and keys takes place. The main mode requires three back and forth exchanges between sender and receiver. In the first step, the two parties agree on algorithms and hashes for that communication. In the second rally, they exchange public keys using the Diffie-Hellman exchange model and they prove their identities to each other. In the final exchange, the sender and receiver verify those identities.

The aggressive mode actually achieves the same end results as the main mode, only it takes two back and forth exchanges of information instead of the three required in the main mode.

Lastly, there is the quick mode, which can be used after an ISAKMP SA has been created using either the main mode or the aggressive mode to create new material for generating keys.

This is also known as a *phase two exchange*. In the quick mode, all packets are already encrypted, so this step is much simpler than the first two modes.

ISAKMP has some competition in the key exchange and management front in the form of the Simple Key management for Internet Protocol (SKIP). SKIP was developed by Sun Microsystems as an automated key management system. It has been submitted to the IETF as a proposed standard, which has caused some tension in the IPSec working group. Several vendors have shown successful implementations of SKIP, and both SKIP and ISAKMP may be used within IPSec. But to get to a standard way of dealing with key management and to ensure interoperability among IPSec implementations, one of these methods has to go, and right now it looks like SKIP is the unlucky one.

SECURELY SPEAKING

Once the IETF gives final sign off on the IPSec suite of protocols, which is expected in early 1998, you can expect the number of products supporting it to increase rapidly. Already there are a number of firewalls and other products that claim to support IPSec, and although many of them may, the immaturity of the standard and interoperability testing is enough to make customers wary for the time being.

Once IPSec finds its way into the security policy of most companies, you can expect to see it used in a variety of applications. It can be used in instances where a network contains information that needs to be kept secure, as well as in cases where only authorized users should be allowed access to network resources.

For example, IPSec can be used when sensitive information is sent over the Internet or when someone from outside the company (a business partner or remote user) wants to come into the network through the Internet.

Perhaps IPSec's most profound impact will be on secure VPNs, which promise to provide a less expensive, yet still secure, method of communication than dedicated private networks. However you look at it, the encryption, authentication, and management capabilities of IPSec make it an encouraging way to extend a private network through the public Internet.

RESOURCES

IPSec Protocol Suite

You can find a white paper on IPSec and related topics at the TimeStep Web site at www.timestep.com/netwsec/index.htm.

To view the IPSec RFCs (1825-1829), go to ftp://nic.mil. Then, click on the RFC/folder and scroll down until you find RFCs 1825, 1826, 1827, 1828, and 1829.

This tutorial, number 115, by Anita Karvé, was originally published in the February 1998 issue of Network Magazine.

Biometrics

Biometrics technology verifies or identifies a person based on physical characteristics. A biometrics system uses hardware to capture the biometric information, and software to maintain and manage it.

The system translates these measurements into a mathematical, computer-readable format. When a user first creates a biometric profile, known as a template, that template is stored in a database. The biometrics system then compares this template to the new image created every time a user accesses the system.

For an enterprise, biometrics provide value in two ways. First, a biometric device automates entry into secure locations, relieving or at least reducing the need for full-time monitoring by personnel. Second, when rolled into an authentication scheme, biometrics adds a strong layer of verification for user names and passwords.

Biometrics adds a unique identifier to network authentication, one that's extremely difficult to duplicate (though not impossible, as I'll discuss in the fingerprint scan section). Smart cards and tokens also provide a unique identifier, but biometrics has an advantage over these devices: a user can't lose or forget his or her fingerprint, retina, or voice.

IDENTIFY VS. VERIFY

It's important to distinguish whether a biometrics system is used to verify or identify a person. These are separate goals, and some biometrics systems are more appropriate for one than the other, though no biometric system is limited to one or the other. The needs of the environment will dictate which system is chosen.

The most common use of biometrics is verification. As the name suggests, the biometric system verifies the user based on information provided by the user. For example, when Alice enters her user name and password, the biometric system then fetches the template for Alice. If there's a match, the system verifies that the user is in fact Alice.

Identification seeks to determine who the subject is without information from, or participation of, the subject. For instance, face recognition systems are commonly used for identification; a device captures an image of the subject's face and looks for a match in its database. Identification is complicated and resource-intensive because the system must perform a one-to-many comparison of images, rather than a one-to-one comparison performed by a verification system.

BIOMETRIC ERRORS

All biometrics systems suffer from two forms of error: false acceptance and false rejection. False acceptance happens when the biometric system authenticates an impostor. False rejection means that the system has rejected a valid user. A biometric system's accuracy is determined by combining the rates of false acceptance and rejection.

Each error presents a unique administrative challenge. For instance, if you're protecting sensitive data with a biometric system, you may want to tune the system to reduce the number of false acceptances. However, a system that's highly calibrated to reduce false acceptances may also increase false rejections, resulting in more help desk calls and administrator intervention. Therefore, administrators must clearly understand the value of the information or systems to be protected, and then find a balance between acceptance and rejection rates appropriate to that value.

A poorly created enrollment template can compound false acceptance and rejection. For example, if a user enrolls in the system with dirt on his finger, it may create an inaccurate template that doesn't match a clean print. Natural changes in a user's physical traits may also lead to errors.

BIOMETRIC TYPES

Finger Scan

Fingerprint scanning is the most common biometric system used today. The human fingerprint is made up of ridges that take the shape of loops, arches, and whorls. Rather than scan each ridge, fingerprint-based biometrics look for minutia, which are the points on a fingerprint where a ridge ends or splits into two. An algorithm extracts the most promising minutia points from an image and then creates a template, usually between 250 to 1,000 bytes in size.

Most fingerprint scanners use an optical reader or silicon-based scanner to acquire the image. Optical systems are the most mature and widely deployed. Optical fingerprint readers are durable, inexpensive, and proven. On the downside, they're larger than silicon devices and susceptible to latent prints—remnants of previous prints left on the plate. In addition, the coating on the plate can wear away, which may affect the image.

In silicon chip-based scanners, a coated chip measures skin capacitance to discover the ridge pattern in the fingerprint. Silicon devices are smaller than optical devices but can generally produce a higher-quality image. Their size also allows them to be more easily integrated into peripherals, desktops, and even smart cards. However, the oils and salt on fingers may degrade the chip's performance over time.

Besides hardware problems, fingerprints themselves can change. Daily wear on fingertips can affect ridge patterns and minutia points. Dirt, sweat, and scars can also distort a print, leading to scanning errors.

Anyone planning a fingerprint scanner deployment should also be aware of recent experiments by Tsutomu Matsumoto, a Japanese researcher at Yokohama National University.

According to a paper published by Matsumoto, he used a fake finger made of gelatin to fool numerous commercially available fingerprint readers—both optical and silicon.

In one experiment, Matsumoto made plastic molds of volunteers' fingers, then filled the molds with gelatin available in grocery stores. The print lifted from the mold was sufficient to fool fingerprint readers up to 80 percent of the time. In a second experiment, Matsumoto captured latent prints left on a glass and made gelatin fingers that also fooled fingerprint scanners.

The uproar that followed the paper was predictable: Some security experts touted Matsumoto's results as a death blow to fingerprint scanners. Vendors of the devices decried the results and insisted on their products's security.

Matsumoto's experiments are interesting, but they aren't cause to abandon fingerprint scanners just yet. To their credit, finger scanning devices are relatively inexpensive (for a biometric system), easily coupled to desktops, and fairly unobtrusive to users. However, his results help underline the fact that no security strategy is 100 percent foolproof, and that layered defenses are necessary for robust protection.

Hand Geometry

Hand scanners use an optical device to measure biometric data such as the length, width, thickness, and surface area of a person's hand and fingers. While accurate, hand scanners don't gather as much biometric data as a fingerprint or eye scan. They aren't suitable for identification (as opposed to verification) because similarities among hands don't allow for a one-to-many database search. Hand injuries may also result in false rejections.

That said, hand scanners are easy for subjects to use, and a hand image is difficult to fake. The template of a hand scan is also much smaller than other biometrics, perhaps 20 bytes or less, allowing for more images to be stored.

Iris and Retina

The iris is the colored ring that surrounds the pupil. A camera using visible and infrared light scans the iris and creates a 512-byte biometric template based on characteristics of the iris tissue, such as rings, furrows, and freckles.

Iris scanning is remarkably accurate, making it suitable for both identification and verification. Peripheral cameras are available for desktop-based network authentication, but iris scanning has traditionally been used for physical access to secure locations.

The retina is a nerve in the back of the eye that senses light. A retina scan creates a template from blood vessels in the retina, the patterns of which are unique to each person. Unlike other biometric characteristics, a person's retinal blood vessel pattern changes very little throughout a person's lifetime (except in cases of severe head trauma or degenerative eye disease), making retina scanning a robust solution. A retina scan is also highly accurate.

Many users find a retina scan to be intrusive. That's because the capture device for a retina scan must be within half an inch from the subject's eye. For an iris scan, the capture device can be as far away as three feet. In addition, glasses may interfere with a proper retina scan, but not an iris scan.

Face Scan

Facial scans capture physical characteristics such as the upper outlines of the eye sockets, the areas around the cheekbones, and the sides of the mouth. Face scanning is appropriate for both identification and verification. For network authentication, PC video cameras are sufficient, but higher quality cameras are necessary for capturing images from greater distances (say in the lobby of a building) and in more variable lighting conditions.

Face recognition systems are becoming popular surveillance methods, in part because they can operate without a subject's knowledge.

Voice and Signature

Voice scanning captures characteristics such as the pitch, tone, and frequency of a subject's voice. Voice biometrics may win a role in network authentication because many PCs include a microphone or can be easily fitted with one. However, background noise or poor-quality microphones can interfere with authentication.

Signature scans capture both the image of the signature and the physical characteristics of the signer, such as the speed and pressure he or she exerts. Signature scanning requires a specialized tablet to capture the biometric data.

At this point, voice and signature biometrics aren't widely deployed, though they may be successful in certain niches, such as point-of-sale transactions. On the positive side, user resistance to these biometrics is low because of their ease of use and familiarity.

ISSUES AND STANDARDS

Administrators must keep several points in mind when considering a biometric solution. Aside from scanner hardware and software, a biometric system has other costs, including the template repository. This repository must be at least as well secured as a password database. Administrators must also consider availability, backup, and general maintenance costs.

Biometric enrollment—when the subject first creates a template—must be handled carefully. A poor enrollment scan can lead to false rejection rates later on. Users also need to be properly trained in the use of the biometric system to prevent unnecessary rejections. Users may also have privacy concerns that will need to be addressed.

Finally, biometric systems traditionally required their own backend systems. Vendors are now introducing products that integrate biometrics into a company's overall backend systems, directories, and Single Sign-On (SSO). These products can also integrate different biometric systems, such as fingerprint scanners and iris scanners, into one system.

Several standards are being developed to help integrate diverse biometric technologies. The BioAPI Consortium's (www. bioapi.org) goal is to promote application integration across numerous biometric systems. Version 1.1 of the consortium's specification for a biometric API is available at the Web site.

The National Institute of Science and Technology (NIST) is promoting the Common Biometric Exchange File Format (CBEFF) standard. Its goal is to develop a common format for exchanging templates among biometric systems. More information is available at www.itl.nist.gov/div895/ isis/bc/cbeff/.

Further CBEFF development may take place under the Biometric Interoperability, Performance, and Assurance Working Group, at www.itl.nist.gov/div895/isis/bc/ bcwg/. This group promotes interoperability and performance metrics within the biometric community.

RESOURCES

The Biometrics Consortium, at www. biometrics.org, is a government-sponsored site for the research, testing, and evaluation of biometric technology.

The International Biometric Group provides a host of good information on biometric technology and biometrics vendors. A long-term market analysis of biometrics is available on the site for a fee. Go to www.biometricgroup.com.

Network Magazine's article "Biometric Devices: The Next Wave" (October 2001, page 48) provides a great overview of the biometric market and the major vendors in each area of biometrics. Go to www.networkmagazine.com/article/NMG20011003S0009.

This tutorial, number 168, by Andrew Conry-Murray, was originally published in the July 2002 issue of Network Magazine.

Firewalls

When your network is connected to a public network, it is exposed to spies, thieves, hackers, thrill seekers, and various other threats. As the public Internet has come to play a critical role in most everyone's networking strategy, protection from undesirable Internet inhabitants is a necessity. One of the key elements of safe networking is a properly designed and configured firewall system.

There are three basic designs for firewall operation. The first, and simplest, is the packet filter. Most routers can be configured to filter packets based on straightforward comparison of packet contents with filter specifications. For instance, particular IP addresses or subnets, particular TCP or UDP port numbers, or combinations of these properties can always be denied passage.

In most modern routers, adding a security-based filter step to the packet forwarding process will add little or no overhead, so packet-filtering firewalls can have very high performance. Unfortunately, would-be intruders have a number of options for defeating simple filtering. For instance, they can spoof packets to appear as if they come from an acceptable source.

The second basic firewall design enhances packet filtering so that it can't be circumvented by these measures. This process, developed and patented by Check Point Software Technologies (www.checkpoint.com), is known as stateful inspection. Stateful inspection extends the packet-by-packet filtering process to include multipacket flows. A connection table tracks individual flows, enabling policy checks that extend across series of packets. For example, TCP ACK packets not preceded by a TCP SYN packet with a correct sequence number can be blocked.

With the correct configuration, a stateful inspection firewall can offset many of the shortcomings of plain packet filtering without serious performance penalties. Like vanilla packet filtering, stateful packet inspection works for all applications because its basic functioning is at the Network and Transport layers. No special client configuration or new client software is required.

The third firewall design is the application proxy. With a pure application proxy, no traffic at all goes through the firewall. Instead, the application proxy behaves like a server to clients on the trusted network and like a client to servers outside the trusted network.

Thus, a Web browser attempting to connect to eBay will speak HTTP and pass the eBay destination URL—and the other information that browsers provide Web servers—to the application proxy firewall. The firewall will then apply its policy rules to the request. If the request is permitted, it will send the request off to eBay. The source IP address on the HTTP packets to eBay will be that of the firewall, not that of the original client.

By operating at the Application layer, application proxy firewalls permit as much granularity as anyone could desire when it comes to rules. For example, lists of specific URLs can be blocked from certain subnets, or FTP clients can be restricted from Puts, but permitted to execute Gets. Added advantages of Application-layer operation include the ability to require strong authentication before connecting and the ability to create detailed logs of security events.

Because the action is at the Application layer, proxies must be provided for each application. A number of traditional Internet applications—including FTP, e-mail, and news—are bundled into common browsers, so they can all be handled by configuring the browser to talk to the firewall. However, custom applications and network applications not bundled into a browser will have to be configured at the firewall individually, assuming they can be adapted to proxy execution at all.

While application proxy firewalls can provide the highest level of security and the finest-grain control, they can also be the most complex to configure. Also, because they act as relay agents for all the clients on the network, their performance can be problematic.

A variant of the application proxy model is the circuit-level gateway. On this model, the firewall relays TCP (and in the most recent implementations, UDP) connections between the trusted and untrusted networks after authenticating the end points. Circuit-level gateways don't access Application-layer information as application proxy firewalls do, so it's not necessary to supply separate proxy processes for each application. The best-known implementations of circuit-level gateways employ an IETF standard protocol called SOCKS. Version 5 of SOCKS includes authentication and UDP flows.

SOCKS 5 (and circuit-level gateways in general) require modifications to either applications or to client TCP/IP stacks. Most common browsers on the most widespread platforms have built-in SOCKS support, and modified protocol stacks are also available for various flavors of Unix, various flavors of Windows, and Macintosh. Nevertheless, a significant administrative effort is needed to implement a circuit-level gateway for a sizable enterprise.

Application proxy firewalls and circuit-level gateways transparently perform Network Address Translation (NAT). The network interface that connects to the untrusted network is all that the untrusted network sees, so the addresses and the structure of the trusted network are shielded from view. In themselves, packet filters and stateful inspection firewalls don't translate IP addresses and port numbers, though there's no reason that NAT can't be implemented separately.

Some makers of NAT devices (mostly those aimed at home networks) describe them as firewalls, but that claim is misleading. NAT does shield internal addresses and their structures from outside view, but it doesn't provide control over outbound traffic, perform authentication, or prevent spoofed inbound packets.

HYBRID DEVICES

In practice, commercial firewalls often make use of multiple protective techniques. Check Point's Firewall-1 has NAT options. Network Associates' Gauntlet, an application proxy firewall, incorporates so-called adaptive proxy technology that essentially hands off packet flows to a stateful inspection engine if the first few packets are passed by the application proxy.

Furthermore, the position of the firewall at the boundary between the trusted network and the untrusted network makes it tempting to build in other services. For example, it may make the most sense to terminate a VPN at the firewall, rather than decrypting the contents of each packet for a security check and then re-encrypting everything that passes the check.

A security check that involves identifying particular traffic flows will have done much of the work required for bandwidth management. It's a small step from providing application proxy services to providing a local cache for common requests, as Novell has done with its BorderManager product.

Firewalls are fundamentally software processes, though they aren't all marketed as software. A firewall can't be any more secure than the operating system underlying it. Most firewalls are deployed on Unix or Windows NT/2000, though there are vulnerabilities with all the variants of these OSs.

Some firewall producers customize an OS—usually Unix—and install their firewall software on that OS in a sealed box. These systems are marketed as firewall appliances, and some would argue that hiding the particulars of the OS contributes an additional layer of security. More importantly, the "hardened" kernels of these customized OSs are stripped down and adapted for firewall functionality instead of general-purpose application execution.

When the trade press began writing about firewalls in the mid-'90s, it was common for tests to be designed around performance, as though these devices were routers or drag racers. Unfortunately, the throughput of a firewall is completely orthogonal to the security it provides. In fact, overly fast performance may once have been an indicator that security was lacking.

Testing for security is inherently difficult—it's impossible to prove that every vulnerability has been eliminated. Security scanner tools that have gathered together all known exploits can be run against firewalls, but that isn't the same as proving invulnerability.

BEYOND TECHNOLOGY

While understanding firewall architectures and technical vulnerabilities is essential to protecting the enterprise network, developing an appropriate security policy is perhaps even more important. Establishing policy requires understanding the value of data, of undisrupted business processes, of various forms of legal and fiduciary liability, and other nontechnical organizational matters.

IT and security people can help ask the right questions, but they aren't the last word on such issues. Once enterprise management determines the fundamental policy guidelines, technical people can implement practices and procedures to carry out those guidelines.

An additional technical matter arises at this point. In some ways, a misconfigured firewall can be more dangerous than none at all. Traditional firewalls were configured with scripts and command-line interfaces. It takes a lot of training and aptitude to mentally convert a list of "deny" and "permit" statements with IP addresses and port numbers into a clear image of the network's protection. Purists may disagree, but I'm convinced that modern GUI-based software, with wizards and graphical representations of procedures, enhances the security of a firewall implementation, in addition to easing the pain.

A related issue is firewall management. Large enterprises will have numerous firewalls to deploy, and will likely want to implement similar policies in many locations. Some enterprises use internal firewalls to protect against potential internal threats, and the configuration of these will perhaps differ substantially from that of the firewalls that protect against threats coming in from the public network. Consistency and flexibility of configuration management will be important for these internal firewalls too.

Firewall logs are critical for anticipating threats, for postmortems after attacks, and for understanding future requirements. Application proxy firewalls, which have the most detailed information about traffic, can provide the most detailed logs. Logs often serve to trigger alarms—you can be paged in the middle of the night when that dreaded Distributed Denial of Service (DDoS) attack finally hits your e-commerce site. Logs provide valuable evidence after successful or unsuccessful firewall activities, and they can certainly play an important role in tracing intruders, in convicting them once they're caught, and in proving that your site took pains to prevent intrusions.

Firewalls can't be expected to do the impossible. If there are back doors into your network, such as desktop or laptop machines with dial-up Internet accounts, or home cable-modem users with access to your network, the firewall can't defend those pathways. Successful firewall operations are dependent on the device being a single point of contact between the untrusted network and the trusted network.

Furthermore, firewalls installed to protect the internal network from outside intruders are no help against insiders. Firewalls don't replace access controls on servers, file systems, databases, or applications. Nor can they protect the network from viruses, unless that capacity is specifically added on.

With well-thought-out configuration and management systems, however, modern firewalls can keep even dedicated attackers from wreaking havoc on your network.

RESOURCES

Network Associates' Gauntlet firewall is part of the company's Pretty Good Privacy (PGP) division. There are white papers on adaptive proxies and other topics at www.nai.com/asp_set/buy_try/try/whitepapers.asp#PGP SECURITIES/.

Check Point Software Technologies' account of stateful packet inspection, along with numerous other technical documents, is available at www.checkpoint.com/products/downloads/fw1-4.0tech.pdf/. Registration is required if you haven't previously registered.

Information Security Policies Made Easy, by Charles Cresson Wood (Baseline Software, Sausalito, CA, 1997, ISBN 1881585042), is a comprehensive starting point for defining organizational security policies.

This tutorial, number 143, by Steve Steinke, was originally published in the June 2000 issue of Network Magazine.

Intrusion Detection

Intrusion detection has its roots in the financial audits of mainframe computers. The scarcity and expense of mainframes meant that access to their computing power had to be tightly controlled (and users had to be accurately billed). In the late 1970s, financial audits were adapted for security purposes, enabling administrators to review logs for anomalies that might indicate misuse, such as unauthorized file changes.

Further developments led to real-time detection capabilities in the mid-1980s, and to network monitoring in 1990. Today, network managers can choose from a variety of solutions and vendors, but the general principles of an Intrusion Detection System (IDS) remain the same.

The idea behind an IDS is simple: an agent monitors file activity on a host or traffic on a network, and reports strange behavior to an administrator (see Figure 1).

The IDS market is divided into two primary groups: host-based and network-based systems. This tutorial examines the basics of host- and network-based detection systems, discusses how to implement an IDS, and outlines a proper incident-response plan.

HOST-BASED IDS

Host-based IDSs add a targeted layer of security to particularly vulnerable or essential systems. An agent sits on an individual system—for example, a database server—and monitors audit trails and system logs for anomalous behavior, such as repeated login attempts or changes to file permissions. The agent may also employ a checksum at regular intervals to look for changes to system files. In some cases, an agent can halt an "attack" on a system, though a host agent's primary function is to log events and send alerts.

The primary benefit of a host-based system is that it can detect both external and internal misuse, something that network monitors and firewalls can't do. The appeal of such a tool is obvious, as security breaches are more likely to come from an internal user than from a hacker outside the network. Host agents are powerful tools for addressing the authorization and access issues that make internal security so complex.

Agents install directly on the host to be monitored, so they must be compatible with the host's OS. Memory requirements and CPU utilization will vary from vendor to vendor, so be sure to learn ahead of time the demands the agent will place on the system.

NETWORK-BASED IDS

A network-based IDS sits on the LAN (or a LAN segment) and monitors network traffic packet by packet in real time (or as near to real time as possible), to see if that traffic conforms to predetermined attack signatures. Attack signatures are activities that match known

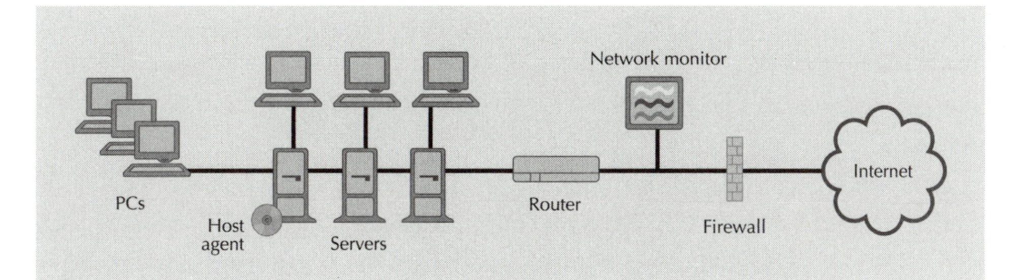

■ **Figure 1: Electronic Eyes.** An Intrusion Detection System (IDS) monitors network traffic or file activity on a host for attacks, anomalous behavior, and misuse. An IDS logs intrusions, sends real-time alerts, and in some situations can halt the attack.

attack patterns. For example, the TearDrop Denial of Service (DoS) attack sends packets that are fragmented in such a way as to crash the target system. The network monitor will recognize packets that conform to the TearDrop signature and take action.

The IDS vendor provides a database of attack signatures, and administrators can also add customized signatures. If the IDS recognizes an attack, it alerts an administrator. In some cases, the IDS can also respond, for example by terminating a connection. In addition to its monitoring and alarm functions, the IDS also records attack sessions for later analysis. Network IDSs can also be linked to other security features, such as firewalls, to make sure those systems haven't been breached.

A network monitor has two main benefits. The first is the real-time nature of the alarm, which can give administrators an opportunity to halt or contain an attack before it does significant harm. This is especially valuable for DoS attacks, which must be dealt with immediately to mitigate damages.

The second benefit is evidence collection. Not only can administrators analyze the attack to determine what damage might have been done, the attack session itself can point out security flaws that need addressing. (This is also true for host-based systems). Because many hackers first scan a target network for known vulnerabilities, a hacker's choice of attack may indicate that such vulnerabilities exist on your network. A simple example is an operating system that has yet to be secured with the latest vendor patch.

Network monitors are OS-independent. Basic requirements include a dedicated node that sits on the segment to be monitored and a NIC set to promiscuous mode. You may also want to set up a secure communications link between the monitor and its management console.

ESTABLISHING AN IDS

The first step in establishing an IDS is to incorporate it into your security policy. (If you don't have a security policy, now would be a good time to develop one. See "Create

Order with a Strong Security Policy," *Network Magazine* July 2000, page 62 for more information.) In brief, a security policy defines the basic architecture of the network, describes how the network will be secured, and establishes a hierarchy of user access to data resources.

When incorporating an IDS into your security policy, you should define how the IDS will fit into the overall security architecture, outline procedures for maintaining and responding to the IDS, and assign resources (software, hardware, and humans to manage the technology).

You'll also have to choose a network- or host-based system, or a combination of both. A combination provides the most comprehensive security; however, this decision will be colored by the level of security you require, the budget at your disposal, and the in-house resources on hand to manage the system.

Generally speaking, network monitors cost significantly more than host-based agents. However, depending on the size of your network, a single monitor can offer substantial network coverage. Conversely, host-based agents cost less, but are limited to a single host.

Other factors play in deciding to implement either or both solutions. For example, network monitors may have difficulty with encrypted traffic. A network monitor functions by reading packet headers and data payloads. If this information is encrypted, the IDS can't detect attacks. Encryption doesn't hinder host agents because the data is decrypted before a host agent sees it.

Network sensors can also become a bottleneck on high-speed LANs, degrading performance and frustrating users. According to an ICSA paper, a network-based IDS can handle up to 65Mbits/sec of traffic before the analysis engine's performance drops (see Resources).

CARE AND FEEDING

Regardless of which solution you implement, you must be prepared to properly maintain the system. IDSs can generate reams of data that have to be reviewed regularly. Most products come with reporting software to help with this task.

Depending on the number of monitors and agents you employ, be prepared to dedicate substantial storage to IDS log files. You also must secure audit logs so that intruders can't tamper with them to erase or obscure evidence of the intrusion.

Like anti-virus products, an IDS's attack signature database must be updated regularly. Vendors will provide new attack signatures, but be sure to query them on the frequency of updates, especially in response to newly discovered attacks. Be aware that slight variations to a known attack may be enough to slip past even an IDS with the most current signatures.

You can also be proactive by monitoring security sites for new attack signatures and exploits.

DEALING WITH INCIDENTS

Once you install your IDS, be prepared for the possibility that it will work! That is, have a plan in place for dealing with intrusions once you detect them.

The first step is to create an incident-handling team to respond to intrusions. The size and capabilities of your team will vary with the size of your organization, but each member of the team should have clearly-defined roles and responsibilities (for example, a Windows NT specialist, a Unix specialist, and so on).

You'll also want to create an incident-handling policy that outlines the response procedures and lists contact information for team members. Procedures include backing up an affected hard drive and determining whether it is necessary to enlist outside expertise or contact law enforcement agencies.

The decision to involve the law is a complicated one, so it's best to have policies in place beforehand. It may not be worth your time to call the cops on a script kiddy who Pings your network.

IS AN IDS RIGHT FOR YOU?

Intrusion detection is still a maturing technology, and not everyone in the security community is convinced of its viability. Some observers have compared intrusion detection to the so-called Star Wars missile defense program—that is, expensive and ineffective.

Of course, "expensive" is a relative term: While an IDS doesn't run cheap, the cost of a network outage from a DoS attack (and the attendant bad press, dissatisfied customers and business partners, and furious executives) can easily justify the vendor's price tag.

As for effectiveness, an IDS is not a "set it and forget it" proposition. Security policies must be in place, attack signature databases must be updated, and logs have to be reviewed regularly to gain the full benefits. If you can meet those requirements, intrusion detection is a valuable tool for protecting your data resources.

RESOURCES

The ICSA's Intrusion Detection Systems Consortium has an informative white paper on Intrusion Detection Systems (IDSs). You can download the paper at www.icsa.net/html/communities/ids/White%20paper/index.shtml.

Intrusion Detection by Rebecca Gurley Brace (Macmillan Technical Publishing, 2000) is an informative and comprehensive guide to the concepts and principles of IDSs.

The IDS vendor Network Security Wizards has a good collection of papers and articles at www.securitywizards.com/library.html. Network managers may be particularly interested in the rebuttal to "50 Ways to Defeat Your Intrusion Detection System."

CERT maintains an IDS checklist at www.cert. org/tech_tips/intruder_detection_ checklist.html.

Network Magazine has published several useful articles on IDSs. Check out "Deploying an Effective Intrusion Detection System" (September 2000, page 60), "Security Reality Check" (July 1999, page 80) and "Can Intrusion Detection Keep an Eye on Your Network's Security?" (April 1999, page 36). In addition, the article "Incident Handling" (January 2000, page 80) provides a good overview of how to build an incident-handling team. Archived article can be found at www.networkmagazine.com.

This tutorial, number 149, by Andrew Conry-Murray, was originally published in the December 2000 issue of Network Magazine.

Vulnerability Assessment Tools

As the complexity of an enterprise network increases, so do its vulnerabilities. Heterogeneous operating systems, each with their own configuration quirks, run myriad devices and applications in a high-speed, highly connected environment. The upshot is a maelstrom of code, with unexpected holes, glitches, and back doors. Oftentimes, network administrators aren't aware such breaches exist until an intruder uses them to gain unauthorized access to network resources.

Scanning is one way to root out possible weak points in your network. A host of software-based scanning tools are available to probe your network for known vulnerabilities in operating systems, applications, passwords, and so on. In fact, would-be intruders use these tools to scope out a network before attacking, so scanning is also a proactive security measure that lets you find the chinks in your armor before someone else does.

Depending on the level of technical expertise available to you, you'll have to decide whether to conduct your own scans or hire a scanning service. If you conduct your own scan, you can choose among commercial, open-source, and freeware tools.

Many of the open-source and freeware tools, such as Nmap, are written by hackers. Why would a legitimate administrator use hacker tools? Good question. A good answer is because such products can be highly effective. You may also find it instructive to learn just how intruders go about casing your network. However, commercial software packages perform the same functions and have easy-to-use interfaces and reporting capabilities.

PINGS AND PORTS

A vulnerability scan takes a hacker's-eye view of your network. Seemingly harmless communication between two machines reveals information pointing to potential vulnerabilities (see "Scan and Deliver," page 34). The scanning tools match this information against a database of exploits to determine which ones may be present on your network.

A vulnerability scan consists of three basic steps. The first is network discovery, which uses the Ping utility to discover active devices on the network. The Ping utility sends Internet Control Message Protocol (ICMP) packets to a target system, looking for a response. A positive response, such as an ICMP ECHO_REPLY (Type 0) means the target is alive. This creates a basic map of live hosts that an intruder can target individually.

The second step is a port scan, which identifies ports in listening mode as well as those that may have exploitable active services. Port scans also identify the operating system on a target, including which service packs or kernel releases have been installed. This information permits an intruder to launch very precise exploits against a target. Be aware that simply finding a listening port doesn't imply vulnerability. Sometimes, an intruder must follow up with further packet manipulation to tease out potentially damaging information. The sidebar at the end of this tutorial describes several of these techniques.

In the final step, the scanner analyzes the data and generates a report detailing potential vulnerabilities and fixes. Because the report data is so crucial, you'll want to choose a solution that displays the results clearly and usefully to you. It's worth your time to review sample reports before buying or using any scanner.

PASSWORD CRACKERS

Password auditing software is another assessment tool available to both intruders and administrators. An administrator uses such a tool to audit his users' passwords to ensure they're following good password policy. A number of open-source password crackers are available on the Internet, and many commercial security scanners include password auditing.

The premier Windows NT password cracking tool is L0phtCrack, developed by the L0pht Heavy Industries group (see Resources).

Windows NT disguises passwords with an encryption function that turns plain text into a string of bytes, known as a hash. The problem is that NT must also support an older and weaker hash algorithm from LAN Manager for compatibility reasons. Because L0phtCrack has reverse-engineered the LAN Manager encryption function, it produces the same hash. Thus, instead of decrypting the password, L0phtCrack merely matches hashes.

The product exploits this flaw in three ways. First, a dictionary attack runs the hash on a collection of words commonly used as passwords. It compares this list to the NT hashes, looking for matches. If it doesn't find a match, the program next adds random characters to each word in its dictionary list. Finally, if still unsuccessful, it runs a brute-force attack, trying every possible combination of letters and numbers until it discovers a match.

While the above information makes it sound as if any old script kiddie running L0pht-Crack can waltz off with your passwords, a would-be cracker must still overcome several barriers. For one, an attacker first has to get the Security Accounts Manager (SAM) file that stores the NT hashes, which requires administrator privileges on the target system. In addition, L0phtCrack must also be run offline because of the time it takes to match the hashes in the SAM file. The further intricacies of password cracking with L0phtCrack are outside the scope of this article, but check out Resources for more information.

SCAN THE SCANNERS

Before purchasing or downloading the latest and greatest network scanner, take some time to search for tools that will meet your objectives.

A good starting point is to determine the product's ease of use. If you prefer GUIs to command line interfaces, you'll likely lean toward commercial products. However, there's nothing stopping you from using both off-the-shelf and freeware tools.

When choosing a product, make sure it prioritizes vulnerabilities. A full-blown scan may generate an 800-page report filled with hundreds of exploits. Addressing such a gargantuan list will be more manageable if you know which ones must be dealt with immediately.

Be aware that a scan will affect your network's performance—to what degree depends on the depth of the scan and the number of devices. Schedule your scans when they're least likely to impact essential business services. Also, look for tools that allow you to target specific systems. You may want to scan particular segments of your network more frequently than others, and there's no sense in blasting every device you own with resource-consuming packets.

Find out how frequently the vulnerability database is updated. Just like viruses, new exploits appear all the time, so your tools should stay abreast of the latest attacks. That said, there are different ways of listing and counting exploits, so don't be dazzled by high numbers. Product A may claim to spot 10,000 vulnerabilities while Product B detects 5,000, but this doesn't mean Product A is a more comprehensive solution.

You may also want to inquire where the vendor gets its vulnerability information. While every vendor makes use of public postings from organizations such as CERT, BugTraq, and the SANS Institute, many also have in-house research teams that alert you to security holes before they're posted at large.

CLEAN SWEEP

Let's say you've scoured your network from top to bottom. You've found the holes, read the reports, and applied the patches. Now your network is one hundred percent secure—at least until an inventive coder discovers an entirely new way to slip packets through your firewalls.

Vulnerability scanning is not a one-time fix. Clever and industrious hackers constantly discover new exploits. In addition, clever and industrious software vendors are constantly releasing new versions of their products; even software that comes fully baked from the shop probably has unforeseen holes, or will interact in unexpected ways with your own network.

The frequency of your scans should depend on your security posture, as well as on the lifecycle of your network devices. If you have a relatively stable architecture with few changes or upgrades, you'll likely require fewer scans (assuming you patched the holes you discovered the first time around). However, particularly sensitive segments of your network, such as the Demilitarized Zone (DMZ), may warrant more frequent check-ups.

Overall, be prepared to invest a good deal of time both for the scan and the clean-up afterwards. It makes no sense to discover vulnerabilities if you simply ignore them.

SCAN AND DELIVER

In a normal TCP communications sequence, a client machine and server must go through a three-step "handshake" to establish a connection. The client initiates the handshake by sending a SYN packet to the server. If the server is available, it acknowledges the communication with a SYN/ACK packet. Finally, the client sends its own ACK packet and makes a connection.

Intruders can manipulate this handshake sequence to glean essential information based on the server's response, including misconfigured operating systems or software versions with known vulnerabilities.

The following port scans are commonly used to case a target host. This list is exerpted, with permission from the publisher, from the book *Hacking Exposed*, Second Edition (Osborne/ McGraw-Hill, 2001), by Joel Scambray, Stuart McClure, and George Kurtz.

TCP Connect scan. This type of scan connects to the target port and completes a full three-way handshake (SYN, SYN/ACK, and ACK). It is easily detected by the target system.

TCP SYN scan. This technique is called half-open scanning because a full TCP connection is not made. Instead, a SYN packet is sent to the target port. If a SYN/ACK is received from the target port, we can deduce that it is in the LISTENING state. If a RST/ACK is received, it usually indicates that the port isn't listening. A RST/ACK will be sent by the system performing the port scan so that a full connection is never established. This technique has the advantage of being stealthier than a full TCP connect, and it may not be logged by the target system.

TCP FIN scan. This technique sends a FIN packet to the target port. Based on RFC 793 (www.ietf.org/rfc/rfc0793.txt), the target system should send back an RST for all closed ports. This technique usually only works on Unix-based TCP/IP stacks.

TCP Xmas Tree scan. This technique sends a FIN, URG, and PUSH packet to the target port. Based on RFC 793, the target system should send back an RST for all closed ports.

TCP Null scan. This technique turns off all flags. Based on RFC 793, the target system should send back an RST for all closed ports.

TCP ACK scan. This technique maps out firewall rulesets. It helps determine if the firewall is a simple packet filter allowing only established connections (connections with the ACK bit set) or a stateful firewall performing advanced packet filtering.

TCP Window scan. This technique may detect open as well as filtered/nonfiltered ports on some systems (for example, AIX and FreeBSD) due to an anomaly in the way the TCP window size is reported.

TCP RPC scan. This technique is specific to Unix systems and detects and identifies Remote Procedure Call (RPC) ports and their associated programs and version numbers.

RESOURCES

Rik Farrow's Network Defense columns in *Network Magazine* provide a host of information on security topics. In regard to vulnerability scanning, check out "ICMP Stands for Trouble" (September 2000, page 98) and "System Fingerprinting with Nmap" (November 2000, page 102). The articles can also be found at www. networkmagazine.com.

The book *Hacking Exposed*, Second Edition by Joel Scambray, Stuart McClure, and George Kurtz, has several in-depth chapters on vulnerability scanning, security exploits, and password cracking. Go to www. foundstone.com for more information.

L0phtCrack is available at www. securitysoftwaretech.com.

You can download the scanning tool Nmap from its creator's Web site at www.insecure.org. The site also has exploit lists and links to other scanning tools and security sites.

A list of commercial scanners is available at http://securityportal.com/research/research.scanners.html.

The magazine *Information Security* ran a four-part series on vulnerability assessment, from July 2000 through October 2000. Back issues are on the Web at www. infosecuritymag. com.

The Web site Windows IT Security discusses how to use L0phtCrack on Windows 2000. Go to www.ntsecurity.net/Articles/ Index.cfm?ArticleID=9186/.

This tutorial, number 153, by Andrew Conry-Murray, was originally published in the April 2001 issue of Network Magazine.

E-Mail Security

Security problems hinder the growth of e-mail as a business tool.

In this installment, I'll cover the security problems that users and organizations face when using e-mail, as well as possible solutions to those problems.

IDENTITY CRISES

E-mail has several inherent security problems (see "What Ails E-Mail"). When placed in the context of business communications, these problems limit e-mail's potential as a serious business tool. For example, one of its limitations is privacy. Normally, e-mail is sent "in the clear," meaning the message is sent in plaintext. So, anyone who can access the e-mail, whether in transit or in storage, can read the message. Clearly, this is a security problem that may prevent companies from using e-mail to convey confidential business information.

WHAT AILS E-MAIL

Here's a list of the main security issues that affect e-mail.

Lack of privacy E-mail is sent in plaintext and can be read by anyone who can access it.

Lack of integrity There is no safeguard to prevent someone from changing the contents of an e-mail message while it's in storage or in transit.

Lack of authenticity Anyone can forge an e-mail message that claims it was written by another individual.

Lack of nonrepudiation Any particular e-mail message can't be bound to its sender, so a sender can deny ever having sent a message to you.

Viruses E-mail messages can contain attachments that are actually viruses in disguise; when you open the attachment, the virus spreads to your PC.

Spam An e-mail account is an open home for spam, those annoying mass e-mail rants and advertisements

Another basic problem revolves around the integrity of a particular piece of e-mail. As mentioned, it's possible for someone to access or intercept a piece of e-mail as it lies in storage or while it's in transit. Since most e-mail messages are in plaintext, anyone who can access the message can also change the contents of the message—without the user knowing that the message had been altered. In this case, the integrity of the message has been compromised.

A more complex security problem is one of authenticity. Currently, there is no method built into e-mail that would let a recipient of a message verify that the sender is actually who he or she claims to be. Combined with the integrity issue, this lack of verification means that e-mail is an untrustworthy system.

A related security problem is a lack of nonrepudiation, in which a sender can deny that he or she ever sent a message. Additionally, there is no way to disprove a sender's claim that his or her message has been tampered with so that its meaning has been changed.

E-MAILER ID

Currently, several schemes seek to address e-mail's security woes, specifically those of privacy, integrity, authentication, and nonrepudiation. These solutions are all rooted in public key cryptography technology. (To learn more about public key technology, read "Public Key Cryptography," following this tutorial.)

If you're not familiar with public key technology, I'll give you a brief overview. In a public key system, a user is assigned a pair of keys that work together. One of the keys is the user's private key, which only the user can possess. The other key is his or her public key, which is freely distributed to the public. Both the keys can encode and decode data. However, what one encodes, only the other can decode. They will not work with any other key.

Here's how the system secures e-mail. When User A wants to send a confidential e-mail message to User B, User A encrypts the e-mail using User B's public key. The only key that can decode the message is User B's private key, which only User B possesses. Consequently, no one else can read the message. This takes care of the privacy and integrity problems.

The system also addresses the issues of authenticity and nonrepudiation. On a basic level, if User A wants to send an e-mail message and assure recipients that the message actually came from him or her, User A can encrypt the message using his or her private key. To read the message, recipients use User A's public key to decode it. Since only User A's public key will decode the message, recipients know that the only key that could encode the message was User A's private key. And, since User A is the only person in possession of that private key, the message must have come from User A.

However, for this system to work, and this is a critical point, users must be able to trust that a user's key is valid. To provide this verification, a trusted entity, generically known as a Certificate Authority (CA), assigns each user a unique digital certificate that assures that a certain public key belongs to a certain user.

THE SOLUTIONS

As mentioned, several schemes for securing e-mail have been developed. All of them are based on public key encryption. However, they differ in the way they implement this technology; specifically, the biggest difference among the schemes is in how they handle key certification.

The hardest aspect of securing e-mail is establishing a valid CA that everyone has access to. Currently, there is no central CA in existence that the general public can use to verify public keys. And, without a governing CA, to whom can a user turn to verify that a particular public key belongs to a particular user? Some organizations that use public key systems internally act as their own CA for their users. However, these organizations can't manage certificates for the public in general.

Privacy Enhanced Mail (PEM), the IETF standard that addresses secure e-mail, proposes using a hierarchy of trusted bodies to reassure users of the validity of a particular e-mail message. At the top level sits the Internet Policy Registration Authority (IPRA), which would be the governing certificate trusted by all. The IPRA would sign certificates for a second layer of trusted bodies called Policy Certification Authorities (PCAs). These in turn would authorize certificates for another layer of bodies called Certificate Authorities (CAs). These CAs would be responsible for authorizing certificates for the public at large.

When a user receives an e-mail message under the PEM model, the e-mail header lists the certificate of the body that authorized the sender's certificate. That way, recipients know that the e-mailer's identity had been verified by the body.

Currently, you can find several products on the market or the Internet that follow the PEM model, including ones from RSA Data Security, Trusted Information Systems, and Michigan State University (called RIPEM). However, because the certificate hierarchy suggested by PEM (the IPRA model) hasn't been established, some of these products use proprietary CA schemes.

Pretty Good Privacy (PGP), a secure e-mail program, offers an alternative take on the CA. Instead of using a hierarchy of authorities, PGP builds something called a web of trust. According to this concept, each user keeps a list of certificates from other PGP users they trust. Each of these certificates contains a list of other trusted certificates. So, when a user receives an e-mail message from someone they don't know, the user can check the chain of certificates listed with the e-mail to see if he or she recognizes a trusted certificate. Or, the user can try to trace a link between certificates listed in the e-mail back to the handful of certificates that the user trusts.

A third secure e-mail scheme is an extension of the PEM standard called MOSS, or MIME Object Security Services, which is still in the debate and development stage within Internet circles. MOSS allows users to secure multimedia messages, using the same principles as PEM.

One difference, however, is that MOSS doesn't require certificates to verify keys, although it does support the certificate structure proposed by PEM. Instead, users must verify keys in any way they can.

A fourth scheme is S/MIME, or Secure MIME, which was developed by RSA Data Security. This specification also provides a way to secure e-mail, including MIME messages. Naturally, S/MIME uses RSA's public key cryptography standards to handle encryption and key pairs. S/MIME uses X.509-based certificates in its model, but it actually leaves the whole matter of setting up a certificate structure to the implementing organization.

AN OPEN DOOR

So far, the security problems I've discussed all stem from the inherent technological limitations of e-mail. However, e-mail also poses a few other security problems. The first problem is one everyone is familiar with: viruses. In the past, viruses were usually spread by floppy disks. E-mail makes the voyage much simpler for viruses. Now, viruses can ride along with an e-mail message as an attachment—right to your desktop. Usually, the virus is disguised as a harmless file. When a user opens the attachment, the virus can spread to the host computer. To combat this problem, you can find several virus scanners on the market that scan incoming e-mail for viruses.

Finally, with the emergence of e-mail comes the birth of spam, otherwise known as junk e-mail. Like junk mail, junk e-mail is a form of advertising in which a person or company sends out a bulk mailing (in this case, via e-mail) to hundreds and thousands of recipients. The contents of the message usually consist of an advertisement peddling some type of service or product, or general ranting and raving.

For businesses, spam is problematic because of the costs it inflicts. On a systems level, spam is an infringement on network resources across the board—including processing power, storage space, network bandwidth, and dial-up and leased lines to the Internet. On an employee level, workers face sorting through junk e-mail to get to their business e-mail. Because of these costs, spam can be seen as a security threat to businesses.

Currently, ISPs are trying to combat spam by filtering out known junk e-mail broadcasts from its systems. Aside from that measure, there are a couple of bills being proposed in the U.S. Congress that seek to limit the actions of spammers. The only solutions businesses have are to install their own filters on their e-mail systems and to educate users about the hazards of handing out their e-mail addresses too freely on the Web and to other parties.

THE MESSAGE IS CLEAR

By itself, standard Internet e-mail is a security liability simply because of the fact that it's sent "in the clear," and is easily forged. With the solutions I've mentioned, you can see that work is underway to make e-mail a more reliable, trustworthy tool for business communication (see "Resources" for Web sites that deal with e-mail security issues). Additionally, there are products available that businesses can implement internally for their employees. However, such a solution doesn't help when trying to secure e-mail being exchanged with other businesses that don't have a secure e-mail system in place.

The real solution will arrive once someone is able to develop a CA structure that everyone, businesses and the public alike, acknowledges. Without this structure, the privacy, integrity, and authenticity of e-mail will always be an issue.

RESOURCES

Here's a list of Web sites you can browse for more information on solutions for shoring up e-mail's security holes.

Privacy Enhanced Mail (PEM)

kekec.e5.ijs.si/security/pem.html

retriever.cs.umbc.edu/~woodcock/cmsc482/proj1/pem.html

ds.internic.net/rfc/rfc1421.txt to 1424.txt (RFCs 1421-1424)

Pretty Good Privacy (PGP)

www.pgp.net

Secure MIME (S/MIME)

www.rsa.com

MIME Object Security Services (MOSS)

ds.internic.net/rfc/rfc1848.txt (RFC 1848)

Virus and Virus Hoax Alerts

www.antivirus.com

www.drsolomon.com

www.mcafee.com

www.symantec.com

Anti-Spam Sites

www.cauce.org (Coalition Against Unsolicited Commercial E-mail)

www.ftech.net/~monark/spam/index.hts

www.mcs.com/~jcr/junkemail.html

This tutorial, number 114, by Lee Chae, was originally published in the January 15, 1998 issue of Network Magazine.

Public Key Cryptography

A secure and simple method of encrypting and decrypting electronic information.

It's been said time and again that Internet commerce won't be viable until tighter security measures are in place and consumers have confidence in them. But on a more basic level, securing computer files and e-mail is of much more concern to users on an everyday basis. After all, if you lose your laptop, your competition can potentially access the sensitive information you were carrying. Also, transferring files back and forth over the Internet or even through proprietary mail systems can present risks if the information is intercepted. To foil the plans of anyone trying to usurp electronic files, many have turned to cryptography, which encrypts and decrypts information such that it's useless to everyone except authorized users.

Traditional cryptosystems are based on the secret key, or symmetric, model (see Figure 1). In this system, each user has his or her own secret key, which is used for both encryption—in which plaintext is converted to ciphertext—and decryption—in which the process is reversed.

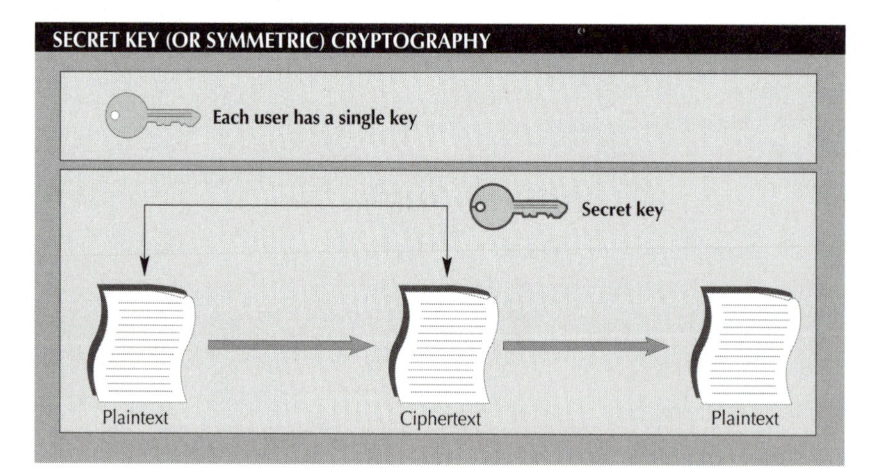

■ **Figure 1:** Secret key, or symmetrical cryptosystems, rely on a single key for all encryption and decryption between parties, which means these systems are fast, but can be easily compromised if the key is lost.

An example of a secret key cryptosystem is the data encryption standard (DES), which was originally developed by IBM and endorsed by the U.S. government in 1977. DES uses a 56-bit key and has a 64-bit block size. The encryption process works on one block of data at a time—in this case, 64 bits—then proceeds to process the next block.

Other symmetric cryptosystems include RC2 and RC4, both of which use variable key lengths and operate on 64-bit blocks of data. RC2 and RC4 were developed by RSA Data Security.

The larger the key size, the more complex the algorithm and the more difficult the cryptosystem is to crack. Within the United States, the sky's the limit in terms of key size, with many organizations going with key sizes as large as 1,024 bits. Currently, however, only keys of 40 or fewer bits can be exported, although this is currently under debate and could change in the near future.

Because secret key cryptosystems rely on a single key for both encryption and decryption, making sure all parties involved have the key, without it being intercepted, becomes a problem. Handing out the key on a disk is a relatively reliable way to go, but even that method can be uncertain.

To avoid having to place trust in a third party and to decrease the possibility of secret keys falling into the wrong hands, a new type of system, public key cryptography, was developed.

PUBLIC PROPERTY

The recognized pioneers in the field of public key cryptography are Whitfield Diffie and Martin Hellman, two researchers whose Diffie-Hellman algorithm was developed in 1976 at Stanford University. Just about a year later, Ron Rivest, Adi Shamir, and Leonard Adleman (the founders of RSA Data Security) invented another public key cryptosystem known as RSA.

Under the public key cryptography model, each user has a pair of keys that are complementary and mathematically related. The keys are generated by a mathematical algorithm that involves very large prime numbers. With a public key system, information encrypted with a particular public key can be decrypted only by using the corresponding private key.

You can hand out your public key to each recipient you communicate with; alternatively, you may choose to post your public key at some central location or on a network directory, so anyone can look it up. To see a working public key directory, visit Pretty Good Privacy's Web site at www.pgp.com. By contrast, you keep your private key, and you should take care to store it in a safe, secure place, whether on a floppy disk or a secured hard drive.

When you want to securely communicate information to another party, you must first generate a key pair, if you don't already have one. Each company's policy on key generation is different, but generally, either each user can generate a key pair using key generation software, or a central authority, such as someone who oversees a company's security policy, creates key pairs for all users.

Once keys have been created, they can be used to send information through e-mail, for example. You first look up the public key of the recipient or have the recipient send the public key directly to you. You then apply the recipient's public key to the plaintext, or unencrypted, message and create a ciphertext, or encrypted, message, which is then sent to the intended

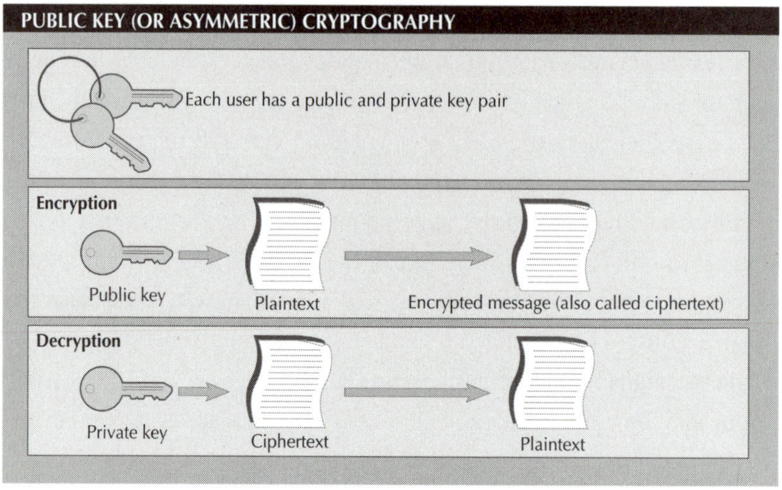

■ **Figure 2:** The use of two mathematically related keys, public key, or asymmetrical cryptosystems, makes messages very hard to crack. The public key can be posted at a central location, while the corresponding private key is kept by the individual at all times.

PUBLIC KEY (OR ASYMMETRIC) CRYPTOGRAPHY

Each user has a public and private key pair

Encryption

Public key

Plaintext

Encrypted message (also called ciphertext)

Decryption

Private key

Ciphertext

Plaintext

recipient. The recipient applies his or her private key to the ciphertext to transform it back to readable plaintext (see Figure 2).

Because Internet mail travels through numerous unknown servers to reach its final destination, it can be intercepted along the way by those with the know-how. However, unless these e-mail hijackers can provide the appropriate private key, all they will see is gobbledygook.

SIGNED, SEALED, DELIVERED

Even though RSA is largely based on Diffie-Hellman, there are several differences between these two schemes of public key cryptography. The Diffie-Hellman model was designed to require a dynamic key exchange between sender and receiver, usually for each session. While this process presents more of a challenge for potential attacks, it also requires a lot of communications overhead.

By contrast, RSA enables each user to have an unchanging key pair, no matter whom the communication is with. The two parties need never actually exchange keys; instead they can post their public keys in a public place, such as a server that can be readily accessed by anyone on the network.

Because Diffie-Hellman requires this dynamic key exchange each time, the keys are susceptible to interception and substitution by a third party, preventing senders and receivers from authenticating one another's identities.

RSA allows a sender to attach a digital signature to a message to show whom the message is from and to ensure the message has not been tampered with during transit. To do this, the sender takes a secure hash function and applies it to the message, which results in a message digest that identifies that message. A hash function is the result of an input of any size that becomes mathematically transformed into a string of a fixed size. The message digest, which represents the message and can be used to detect changes in the message, is then encrypted with the sender's private key, which results in a digital signature. The message digest and message itself are then routed to the appropriate recipient.

The recipient uses the sender's public key to decrypt the digital signature and see the message digest. Next, the recipient applies to the message the same hash function the sender used, and the hash function is then compared with the decrypted message digest. The signature is verified if both are the same. If, for some reason, they are not the same, the message either didn't come from the supposed sender or it has been changed in transit. This process is similar to when a store clerk compares your handwritten signature on a credit card sales slip with the signature on the credit card itself.

In addition to RSA's algorithm for creating a digital signature, the U.S. government has its own standard, the Digital Signature Standard (DSS). DSS uses the Secure Hash Algorithm, which converts a message into a 160-bit hash of the message. The hash value, which is, in essence, the contents of the message, is combined with the sender's private key, resulting in a digital signature. DSS employs the Digital Signature Algorithm, which takes the hash value and the private key and generates a 320-bit signature from the two.

In some instances, a digital signature is simply not enough. Although it ensures the message hasn't been altered in transit and it shows the signature of the sender, unless you know the sender, there really is no way to prove that the sender really is the person he or she claims to be. To provide further proof of identity, you can register your public key with a Certificate Authority (CA), an in-house or third-party group that verifies information about the person who holds a particular set of keys and then issues a digital certificate. Consider the certificate as similar to what a notary public issues to vouch for and verify the validity of a paper-based document. When this digital certificate is applied to an encrypted message, the recipient knows for sure the message did indeed come from the sender.

As we've seen, public key cryptography has several advantages over secret key cryptography. First, half of the cryptographic scheme (private keys) is never divulged, thereby increasing security. Also, public key cryptography allows for digital signatures, which enable sender and recipient to feel more confident about communicating with one another. Public key cryptography also has its drawbacks, most notably the fact that it's much slower than the secret key models. Because both types of cryptography have plenty of pros and cons, a combination of the two methods may be the most practical solution for electronic communications. You could encrypt private files, such as those on a hard drive, using the secret key method because technically those files aren't communicated to others. Then, for environments in which many people potentially have access to information, such as e-mail, you could use public key cryptography to secure information.

Whatever you choose, remember that the rapid growth of the Internet also extends invitations to unwanted hackers who will try to intercept messages and break into your network. Some form of cryptography, whether it's secret key or public key, will keep the wolves at bay.

RESOURCES

For an informative 10-part cryptography FAQ listing, visit www.cis.ohio-state.edu/hypertext/faq/usenet/cryptography-faq/top.html.

For a comprehensive look at cryptography, which includes a search engine, see www.rsa.com/rsalabs/newfaq.

This tutorial, number 104, by Anita Karvé was originally published in the April 1997 issue of LAN Magazine/Network Magazine.

Proxy Servers and Caching

Old ideas aren't the trendiest, but they're often the best. Take the example of Yahoo!, whose founders Jerry Yang and David Filo saw that despite the merits of search engines, the venerable concept of card catalog subject headings still had plenty of merit. By a process of reinvention, and without bothering to hire any actual librarians, Yang and Filo quickly proceeded to fortune and fame.

Or take the example of the proxy server. This is a concept that sounds trendy and cutting-edge, but its roots are also in dusty old library science. Remember in college when you first needed to check out a book housed in your university's locked stacks? Since you weren't allowed to go into this secured part of the library, a staff member acted as your proxy and retrieved the book for you.

All too often, of course, this process took longer than if you'd been able to go to the shelf and get the book yourself. But suppose that each time librarians retrieved a book for one student, they also made several copies, keeping them at the front desk for other users who requested the same title. The result would've been an ideal blend of fast service and airtight security.

This analogy explains the two main functions of a proxy server. First, the proxy server acts as an intermediary, helping users on a private network get information from the Internet when they need it, while ensuring that network security is maintained. Second, a proxy server may store frequently requested information in a local disk cache, rapidly delivering it to multiple users without having to go back to the Internet to get it.

THE LAYERED APPROACH

A proxy server is usually just one component of software that provides a variety of other services, such as a gateway to connect the local network to the Internet or a firewall to provide protection from outside intrusion.

Because proxy servers and firewalls are so often bundled together, people often confuse the two. However, a packet-filtering firewall operates at the Network layer of the OSI model, while proxy servers work at the Application layer. Packet filters use routers to filter information coming to and from a network. Because routers check each packet against some sort of access-control table (listing, for example, the IP addresses of trusted servers), they make it easy to block traffic that's not trusted. Firewalls can also screen packets based upon TCP and UDP port numbers; therefore, you can permit certain types of connections (Telnet or FTP, for example) to only certain trusted servers.

Packet-filtering firewalls have the advantage of speed, and they require no special configuration on the part of end-user applications. On the other hand, creating complex access rules can be difficult. Further, all packet filters can do is grant or deny access based on a packet's apparent source or destination address. Hackers can fool such firewalls by forging source addresses via IP packet spoofing. Since client-server connections are direct, hackers can also use packet sniffers to discern a network's address structure with relative ease.

In our library analogy, the equivalent of a packet-filtering firewall would be the librarian keeping a list of trusted students, then allowing only those individuals into the locked stacks to retrieve books. This might make book retrieval faster, but it would require that a list be created and maintained. It would also be vulnerable to impostors who turn up at the front desk bearing fake IDs.

Proxy servers are different. They break the direct link between client and server (or, if you will, between the student and the valuable book). They start by performing network address translation, mapping all of a network's internal IP addresses to a single "safe" IP address. Since the latter is the only address the untrusted network is aware of, spoofing attacks are no longer possible.

Because they operate at the Application layer of the OSI model, proxy servers can do a lot more. Any given proxy server includes a collection of application-specific proxies: an HTTP proxy for Web pages; an FTP proxy; an SMTP/POP proxy for e-mail; a Network News Transfer Protocol (NNTP) proxy for news servers; a RealAudio/RealVideo proxy; and more. Each of these proxies accepts only packets generated by services it is designed to copy, forward, and filter.

Application-specific proxies are almost infinitely configurable. For example, they can be set to block access to certain Web servers at all times, let only certain users play RealAudio files, permit FTP downloads but not uploads, or keep employees from logging on to their personal America Online accounts until after 5 p.m. Proxy servers can also bar specific MIME types and, in conjunction with a third-party plug-in such as SurfWatch, even filter content.

Proxy servers also do a superior job of logging network traffic, and can ensure that connectivity is always available for certain traffic types. For example, a small office might be connected to the Internet at all times for Web browsing via a single dial-up connection; a proxy server could automatically bring up a second dial-up connection when a user starts a long download via FTP.

As usual, though, the flip side of extensive configurability is complexity. Client applications such as Web browsers and RealAudio players must often be reconfigured to be made aware of proxy servers. In addition, as new Internet services become available and use new protocols and ports, new proxies must be written to support them. The process of adding users and defining permissions can also be complicated, though some proxy servers ease this task by working with Lightweight Directory Access Protocol (LDAP) information.

ESTABLISHING A VIRTUAL CIRCUIT

Circuit-level proxy servers were devised to simplify matters. Instead of operating at the Application layer, they work as a "shim" between the Application layer and the Transport layer, monitoring TCP handshaking between packets from trusted clients or servers to untrusted hosts, and vice versa. The proxy server is still an intermediary between the two parties, but this time it establishes a virtual circuit between them.

With a circuit-level proxy, client software no longer needs to be configured on a case-by-case basis. With Microsoft's Proxy Server, for example, once WinSock Proxy software has been installed onto a client computer—a one-time procedure—client software such as the Windows Media Player, Internet Relay Chat (IRC), or Telnet will perform just as if it were directly connected to the Internet.

The downside of a circuit-level proxy is that it cannot examine the Application-layer content of the packets it passes. Also, some computers (such as Macintosh) may not have the required client software available to them. (In such cases, Web browsers and the like may still operate, but they must be configured manually.) This problem has been addressed by the software technology known as SOCKS.

SOCKS was originally developed in 1990 and has currently reached version 5 (defined in RFC 1928). It provides a cross-platform standard for accessing circuit-level proxies. These may be accessed either by a single "SOCKSified" application on an otherwise-unmodified client computer, or by any application running on a computer that has had a SOCKS shim (shared or dynamic link libraries) put onto it.

Apart from standardization, SOCKS has other advantages. Version 5 supports both username/password (RFC 1929) and API-based (RFC 1961) authentication. It also supports both public and private key encryption.

It is historically difficult to proxy UDP-based services, since these are not connection-based; each packet is sent as a separate message. SOCKS 5 is capable of solving this problem by establishing TCP connections and then using these to relay UDP data.

Finally, aspects of packet-filtering firewalls, application-level proxies, and circuit-level proxies are combined by stateful-inspection firewalls. These devices are capable of intercepting and examining all of the packets they pass, using algorithms to recognize Application-layer data. Unlike Application-layer proxies, stateful-inspection firewalls do not break the client-server model in order to analyze data.

CACHE AS CACHE CAN

Though I have focused on architectural and security issues so far, most users are interested in proxy servers for a single reason—caching. Though theoretically optional, this feature has been closely associated with Web proxy servers ever since they were described at the first International World Wide Web Conference (Geneva, April 1994).

A proxy server's basic caching function works much like what's built into a Web browser, with the exception that the contents of the proxy server cache are available to multiple users. Whenever one user on the local network retrieves pages from the Internet, the pages are stored locally, which dramatically speeds access (see Figure 1). For example, Novell claims that when its BorderManager FastCache is configured to run from RAM, it is capable of processing more than 5,000 hits per second.

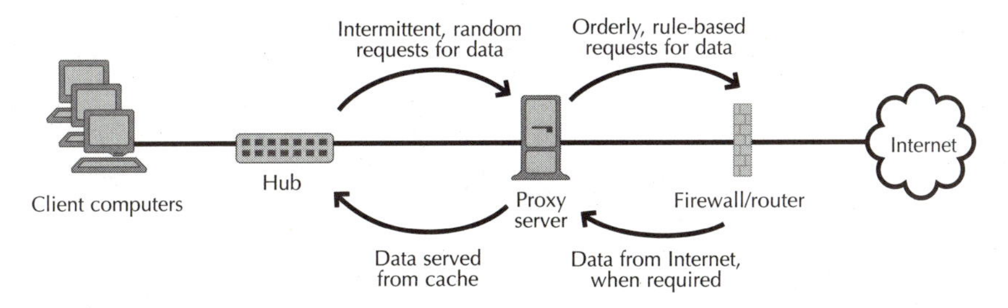

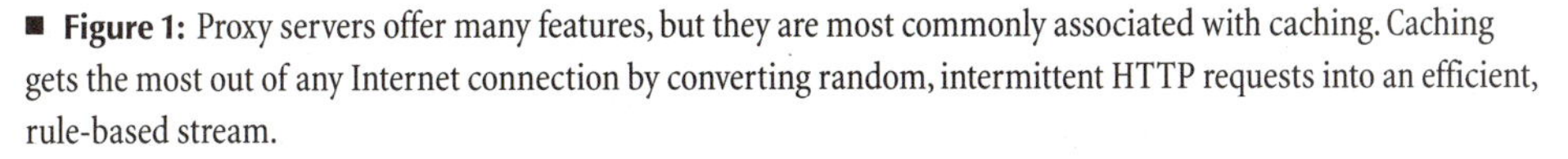

Figure 1: Proxy servers offer many features, but they are most commonly associated with caching. Caching gets the most out of any Internet connection by converting random, intermittent HTTP requests into an efficient, rule-based stream.

Some proxy servers offer read-ahead caching, which is capable of loading images and other objects embedded on a Web page into a cache before a Web browser has requested them. Caches may also be preloaded via a mechanism known as the last-modified multiplier. With the last-modified multiplier, a proxy server examines the creation dates of frequently requested pages, learning when updates are likely to occur and retrieving the pages when appropriate. And of course proxy servers also let administrators schedule batch retrieval of Web pages during any time of day when network traffic is known to be light.

Reverse caching is an additional feature of some proxy servers. In reverse caching, the cache server not only stores pages from the Internet for the benefit of local users, but it also stores local pages for the benefit of Internet users.

LINKING MULTIPLE CACHES

No matter how large and speedy it is, no single cache server can store everything. Inevitably the time will come when some user requests uncached data, which then has to trek slowly across the Internet. However, it is possible to ameliorate this problem by linking multiple caches together so they can draw information from one another. RFC 2187 describes the Internet Cache Protocol (ICP), which permits the hierarchical connection of caches.

In a cache hierarchy (or mesh), one cache establishes peering relationships with other caches. There are two types of relationships: parent and sibling. When one cache does not hold a requested object, it performs an ICP query to ask whether any of its siblings has the object. If a sibling does have it, the original cache requests it. If no siblings have it, then the request is forwarded to the parent or to the origin server. Figure 2 shows a typical cache hierarchy.

Although ICP lets cache servers be linked, it does have some problems. One is that ICP queries generate extraneous network traffic as they attempt to locate cached information. The more cache servers in the array, the more traffic there is, which results in negative scalability.

Another problem with ICP is that arrays become redundant over time. Each server tends to wind up holding duplicate copies of the most frequently requested URLs. For these reasons, ICP is gradually being replaced by the Cache Array Routing Protocol (CARP), originally devised by Microsoft.

With CARP, cache servers are tracked via an "array membership list," which is automatically updated via a Time-to-Live (TTL) function that regularly checks for active servers. A hashing algorithm is then used to determine which of the members of the array should be the receptacle of a particular URL request.

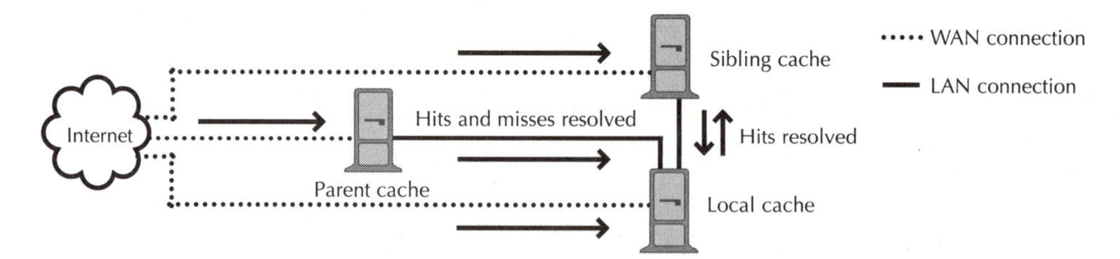

■ **Figure 2:** The Internet Cache Protocol (ICP) links multiple cache servers together in a sibling-parent hierarchy. The local cache can retrieve hits from sibling caches, hits and misses from parent caches, and misses from origin servers directly.

CACHING UNPLUGGED

Cache servers used to be viewed as nice-to-have items you got for free when you purchased a proxy server. Now that the Internet is growing steadily more congested and more and more clients have broadband connections, the terms "cache server" and "proxy server" may not be used quite so interchangeably.

Proxy servers will continue to offer caching as one of their features. However, the increasing demand for specialized caching means that cache servers will gain more visibility as separate products. For example, the CacheQube from Cobalt Networks (Mountain View, CA) is an appliance that can simply be connected between a LAN and a router to provide transparent caching. The Streaming Media Cache from Inktomi (San Mateo, CA) and MediaMall from InfoLibria (Waltham, MA) are caches designed specifically for handling streaming audio and video.

RESOURCES

The Internet Caching Resource Center, at www.caching.com, offers a variety of information and articles about caching.

A SOCKS 5 white paper is available from www.aventail.com/index.phtml/solutions/white_papers/sockswp.phtml. NEC also offers an introduction to SOCKS at www.socks.nec.com/introduction.html. RFC 1928 is readable at http://info.internet.isi.edu:80/in-notes/rfc/files/rfc1928.txt/.

"Patrolling the Borders of Your Network," though written to promote Novell's BorderManager, is of general background interest. You can find it at www.novell.com/bordermanager/bmgr3_wp.html.

You'll find an introduction to CARP at www.microsoft.com/proxy/guide/CarpWP.asp/.

Finally, a well-regarded book is Ari Luotonen's *Web Proxy Servers* (Prentice-Hall, 1998, ISBN 0-13-680612-0). Chief architect of the Netscape Proxy Server, Luotonen was also co-developer of the first Web proxy server (CERN, 1994).

This tutorial, number 129, by Jonathan Angel, was originally published in the May 1999 issue of Network Magazine.

The Virus Threat

By most accounts, in October 1988, only three DOS computer viruses were known. By October 1991, McAfee Associates identified some 900 computer virus strains. At the 18th Computer Security Institute Conference, Scott Charney from the U.S. Justice Department indicated that the government expected to see an additional 600 viruses and mutant strains introduced during 1992.

Peter Tippett, president of Certus, reported a new virus was discovered about every six days in January 1990; by June 1990, a new virus was discovered about every four days; and in September, 1990, one was found every three days. According to Charney's prediction, we would have discovered 1.6 new viruses daily in 1992.

THE COST OF A VIRUS

Although many viruses are labeled benign (more annoying than actually causing damage to the system or data), a virus usually causes at least some inconvenience, some loss of system-access time and, at the worst, loss of data.

Recent virus cleanup figures at a large corporation found an average of one hour of technician time was required to locate and remove a virus from each computer. Data from Certus support staff suggests that a reasonable figure for direct technician time to resolve a disastrous computer virus is approximately $250 per computer or network workstation.

WHAT IS A VIRUS?

A virus is a program that has the ability to reproduce by modifying other programs to include a copy of itself. It may contain destructive code that moves into multiple programs, data files, or devices on a system and spreads through multiple systems in a network. Viral code may execute immediately or wait for a specific set of circumstances. Viruses are not distinct programs; they need a host program executed to activate their code.

Many distinct programmed threats have been lumped together as viruses; however, most are not actually viruses at all. To better understand the threat, we will identify and define these other techniques.

Bacteria, also known as rabbits, are programs that do not explicitly damage any files. Their sole purpose is to reproduce themselves. A typical bacteria program does nothing more than reproduce itself exponentially, eventually eating up all the processor capacity, memory, or disk space, denying the user access to those resources. This kind of programming attack is one of the oldest forms of programmed threat.

A *logic bomb* is a program or a section of code built into a program that lies dormant until a predefined condition is met. When that condition occurs, the bomb goes off with a result that is neither expected nor desired. Time bombs explode frequently. Depending on who authored the bomb and how many generations of backups it contaminated, the recovery effort ranges from mildly inconvenient to nearly impossible.

A *password catcher* is a program that mimics the actions of a normal sign-on screen but stores supplied ID and password combinations, usually in a hidden file accessible only to the author. The collected ID/password combinations are later used to attack the system. Running for three or four days will generally yield more than 90 percent of a site's active system users. Truly clever ones cause no perceptible delay and force no repeated sign-ons.

A *repeat dialer* is a program that continually calls the same number, thus placing it virtually out of service for any other caller. This technique has been used against TV preachers during fund drives to stop pledge calls. As would-be donors keep receiving busy signals, they tend to get frustrated and stop trying.

Trapdoors, also known as backdoors, are undocumented entry points into an otherwise secure system. During software development, programmers often create entry points into "their" programs to aid in debugging and adding final enhancements. These trapdoors are supposed to be closed before the program goes through the promotion to production process. Many times, however, they are not. This breach leaves an unadvertised but very real hole in the system's security.

A *Trojan horse* is a program that appears to perform a useful function, and sometimes does so quite well, but also includes an unadvertised feature that is usually malicious in nature. Viruses and logic bombs can be hidden in Trojan horses. The code in a well-constructed Trojan horse can perform its apparent function long and admirably before encountering the triggering condition that prompts it to let loose its secret agenda.

A *war or demon dialer* is a program run from outside an organization's environment and control, usually by hackers, to find dial-up ports into computer systems. Such a program identifies numbers in a given range that connect to a computer. From a modem-equipped PC, the hacker enters a starting phone number and an ending value to the resident dialer program. The program then sequentially dials each number in the range, seeking a computer tone. Copies of this type of program are found on bulletin boards everywhere.

A *worm* is a program that scans a system or an entire network for available, unused disk space in which to run. Originally, worms were developed by systems programmers searching for fragments of core in which to run segments of large programs. They tend to tie up all computing resources in a system or on a network and effectively shut it down. Worms can be activated at boot-up or submitted separately. Probably the most well-known worm was the November 2, 1988, Internet incident. In two days, an estimated 6,200 Unix-based computer systems on the network were infected.

HOW DOES A VIRUS SPREAD?

A computer virus, like its human counterpart, does not spread through the air. Humans become infected by a virus by coming in contact with someone who is infected. So it is with computers. They must come in contact with some contaminated source. The virus infects any form of writable storage, including hard drives, diskettes, magnetic tapes and cartridges, optical media, and memory. The most frequent sources of contamination include:

- physical or communication contact with an infected computer,
- copying an unknown disk containing a carrier program,
- downloading a file from a bulletin board system,
- running an infected network program,
- booting with an infected disk,

- infected software from a vendor,
- overt action by individuals, and
- e-mail attachments.

VIRUS PROTECTION

A number of clues can indicate that a virus has infected or attempted to infect a system, even before any damage is done. Unexplained system crashes, programs that suddenly don't seem to work properly, data files or programs mysteriously erased, disks becoming unreadable—all could be caused by a virus.

Here are some indicators that may confirm the presence of a virus. Most viruses use information provided by the command DIR, which lists a disk's directory, and CHKDSK, which snapshots disk and memory usage.

File size increase. A file's size may increase when a virus attaches itself to the file.

Change in update timestamp. When a virus modifies another program—even one such as COMMAND.COM (which is part of the operating system)—the "last-update" date and time are often changed. Since most programs are normally never modified (except when they're upgraded), periodically checking the last-update timestamp, with the DIR command, can alert the user to the presence of a virus. Another danger sign is when many programs list the same date and/or time in their last-update field. This occurrence indicates that all have been modified together, possibly by a virus.

Sudden decrease of free space. When running a new program, particularly if it is freeware or shareware, be alert for a sudden, unexpected decrease in disk space or memory.

Numerous unexpected disk accesses. Unless a program is exceptionally large or uses huge data files, it should not conduct a high number of disk accesses. Unexpected disk activity might signal a virus.

PREVENTING INFECTION

Preventing a virus infection is the best way to protect your organization against damage. If a virus cannot establish itself within your systems, then it cannot damage your programs or data. The following steps can help keep a clean system from becoming infected with a virus.

Awareness training. All employees having access to computer systems should be required to attend a training session on the virus threat. It is crucial that employees realize how much damage a virus can inflict.

Policies and procedures. The organization should prepare a policy on virus control to address the following issues: tight control of freeware and shareware; a control process that includes running anti-virus software regularly by each department; a virus response team and methods for contacting the team; control of the infection once it is detected; and recovery from the virus, including backup and dump policies.

For two very important reasons, the user community should be made aware of the risks of sharing software. The primary cause of the spread of virus infections is through the uncontrolled use of diskettes being introduced into computer systems. The other reason is the possibility of the illegal use of copyrighted software.

If your organization lets employees transport diskettes out of the work facility, a quarantined system to test diskettes and software before their introduction into the system should be in effect. This quarantine system should test all diskettes for the possibility of virus contamination.

In the network environment, avoid placing shareware in a common file server directory, thereby making it accessible to any PC in the network. Only allow the network administrator to sign on to the file server node.

The most prudent precaution is to carefully make, store, and routinely check backup copies of files and programs—all on an established schedule. And control access to backups to guarantee integrity.

VIRUS-PROTECTION PACKAGES

Several commercially available programs can now help detect viruses and provide some degree of protection against them. However, if you use such programs, be careful that they don't cause greater problems than they solve. Some anti-virus programs interfere with the normal operations of programs they are supposed to protect (such as blocking a disk formatting utility).

Also, an anti-virus program may warn of a suspected infection when none has actually taken place. Because of the differences in anti-virus packages, it's important to standardize testing procedures and analytical tools, so results can be compared on a consistent basis.

Unfortunately, malicious code is now a fact of life. Computer viruses appear to be a long-term threat. Systems and data will continue to be vulnerable until a proactive preventive and corrective action is established. In the short term, caution in testing and using unfamiliar software, as well as carefully made backups, are your best safeguards. Your slogan should be, "Don't accept software from strangers."

This tutorial, number 45, by Thomas Peltier, was originally published in the April 1992 issue of LAN Magazine/Network Magazine.

Trojan Horses

No matter what security measures you have in place, every network suffers from one serious weakness: human gullibility. Trojan Horses take advantage of this, hiding a malicious program inside something apparently harmless. If software has been installed in good faith, it can get around almost any firewall, authentication system, or virus scanner.

Trojans vary in the nefarious acts they perform once inside a machine. They can be harmless pranks that display an obscene or political message, or logic bombs that erase data and try to damage hardware. Some are coupled with viruses, spreading between systems by e-mail. The most insidious are stealthier, and often have a purpose beyond wreaking havoc. As well as hacking, Trojans have been used to spy on people, and have acted as the culprits in some spectacular frauds.

No one is safe. In fall 2000, Microsoft suffered a much-publicized attack in which hackers downloaded, and perhaps changed, the source code of a future operating system. This was the result of a Trojan concealing a worm—a program that copies itself onto other machines throughout a network. Once installed on a Microsoft machine, the code spread until it found a computer containing secrets worth stealing. The Trojan then signalled its presence to a hacker, opening a backdoor to the network.

So, how can you avoid becoming the next Microsoft? Short of banning all users from your network, you can't. But there are ways to minimize the risk, starting with vigilance and education. Regular backups are a must to undo the damage caused by those that only delete data. So is running a full suite of security software, as firewalls and virus scanners can catch some of the best-known offenders. Most importantly, you need to teach your users and yourself about Trojans. Find out their effects, and what kind of programs they hide inside. Then learn to how to distinguish a Trojan from a real gift horse, before it gets inside your network.

WORMING HORSES

Most Trojans conceal viruses or worms, both of which exist primarily to replicate themselves, but may also cause damaging effects (see Malware Taxonomy). Trojans have become increasingly important to viruses, because most are now sent as e-mail attachments. A user must open an e-mail attachment, whereas earlier floppy disk-based viruses were loaded automatically when a PC booted.

With the exception of Bubbleboy, which was very rare and exploited a now-fixed security hole in Microsoft Outlook, it's impossible to catch a virus simply by reading an e-mail message. Users need to be tricked into running an attached file, something that virus writers have found to be embarrassingly easy. Many people automatically double-click everything that arrives by e-mail, and must be educated otherwise.

Most IT staff should already know that Windows files ending in .com (command), .exe (executable) and .dll (dynamic link library) are programs. They have the potential to do literally anything to a system, and so should be treated with extreme caution: run them only if you trust their source completely, and you know what the program actually does. The fact that a program was e-mailed to you by a friend or colleague is not reason enough to run it. A Trojan could have commandeered your friend's backdoor mail system and spammed itself to an entire address book.

To prevent infection, many organizations have a policy against users installing unauthorized software. However, this is often impossible to enforce, and can prevent employees from using the best tools available to do their jobs. Whether or not you do implement such a policy, it's important to make users aware of the dangers. If people are allowed to download software, they should know what is likely to be most dangerous; if they aren't, they're more likely to respect the rules if they understand the reason for them.

The most serious risk comes from pirated software, because its source is almost by definition untrusted. Angry programmers have been known to wreak revenge on pirates, distributing Trojans that claim to be illegal software. The first attack on the Palm platform fell into this category, with a program that claimed to be a popular GameBoy emulator called Liberty. Instead, it deleted all files and applications.

The list of file extensions used by programs is growing all the time, making it difficult for virus scanners to keep up. Most anti-virus software checks around 30 different types, but was still caught out by the .vbs (Visual Basic Script) files used in the Love Bug of 2000. If your anti-virus software is older than this, manually set it to scan all file types, or consider an upgrade. Automatic updates provided over the Internet usually only list new viruses and fixes, not new file types where they may hide.

The most dangerous file type is the shell scrap object, which seems to be designed as a Trojan. Though it is supposed to have the .shs or .shb extension, this remains hidden under Windows 98 and Me, disguising it as any other file type. The first program to take advantage of this vulnerability was the Stages worm, which struck in June 1998. Appearing to be a harmless text file, it was, in fact, a VB-Script that e-mailed itself to all a user's contacts.

Shell scrap objects are so dangerous that Symantec's Anti-Virus Research Center recommends not using them at all. They have so few legitimate applications that many users might want to disable them entirely, by deleting the file schscrap.dll from the Windows/system directory on every PC. Less drastically, they can be forced out of hiding by deleting the registry entry for HKEY_CLASSES_ ROOT\ShellScrap.

PULLING THE REINS

As threatening as viruses and worms are, they're perhaps the least dangerous payload that can lurk within a Trojan. Many are instead designed to gain access to your network, concealing small server programs that run almost unnoticed. These can let a hacker spy on your secrets, or even take control of your PC.

The most infamous hacking tool is Back Orifice 2000, often known simply as BO2K, and produced by hacker collective the Cult of the Dead Cow (www.cultdeadcow.com). The authors describe the program as a "remote administration tool," which just happens to be able to administer a computer without its user's knowledge or consent. It can run almost undetected under any version of Windows, allowing an outsider almost unrestricted access to a system. As well as copying or altering files, hackers equipped with BO2K can record a user's every keystroke, and even receive a live video feed of their screen.

In an ironic twist, the Cult of the Dead Cow itself has fallen victim to a Trojan. The first Back Orifice 2000 CD-ROMs to be distributed were infected with Chernobyl, a nasty virus that can cause permanent damage to hardware. Aspiring hackers at 1999's DefCon convention found that far from gaining control over other people's computers, they lost control of their own as hard disks were over-written and BIOS chips erased.

The Microsoft hack in fall 2000 used a Trojan called QAZ, which disguises itself as the Notepad utility, the file notepdad.exe. Notepad itself is still available, but renamed note.exe, so that users won't notice a change. An administrator trying to fix the problem might know that this file is not part of the standard Windows installation and remove it, a course of action which would stop Notepad from working, but leave the Trojan intact.

Even if intruders aren't interested in your data, gaining control of a computer can still be a serious coup. The Distributed Denial of Service (DDoS) attacks that brought down leading Web sites in early 2000 were all accomplished by Trojan Horses. These attacks rely on thousands of computers all working together, and so can't normally be launched by any one individual. However, the attack becomes possible if that individual first gains control of thousands of computers.

Participation in a DDoS attack means more than just being a bad Netizen and opening your organization to lawsuits. Though the headlines told of Yahoo and eBay being knocked out, the thousands of individuals and businesses whose computers relayed the attacks also suffered. If your mail server is busy launching an attack, it won't be available for its intended purpose.

Any PC connected to a phone line is a valuable target for a financially motivated attack, because its modem can be reprogrammed to dial premium-rate numbers. Many Trojans change a user's dial-up networking settings to an international number, often charged at several dollars per minute. If the number actually connects to an ISP, victims may notice nothing until they receive their phone bills.

This kind of Trojan first struck in 1998, when thousands of European users downloaded a pornography slide-show, only to find their modems calling an expensive number in Ghana.

It has since moved to number three on the Federal Trade Commission list of Internet frauds, and is considered more dangerous than phone slamming and pyramid schemes.

SHUTTING THE STABLE DOOR

Most Trojans signal their presence to a hacker using a preset TCP port, so a properly configured firewall may be able to detect or block them. Lists of ports used by popular Trojans are published on several Web sites (see Resources), and some will even scan for them automatically. However, the latest versions of many Trojans can vary their port, making detection more difficult.

Anti-virus software also detects Trojans, though it can pose risks. It needs regular updates, which gives the anti-virus company access to your network. In November 2000, an update for Network Associates' McAfee VirusScan caused certain versions of the software to crash their systems, losing unsaved data. This was due to a bug, rather than a deliberate act, but with already-compromised companies, such as Microsoft, moving into the anti-virus space, there is a risk that some Trojans may use this method of attack.

The German government believes that Windows 2000 may already harbor a Trojan Horse. It went so far as to threaten a ban on the software unless Microsoft removed the Disk Defragmenter utility, where the offending code allegedly hides. Microsoft refused to do this, but posted detailed instructions on its German support site for users to remove it themselves. Concerned managers should note there is no evidence that this Trojan actually exists. Indeed, the U.S. government is so confident of Windows 2000's security that it uses the program in many agencies, including the armed forces.

MALWARE TAXONOMY

Though the media and some users often describe every damaging program as a "virus," security experts know otherwise. Here's a recap of the three most common malicious software types, any and all of which can hide inside a Trojan:

Viruses are technically self-replicating code that is attached to another file, in the same way that real viruses attach themselves to cells. Viruses originally targeted .com or .exe programs, but scripting languages have enabled them to hit office documents and even e-mail messages.

Worms are standalone programs that replicate, usually by copying themselves to another computer on a network. They are sometimes known as bacteria because they don't rely on other programs. The most widespread is happy99.exe, which paralyzed many computers two years ago, and still strikes occasionally—particularly in the New Year.

Logic Bombs don't replicate, but can be very damaging. They are simply programs that perform a harmful function, such as deleting a user's files, when a given condition is met.

RESOURCES

HackFix, a nonprofit organization dedicated to fighting Trojans, has some useful tools and information at www.hackfix.org. It specializes in Back Orifice and NetBus, but also deals with other threats.

If you're worried that a hacker might already be inside your network, go to www.trojanscanner.com. This site offers a free online scanner that can probe every TCP port on your system, and notify you of any that may be used by an intruder.

Want to break into (or as its authors claim, "remotely administer") a PC running Windows? Go to the official Back Orifice Web site, www.bo2k.com, to download the latest version.

The Federal Trade Commission has more information on "dot cons," including the premium rate dialer masquerading as a porn viewer, at www.ftc.gov/bcp/conline/edcams/dotcon/.

This tutorial, number 150, by Andy Dornan, was originally published in the January 2001 issue of Network Magazine.

Security and 802.11 Wireless Networks

Wireless LANs (WLANs) conforming to the IEEE's 802.11b specification have become popular, and there's every reason to think that 802.11a and 802.11g networks will also be widely deployed in the next few years. One attraction of these wireless networks is how easy they are to implement. Unfortunately, 802.11b networks suffer from various security shortcomings. Coping with these security problems is complex and potentially costly—possibly even negating the value of such networks.

Traditional LANs shared a single medium—a copper cable and passive hubs or concentrators. These hub ports and cable taps were almost always located within a facility with some physical security that made it nontrivial for an attacker to tap into. Many modern LANs associate a single switched port to each user, limiting the span of even an authenticated internal user, much less an outside attacker. By contrast, WLANs share an ill-defined medium in free space, which almost certainly includes locations outside the physical control of WLAN administrators, such as the company parking lot, other floors of the facility, or nearby high-rise buildings. For this reason, a wireless network is fundamentally less secure than a wired one.

Acknowledging the inherent security deficiencies of WLANs, the 802.11 committee adopted an encryption protocol called Wired Equivalent Privacy (WEP). Note the rather mealy-mouthed terminology: WEP isn't positioned to provide real privacy—just privacy comparable to otherwise unprotected wired networks. Furthermore, WEP isn't positioned to provide authentication, access control, or data integrity. However, the authentication and data-integrity capabilities provided by 802.11 networks are built on a WEP foundation, so if WEP is broken, so are these mechanisms. It turns out that authentication, data integrity, and access control on 802.11 networks can be broken without breaking WEP, but WEP's failure as an encryption protocol is nevertheless a serious problem.

What's at stake if WEP encryption can be defeated? An eavesdropper can, among other things, watch and intercept traffic flows, including e-mail, browsing, file transfers, and remote terminal sessions. An eavesdropper can also map and capture all conversations on the network, including management and configuration processes, as well as end-user data; or capture IDs and passwords that users employ to log in to other networks and resources.

WEP DEFICIENCIES

How serious are WEP's deficiencies? Before answering that question, I'll first have to discuss how WEP operates. WEP uses a stream cipher named RC4, which means that it uses a shared secret key to generate an arbitrarily long sequence of bytes from a pseudorandom number generator. This stream is XORed with the plaintext to produce the encrypted ciphertext. (RC4 encryption works successfully with Secure Sockets Layer (SSL), the encryption protocol that lets you breathe easier when you use credit cards to makes purchases on the Web.)

Early 802.11b networks used 40-bit keys because of the federal government's restrictions on encryption in those days, but most current components use 104-bit keys. Hackers can crack a 40-bit key with a brute-force attack in just hours with modern PCs, but for now, a brute-force attack on a 104-bit key would take longer than the current age of the universe, so there's little to worry about on that front.

It's easy to break RC4 encryption if a second instance of encryption with a single key—a keystream reuse—can be isolated. The WEP designers were aware of this problem, and they built into WEP a so-called Initialization Vector (IV), a 24-bit value that changes with each packet and is appended to the unchanging shared secret key to minimize the likelihood of "key collision."

The IV is carried in the clear in each packet—otherwise the receiver couldn't set up the RC4 engine for decryption. Because 224 is 16,777,216, it initially appears that an eavesdropper would have to capture many millions of packets to identify keystream reuse instances, but thanks to the "birthday paradox," a key collision is likely to occur after only 5,000 or so packets. (The birthday paradox uses elementary probability techniques to demonstrate that the odds are greater than even that, out of a group of 23 randomly chosen people, at least two of them will have the same birthday.)

If this problem weren't serious enough, Fluhrer, Mantin, and Shamir, in "Weaknesses in the Key Scheduling Algorithm of RC4" (see Resources), identified a further problem with RC4. This paper demonstrates that a fraction of the keys in RC4 are weak, revealing more of the structure of early bytes in the output than they ought to. By exploiting the statistical properties of this weakness, an attacker can crack any message in hours, independently of other attacks. AirSnort (http://airsnort.shmoo.com/), one of the best-known WEP cracking tools, employs this attack.

Even assuming a less than fully loaded network, an attacker can create a full keystream dictionary in just a few days. Various vendor-specific implementation choices can reduce this crack time substantially. The 802.11 standard doesn't specify a key distribution method, so vendor algorithms generating shared secret keys from passwords might be subject to simple dictionary attacks that greatly reduce the problem of guessing the key. The general reliance on out-of-band—usually manual—key distribution ensures that keys won't be changed often. Attackers capable of sending data into the network can speed up the key-cracking process in various ways. (See "Intercepting Mobile Communications: The Insecurity of 802.11," at www.isaac.cs.berkeley. edu/isaac/mobicom.pdf for more details.)

AUTHENTICATION ISSUES

Authentication failures result in unknown users on the network. An attacker who intends to crack WEP or misuse network resources must first authenticate successfully. (A large fraction of 802.11 networks are configured with the authentication system turned off, so a person who wants to attack one of these systems wouldn't even have to take the simple steps listed here.)

The 802.11 specification supports a two-step form of authentication. Potentially participating stations must respond correctly to a cryptographic challenge (the authentication step) and then associate with an access point by submitting the access point's Service Set Identifier (SSID.) The association step adds little security to the system—some vendors provide clients with a list of SSIDs to choose from. But all vendors broadcast the SSID values in the clear, so a protocol analyzer with a wireless card can find these values in seconds.

The authentication step relies on RC4 encryption, as WEP does. The problem isn't the insecurity of WEP as such, or of RC4 in itself. Again, it's an issue of implementation. An access point issues a cryptographic challenge by encrypting a random string with the shared secret key using RC4. The initiator must decrypt the challenge and send the plaintext back to the access point, which compares the decrypted plaintext with the original random string. If they match, the initiator is authenticated.

By capturing only two frames—the challenge frame and a successful response frame—an attacker can easily derive a keystream that will successfully decrypt future challenges. An integrity check is built into WEP systems, presumably to prevent this replay attack. But the integrity check is based on the Cyclic Redundancy Check (CRC) mechanism that many data-link protocols use, and CRC doesn't de-pend on a cryptographic key, so it's easy to get around this obstacle.

In addition, attackers can use a well-understood method to make arbitrary changes to a message, so the checksum of the changed message is the same as that of the original. This data-integrity failure not only implies that an attacker can modify any content—for example, the position of a decimal point in a financial document—but it also lets attackers use the checksum to assess the correctness of their decryption attempts.

Properly authenticated and associated clients are often given full access to the wireless network. Even without cracking WEP encryption, attackers can access wired networks connected to the wireless one, and perform illegal, embarrassing, or otherwise undesirable acts that reflect badly on the network administration. Attackers can also spread viruses, Trojan Horse programs, and perform local or remote Denial of Service (DoS) attacks.

The 802.11 and WEP mechanisms say little about enhanced access control. Some access-point vendors build in a MAC address table that can serve as an access-control list, accepting traffic only from clients whose MAC address appears on the list. The problem is that MAC addresses are necessarily transmitted in the clear, so a wireless protocol analyzer can pick them up immediately. In general, you can configure wireless NICs with different MAC addresses, so a spoofing attack on this form of access control is trivial.

PROTECTIVE RESPONSES

It should be clear that the 802.11 families of wireless products share serious deficiencies in privacy, confidentiality, data integrity, and provisions for safety from various other attacks. That isn't to say that WEP is useless. A home network that isn't connected to an enterprise network, isn't part of a commercial operation, and has no confidential, illegal, or embarrassing online content will provide little incentive for an attacker to defeat WEP authentication and encryption. Every other 802.11 network should be accessible only via VPN login or equivalent mechanisms.

Network managers should periodically audit their facilities for rogue 802.11 networks. Wireless traffic should be outside the enterprise firewall or within the Demilitarized Zone (DMZ). User-initiated wireless networks probably won't be outside the secure perimeter and offer a juicy target for attackers, whatever their motives. In fact, user-initiated networks are likely to be implemented without authentication and WEP, which is as big a security hole as can be imagined.

The principal deficiencies of 802.11 security are the result of security implementation by software and hardware engineers who lack sufficient understanding of real-world security. The security experts who criticized WEP implementation from the beginning weren't simply tooting their own horns. Some of the choices seem elementary, such as the use of CRC for integrity assurance.

Similarly, security textbooks emphasize that RC4 keys ought never to be repeated, not just limited to once every 16 million packets. No doubt there were performance, power consumption, and cost constraints that entered into the poor decisions. Every packet of an 802.11 wireless network requires a lot of processing work, and radios are often finicky performers. Wireless LAN implementations prior to 802.11b had lousy performance and no security, so even WEP was a step in the right direction.

The 802.11 committees are aware of these shortcomings. The 802.1X committee has defined a standard for authentication and key management for Ethernet and other 802-numbered data-link technologies that should eventually be integrated with 802.11 systems. Proposals to strengthen or replace WEP are also in discussion. All the problems identified so far can be addressed, but delays of months or years are almost certain where standards organizations are concerned. In the meantime, the need to add security features beyond those built into the 802.11 protocols will undermine the attractive simplicity of wireless networks.

RESOURCES

The Unofficial 802.11 Security Web Page, www.drizzle.com/~aboba/IEEE, has a comprehensive list of relevant materials.

"Intercepting Mobile Communications: The Insecurity of 802.11," by Nikita Borisov, Ian Goldberg, and David Wagner, can be found at www.isaac.cs.berkeley.edu/isaac/mobicom.pdf.This paper is one of the most clear and understandable descriptions of cryptographic subjects to be found anywhere. It explains WEP thoroughly and precisely and presents the problems associated with keystream reuse.

"Weaknesses in the Key Scheduling Algorithm of RC4," by Scott Fluhrer, Itsik Mantin, and Adi Shamir, can be found at www.drizzle.com/~aboba/IEEE/rc4_ksaproc.pdf. The details provided in this paper are probably beyond the ability of security civilians, but these findings provide the basis for AirSnort, a widely available cracking tool. You can find out more about AirSnort and download the software at http://airsnort.shmoo.com/.

With a laptop outfitted with NetStumbler (www.netstumbler.com/), you can find unauthorized wireless networks, assess the reachability of authorized networks, or go war driving and find all the wireless access points in the neighborhood.

The protocol analyzers from Network Associates (www.nai.com), WildPackets (www. wildpackets.com), and Network Instruments (www.networkinstruments.com), among others, can decode wireless traffic when outfitted with supported wireless interface cards.

This tutorial, number 167, by Steve Steinke, was originally published in the June 2002 issue of Network Magazine.

Coping with Home Network Security Threats

While there are thousands of exploitable vulnerabilities in network-connected home systems, there's a short list of basic types of attacks, which can help make securing those systems a manageable project. These attacks can result in a short list of bad results: loss of data; loss of confidentiality; impersonation or a similar abuse of the system that can be traced back to and blamed on you; and Denial of Service (DoS). Loss of data can include loss of data integrity as well as formatted disk drives and deleted files. Loss of confidentiality can include the exposure of embarrassing or compromising information as well as intercepted secrets. Impersonation can take the form of your system being employed as a "zombie" in attacks on other users, as well as online purchases with your credit card. DoS may simply tie up your computer, but it may also include overloading your access link to the Internet.

The root cause of these threats is that someone else gains the capability to execute software on your system. The most likely means of an attacker's accomplishment of this goal is via a virus or a worm. Viruses duplicate themselves in the file system on executable files so simply shutting down the computer can't eliminate them. On traditional unprotected desktop OSs, including the consumer versions of Windows up to and including Windows ME and MacOS versions through OS 9, viruses that are successfully propagated onto a computer can format drives, erase files, send e-mail, and attack other systems. Viruses can also: install back doors that allow the ready return of control by the attacker; capture passwords and credit card numbers; and basically accomplish any of the bad outcomes described earlier in this article.

Worms differ from viruses in that they spread across networks without piggybacking on an executable host file. The most common vehicles are e-mail attachments and openly shared files. You can stop a pure worm by shutting down any computers it's running on, though some recent worms, such as Code Red and Nimda, include viral components that permit destructive code to start up again after a shutdown. A worm running on your system can perform any of the deleterious actions that a virus can.

A third category of dangerous software is the Trojan Horse, which disguises itself either as something useful—a network login window—or as something interesting—an online game or other form of entertainment. But the software actually captures your password for a later retrieval, or installs other software that allows the attacker to re-contact your computer and take full remote control. Some viruses and worms install Trojan Horses on the computers they infect.

COUNTERMEASURES FOR EVIL EXECUTABLES

Up-to-date anti-virus software will protect against viruses, worms, and Trojan Horses that the anti-virus software providers have identified and neutered. If the anti-virus developers provide solutions before new malware becomes widespread, the risk of infection is minimal. However, viruses or worms that spread quickly may arrive before you get the updated anti-virus version that protects against the latest threat. In that case, your security settings and ultimately the common sense of your users will be the final protective barriers.

At one time, Microsoft Outlook Express' default settings allowed executable e-mail attachments to run automatically when the message was opened. The destructive ILOVEYOU, or VBS.LoveLetter, virus demonstrated the foolishness of those defaults. Early versions of Microsoft Word had automatic macro execution enabled by default, and that loophole was closed only after early macro viruses became widespread.

Browser settings can also help reduce the risk of infection. Internet Explorer has a granular set of options for coping with different executable files, based on the sites that provide them and whether the files have been signed and certified by credible authorities. Windows XP, Windows 2000, MacOS X, and Linux offer administrative options that can prevent the installation and execution of unknown or forbidden software. It's probably impossible to configure earlier versions of MacOS and non-NT Windows in such a way that a well-informed user could be kept from installing arbitrary executable code and defeating any protective measures.

If the OS can't be locked down and the other protective mechanisms fail, the users' common sense is the last barrier. Users and their roommates and family members need to understand the risks of installing software from unknown sources, downloading files, double-clicking on e-mail attachments, deactivating protective software, and changing the system configuration to an insecure state. An enterprise that needs to secure work-at-home and mobile users should quickly migrate its users to OSs with options for tight protective measures. This is because the entire population, including grandmothers, teenage boys, and people who simply aren't interested in computer security, will never learn enough about the subject to protect themselves adequately.

OUTSIDE INTRUSION

Viruses and worms are created by their lovable authors and launched into the world. They're designed to spread on their own, without further activity on the author's part. Good practices by end users will almost always prevent damage. Intrusion over the Internet is a rather different, scarier threat. In these attacks, the attacker is targeting your system rather than simply launching a destructive bit of software into cyberspace. It's the difference between someone who's rattling your doorknobs or using bolt cutters on your padlocks and someone who arbitrarily sets out land mines, or perhaps whoopee cushions, with no specific target in mind.

Intruders may look for vulnerabilities at random IP destinations or they may scan specific blocks of IP addresses looking for likely targets. It's a no-brainer to use whois to find what the addresses of @Home's cable modem users or SBC's DSL users are. Systems on these hacker-popular networks are probably scanned and Pinged several times a day, though these initial probes cause no harm.

The good news is that external intrusions only succeed against targets that run insecure server processes. Hackers can try all their tools against your system, but if you don't have a server process waiting to answer incoming TCP or UDP requests, or if the server process (or the quivering remnant that's left after a successful takeover attack) neither hands over files, passwords, configuration settings, or other data that should be secure, nor allows itself to be perverted into an attack avenue, there's nothing to worry about.

The bad news is that there are good reasons for home users to run server processes, and it's not always obvious that a process is a server process. One obvious type of server is peer-to-peer file sharing. A standalone home PC or the PCs on a home network can be configured to share files over the Internet. If you've enabled file sharing for sensitive files without protecting them with a strong password, it won't take any kind of hacker skills to read, copy, or delete your files. Intruders who can write files to your computer can install software and ultimately do what they wish.

Some people writing about Windows security over-cautiously recommend disabling file sharing tout court. There are four separate layers of protection even if file sharing is turned on. First of all, if file sharing isn't bound to TCP/IP, no one on the Internet will see your directories or be able to discern that file sharing is occurring. If you need to share files only on the local network, NetBEUI or IPX/SPX will work fine and remove the temptation of a potentially open file share. Second, you must actively indicate that a drive, directory, or file is to be shared. Items not explicitly marked as shares will not be visible to other clients, though it's possible to enable sharing of an entire volume by selecting it at the root level—most likely an inadvisable practice. Third, setting a Scope ID will make a share invisible to an intruder who doesn't know it. Finally, passwords can (and should) be assigned to shared resources.

File and Printer Sharing for Microsoft Networks turns a Windows 9x system into a file server. Given the installation of updates and the use of strong password choices, even running this server process is reasonably secure. The only other explicit server that comes with consumer Windows OSs is the Personal Web Server. Apparently this program has sufficient code in common with Internet Information Server (IIS), the NT/2000 Web server, that some of the exploits that threaten IIS also require patching on the Personal Web Server.

The most dangerous disguised server is the remote control application. pcANYWHERE,

Carbon Copy, Timbuktu, and LapLink are some of the better-known commercial remote control packages. Back Orifice, SubSeven, and NetBus are three of the better known stealth remote control applications. Trojan Horse programs typically install the stealth applications because they give a remote intruder complete control over the system. The commercial products can be protected with passwords, and the power of these systems is so great that there's a high incentive to create strong passwords.

Chat and Instant Messaging (IM) programs can execute server processes on home computers. Internet Relay Chat (IRC), the Internet forerunner of IM, is the source of numerous destructive exploits. It serves as the elementary school for script kiddies, as well as the neighborhood watering hole for more experienced intruders. Napster and its decentralized offspring are essentially file servers with more or less restricted realms of operation. The Gnutella derivatives can be configured to share files almost as profligately as Windows File and Print Sharing. While I don't know of specific exploits affecting home users in these areas, I suspect that watertight security is rarely a high priority for the developers of these not-so-obvious server processes.

While a wireless access point doesn't count as a server in the OSI mindset—it's a layer-2 bridge, after all—a war-driving intruder could pose just as big a threat to corporate data as a hacker who installs SubSeven. Employees installing 802.11 networks can avoid drilling holes in their baseboards, but they may be making their network, and the company data they access, visible to anyone who cares to look for it.

INTRUSION COUNTERMEASURES

The first line of defense against intruders is a well-patched OS. It's rare that a class of attacks is actually employed before the OS vendors make their patches available. The next preventative step is to understand the implications of configuration choices for any explicit or veiled server processes that run on your computer and configure them as safely as possible, with well-chosen passwords.

The next level of security escalation for broadband households that wish to share Internet access among multiple PCs is to install a Network Address Translation (NAT) router. By presenting a single IP address to the outside world and mapping that address and a particular port to a non-routable address inside the router, the inside machines become invisible to Internet-based attackers. These NAT routers often come with filtering capabilities or even stateful-inspection firewalls that provide an additional level of protection at the price of some complicated installation procedures.

Personal firewalls, which often incorporate some kinds of intrusion detection, are perfect for individual machines connected to the Internet without a NAT router. They can also be installed on each PC in a local network, located outside the router, or run on a dedicated PC between the router and a hub or switch that connects to the client PCs. Some of these software products will detect outbound traffic from stealth remote control processes, providing a valuable backup to preventative efforts. Finally, enterprises whose remote employees have access to the most crucial, sensitive data will probably want to install firewalls at their employees' residences.

Note that certain commonly cited security threats—having a broadband connection and having an always-on Internet connection—are actually not threats at all. If you properly configure and patch the OS and file-sharing applications; guard against viruses, worms, and rogue software they may install; and insulate your home system with a NAT router or a firewall, then the length of time you're connected and the speed of your connection have no impact on your security.

RESOURCES

One of the most sensible and comprehensive sites for safely configuring broadband networks is the Navas Cable Modem/DSL Tuning Guide at http://cable-dsl.home.att.net/#security. This site has especially good advice on File and Printer Sharing for Microsoft Networks, but also discusses Macintosh security, OS/2 security, and the pros and cons of personal and hardware-based firewall products.

Hacking Exposed, Second Edition, by Joel Scambray, Stuart McClure, and George Kurtz, Osborne-McGraw Hill, 2001, ISBN 0-07-212748-1, is the definitive compendium of intruder practices and tools.

This tutorial, number 162, by Steve Steinke, was originally published in the January 2002 issue of Network Magazine.

Tools for Securing Home Networks

The first and most efficacious step toward securing home networks, including those that connect with enterprise networks, is to ensure that each computer has capable, up-to-date anti-virus protection. The anti-virus system should not only scan the system for executable files with viruses and monitor executable files arriving from the Internet or from removable media; it should also be capable of scanning e-mail attachments and files with embedded macros. Because connecting via a VPN will enable the enterprise network to seem to be part of the local network, home users need to be assured of good virus protection on the enterprise network.

Home computers may become infected through several routes with Trojan Horse software that provides remote access to attackers. The Trojan Horse may arrive before the anti-virus vendor's update is installed. The offending remote control code may be part of a game or file swapping program that a user installs or discovers intentionally. An attacker who installs the remote control program may take over an inadequately patched home-based server. Once NetBus or SubSeven or Back Orifice is installed, anti-virus software will, at best, only find it when it does a full system scan. In the worst case, it will fail to realize that any infection exists.

This is when a personal firewall product with outbound application control capabilities is essential. These products prompt users the first time an application needs to send traffic across the network. The name of the program (or service), perhaps the pathname of the application, and additional information may be provided, and the user will be prompted to specify whether or not to block the traffic. If the user chooses to block the traffic for an application, rules will be created in the firewall to prevent future traffic. Some of these products have a database of known illegitimate remote control programs and automatically block their ability to transmit, which is a good idea for situations where a user might not recognize the names of dangerous executable files. In some products, the firewall silently blocks traffic it identifies as dangerous. This approach has the advantage of minimizing confusion on the part of users who have no basis to decide what applications should be allowed to communicate, but it increases reliance on the vendor to keep the database of dangerous applications up to date.

As we've pointed out in previous tutorials (see "Coping with Home Network Security Threats," the previous tutorial), it's easy to overestimate the vulnerability of home computers and home networks to direct outside attacks. A single computer, even one connected to an always-on, high-speed access service, will rarely be running a Web server, mail server, FTP server, or other traditional hacker target, and therefore usually will not be vulnerable to a direct exploit. A home network that uses network address translation (NAT) to share a single IP address among multiple devices generally isn't vulnerable to outside attack, provided none of the devices is acting as a crackable server. Some vendors even describe NAT gateway products as "firewalls," though that's an overstatement. NAT boxes maintain a table that matches incoming address and port number combinations with more or less arbitrary local address and port number combinations, while packet filtering firewalls maintain tables of rules for forwarding or denying packets on each port.

Setting the inbound rules on a personal firewall is only crucial sometimes—in particular, if you're running server processes for remote clients. Of course, not all services are obviously services. Peer-to-peer software, including Instant Messaging (IM) and teleconferencing require running server components. A recently uncovered and patched vulnerability in Windows XP attacked a service in the Universal Plug and Play (UPnP) system, designed to support automatic installation of remote devices. It's not intuitive to imagine that being able to connect with a printer across the Internet requires your desktop machine to operate a service on ports 1900 and 5000.

Services aren't a danger if they're implemented properly—the UPnP vulnerability was based on an input buffer overflow, as many other attacks are. Implemented properly, UPnP represents no vulnerability to an external takeover attack.

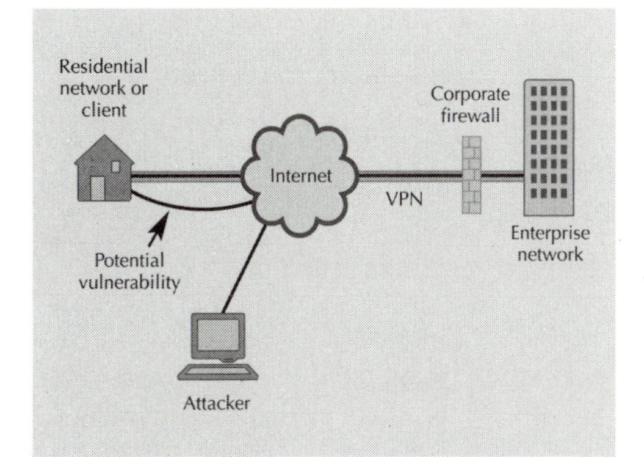

Potential Vulnerablity of VPN. An external attacker with remote control access to a residential client may be able to access the enterprise network with the employee's access rights.

Another issue regarding personal firewalls is peace-of-mind. Assessing a home network's vulnerability is more complex than most configuration jobs for home PCs, and the stakes can be high for workers who have enterprise data on their home network and for remote workers who connect with their offices. Enterprise network managers who define the standards for remote connections, as well as individuals who set up their own remote connections to the office, would be well advised to install personal firewalls that block unsolicited incoming connections and unintentional outbound remote control connections. (Note that Microsoft's Internet Connection Firewall, included with Windows XP, isn't designed to block this sort of dangerous outbound traffic.) For enterprises looking for consistency and manageability, some personal firewall products have central configuration options. The real trade-off for enterprises with remote workers is the safety net provided by personal firewalls versus the complication, support requirements, and perhaps a performance penalty that comes with the personal firewall.

When the stakes are really high, software-based personal firewalls will likely be perceived as incommensurate with the task of securing remote workers. In these cases, the enterprise will probably install centrally controllable standalone firewalls and dedicate the computer and perhaps even the network access connection to secure applications.

FITTING IN WITH THE VPN

A remote-access VPN is designed to authenticate users accessing their office network over the Internet and to encrypt all the traffic over the link. One way to look at a VPN is as an opening through the enterprise firewall. If an attacker can install remote control software on a home computer, the VPN may serve as a path into the enterprise network. On the other hand, the home firewall needs to recognize that VPN traffic is legitimate and to stay out of the way. Most common personal firewall products can be configured to coexist with VPNs.

You might think of firewalls as the walls of a building, with locked doors corresponding to particular port and protocol combinations that permit certain types of entry and exit.

Extending this image, Intrusion Detection Systems (IDSs) listen for someone to jiggle the locks and try the doorknobs to see if unauthorized entry is possible. Where firewalls apply rules to particular ports and protocols, network-based IDSs monitor all inbound traffic to identify "attack signatures."

Both firewalls and intrusion detectors create events when they detect violations. The least common denominator destination for events is the log file. Various products can log some or all of the following data: the source, destination, and port addresses of particular packets and the rules they broke; the complete contents of offending packets; inferences about the type of attack that the offending traffic indicates; and the configuration history of the software. Products that emphasize intrusion detection often display onscreen alerts when they identify attack conditions. Some products send e-mail notices of alert conditions.

For some people, flashing alerts whenever an automated port scan or nmap episode occurs is more than they care to know about the dangers of the Internet—they're content to go about their business as long as they're protected. Others take script-kiddie attacks to heart, and feel obligated to run whois and trace the ISP of the source address of every log event. It's important for the future self-government on the Internet that irresponsible users be held in check, but there's a certain amount of danger in auto-responding to every alert with e-mails to the abuse departments of ISPs. The primary reason for caution is the presence of "false positives" in firewall and intrusion detection logs.

Lawrence Baldwin, who operates www. mynetwatchman.com, has closely studied personal firewall logs over the last year and a half. A nice compromise between ignoring events that appear to be attacks and blasting away at ISPs a hundred times a day when a Code-Red-type siege is underway is to follow the instructions at the myNetWatchman site and send your logs to them. Their scripts filter out the false positives, identify spoofed source addresses, and track down the real sources of attack traffic. Then they pack up all the evidence and send it to ISPs in a form that's hard to ignore.

FALSE POSITIVES

Here are some of the most common causes of false positive events for residential clients as identified by myNetWatchman: If you request a response from a sufficiently slow server, your firewall may time out and no longer maintain the connection in its table. The slow-to-arrive response now looks like an unsolicited transmission, requiring an alert to the user. Furthermore, Web pages often have components that must be collected from multiple geographically distinct servers, including Content Delivery Networks (CDNs), such as Akamai's, and advertising servers, such as doubleclick's. Thus, any of a number of slow sources may falsely appear to be an attack.

A second common false positive is the "proximity probe." When you start to request data from a large, widely distributed Web site that employs a load balancer with a DNS lookup, multiple sites issue proximity probes. The ICMP normal responses to these probes help identify the closest or fastest server for providing the requested data. This traffic arrives at the personal firewall from addresses that no request was sent to, so they appear to be probe events on TCP port 53.

Another common source of false positives is the stale IP cache. If you have a dynamic IP address, you inherit all the cached baggage of the address' former lessee. Certain games and peer-to-peer file sharing programs are notorious for blasting out thousands of requests to expired addresses. Each of these requests hits the personal firewall log as an attempted attack.

Another type of false positive is associated with Microsoft's Internet Information Server (IIS), the Web server built into all NT-based versions of Windows. Apparently when NetBIOS over TCP/IP is enabled on a public-facing interface, which is common, though it hardly seems useful, IIS attempts to identify the NetBIOS names of clients that make HTTP requests. These queries will show up as hostile probes on the personal firewalls.

There's a lot of scanning and port mapping and scripted exploit execution underway at any given moment. It's relatively painless to protect against such activity—in fact, to be equipped with not only a belt but also a couple of pairs of suspenders. Some reasonably good personal firewalls are available free for the download, and the most expensive ones are just $50. Popular and reliable NAT routers with firewall capability included are available for less than $100. The real danger may be the amount of time users spend installing software, defining firewall rules, trouble-shooting dicey installations, waiting for tech support, and flaming unresponsive ISPs that don't shut down obvious miscreants. One last word of advice: Try to be as well-informed as possible before asking hardware, software, or ISP help desk people about the potential sources of rogue Internet traffic. You don't want to be classified as a GWF, short for "Gomer with Firewall."

RESOURCES

Lawrence Baldwin's useful discussion of firewall false positives can be found at www.dslreports. com/forum/remark,2169468.

The Home PC Firewall Guide (www.firewallguide. com) has high-quality background information, along with numerous pointers to comparative reviews of these products. One good one is at www.boran.com/security/sp/pf/pf_main20001023.html, written by Seán Boran.

Another valuable comparison, "Getting Personal with Firewalls," was written by Curtis Dalton in the January 2001 issue of Network Magazine (page 100, www.networkmagazine. com/article/NMG20010103S0010/1).

If you want to see the center of the Gomer-with-firewall universe, tune in to the grc.security newsgroup at news.grc.com. The postings aren't always accurate and reliable, so you'll have to sort through them with a wary eye.

This tutorial, number 164, by Steve Steinke, was originally published in the March 2002 issue of Network Magazine.

Virtual Private Networks

Although the Internet can't quite do our laundry yet (or even cook a decent meal for that matter), it has changed the way we are able to transact business. One of its latest offerings for organizations that are motivated to reduce costs and increase services is the Virtual Private Network, or VPN.

In a nutshell, a VPN is a private connection between two machines or networks over a shared or public network. In practical terms, VPN technology lets an organization securely extend its network services over the Internet to remote users, branch offices, and partner companies. In other words, VPNs turn the Internet into a simulated private WAN.

The appeal is that the Internet has a global presence, and its use is now standard practice for most users and organizations. Thus, creating a communications link can be done quickly, cheaply, and safely.

HOW IT WORKS

To use the Internet as a private wide area network, organizations may have to overcome two main hurdles. First, networks often communicate using a variety of protocols, such as IPX and NetBEUI, but the Internet can only handle IP traffic. So, VPNs may need to provide a way to pass non-IP protocols from one network to another.

Second, data packets traveling the Internet are transported in clear text. Consequently, anyone who can see Internet traffic can also read the data contained in the packets. This is clearly a problem if companies want to use the Internet to pass important, confidential business information.

VPNs overcome these obstacles by using a strategy called tunneling. Instead of packets crossing the Internet out in the open, data packets are first encrypted for security, and then encapsulated in an IP package by the VPN and tunneled through the Internet (see Figure).

To illustrate the concept, let's say you're running NetWare on one network, and a client on that network wants to connect to a remote NetWare server.

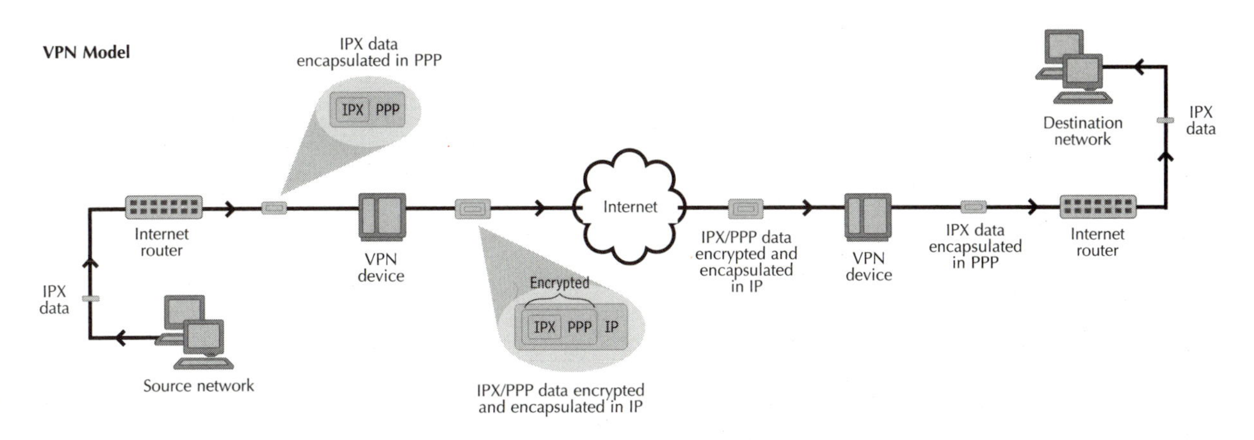

■ **How Tunneling Works:** When a VPN device receives instructions to transmit a packet over the Internet, it negotiates encryption with the VPN device on the destination network, then encrypts the packet accordingly. Next, it encapsulates the encrypted packet in an IP packet and sends it over the Internet to the destination network. Once the packet arrives, the receiving VPN termination device reverses the process and lets the packet continue to its destination on the internal network.

The primary protocol used with traditional NetWare is IPX. So, to use a generic layer-2 VPN model, IPX packets bound for the remote network reach a tunnel initiating device—perhaps a remote access device, a router, or even a desktop PC, in the case of remote-client-to-server connections—which prepares them for transmission over the Internet.

The VPN tunnel initiator on the source network communicates with a VPN tunnel terminator on the destination network. The two agree upon an encryption scheme, and the tunnel initiator encrypts the packet for security. (For better security, there should be an authentication process to ensure that the connecting user has the proper rights to enter the destination network. Most currently available VPN products support multiple forms of authentication.)

Finally, the VPN initiator encapsulates the entire encrypted package in an IP packet. Now, regardless of the type of protocol originally being transmitted, it can travel the IP-only Internet. And, because the packet is encrypted, no one can read the original data.

On the destination end, the VPN tunnel terminator receives the packet and removes the IP information. It then decrypts the packet according to the agreed upon encryption scheme, and sends the resulting packet to the remote access server or local router, which passes the hidden IPX packet to the network for delivery to the appropriate destination.

THE METHODS

Currently, there are a handful of VPN protocols rising to the surface in the industry—namely L2TP, IPsec, and SOCKS 5. Because they provide tunneling functions, these protocols are the building blocks used to create VPN links. Some of the protocols overlap in functionality, and some offer similar but complementary functionality. Each of the protocols requires further investigation when shopping for a solution. In the meantime, here's a quick summary of the protocols.

Also known as the Layer-2 Tunneling Protocol, L2TP is the combination of Cisco Systems' Layer-2 Forwarding (L2F) and Microsoft's Point-to-Point Tunneling Protocol (PPTP). L2TP supports any routed protocol, including IP, IPX, and AppleTalk. It also supports any WAN backbone technology, including frame relay, ATM, X.25, and SONET.

One key to L2TP is its use of PPTP. This Microsoft protocol is an extension of PPP and is included as part of the remote access features of Windows 95, Windows 98, and Windows NT. So, in the big picture, most PC clients come equipped with tunneling functionality. PPTP provides a consistent way to encapsulate Network-layer traffic for remote access transmission between Windows clients and servers. The protocol doesn't specify a particular encryption scheme, but the remote access functions included in the Microsoft stable of operating systems are supplied with Microsoft Point-to-Point Encryption (MPPE).

The L2F portion of L2TP lets remote clients connect and authenticate to networks over ISP and NSP links. Besides the basic VPN capability, L2TP can create multiple tunnels from a single client. In practice, a remote client can create tunneled connections to various systems simultaneously—for instance, to a corporate database application and to the company's intranet.

As for IPsec, the full name for it is Internet Protocol Security, and it's basically a suite of protocols that provide security features for IP VPNs. As a layer-3 function, IPsec can't perform services for other layer-3 protocols, such as IPX and SNA. IPsec provides a means of ensuring the confidentiality and authenticity of IP packets. The protocol works with a variety of standard encryption schemes and encryption negotiation processes, as well as with various security systems, including digital signatures, digital certificates, public key infrastructures, and certificate authorities.

IPsec works by encapsulating the original IP data packet into a new IP packet that's fitted with authentication and security headers. The headers contain the information needed by the remote end, which took part in the security negotiation process to authenticate and decrypt the data contained in the packet.

The appeal of IPsec is its interoperability. It doesn't specify a proprietary way to perform authentication and encryption. Instead, it works with many systems and standards. IPsec can complement other VPN protocols. For instance, IPsec can perform the encryption negotiation and authentication, while an L2TP VPN receives the internal data packet, initiates the tunnel, and passes the encapsulated packet to the other VPN end point.

Another approach to VPNs is SOCKS 5, which was first developed by Aventail. SOCKS 5 is a bit different from L2TP and IPsec: It follows a proxy server model and works at the TCP socket level. To use SOCKS 5, systems must be outfitted with SOCKS 5 client software. Furthermore, your organization needs to be running a SOCKS 5 server.

Here's how the SOCKS 5 model works. First, a client request for services is intercepted by the SOCKS 5 client. The request is sent to the SOCKS 5 server, which checks the request against a security database. If the request is granted, the SOCKS 5 server establishes an authenticated session with the client and acts as a proxy for the client, performing the requested operations.

The upside to SOCKS 5 is that it lets network managers apply specific controls on proxied traffic. Because it works at the TCP level, SOCKS 5 lets you specify which applications can cross the firewall into the Internet, and which are restricted.

WHY IT'S APPEALING

VPN vendors can recite a litany of benefits that the technology provides, and more will emerge as VPN products mature.

Perhaps the biggest selling point for VPNs is cost savings. If you use the Internet to distribute network services over long distances, then you avoid having to purchase expensive leased lines to branch offices or partner companies. And, you escape having to pay for long distance charges on dial-up modem or ISDN calls between distant sites. Instead, users and systems simply connect locally to their ISP and leave the rest of the journey to the vast reach of the Internet. On another cost-related note, you can evade having to invest in additional WAN equipment and instead leverage your existing Internet installation.

Another benefit of VPNs is that they are an ideal way to handle mobile users. VPNs allow any user with Internet access and a VPN client to connect to the corporate network and to receive network services. Since Internet access is now widespread, you won't have to juggle users and locations when setting up remote, mobile access.

In the same vein, because Internet use is commonplace, you can deploy a network-to-network arrangement quickly and cheaply. There's no need to order and configure data lines and WAN interfaces for each site. Again, you just leverage each site's Internet connection to form the link. This is especially advantageous in the current business environment, where partner companies are connecting networks to improve the speed and efficiency of shared business operations.

This tutorial, number 123, by Lee Chae, was originally published in the October 1998 issue of Network Magazine.

IP VPN Services

An IP VPN is commonly defined as a routed link between two or more points across a heterogeneous network topology with various degrees of security that ensure privacy for all parties. The idea behind the IP VPN is to leverage the Internet's reach—and low cost—to eliminate the more expensive dedicated links common today. Some industry experts also claim IP VPNs will guarantee secure data transmission for businesses and, in the process, allow service providers to offer more-profitable value-added services.

You can set up VPNs in several ways, but Customer Premises Equipment (CPE)-based IP VPNs and network-based IP VPNs are the most common approaches. The difference lies in the network architecture: In CPE-based VPNs, the routing intelligence resides at an end-user site, while in carrier-based VPNs, it resides at the provider's edge, where it can be extended out to many end-user locations.

WHAT KIND OF VPN DO YOU NEED?

There are three types of VPNs. First, remote-access VPNs allow telecommuters or home workers who use DSL, cable, dial up, or wireless to access their corporate data networks. Second, site-to-site VPNs connect remote offices over the Internet. Site-to-site VPNs use secure point-to-point connections in a mesh topology overlaid on the Internet or even on a single provider's network.

Third, extranet VPNs connect a "community of interest," such as a company, its partners, suppliers, and customers, and so on, to an enterprise network and perhaps other relevant destinations. In extranet VPNs, an authorized third-party user arrives at the enterprise firewall after traversing the public network, and the destination (often the home office) VPN gateway terminates the traffic and grants trans-firewall access where appropriate.

When it's time to engineer your VPN, you need to consider many issues, such as the timeline for deployment and what applications the VPN will serve. For intranet and extranet users, you need to consider what type of data is involved and whether you ought to set priority levels based on the type of traffic. You need to know the number of users, how many classes of users you have, and what specific restrictions apply to these user groups. These concerns might include limiting user access to a specific location or during certain times of the day or week. You might also consider the type of resource access to grant, such as connectivity to a particular server, to a subnet, or to the whole network, as well as assigning QoS levels.

Next, decide whether to use a CPE- or network-based solution. Each approach has pros and cons (see "IP VPNs: The Next Wave," *Network Magazine* February 2001, page 96), but remember that most first-generation network-based solutions encrypt data only from service provider edge to service provider edge—the access link isn't secured. Of course, it's also

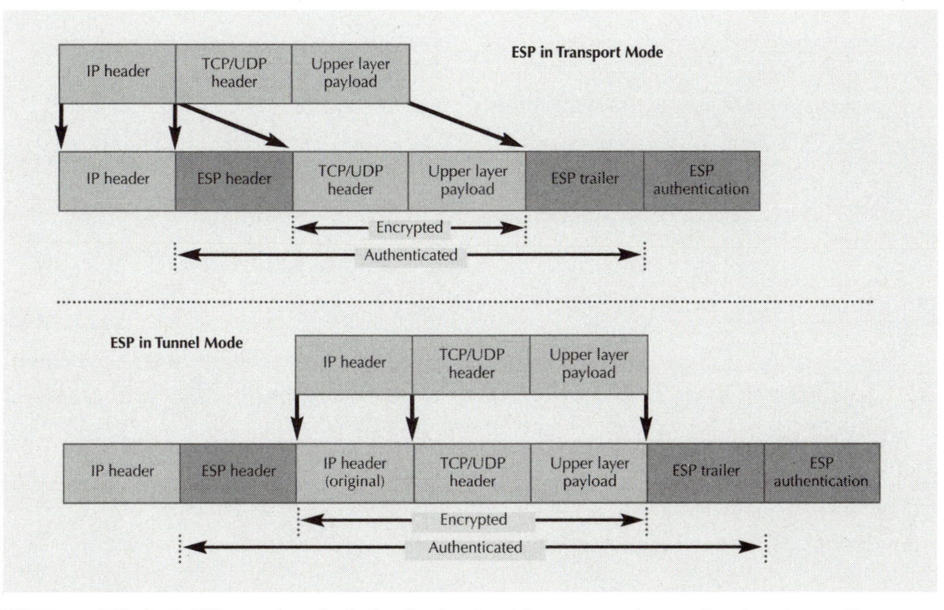

ESP Tunnel Mode. In ESP tunnel mode, the header data is neither encrypted nor authenticated, so traffic can traverse a Network Adress Translation (NAT) device.

possible for service providers to install devices they own and control on the customer premises. Alternatively, some software overlay VPNs (see "VPN Overlay Networks: An Answer to Network-based IP VPNs?" *Network Magazine* June 2001, page 48) offer a mix of CPE- and network-based equipment or service options.

TYPES OF VPN

IPSec is the predominant standard for creating IP VPNs. IPSec handles these three types of VPNs by using either a gateway-to-gateway (GTG) or client-to-gateway (CTG) approach to create an overlay network on top of the Internet. IPSec requires each site to run a gateway that routes data to the Internet in a virtual tunnel that both verifies the authenticity of the parties at both endpoints and encrypts the data. GTG is used for site-to-site VPNs and designates tunnel endpoints at the WAN interface of the VPN gateway. Traffic between these endpoints is encrypted and authenticated, but the LANs behind the gateway are left alone, since these are supposedly already secured subnets.

CTG configurations are designed for mobile remote workers or telecommuters who need secure access to corporate data. This approach moves one of the tunnel endpoints onto the end user's machine. The client software supports the same IPSec authentication, encryption, and security association mechanisms as the VPN gateway.

You'll want to consider the network topology of your VPN overlay network, too. Scalability and performance are the key issues that determine whether to use a partially or fully meshed, distributed topology or a hub-and-spoke (HAS) topology, or some combination of these. A fully meshed network is hard to scale because it requires a link between each device and every other one with attendant encrypted-vs.-clear-traffic processing issues, configuration complexity, high availability requirements, high demand for knowledge of related security features, and the need to implement numerous QoS parameters. All of these factors may vary by user. It will often be better to use a partially meshed network, which establishes inter-spoke connections as needed. In both cases, the key is to keep the number of tunnels well within the VPN gateway's performance capabilities. One way to simplify configuration and improve scalability is to have the gateways use a dynamic tunnel endpoint discovery mechanism.

The most scalable topology is the HAS model, since the headend or hub can grow if spoke end-user capacity requirements increase. All traffic goes through the hub site, so spokes can interconnect with other spokes without using a direct link, but this may require fat pipes to provide enough bandwidth for spoke-to-spoke, as well as spoke-to-hub traffic. This approach may be too costly, especially since some headend VPN devices don't support direct spoke-to-spoke intercommunication. In cases where endpoints are more spread out across regions, or most traffic doesn't demand access to networks through the hub site, a distribution layer to decrease the bandwidth requirements at the headend and improve scalability might make sense.

WHERE SHOULD THE VPN TERMINATE?

Another knotty question is where to terminate a VPN—at the corporate or branch office firewall, or on the end user's private network? The answer depends on the end user's security needs and how this relates to the network architecture. In a remote access application where end users are tunneling into the corporate network, VPN devices commonly sit parallel to an existing firewall, or behind that firewall. Less common approaches include terminating sessions in front of the firewall or on the firewall itself.

A parallel VPN device/firewall approach means the firewall doesn't have to be reconfigured for the VPN traffic, but it does add to the number of entry points into the private network. It's imperative that VPN devices block all non-VPN traffic in order to minimize the additional risk. This may require the VPN device to perform some degree of address translation, or to redirect such traffic to the corresponding firewall.

If you put the VPN device behind the firewall, that firewall will require modification and the ability to filter and forward the VPN traffic to the appropriate gateway. But this setup may allow you to use only one of the Ethernet ports on the VPN gateway, which reduces some complexity.

On the other hand, if you put the gateway in front of the firewall, the secured traffic must terminate in a public zone. This forces the network manager to assign specific static IP addresses. This allows the manager to control the traffic's destination through the existing firewall, but most VPN gateways can also perform this task.

Finally, running a VPN gateway on an existing firewall has the appeal of simplicity by keeping all security access functions at the network perimeter, but this requires the firewall to take on a heavy processing burden that may be beyond the firewall's capabilities. This approach is significantly less popular than a parallel or behind-the-firewall architecture.

IP SEC AND NAT

Using Network Address Translation (NAT) with IPSec VPNs can also be tricky. One trouble spot is the Authentication Header (AH) protocol in IPSec, which computes a hash value for outgoing packets that includes both the data payload and the address headers, and inserts this value into the packet.

A NAT device between the IPSec endpoints will change the addresses and port numbers to those appropriate for the end network. In this case, the receiving VPN device, which is unaware of any NAT activity, will drop the incoming packets—the receiving device computes a hash value based on the incoming packet, and then tries to match that value to the hash value in the packet. No match, no access, no VPN.

The solution is to use the Encapsulating Security Payload (ESP) protocol in tunnel mode (see figure). This mode encrypts both the packet's data and its headers in a new packet with its own headers. In other words, the newly created packet's source address is the sending gateway and its destination address is the receiving VPN gateway. By using ESP with authentication, the whole packet is now encrypted. But the header values changed by NAT aren't locked in by the integrity-checking mechanism and the traffic can flow freely.

One tricky area remains, however. The third major component of IPSec, after AH and ESP, is the Internet Key Exchange (IKE). IKE, formerly referred to as the Internet Security Association and Key Management Protocol (ISAKAMP), with the arbitrary name "Oakley" attached—ISAKAMP/ Oakley—manages the creation, management, and distribution of cryptographic keys. (Security Associations [SAs] are the endpoints of IPSec links.) Before any data can be exchanged, IKE must first establish an SA to serve as an initial secure channel for exchanging keys. This initial IKE SA establishment is referred to as Phase 1, and can be accomplished with a main mode exchange or an aggressive mode exchange. Phase 2 is the establishment of an IPSec SA, and is known as quick mode. Possible main mode establishment mechanisms include X.509 digital certificates, preshared keys, and encrypted nonces. Preshared keys are symmetrical keys installed in advance on VPN endpoints.

Encrypted nonces involve the generation of public/private key pairs on each endpoint and the manual copying of public keys to every other endpoint. As with AH traffic, NAT will bring down IKE if the keys or certificates exchanged by IKE bind a gateway's identity to its IP address.

IPSec would undoubtedly be much more widely used if PKI were more widely adopted. PKI digital certificates are the most straightforward way to create IPSec SAs. As digital certificates and certificate authorities become more established, we can expect to see more IP VPNs using IPSec technology.

RESOURCES

Cisco Systems has a comprehensive discussion of VPN technology at: www.cisco.com/warp/public/ cc/so/cuso/epso/sqfr/safev_wp.htm.

A good, concise VPN FAQ can be found at: www.zyxel.com/support/supportnote/zywall10/faq/vpn_faq.htm.

Check out Virtela Communications' "IPSec VPN Tutorial" at www.virtela.com for more information on IPSec.

Numerous white papers address more general applications for VPNs at www.aplion.com. This company offers good high-level views of how VPNs should be used in the future.

This tutorial, number 165, by Doug Allen, was originally published in the April 2002 issue of Network Magazine.

Layer-2 VPNs

Hold on to your toolbelt. Over the next eight months, new layer-2 (L2) Multiprotocol Label Switch (MPLS)-based VPN services are expected to deliver ATM- and frame-like connectivity at a fraction of ATM and frame prices.

Unlike most existing MPLS VPN services, these new L2 VPNs afford networks the same raw L2 protocol that any data service might provide, but they deliver it over an integrated backbone capable of supporting both voice and data. The result, say proponents, is a huge cost savings—one that, at least in part, is expected to be passed back to the consumer. L2 VPNs will also offer networkers better control and a smoother migration path to an MPLS backbone than is possible with existing layer-3 (L3) VPN offerings.

Of course, with more choices comes more confusion. Not only will companies need to select between L2 and L3 VPNs, they'll likely have to choose between two types of L2 VPNs. To get ready for the CEO's and CTO's inevitable purchasing questions, start by defining VPNs for them: a closed user group where admittance is based on an address, such as the Ethernet or frame address for L2 VPNs, and the IP address for L3 VPNs. Then explain the differences between L2 and L3 VPNs, as well as the differences between L2 VPN approaches.

TODAY'S VPNS

Most of today's MPLS-based VPNs are L3 VPN services, well suited to IP-centric corporate networks with simple routing architectures. But they also include companies with offices connected via a mix of frame relay and ATM, or Ethernet and leased line connections, as well as companies looking for more advanced services, such as filtering and Class of Service (COS).

The basis for these VPN services (RFCs 2547 and 2547bis), though, imposes significant scalability, reliability, and operational challenges on providers, even as the RFCs enable them to multi-source their equipment. While only the biggest networks should be impacted by the scalability limitations of L3 VPNs, VPNs can still span about 200 sites, and router instability is another matter. With L3 VPNs, customer routers can flap their routes, or very rapidly turn their routes on and off, causing instability in the provider's edge router or even the carrier's entire network (see "Inside RFC 2547bis" for a more in-depth analysis of the RFC specification at www. networkmagazine.com).

L2 VPNs offer a compromise. They enable networkers to connect new sites with native IP connections, while migrating existing frame relay or ATM sites to MPLS by providing native L2 connections—similar to AT&T's original IP Enabled Frame Relay or IP Enabled ATM services, but based on approved standards. Existing sites might continue connecting with one another via frame relay, while new sites might access the corporate network via Ethernet, all the while carrying a wide range of protocols across the MPLS connection. As for L2 reliability, at worst a Customer Edge (CE) router might destabilize its own line, but it wouldn't destabilize another customer's connection.

THE L2 SOLUTIONS

Pundits proposed L2 VPN drafts in two groups within IETF: the Pseudo Wire Emulation Edge to Edge (PWE3) group and the Provider-Provisioned VPN (PPVPN) group. The PWE3 group is working on the Martini draft, named after Luca Martini, a senior architect at Level 3 Communications (www.level3.com) and a major contributor to the draft. The draft has garnered broad industry support from many companies including Cisco Systems and Juniper Networks (www.juniper.net), and defines a provider-provisioned point-to-point service called Virtual Private Wire Service (VPWS). Within the PPVPN group, the less-popular Kompella draft, named after Kireeti Kompella, a distinguished engineer at Juniper, defines a provider-provisioned VPWS and a point-to-multipoint service called a Virtual Private LAN Service (VPLS), previously known as a Transparent LAN Service (TLS).

Both drafts specify a way to group individual connections at different routers into a flat network. Those connections may be a single physical segment, as in the case of Ethernet, or virtual channels over a particular port, such as VLANs in Ethernet or virtual circuits in the case of frame relay or ATM.

Under Martini, providers build VPNs from VC IDs, tags that identify the virtual channels running between the specific ports on a CE router and the locally connected Provider Edge (PE) router. Provisioning a link involves configuring an MPLS tunnel between a VC ID on one PE router to the VC ID on another PE router. The provider enters the remote VC ID information at each PE router, associates the ingress and egress VC IDs with one another, and then relies on MPLS's Label Distribution Protocol (LDP) to distribute the necessary label information among the interior routers to carry the L2 packet through an MPLS tunnel between the two points.

With Kompella, however, providers build VPNs with a BGP attribute, the Route Target community. With BGP, PE routers can learn the VPN membership, freeing up the providers from supplying all of the information at every router. To do that providers must first configure PE routers with a list of the locally connected CE routers (identified through a CE-id) participating in each VPN, the specific Route Target community, and the physical interface over which those customers connect to the PE router (called the Interface Index). The router derives the Label Block, a block of MPLS labels mapped to the VCs connecting to the CE, on its own.

Using BGP, PE routers then announce all of this information to the other PE routers. The PE routers receive the announcement and check whether they belong to the particular Route Target community. If they do they add the CE ID and Label Block to their databases for the VPN, and create the necessary routes in their MPLS tables. The L2 address is then associated with an MPLS tunnel.

The two drafts are almost identical during the packet's movement across the MPLS network and in the way PE routers encapsulate ATM, frame relay, or Ethernet signals inside MPLS. When a packet reaches the ingress PE router, it determines the appropriate tunnel and constructs a VC Label. The VC Label is determined from the ingress Interface Index and the VC ID or CE ID (depending on if it's Martini or Kompella), and contains information about the type of circuit (VC Type), as well as specifics about the egress interface. A separate field, called the sequencing control word, provides the circuit's properties, such as the sequential packet propagation, padding, and control bit propagation. The router constructs and appends the VC Label and the requisite MPLS labels to the L2 packet. Depending on the connection, the packet will also get a sequencing control word. Under Kompella, if the network runs IP Interworking, the PE router also strips off the L2 header (see Figure 1).

ANATOMY OF A PACKET

L1 encapsulation	Transport label	VC label	Optional control word	Encapsulated L2 frame	L1 encapsulation

■ **Figure 1:** While Kompella and Martini may differ in their signaling, their generalized packet configuration is very similar. The L1 encapsulation is the additional L1 information, most likely Sonet or SDH, needed to move data across the carrier's infrastructure. The Transport Label is the MPLS label that identifies the MPLS tunnel transporting the encapsulated L2 frames or cells through the MPLS network. The VC label is an MPLS label that identifies the particular L2 virtual connection, such as a Frame Relay DLCI, that is being transported through the MPLS tunnel. The control word contains information about the connection. It may be optional or mandatory depending on the network configuration. The L2 frame or cell is the L2 frame presented to the provider's edge router.

VPN OPERATION

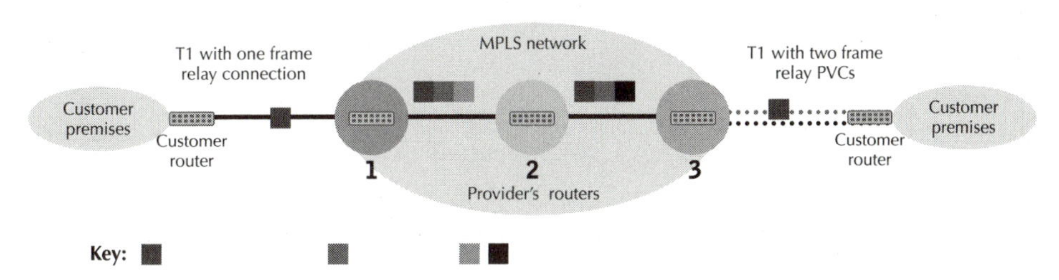

■ **Figure 2:** With L2 VPNs, the provider's edge router encapsulates the L2 packet within an MPLS frame and adds a a special MPLS label, the VPN Label, that designates the destined port and virtual circuit (1). The packet traverses the MPLS network, with each MPLS router swapping labels (2). The final router removes the VPN Label exposing the L2 packet to the customer edge (CE) router (3).

The packet then traverses the MPLS network until reaching the penultimate router or the egress router, where, depending on the network configuration, one of the two strips off the last MPLS label and exposes the VC Label. Either way the egress PE router reads the VC Label, determines the outgoing channel and interface, and sends the L2 packet to the receiving CE device (see Figure 2).

DIFFERENCES

Martini and Kompella share much in common vis-à-vis the packet formats and their ability to connect customer premises with a range of L2 protocols including frame relay, ATM, Ethernet, and leased lines. Their respective strengths and weakness are largely derived from their use of BGP, in the case of Kompella, or LDP, in the case of Martini, to distribute VPN information.

Kompella argues that configuring a Martini network is more complicated in two ways. First, Martini lacks a distribution scheme for spreading configuration information across the network, requiring service providers to provision both ends of the circuit. Martini networks only support point-to-point circuits, requiring providers to configure more circuits in a fully meshed network than Kompella requires. Kompella permits both point-to-multipoint and point-to-point circuits.

However, Luca Martini agues that there's likely to be little difference as provisioning systems will automate circuit establishment under either draft. And the use of point-to-multipoint circuits may be a mixed blessing. However, carriers typically want partially meshed networks, he contends, and establishing those under Kompella is greatly complicated by the PE router's need to filter BGP advertisements.

Kompella argues that troubleshooting a VPN under the Kompella draft is simpler than under Martini. Configuring both ends of the links introduces additional operational complexity of the network, especially if a full mesh is required. With a 40-site VPN, Martini would require 1,600 configuration statements while Kompella would require just 40. A separate draft defining single-sided provisioning for Martini would help the matter, reducing the number of statements to 800.

Martini says that the number of such configuration statements is irrelevant because it only affects the router at boot-up time. Even when individuals need to identify specific lines of configuration code, major equipment providers offer tools to zoom in on a specific point in the configuration, leaving the number of statements irrelevant, says Martini.

Finally, Kompella points out that the existing Martini draft inherently limits VPNs to a single Autonomous System (AS) as the draft relies on LDP, an intra-AS protocol. The Kompella-draft's use of BGP for label distribution enables VPNs to work between ASs.

Even so, Martini argues that LDP can be used between ASs, and that according to Nasser El-Aawar, director of engineering at Level 3 and a coauthor of the Martini draft, the company chose LDP because it offers better protection from Internet attacks on L2 infrastructures, and because LDP reduces the long convergence time inherent in BGP. Regardless, networkers should drill service providers about support for such a draft and how they'll interconnect offices that might exist on other providers' networks or in different ASs.

Additional thanks go out to Andrew Malis, chief technologist at Vivace Networks, and Dr. Vijay Srinivasan, senior vice president of technology, and Chandru Sargor, principal engineer, at Cosine Communications for lending their expertise in the formation of this article.

This tutorial, number 171, by David Greenfield, was originally published in the October 2002 issue of Network Magazine.

Fibre Channel SAN Security

Like so many terms in the networking industry, "Storage Area Network" (SAN) boasts a bevy of definitions. Semantics aside, whatever form of SAN you construct must be secure enough to avoid the risk of losing or compromising critical data.

Here, "SAN" refers to a network of devices (typically storage devices and servers) that communicate using a serial SCSI protocol such as Fibre Channel or iSCSI. This article focuses on Fibre Channel-based SAN security.

In their brief history, Fibre Channel SANs have been perceived by many as inherently secure compared to more traditional storage technologies. This is partially due to the fact that SANs are dedicated networks typically devoted to enabling communication between storage devices and computers. This has contributed to the Fibre Channel SAN's image of being less vulnerable to security breaches on the enterprise network. In addition, Fibre Channel SANs are based on optical fiber, which is more resistant to sniffing than copper cabling. And many would argue that traditional Fibre Channel SANs that aren't linked to the Internet are less likely to be compromised than IP-based (read: Internet-connected) networked storage systems.

But as Fibre Channel SANs become larger and more complex, ensuring the security of the data they contain becomes more difficult. The more devices, servers, data, and users become intertwined, the greater the possibility of a security glitch. In addition, there are relatively few standards pertaining to the security of Fibre Channel SANs. (For more information on this topic, see Resources.)

MEMBERS ONLY

The primary problem is the potential for unauthorized access. For example, while a hacker may not have a direct line into a Fibre Channel SAN, he or she might be able to compromise a server that has access to that SAN—perhaps by gaining administrator rights to the system. From this standpoint, the SAN is potentially subject to security problems stemming from servers attached to it. Once the SAN is compromised, an unauthorized user from, for example, the engineering department might obtain access to sensitive information in the form of HR, accounting, or marketing department files that he or she doesn't have rights to view. An even worse scenario could occur if the user were able to obtain rights to alter that data, or if the user decided to confiscate competitive corporate data and sell it to the highest bidder.

A less likely, but certainly possible, development would be a Denial of Service (DoS) attack that bombards the SAN with so many requests that it basically goes out of commission. To accomplish such a feat, the hacker would require fairly sophisticated capabilities in terms of generating detailed device driver code. There simply aren't that many individuals with this level of skill now, but—as with death and taxes—it's just a matter of time.

For these reasons, the SAN must be designed so that only authorized users or systems can access storage resources. Ensuring proper execution of access rights and authorization procedures is critical to protecting the Fibre Channel SAN.

LINES OF DEFENSE

There are many layers of security within a SAN. Firewalls, Intrusion Detection Systems (IDSs), and other basic security mechanisms on the network serve as the first line of defense. An additional layer of security lies in the OS software, which should (among other things) help secure servers attached to the SAN. But these are far from bulletproof; security must be implemented at much deeper, more granular levels within the SAN to make the storage network as airtight as possible.

Although it's far from a panacea, well designed and properly used storage management software can go a long way toward ensuring Fibre Channel SAN security. Unfortunately, the management interface itself can pose a security risk: If an unauthorized user with ill intent were to breach the system, he or she could use the management interface to manipulate a number of variables, and could then reassign storage resources, redefine policies, or otherwise compromise the security and integrity of the data on the SAN.

Because of the amount of damage that could occur if an unauthorized user were to gain control of the management software, access rights to such software should be carefully controlled through mechanisms such as strictly enforced use of user IDs and passwords (preferably encrypted) for login. Using Secure Sockets Layer (SSL) or comparable encryption methods for communication across the network is also helpful.

DIVIDE AND CONQUER

One of the primary ways to help ensure Fibre Channel SAN security is to segment, or partition, storage resources so that only authorized users or departments can view them. You can start by separating sensitive information from less critical data on the SAN.

There are more granular methods of partitioning storage resources as well. Zoning is one way to limit the visibility of storage resources to specific users or departments. Zoning can be implemented in a variety of ways, one of which is switch zoning. This approach is divided into two subcategories: hard zoning and soft zoning. A zone can be relatively small, or it can encompass multiple switches in a SAN fabric. Servers and storage systems can be members of multiple zones, depending on the configuration.

In hard (or port) zoning, zones are defined on the basis of the Fibre Channel switch's ports. These zones typically include components such as servers, storage devices, subsystems, and Host Bus Adapters (HBAs). In this case, the zones are based on the physical port attachment of the devices to the switch. Members of individual zones are only allowed to communicate with other systems within the same zone (see figure).

As the name implies, hard zoning occurs within the switch's hardware (specifically, within its circuitry), which reads the destination address of frames entering the system to determine which output port they should be sent to. The switch contains a table of port addresses that are allowed to communicate with each other. If a port tries to communicate with a port in a different zone, the frames from the nonauthorized port are dropped, and no communication can occur.

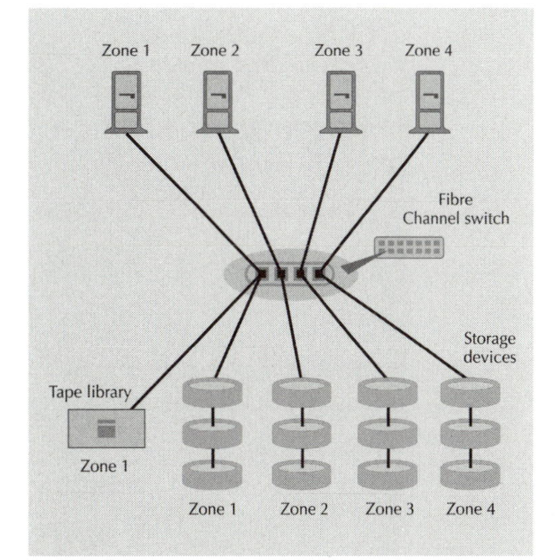

Zoned Out. In Fibre Channel SANs, one way to partition storage resources for additional security is through zoning. Switch zoning involves dividing systems such as servers, storage devices, subsystems, and host bus adapters into groups, or zones. These zones can be based on switch port connectivity (hard zoning), or the switch can read incoming frames to ensure that source and destination addresses are within the same zone (soft zoning).

Because it's based on hardware, hard zoning is more secure than its software-based counterpart. It also doesn't have the performance hit that soft zoning entails (more on this later). However, hard zoning is less flexible than soft zoning. Because the zone assignment remains with the port as opposed to the device, keeping track of configuration changes is more difficult. If a device is moved from one port to another, the network manager or administrator must reconfigure the zone assignment, which can result in a significant amount of overhead. This approach can be particularly cumbersome in dynamic environments in which frequent configuration changes are required.

Soft zoning is based on the use of World Wide Names (WWNs). A WWN is a unique identifier assigned to each Fibre Channel device. In soft zoning, the switch reads incoming frames and ensures that the source and destination addresses (WWNs) have been assigned to the same zone. If these addresses don't correspond, the switch discards the offending frame.

A major benefit of soft zoning is flexibility. If a device needs to be moved from one switch port to another, its current zone membership(s) isn't altered. This can save a lot of administrative time in SANs that have frequent configuration changes involving devices such as servers and storage subsystems.

But while soft zoning is more flexible, it's also less secure than hard zoning. With soft zoning, a hacker could pull off address spoofing by altering frame headers and making his or her way into a switch zone that's off limits. Another downside is that soft zoning can introduce latency and thus impair a Fibre Channel switch's throughput.

IT'S ONLY LOGICAL

Another SAN security mechanism is Logical Unit Number (LUN) masking. This technique accomplishes a similar goal to zoning, but in a different way. To understand this process, you must know that an initiator (typically a server or workstation) begins a transaction with a target (typically a storage device such as a tape or disk array) by generating an I/O command. A logical unit in the SCSI-based target executes the I/O commands. A LUN, then, is a SCSI identifier for the logical unit within a target. In Fibre Channel SANs, LUNs are assigned based on the WWNs of the devices and components. LUNs represent physical storage components, including disks and tape drives.

In LUN masking, LUNs are assigned to host servers; the server can see only the LUNs that have been assigned to it. If multiple servers or departments are accessing a single storage device, LUN masking enables the network manager or administrator to limit the visibility of these servers or departments to a specific LUN (or LUNs) to help ensure security.

LUN masking can be implemented at various locations within the SAN, including storage arrays, bridges and routers, and HBAs.

When LUN masking is implemented in an HBA, software on the server and firmware in the HBA limit the addresses from which commands are accepted. The HBA device driver can be configured to restrict visibility to specific LUNs. One characteristic of this technique is that its boundaries are essentially limited to the server in which the HBA resides.

LUN masking can also be implemented in a RAID subsystem—typically a disk controller(s) that orchestrates the operation of a set of disk drives. In this scenario, the subsystem maintains a table of port addresses via the RAID subsystem controller. This table indicates which addresses are allowed to issue commands to specific LUNs; certain LUNs are masked out so specific storage controllers can't show them. This form of LUN masking extends to the subsystem in which the mapping is executed.

If a RAID doesn't support LUN masking, you can implement this functionality via a bridge or router placed between the servers and the storage devices and sub-systems. In this case, you can configure the system so that only specific servers are allowed to see certain LUNs.

A MOVING TARGET?

The techniques described in this tutorial represent the primary means of securing a Fibre Channel SAN, but there are additional approaches (see Resources). It's important to keep in mind that none of the security measures described here are bulletproof. Whenever possible, multiple techniques should be used to obtain the highest level of security achievable for a SAN given the characteristics of the network, availability requirements, the level of flexibility needed, and so on.

The security of Fibre Channel SANs will continue to evolve out of necessity. Vendors are working on storage processors that encrypt and compress data on storage networks, and storage security appliances that can perform authentication and encryption, and access data across Fibre Channel SANs at wire speed. But regardless of the possibilities such devices might hold, nothing obviates the need to observe the less exciting but time-tested security practices for Fibre Channel SANs.

RESOURCES

For a free white paper titled "SANs Heighten Storage Security Requirements," by James Bannister and Dennis Martin of the Evaluator Group, go to www.evaluatorgroup.com.

Information on Fibre Channel technologies and standards can be found at: The Fibre Channel Industry Association www.fibrechannel.com. Storage Networking Industry Association (SNIA) www.snia.org. Technical Committee T1 www.t11.org/index.htm

This tutorial, number 170, by Elizabeth Clark, was originally published in the September 2002 issue of Network Magazine.

Tape Backup

Installing and testing a tape backup system is one of those tasks that is either a breeze or a nightmare. If you don't run into any surprises, the process will take a few stress-free hours—not including, of course, the time spent carefully planning the system before even ordering the hardware. Installations don't always go smoothly, however. At the first device unavailable message, you might as well call home to say you'll be working late.

At first glance, you wouldn't expect a tape drive installation to be any more fraught with pitfalls than installation of a hard drive. However, because the tape drive is often added to a fully configured system—usually as something of an afterthought—you're more likely to bring any hidden problems to the forefront. Also, given the different nature of tape drives and the fact that you're more likely to disconnect and reconnect them, the devices can strain a SCSI chain that's already marginal.

In this tutorial, we'll look at some of the dangers lurking in the back of the SCSI bus (or channel), and in the process, get an overview of SCSI fundamentals.

THE SCUZZY SPEC

The Small Computer Systems Interface (SCSI, pronounced SKUH-zee) specification started as an attempt to provide a device-independent way to connect multiple devices to a computer, without requiring a separate device driver customized for each type of hard disk or tape device. While the goal of eliminating separate drivers remains elusive, SCSI does promote the use of much more intelligent devices that handle many of the details of data writing and reading and defect management. Thus, the software on the computer doesn't have to include separate logic to handle each different type of drive.

A SCSI bus can address up to eight devices, each of which can be classified as initiator (a device giving commands to other devices) or target (a device carrying out those commands). Most often, the host adapter will take the role of initiator and the attached drives will be the targets. But, some SCSI devices can alternate between being initiators and targets.

Total SCSI cable length should be less than 6 meters, or about 19.7 feet. When measuring cable length, include cables inside the computer and within external devices, such as autoloaders. Cables with different impedance values should not be used in the same chain or the signal may tend to be reflected at the junction. You'll save yourself trouble by using high-quality cables and keeping the total length of the SCSI chain as short as possible—much shorter than the 6 meters specified in the standard.

TERMINATORS

Each end of the SCSI chain should be terminated. Terminating resistors keep the signal from reflecting back into the cable. Some of the hardest-to-find problems spring from improper termination; having too many, too few, or incorrect placement of terminators can cause unreliable operation. Also, bear in mind that improperly terminated SCSI chains may work fine for a time, but then fail when you add another device to the chain or use applications that stress the limits of the SCSI data transfer rate.

A SCSI bus with no termination isn't likely to work at all. A bus that's missing just one terminator will work in most cases, but it won't be as reliable with long cables or in an environment with a lot of electrical noise. The same holds true for a SCSI bus in which the terminators aren't at the very ends of the chain.

Remember that for proper termination, only the devices (drives or host adapter) at the ends of the daisy chain must have terminators installed. Note that many drives come with terminating resistor packs installed. If you use drives in external cabinets, make sure that the drives installed in those cabinets are properly configured. Usually, it's easiest to remove the termination from the drive itself and install a plug-type terminator on the external box.

If the host adapter controls one SCSI channel and both internal and external SCSI devices are installed, the devices at each end of the chain need to be terminated and the host adapter itself should not be terminated. Some SCSI host adapters simplify configuration by including separate internal and external SCSI channels. With these cards, the adapter and the last drive in each chain should be terminated.

The original SCSI specification called for passive terminators—a pair of resistors for each signal line on the SCSI bus. With its faster data transfer rates and tighter timings, the SCSI-2 specification introduced an alternate, more reliable form of termination called active termination. Incorporating one or more voltage regulators, active terminators provide much more consistent termination and are less susceptible to noise. Active terminators will work with SCSI or SCSI-2 channels. So, if you are experiencing hard-to-trace problems with SCSI devices, you may want to try replacing your passive terminators with active ones.

Improper termination isn't always the culprit when SCSI channels exhibit intermittent problems. The power supplies of some external SCSI cabinets can occasionally malfunction, causing unreliable operation of the drive and mimicking termination or SCSI ID problems with the SCSI channel. Other problems can arise from incorrect settings on the drives themselves.

Sometimes, adding a tape drive to a SCSI bus can suddenly cause the entire chain to operate more slowly. If you have a tape unit, CD-ROM, or other device attached to the SCSI bus, you should enable the disconnect feature on those devices. This allows the host-based adapter (HBA) to issue a command or series of commands to that device and then disconnect and return to servicing disk requests. If the disconnect feature is not enabled, the SCSI bus must wait for the device to return with a completion code, which may make the bus unable to fulfill other requests to talk to the drive.

SCSI IDS

Up to eight devices, including the host adapter, can attach to one SCSI channel. The SCSI ID must be different for each device on the same channel. The controller itself usually has ID 7. A bootable hard drive is usually set to ID 0. Other devices on the chain should have IDs that correspond to their priority; 0 has the lowest priority, and 7 has the highest.

This makes intuitive sense in the case of the controller, because it would obviously need the highest priority. However, with bootable hard drives, device priority may be confusing. The BIOS on many controllers will only make a drive with SCSI ID 0 bootable; this seems problematic because the device needing the highest priority is assigned the lowest. But, it turns out that SCSI priority isn't very important in normal operation. Most devices cooperate to keep the bus as free as possible, so arbitration priority is needed only as a last resort during device contention. In addition, arbitration is needed only at the beginning of a transaction between two devices, usually the host adapter and another SCSI device. While the two devices communicate, no other devices can access the SCSI bus.

SCSI ID is very important, even if arbitration priority isn't always a serious issue. Remember that each device, even internal drives, must have a unique ID. The ID also serves as the device address, a three-bit sequence that's included with each command issued to a device. Duplicate device IDs are a very common source of errors, and the problems they cause can be intermittent and hard to track down. This is especially true if you install drive mechanisms into do-it-yourself external SCSI cabinets and have incomplete information about the drive-select jumpers.

Installing a SCSI host adapter with multiple channels may ease the process of selecting unique IDs. Some SCSI cards use different channels, each supporting up to seven devices, for the internal and external connectors. Thus, you wouldn't need to worry about the IDs of internal devices when you install external SCSI devices. Separate channels can also help you segregate fast hard drives from slower devices, such as CD-ROMs and tape drives.

It's a good idea to keep tape drives on a different SCSI channel than your hard drives. This keeps fast hard drives from becoming burdened by slower devices. It's also a much safer configuration for emergency surgery on a tape drive. Although disconnecting a tape drive from the SCSI chain with the server power on is never recommended, in an emergency, you may have to. Having the tape drive segregated from the hard drives will at least ensure that you don't bring the whole file system down when you replace a jammed tape drive. Better yet, get a server that actually supports hot-swapping of SCSI devices.

LAST WORDS

Some 8mm tape drives default to asynchronous transfer mode, while many host adapters default to synchronous transfer mode. If you power down the tape drive while the server is still running and then power it back on, the device will attempt to use its default asynchronous mode without renegotiating the transfer mode with the host controller. The next tape operation is then likely to lock up the workstation or server.

If there's no bootable hard disk on the SCSI bus, disable the SCSI adapter's BIOS and let the software driver control the SCSI devices.

Make sure that you are loading the latest version of the drivers for your host adapter. Tape backup programs will generally require ASPI (Advanced SCSI Programming Interface) drivers. The interface consists of two parts: a low-level, hardware-specific ASPI Manager that accepts ASPI commands and translates them into commands the adapter hardware understands; and an adapter-independent ASPI module. Although the ASPI module is adapter independent, it is generally customized to support a particular type of SCSI device, such as tape drives, CD-ROMs, or hard drives.

FEELING KIND OF SCSI: SCSI'S NEW GENERATION

In the face of increasing speeds offered by hard drives, the SCSI specification has been updated. The SCSI-2 specification included methods for faster and wider data transfer and for more robust termination. The SCSI-3 specification, which is currently being developed, is intended to solidify some of the improvements needed to handle even faster data transfer, such as support for fiber optic interfaces. It will also add support for other types of peripherals.

Here are some of the SCSI-related terms you're likely to hear:

Differential SCSI vs. single-ended SCSI. These terms were defined by the original SCSI specification to support different types of SCSI interface cables. The most commonly encountered type is single-ended SCSI, which has one wire for each signal to be transmitted. Differential SCSI, in contrast, uses a pair of wires for each signal, providing greater immunity to electrical noise and supporting greater cable lengths.

A single SCSI bus supports either single-ended or differential devices; the host determines which type. You can't combine both types on the same chain. Most generic SCSI drives are single-ended SCSI devices.

Synchronous vs. asynchronous transfer. Defined in the original SCSI specification, synchronous transfers allow data rates of up to 5MBps on SCSI devices by allowing devices to transmit without waiting for acknowledgment from the other device—the acknowledgments should eventually catch up, but the transmitting device doesn't have to wait for each one. SCSI-2 expanded this concept to support rates of up to 10MBps. This 10MBps rate is generally called Fast SCSI.

Fast SCSI. By reducing some of the timing margins in the SCSI specification, SCSI-2 devices can transfer data faster than SCSI devices— 3MBps for asynchronous SCSI-2 transfers vs. 1.5MBps for SCSI, and up to 10MBps for SCSI-2 synchronous transfers. Fast SCSI generally refers to SCSI-2 devices capable of performing synchronous transfers faster than 5MBps.

Wide SCSI. The original SCSI specification called for a single cable serving as the backbone of the SCSI bus. This cable supported a data width of eight bits, or one byte. SCSI-2 also allows for transfers using wider data paths of 16 bits or 32 bits. However, this wider path entails the use of an additional 68-pin cable called the B-Cable. Not surprisingly, the original SCSI cable is called the A-Cable. Because wide transfers are automatically negotiated between devices, you can mix devices using different data widths on the same SCSI bus.

Because the B-Cable carries only data signals, Wide SCSI devices conforming to the SCSI-2 standard still use only eight bits for command, status, message, and arbitration lines. SCSI-3 will add a specification for a new type of cable, the P-Cable, that supports 16-bit arbitration, as well as 16-bit data transfers.

This tutorial, number 84, by Dave Fogle, was originally published in the August 1995 issue of LAN Magazine/Network Magazine.

Fault-Tolerant Systems

Network downtime is not only frustrating to users and network administrators, it can become downright expensive. A nonoperational network can cost an organization upwards of $50,000 per hour, depending on the application and marketplace (according to the findings of a downtime cost survey performed by market research firm The Yankee Group). Take the case of a reservation system—an airline or theater-ticket agency. During a two-hour period in which such a system is down, an agency dependent on such a network could lose thousands or—in the case of an airline—millions of dollars in revenue.

As a result, there is a strong demand among end users for products that protect their network systems against loss of data; these include magnetic tape-based backup-and-recording systems, uninterruptible power supplies, and fault-tolerant systems.

Most network managers rely on some form of regular backup procedures, which create archival files, and uninterruptible power sources, which provide battery-supplied electricity that takes over operation of the network automatically when regular electrical service fails. As a result, these are fairly well-understood and implemented technologies.

Fault-tolerant products such as Novell's System Fault Tolerant (SFT) NetWare are not so well understood and, as a consequence, there are relatively few in use. Because they are based on hardware redundancy that provides two identical copies of data and program files—fault-tolerant systems also go by the nickname of mirroring devices.

The term mirroring is apt in one sense: Fault-tolerant products rely on two mass-storage devices—usually, a server or hard disk—that work in tandem to support a mirror image of each other. That is, they contain identical formatting, applications, and data files. The analogy isn't exact, however—a true mirrored image is backward from its original, while the formatting and files on fault-tolerant systems, as one might expect, are identical, not backward.

Still, the mirroring analogy is helpful in describing and understanding the concept of fault-tolerant systems. Fault-tolerant products offer a measure of security that goes beyond the backup-and-recovery process, which provides a static, or time-specific, record of the data stored on a network's hard disk drives.

IDENTICAL DATA STORES

Fault-tolerant systems prevent data loss and network downtime by giving the network operating system real-time, immediate access to two identical and dynamically changing copies of the information stored on the networks. A fault-tolerant system thus relies on hardware redundancy, either with two identical hard disks or servers.

In a fault-tolerant system, the failure of one mirrored component—for example, a hard disk—doesn't bring on a catastrophic collapse of the network: The duplicate, or secondary, device, which is running concurrently with the primary device, merely takes over the operation of the tasks the primary component was handling, and the user isn't aware that his or her network has experienced trouble.

In a disk-mirroring system, a network server contains hard disks in shadowed pairs of primary and secondary drives. When network users store data to the network, the server writes it to both drives, thus creating mirrored images on the separate devices. Should either hard disk fail, network operation can continue uninterrupted, since the NOS automatically makes all reads and writes to and from the remaining hard disk.

Some disk-mirroring packages offer specific options to fine tune a system. Novell's SFT NetWare, for example, lets network administrators duplicate directories and file-allocation tables while providing a read-after-write verification process. And if a NetWare SFT server fails to read a block of data from one mirrored disk, the server automatically looks to the secondary disk to fetch the data. Moreover, SFT NetWare marks that bad area on the disk unusable, then repairs the file by copying the valid data from the secondary disk to a known-usable area on the primary disk.

MIRRORING SERVERS

Similarly, a server-mirroring system, composed of primary and secondary servers, operates with the two servers running in parallel. The primary server handles all network activity while the secondary server operates concurrently— in the background, as it were.

Mirrored servers generally are linked via a special cable and dedicated interface adapters, one of which must be plugged into each server's internal bus. Depending on vendor, these links can be made via RS-232, parallel, or SCSI connections.

Each server continuously monitors the operation of the other, so when the primary server fails, the secondary server automatically takes control of the network. Because the secondary server's hard disks contain mirrored images of those on the primary server, users don't lose data in the exchange. Should the primary server failure be limited to just a hard-disk crash, then the primary server automatically switches disk I/O to the secondary server's mass-storage system.

DISK DUPLEXING

Another form of fault tolerance is disk duplexing, in which two disk controllers rather than one are used within a server. This provides nonstop operation should a disk controller, disk interface or disk power supply fail. Duplexing can improve a system's performance by creating two data channels. If any component within one channel fails, the second channel takes over automatically, again without loss of data.

In this type of system, the server's processor can receive data from whichever disk channel responds first. This often improves performance because the majority of requests across a network are disk reads. In addition, a server with duplexed disk drives can read from one disk drive, write to a second, then, when the process is completed, create a mirror image of the most-recently stored data.

After a disk or server has been taken out of operation and is ready to be put back online, a fault-tolerant system must provide a synchronization process that puts the redundant components back in sync with each other.

COSTLY OVERHEAD?

All fault-tolerant products offer drawbacks, the most obvious of which are costs of the redundant hardware and the software to run the hardware. Another factor is overhead from the executable code that controls the mirroring or duplexing process.

Monetary costs are easy to figure out. You'll pay for two of the most expensive components on a network—a server or hard disk (or, in duplexing, a disk controller)—not just one.

Determining whether such hardware and software costs are justifiable in any particular installation is a much more complex matter. This risk analysis process involves evaluating the loss potential, determining the equipment necessary to reduce the risk, and forming effective management procedures. (Numerous network vendors can provide worksheets and formulas that help in determining whether a particular installation would benefit from the installation of fault-tolerant products.)

Software overhead is a key issue in server and disk mirroring. Not only must the fault-tolerant product provide the code to handle reads and writes to and from redundant disks or servers, it must also be capable of determining when one of the disks or servers has failed, then put the secondary component in charge. This, of course, can consume quite a bit of RAM and CPU cycles on the server and traffic on the wire.

In the final analysis, however, many network managers believe that the security that fault-tolerant products provide is well worth the costs.

This tutorial, number 18, was originally published in the January 1990 issue of LAN Magazine/Network Magazine.

Glossary

1Base5. 1Base5 is the implementation of 1Mbit/sec StarLAN, which is wired in a star topology.

10Base2. 10Base2 is the implementation of the IEEE 802.3 Ethernet standard on thin coaxial cable. Thin Ethernet or thinnet, as it's commonly called, because the cable is half the diameter of 10Base5 Ethernet cable, runs at 10Mbits/sec. Stations are daisy-chained along a terminated bus topology, and the maximum segment length is 185 meters.

10Base5. 10Base5 is the implementation of the IEEE 802.3 Ethernet standard on thick coaxial cable. Thick, or standard Ethernet, as it's commonly called, runs at 10Mbits/sec. It uses a bus topology, and the maximum segment length is 500 meters.

10BaseF. This is the specification for running IEEE 802.3 Ethernet over fiber-optic cable. It specifies a point-to-point link.

10BaseT. 10BaseT is the implementation of the IEEE 802.3 Ethernet standard on unshielded twisted-pair wiring. It uses a star topology, with stations directly connected to a multiport hub. It runs at 10Mbits/sec, and it has a maximum segment length of 100 meters.

100BaseT. The 100Mbit/sec Ethernet standard, 100BaseT, is defined by the IEEE 802.3 committee for two pairs of unshielded twisted pair (100BaseTX), for four pairs of unshielded twisted pair (100BaseT4), and for fiber-optic cable (100BaseFX).

3+. 3+ was 3Com's network operating system that implemented Microsoft MS-Net file sharing and Xerox's XNS transport protocols. 3Com no longer sells 3+.

3+Open. 3+Open was 3Com's network operating system based on Microsoft's OS/2 LAN Manager. 3Com no longer sells 3+Open.

access method. An access method is the set of rules by which the network arbitrates access among the nodes. Carrier Sense Multiple Access with Collision Detection and token passing are two access methods commonly used in LANs.

Address Resolution Protocol (ARP). Within TCP/IP, ARP is the protocol that matches a MAC address to an IP address. Without ARP, packets couldn't take the final step to their destinations on local networks, because Ethernet and Token Ring NICs don't recognize IP addresses.

address. An address is a unique identification code that is assigned to a network device, so it can independently send and receive messages.

Advanced Peer-to-Peer Networking (APPN). APPN is the network architecture within IBM's Systems Application Architecture that provides for peer-to-peer access among computers. Under APPN, a mainframe host is not required. It also implements concepts such as dynamic network directories and routing in SNA. Advanced Peer-to-Peer Networking is an extension of SNA that is based on LU6.2 and that provides additional distributed network control. APPN also automates resource registration and directory lookup. There are two types of APPN devices: EN, or End Nodes, which contain client and server applications, and NN, or Network Nodes, which provide routing and network management services. See LU, SNA.

Advanced Program-to-Program Communications (APPC). APPC is the protocol suite within IBM's Systems Application Architecture that provides peer-to-peer access, enabling PCs and midrange hosts to communicate directly with mainframes. APPC is key for distributed computing within an IBM environment. APPC can be used over an SNA, Token Ring, Ethernet, or X.25 network. The term Advanced Program-to-Program Communication is often used synonymously with LU6.2, in particular to refer to products based on the LU6.2 architecture. See LU.

American National Standards Institute (ANSI). ANSI is the principal group in the United States for defining standards. ANSI represents the U.S. in ISO, the international standards-making body. Fiber Distributed Data Interface, a 100Mbit/sec network, is one network standard developed by ANSI.

analog. An analog signal or representation continuously indicates some value or quantity, while a digital signal or representation indicates a limited number of discrete values, most commonly the binary distinction between on and off or 0 and 1.

APPC See Advanced Program-to-Program Communication, LU.

AppleShare. Apple Computer's network operating system is designed to run primarily with Macintoshes, but also accommodates DOS and Windows PC clients. AppleShare Pro runs under A/UX, Apple's version of Unix, and is a high-performance version of the network operating system.

AppleTalk. AppleTalk is the name of Apple Computer's networking specification. AppleTalk includes such Physical layer specifications as LocalTalk, EtherTalk, and TokenTalk; network and transport functions such as Datagram Delivery Protocol and AppleTalk Session Protocol; addressing such as Name Binding Protocol; file sharing such as AppleShare; and remote access such as AppleTalk Remote Access.

Application layer. The seventh and uppermost layer of the OSI model, the application layer allows users to transfer files, send mail, and perform other functions where they interact with the network components and services. It is the only layer that users communicate with directly, though many Application layer services are provided to programs or processes, and are not intended for direct consumption by people.

application programming interface (API). An API is a set of programming functions, calls, and interfaces that provide access to services, such as messaging, text formatting, or the functions of a particular Network layer. Programmers access the functionality of operating systems, utilities, and other packaged functions through APIs.

APPN See Advanced Peer-to-Peer Networking

ARCnet. Datapoint designed this 2.5Mbit/sec token-passing, star-wired network in the 1970s. Its low cost and high reliability has made it attractive to companies on a tight network budget. ARCnet-Plus is a proprietary product of Datapoint that runs at 20Mbits/sec. TCNS is a 100Mbit/sec version of ARCnet over fiber-optic cabling developed by the Thomas Conrad Corporation.

ASCII (American Standard Code for Information Interchange). The ASCII schema represents 128 characters—the upper and lower case alphabetical characters, 10 numerals, common punctuation marks, and certain printer commands—using the numbers that can be formed with seven binary digits. Other representation codes are in use, such as EBCDIC in the IBM mainframe world, and it is ostensibly the responsibility of layer 6 of the OSI model, the Presentation layer, to handle character code conversions if necessary.

ASN.1 The Abstract Syntax Notation is a formal language—that is to say, a machine-readable or compilable format—defined by CCITT X.208 and ISO 8824. Under both CMIP and SNMP, ASN.1 defines the syntax and format of communication between managed devices and management applications. See CMIP, SNMP.

asynchronous communication server (ACS). An asynchronous communication server is some combination of a computer motherboard, asynchronous modems, and software that enable multiple people to dial out of a LAN. ACSs also provide dial-in service, where users not in the office can use modems to call up their network services in the office. ACSs are also called dial-in/dial-out servers or modem servers.

Asynchronous Transfer Mode (ATM). ATM is a method of data transmission used by Broadband ISDN. It is specified as 53-octet fixed length cells that are transmitted over a cell-switched network. Speeds up to 10 gigabits per second and higher are possible, and it is capable of carrying voice, video, and data. ATM has been embraced by the LAN and WAN industries, who have proclaimed it as the solution to integrating disparate networks across a large geographic distance. It is also called cell relay.

attenuation. Attenuation is the amount of power that is lost as a signal moves over a medium from the transmitter to the receiver. It is measured in decibels (dBs).

backbone. A backbone is the main "spine" or segment of a building or campus network. Departmental networks are attached as "ribs" to the central backbone. The long-haul intercity segments of public carrier networks are also referred to as backbones.

bandwidth on demand. A concept in wide area networking that allows a user or application to agglomerate additional WAN bandwidth as the application warrants. It enables users to pay only for bandwidth that they use, when they use it. Implementing bandwidth on demand requires switched services, such as ISDN or Switched 56 lines.

bandwidth. Bandwidth is the difference between the highest and lowest frequency a channel can conduct, measured in Hz. The bandwidth of a voice-grade telephone line is about 4KHz, while the bandwidth of a broadcast TV channel is 6MHz. The term bandwidth is often used informally to refer to a channel's throughput, which is typically measured in Kbits/sec or Mbits/sec. All other things being equal, a channel with twice the bandwidth of another channel can carry twice as much traffic—that is, it can have twice the throughput. See throughput.

Basic Rate Interface (BRI). BRI is an ISDN service that offers two "bearer" (B) channels with 64Kbits/sec throughput that can be used for bulk data transfer plus a "data link" (D) 16Kbits/sec channel for control and signaling information.

blackout. A blackout or power outage is an interruption or total loss of commercial electrical power.

Uninterruptible power supplies provide battery-backed up power that will supply electricity during a blackout (while their batteries last).

bridge. A bridge connects two networks of the same access method, for example, Ethernet to Ethernet or Token Ring to Token Ring. A bridge works at the OSI's Media Access Control layer, and is transparent to upper-layer devices and protocols. Bridges operate by filtering or forwarding packets according to their destination addresses. Most bridges automatically learn where these addresses are located, and thus are called learning bridges.

Broadband ISDN (B-ISDN). A class of emerging high speed data and voice services for the wide-area network. Switched Multimegabit Data Services and Asynchronous Transfer Mode are two emerging B-ISDN services that were designed to provide megabits and gigabits of bandwidth across a wide area network.

broadcast storm. In a broadcast storm, network congestion occurs when excessive numbers of frames are broadcast.

broadcast. A broadcast message is addressed to all stations on a network.

brouter. A brouter is a device that can transparently bridge protocols as well as route them. It is a hybrid of a bridge and a router.

brownout. A brownout is an abnormally low voltage on commercial power distribution lines. Power utilities may intentionally produce a brownout when there is near overload demand for power, or natural conditions, such as storms, fires, or accidents, may cause a brownout.

bus topology. A bus topology is a network architecture in which all of the nodes are connected to a single linear cable.

campus network. A campus network connects LANs from multiple departments within a single building or campus. Campus networks are typically local area networks; that is, they don't include wire-area network services, though they may span several miles.

campus wiring system. A campus wiring system is the part of a structured wiring system that connects multiple buildings to a centralized main distribution facility, local exchange carrier, or other point of demarcation. It is also referred to as a backbone.

Carrier Sense, Multiple Access with Collision Detection (CSMA/CD). Ethernet and 802.3 LANs use the CSMA/CD access method. In CSMA/CD, each network device waits for a time when the network is not busy before transmitting—they detect transmissions already on the wire from other stations.

cascaded star. A cascaded star topology is a network configuration in which multiple data centers or hubs are constructed for the purposes of redundancy. It is also called a tree topology.

Category 1. The Electronics Industry Association/Telecommunications Industry Association (EIA/TIA) specifies a five-level standard for commercial building telecommunications wiring. Category 1 wiring is old-style unshielded twisted-pair telephone cable, and it is not suitable for data transmission.

Category 2. The EIA/TIA 568 standard certifies Category 2 UTP for use up to 4MHz. Category 2 UTP is similar to the IBM Cabling System Type 3 cable.

Category 3. The EIA/TIA 568 standard specifies Category 3 UTP for speeds up to 10MHz, and it is the minimum-performance cable required for 10BaseT. The wire pairs should have at least three twists per foot, but no two pairs should have the same twist pattern.

Category 4. The EIA/TIA 568 standard specifies Category 4 as the lowest grade UTP acceptable for 16Mbit/sec Token Ring.

Category 5. The EIA/TIA 568 standard specifies that Category 5 is certified up to 100MHz. It is suitable for FDDI over copper, 100BaseT and other high-speed networks.

cell relay. Cell relay is a form of packet transmission used by Broadband ISDN networks. Also called ATM, cell relay transmits 53-octet fixed-length packets over a packet-switched network. ATM is important because it makes it possible to use a single transmission scheme for voice, data, and video traffic on LANs and WANs.

cell. A fixed-length packet. For example, Asynchronous Transfer Mode (ATM) uses 53-octet cells.

CIF With resolution of 352 horizontal pixels by 288 vertical pixels, the Common Intermediate Format is a popular size for video conferencing images. For lower bandwidth applications, video systems often use either QCIF (Quarter CIF), which displays images at 176 pixel by 144 pixel resolution, or SQCIF (Sub-Quarter CIF), which is actually one-ninth of CIF's resolution, at 128 pixels by 96 pixels. High-bandwidth video can be described as 4CIF (704 pixels by 576 pixels) or 16CIF (1,408 pixels by 1,152 pixels). See H.261.

client. A client is a computer that requests network or application services from a server. A client has only one user; a server is shared by many users.

CMIP The Common Management Information Protocol is the network management standard for OSI networks. It has some features that are lacking in SNMP and SNMP-2, and is more complex. CMIP has a far smaller mind share and market share than SNMP in North America, though support for this standard is sometimes mandated, especially in Europe. See MIB, SNMP.

coaxial cable. Coaxial cable has an inner conductor made of a solid wire that is surrounded by insulation and wrapped in metal screen. Its axis of curvature coincides with the inner conductors, hence the name coaxial. Ethernet and ARCnet can use coaxial cable. It is commonly called coax.

common carrier. A common carrier is a licensed, private utility company that provides data and voice communication services for a fee. For example, Sprint and MCI are common carriers.

Common Management Information Protocol (CMIP). CMIP is the OSI management information protocol for network management. It is not as widely implemented as SNMP, the IETF management protocol. See CMIP.

compression. A technique to "squash" files, making them smaller to optimize bandwidth utilization. Compression is important for WAN transmission and disk and tape storage.

concentrator. A concentrator is a multiport repeater or hub that brings together the connections from multiple network nodes. Concentrators have moved past their origins as wire concentration centers, and often include bridging, routing, and management devices.

Connectionless Network Protocol (CLNP). Of the two OSI transport protocols—CLNP and Connection-Oriented Network Service (CONS)—CLNP is more efficient for LANs. Like TCP/IP, it uses datagrams to route network messages by including addressing information in each.

Connection-Oriented Network Service (CONS). Of the two OSI transport protocols—CLNP and CONS—CONS is more efficient for WANs. CONS allows the transport layer to bypass CLNP when a single logical X.25 network is used.

Consultative Committee for International Telegraphy and Telephony (CCITT). The CCITT defines international telecommunications and data communication standards. In March of 1993, the group changed its name to ITU-TS.

Controlled Access Unit (CAU). A CAU is a managed Multistation Access Unit (MAU), or a managed multiport wiring hub for Token Ring networks. Management features include turning ports on and off.

CTI Computer-Telephony Integration relates to the implementation of traditional telephone-based audio (and sometimes video) services over a data network. CTI may be implemented over systems that guarantee bandwidth, such as ATM, or over frame-based networks like Ethernet or frame relay.

Data Access Language (DAL). DAL is Apple's database query language that is based upon SQL, but it provides far greater functionality.

data dictionary. In a distributed database, a data dictionary keeps track of where the data is located and stores the necessary information for determining the best way to retrieve the data.

Data Encryption Standard (DES). DES is the United States government's standard for encryption, in which data is scrambled using security codes called keys, so data cannot be deciphered by unauthorized users.

database server. A database server is a database application that follows the client-server model, dividing an application into a front end and a back end. The front end, running on the user's computer, displays the data and interacts with the user. The back end, running on a server, preserves data integrity and handles most of the processor-intensive work, such as data storage and manipulation.

Data-link layer. The Data-link layer is the second layer of the OSI model. It defines how data is framed and transmitted to and from each network device. It is divided into two sublayers: medium access control and logical link control.

DECnet. Digital Equipment Corporation's network system for networking personal computers and host computers. DECnet can use TCP/IP and OSI, as well as its proprietary protocols.

departmental LAN. A departmental LAN is a network that's used by a small group of people laboring toward a similar goal. Its primary goal is to share local resources, such as applications, data, and printers.

directory services. Directory services provide a white pages-like directory of the users and resources that are located on an enterprise network. Instead of having to know a device or user's specific network address, a directory service provides an English-like listing for a user. The OSI's X.500, Novell's NetWare Directory Services (NDS), Microsoft's Active Directory, and Banyan's StreetTalk are examples of directory services.

distributed computing. In a distributed computing architecture, portions of the applications and the data are broken up and distributed among server and client computers. In the older model, all applications and data resided on the same computer.

distributed database. A database application where there are multiple clients as well as multiple servers. All databases at remote and local sites are treated as if they were one database. The data dictionary is crucial in mapping where the data resides.

Distributed Queue Dual Bus (DQDB). The medium access method of the IEEE 802.6 standard for metropolitan area networks.

downsizing. Downsizing or rightsizing is the process of porting mission-critical applications from a mainframe to a minicomputer or PC LAN or from a minicomputer to a PC LAN.

dual homing. In FDDI, dual homing is a method of cabling concentrators and stations in a tree configuration that permits an alternate path to the FDDI network in case the primary connection fails.

Dual-Attached Station (DAS). In FDDI, a DAS connects to both of the dual, counter-rotating rings. Concentrators, bridges, and routers often use DAS connections for fault tolerance. In contrast, a single-attached station is connected to only one ring.

Dynamic Data Exchange (DDE). DDE is Microsoft's specification for Windows 3.1 and earlier versions that enables applications to communicate without human intervention.

E1. In Europe, E1 is the basic telecommunications carrier, and it operates at 2.048Mbits/sec. In the U.S., the basic carrier is T1, which operates at 1.544Mbits/sec.

EBCDIC Extended Binary Code for Data Interchange is IBM's standard for representing alphanumeric characters in an 8-bit field; it can be compared with the more familiar ASCII encoding scheme. However, EBCDIC (pronounced "eb'-sidik") is different from ASCII, and therefore translation is a common function of SNA gateways and terminal emulators. In EBCDIC, for example, an uppercase *a* is hex C1, a lowercase *a* is 81, and the number 5 is F5; in ASCII, those characters are hex 41, 61, and 35, respectively. See IBM 3270, IBM 5250.

electromagnetic interference/radio frequency interference (EMI/RFI). EMI and RFI are forms of noise on data transmission lines that reduce data integrity. EMI is caused by motors, machines, and other generators of electromagnetic fields. RFI is caused by radio waves.

Electronic Data Interchange (EDI). EDI is a method of electronically exchanging business documents, such as purchase orders, bills of lading, and invoices. Customers and their suppliers can set up EDI networks. EDI can be accomplished through OSI standards or through proprietary products.

electronic mail. E-mail is an application that enables users to send messages and files over their computer networks. E-mail can range from a simple text-based system to a messaging system that accommodates graphics, faxes, forms-processing, workflow, and more.

encapsulation. Encapsulation or tunneling is the process of encasing one protocol into another protocol's format. For example, AppleTalk is often encapsulated into TCP/IP for transmission over a WAN because TCP/IP is more efficient over a WAN.

End System To Intermediate System (ES-IS). ES-IS is an OSI routing protocol that provides the capabilities for hosts (or end systems) and routers (or intermediate systems) to find each other. ES-IS does not handle the router-to-router protocols; the Intermediate System to Intermediate System protocol does.

end system. In Internet terminology, an end system is a host computer.

Enterprise Management Architecture (EMA). EMA once was Digital Equipment Corp.'s umbrella architecture for managing enterprise networks. EMA is a distributed approach.

enterprise network. An enterprise network is one that connects every computer in every location of a company and runs the company's mission-critical applications.

Enterprise RMON A proprietary extension of RMON and RMON-2, Enterprise RMON was developed by NetScout Systems (formerly Frontier Software Development) and is supported by several other vendors, including Cisco Systems. Enterprise RMON's extensions monitor FDDI and switched LANs. See RMON, RMON-2.

Ethernet. Ethernet is a CSMA/CD network that runs over thick coax, thin coax, twisted-pair, and fiber-optic cable. A thick coax Ethernet and a thin coax Ethernet use a bus topology. Twisted-pair Ethernet uses a star topology. A fiber Ethernet is point-to-point. DIX or Blue Book Ethernet is the name of the Digital Equipment Corp., Intel, and Xerox specification; 802.3 is the IEEE's specification; 8802/3 is the ISO's specification.

EtherTalk. EtherTalk is Apple Computer's implementation of Ethernet.

fast packet. Fast packet is a technique for asynchronously transferring data across the network.

fault management. Fault management, one of the five categories of network management defined by ISO, is the detection, isolation, and correction of network faults.

fault tolerance. Fault tolerance is the ability of a system to continue operating in the event of a fault. You can implement fault tolerance in many places in a network, including in file servers with Novell's NetWare SFT III, in disks with RAID, and in bridges with the spanning-tree algorithm.

Fiber Distributed Data Interface (FDDI). FDDI is the ANSI X3T9.5 specification for a 100Mbit/sec network that is logically implemented as dual, counter-rotating rings. A fiber FDDI network can support up to 500 stations over 2 kilometers. FDDI, originally specified to run over fiber, can also operate over shielded and unshielded twisted pair, although the distances are greatly shortened.

fiber-optic cable. Fiber-optic cable can be used to transmit signals in the form of light. Glass fiber is composed of an outer protective sheath, cladding, and the optical fiber. It comes in single mode and multimode varieties. Single-mode fiber is more often used in the public-switched telephone network; multimode fiber is more often used in local area networks. Single-mode fiber uses lasers to transmit the light; multimode uses light-emitting diodes.

File Transfer Protocol (FTP). FTP is the TCP/IP protocol for file transfer.

File Transfer, Access, and Management (FTAM). FTAM is the OSI protocol for transferring and remotely accessing files on other hosts also running FTAM.

filtering. Filtering is the process by which particular source or destination addresses can be prevented from crossing a bridge or router onto another portion of the network.

firewall. A firewall is an impermeable barrier through which certain types of packets cannot pass.

forwarding. Forwarding is the process by which a bridge copies a packet from one segment or ring to another.

fractional T1. In fractional T1, the 1.544Mbit/sec T1 capacity is divided into 64Kbit/sec increments.

Users can order as many channels as they need, but they are not required to purchase the entire 1.544Mbits/sec from the service provider.

fragmentation. Fragmentation is the process in which large frames from one network are broken up into smaller frames compatible with the network to which they'll be forwarded.

frame relay. Frame relay is the ITU-T standard for a low-overhead packet-switching protocol that provides dynamic bandwidth allocation at speeds up to 2Mbits/sec or more. It is considered a second generation X.25 in that it is more efficient.

front-end application. Users present, manipulate, and display data via front-end or client applications. These applications work with back-end applications, such as a mail or database engines.

G.711 One of the major ITU-T codec (coder-decoder) standards for audio (voice and music), G.711 can be incorporated into broader multimedia standards, such as H.320 and H.323, or used on its own for computer telephony. G.711 specifies an audio signal with a 3.4KHz bandwidth (that is, an ordinary analog voice signal) over a 64Kbit/sec data path. Related standards include G.722, which defines a 64kbit/sec 7.0KHz bandwidth audio stream, and the lower-bandwidth G.728 standard, which defines a 16Kbit/sec 3.4KHz audio stream. For computer-based audio over narrow-band phone lines, there's G.723, which supports a compressed 3.4KHz signal over a POTS (Plain Old Telephone System) line and is used in the H.324 multimedia standard. See H.320, H.323, H.324.

gateway. In OSI terminology, a gateway is a device or process that connects two dissimilar systems, such as a PC on a LAN and a mainframe. It operates on all seven layers of the OSI model. In Internet terminology, a gateway is another name for a router.

global network. A global network spans all departments, campuses, branch offices, and subsidiaries of a corporation. Global networks are international, and bring with them the problems of dealing with multiple languages, cultures, standards, and telephone companies.

Government OSI Profile (GOSIP). GOSIP is the U.S. government's specification for OSI conformance. Some level of GOSIP support has been required for all bids made on government projects, though it is unclear that it will be required in the future.

H.233 The ITU-T's data-encryption standard for real-time multimedia, H.233 is supported across a wide range of standard services, including H.320, H.323, and H.324. A related standard is H.234, which specifies how encryption keys are handled. See H.320, H.323, H.324.

H.261 An ITU-T video-compression codec (coder-decoder) standard, H.261, which is supported by H.320, H.323, and H.324, supports CIF and QCIF images. H.261 was designed for use with ISDN and assumes data rates in multiples of 64Kbits/sec. A newer standard, H.263 (supported by H.324), improves H.261's efficiency and adds support for SQCIF-, 4CIF-, and 16CIF-sized images. See CIF, H.320, H.323, H.324.

H.320 One of the major ITU-T standards for real-time multimedia, H.320 is the standard for video conferencing over narrow-band circuit-switched WAN services such as ISDN. This standard includes specifications for T.120-based data, G.711- and G.728-based voice, H.261-based video, and H.233- and H.234-based encryption. The basic H.320 standard has been enhanced in H.323 to include basic packet-switched networks. Related standards for real-time multimedia that are not broken out separately in this glossary are H.321, for broadband ISDN and ATM; H.322, for guaranteed-bandwidth packet-switched networks; and H.310, for higher-resolution multimedia over ATM. See G.711, H.233, H.261, H.323, T.120.

H.323 An extension of the older H.320 standard, the ITU-T's H.323 covers video conferencing not only over narrow-band WAN services, but also on packet-switched networks such as corporate LANs and the Internet. H.323 is based on the Internet Engineering Task Force's (IETF's) Real-Time Protocol (RTP), H.261-based and H.263-based video, and T.120. See H.261, H.320, RTP, T.120.

H.324 The ITU-T's standard for real-time multimedia over standard POTS (Plain Old Telephone System) lines using 28.8Kbit/sec V.34 modems or better, H.324, like H.320, incorporates the T.120 standard for data sharing, H.261-based video compression, and the H.233 and H.234 encryption standards. Unlike H.320, H.324 uses the G.723 audio standard. See G.711, H.233, H.261, H.320, T.120.

Heterogeneous LAN Management (HLM). HLM is an IEEE 802.1 specification for jointly managing mixed Ethernet and Token Ring networks with the same objects.

heterogeneous network. A heterogeneous network is made up of a multitude of workstations, operating systems, and applications of different types from different vendors. For example, a heterogeneous network might contain 3Com Ethernet adapter cards, Dell 486 PCs, Compaq SystemPros, Novell NetWare, FTP TCP/IP, and an HP 9000 Unix host.

High Level Language API (HLLAPI). HLLAPI is a set of tools developed by IBM to help developers write applications that conform to its Systems Application Architecture.

High-level Data Link Control (HDLC). HDLC is an ISO standard for a bit-oriented, link-layer protocol that specifies how data is encapsulated on synchronous networks.

High-Speed Serial Interface (HSSI). HSSI is a standard for a serial link up to 52Mbits/sec in speed over WAN links.

homogeneous network. A homogeneous network is made up of identical or similar components—hardware, operating systems, protocols, databases, and applications.

horizontal wiring subsystem. This part of a structured wiring system connects the users' computers in the departments. It is attached to the vertical wiring system. The horizontal wiring system is often copper cable, such as twisted pair or coax.

hub. A hub or concentrator is a multiport repeater that brings together the connections from multiple network nodes. Modular concentrators have moved beyond their origins as wire concentration centers and often house bridges, routers, and network-management devices. Some hubs are stackable, which usually implies that they have an external backplane that permits multiple units to be connected together without taking up a "repeater hop" or using one of the ports.

IBM 3270 The ubiquitous mainframe terminals and their controllers belong to the 3270 family. Common terminal models are the 3278, which supports color text, and the 3279, which supports graphics. Other members include the modern 3290 and 3187 terminals and the 3287 printer. The 3270 devices operate in block mode, where an entire page of text is transmitted at once, rather than character-by-character as keyed by the user. From the SNA perspective, these are LU2 devices. IBM 3270 terminals are emulated in the TCP/IP environment with the TN3270 service, which is an extension of telnet that is designed to handle ASCII-EBCDIC translation, keyboard mapping, and block-mode operation. See EBCDIC, IBM 5250, LU.

IBM 5250 IBM's LU7 terminals for midrange systems, such as the AS/400, constitute the 5250 family. Like IBM 3270 terminals, these are EBCDIC devices, which transmit information a page at a time. In the TCP/IP environment, the 5250 is emulated by the TN5250 service. See EBCDIC, IBM 3270, LU.

IEEE 802 The main IEEE standard for local area networking (LAN) and metropolitan area networking (MAN), including an overview of networking architecture, approved in 1990. According to the IEEE, the numbering for IEEE's 802-series LAN standards follows a unique pattern. If the number is followed by a capital letter, the designation refers to a stand-alone standard. If it is followed by a lower-case letter, it is a supplement to a standard or part of a multiple-number standard.

IEEE 802.1B Standard for LAN/WAN management, approved in 1992 and, along with 802.1k, became the basis of ISO/IEC 15802-2.

IEEE 802.1D Standard for interconnecting LANs using MAC bridges. Approved in 1990, it became the basis of ISO/IEC 10038.

IEEE 802.1E Standard for LAN and MAN system load protocols. Approved in 1990, it became the basis of ISO/IEC 15802-4.

IEEE 802.1F The standard for defining management information specified in 802, approved in 1993.

IEEE 802.1g A proposed standard for remote MAC bridging.

IEEE 802.1H Recommended practices for MAC bridging of Ethernet 2.0 LANs, approved in 1995.

IEEE 802.1i Standard for using Fiber Distributed Data Interface (FDDI) as a MAC bridge. The standard was approved in 1992 and included in ISO/IEC 10038.

IEEE 802.1j A supplement to 802.1D, this standard covers LAN connectivity using MAC bridges. It was approved in 1996.

IEEE 802.1k Standard for LAN and MAN networks' discovery and dynamic control of event forwarding. It was approved in 1993 and, along with 802.1B, became the basis of ISO/IEC 15802-2.

IEEE 802.1m Conformance statement for 802.1E. It covers the managed-object definitions and protocols for system load protocol. It was approved in 1993 and incorporated into ISO/IEC 15802-4.

IEEE 802.1p Proposed standard for LANs and MANs that deals with expediting traffic and multicast filtering using MAC bridges.

IEEE 802.1Q Proposed standard for virtual bridged LANs.

IEEE 802.2 Standard for logical link control in LAN and MAN connectivity, mainly using bridges. It is the basis of ISO/IEC 8802-2. The current version, approved in 1994, replaced an earlier 802.2 standard that was approved in 1989.

IEEE 802.3 Standard for LAN-based Carrier Sense Multiple Access with Collision Detection (CSMA/CD) access methods and Physical layers, as well as the basis of ISO/IEC 8802-3. This is sometimes referred to as the "Ethernet standard." It was revised in 1996.

IEEE 802.3b Standard for broadband media attachment unit and specifications for 10Broad36. It was approved in 1985 and incorporated into ISO/IEC 8802-3.

IEEE 802.3c Standard for 10Mbit/sec baseband network repeaters. It was approved in 1985 and incorporated into ISO/IEC 8802-3.

IEEE 802.3d Standard for media attachment units and baseband media specifications over fiber-optic repeater links. It was approved in 1987 and incorporated into ISO/IEC 8802-3.

IEEE 802.3e Standard for Physical signaling, media attachment, and baseband media specifications for a 1Mbit/sec network—that is, 1Base5. It was approved in 1987 and incorporated into ISO/IEC 8802-3.

IEEE 802.3h Standard for layer management in Carrier Sense Multiple Access with Collision Detection (CSMA/CD) networks. It was approved in 1990 and incorporated into ISO/IEC 8802-3.

IEEE 802.3i Standard covering two areas: multisegment 10Mbit/sec baseband networks and twisted-pair media for 10BaseT networks. It was approved in 1990 and incorporated into ISO/IEC 8802-3.

IEEE 802.3j Standard for 10Mbit/sec active and passive star-based segments using fiber optics—that is, 10BaseF. It was approved in 1993 and incorporated into ISO/IEC 8802-3.

IEEE 802.3k Standard for layer management for 10Mbit/sec baseband repeaters. Approved in 1992, it was incorporated into ISO/IEC 8802-3.

IEEE 802.3l Conformance statement for the 10BaseT media attachment unit protocol. It was approved in 1992 and incorporated into ISO/IEC 8802-3.

IEEE 802.3p Standard for the 10Mbit/sec baseband media attachment units' layer management. It was approved in 1993 and incorporated into ISO/IEC 8802-3.

IEEE 802.3q Guidelines for the development of managed objects. It was approved in 1993 and incorporated into ISO/IEC 8802-3.

IEEE 802.3r The standard for the Carrier Sense Multiple Access with Collision Detection (CSMA/CD) access method and Physical layer specifications using 10Base5. It was updated in 1996.

IEEE 802.3t Standard for supporting 120-ohm cables in 10BaseT simplex link segments. It was approved in 1995 and incorporated into ISO/IEC 8802-3.

IEEE 802.3u Supplement to 802.3 covering MAC parameters, the Physical layer, and repeaters for 100Mbit/sec operation—that is, 100BaseT, generally known as Fast Ethernet. It was approved in 1995.

IEEE 802.3v Standard for supporting 150-ohm cables in 10BaseT link segments. It was approved in 1995 and incorporated into ISO/IEC 8802-3.

IEEE 802.3w Proposed standard for enhanced MAC algorithms.

IEEE 802.3x Proposed standard for 802.3 full-duplex operation.

IEEE 802.3y Proposed Physical-layer specification for 100Mbit/sec operation on two pairs of Category 3 or better balanced twisted-pair cable—that is, 100BaseT2.

IEEE 802.3z Proposed standard for Physical-layer, repeater, and management parameters for 1,000Mbit/sec operation; often referred to as "Gigabit Ethernet."

IEEE 802.4 Standard for token-passing bus access methods and Physical-layer specifications. It was approved in 1990.

IEEE 802.5 Standard for token-ring access methods and Physical-layer specifications—that is, common Token Ring architecture. It became the basis of ISO/IEC 8802-5. The current version was approved in 1995.

IEEE 802.6 The family of standards for a LAN's Distributed-Queue Dual-Bus (DQDB) subnetwork. It was approved in 1990.

IEEE 802.9 Standard for Integrated Services LAN (ISLAN), designed to connect 802.x LANs to publicly and privately administered backbone networks such as FDDI or ISDN. It was approved in 1994 and is the basis of ISO/IEC 8802-9.

IEEE 802.10 Standard for Interoperable LAN Security, also known as SILS. It was approved in 1992.

IEEE 802.11 Standard for wireless LAN MAC and Physical-layer specifications. Current drafts focus on the 2.4GHz band.

IEEE 802.12 Standard for 100Mbit/sec demand-priority access method Physical-layer and repeater specifications, also known as 100VG-AnyLAN. It was approved in 1995.

IEEE Institute for Electrical and Electronics Engineers; for information, browse www.ieee.org . Many IEEE-approved networking standards became the basis of the international networking standards specified by the International Organization for Standardization (ISO) and the International Electrotechnical Commission (IEC). See ISO/IEC.

impedance. Impedance is the resistance equivalent for AC, and it affects a network's propagation delay and attenuation. Each protocol and topology has its own impedance standards. For example, 10BaseT UTP cable has an impedance of 100 ohms to 105 ohms, while 10Base2 coaxial cable has an impedance of 50 ohms.

IND$FILE This program runs on mainframes to support file transfers between the mainframe and LU2 terminal devices, such as 3270 terminals or PCs running TN3270. IND$FILE is analogous to ftp under TCP/IP. TN3270 and TN5250 emulators often use IND$FILE to transfer files between PCs and mainframe and midrange computers. See IBM 3270.

infrared. Infrared electromagnetic waves have frequencies higher than microwaves but lower than the visible spectrum. Infrared transmission is used for wireless LANs, as well as for point-to-point communications with portable devices.

Infrared may also be used as a wireless medium, and has greatest applicability for mobile applications due to its low cost. Infrared allows for higher throughput—measured in megabits per second—than spread spectrum, but it offers more limited distances. Infrared beams cannot pass through walls.

Institute of Electronics and Electrical Engineers (IEEE). The IEEE is a professional society of electrical engineers. One of its functions is to coordinate, develop, and publish data communications standards for use in the United States. See IEEE.

Integrated Services Digital Network (ISDN). ISDN is the ITU standard for carrying voice and data to the same destination. Although ISDN has not been popular in the United States, it is commonly available in Europe (especially in the U.K., Germany, and France) and in Japan.

intermediate system. In OSI terminology, an intermediate system is a router.

Intermediate System-to-Intermediate System (IS-IS). IS-IS is an OSI routing protocol that provides dynamic routing between routers or intermediate systems.

International Organization for Standardization (ISO). ISO is a multinational standards-setting organization that formulates computer and communication standards, among others. ISO defined the OSI reference model, which divides computer communications into seven layers: Physical, Data-link, Network, Transport, Session, Presentation, and Application.

Internet Activities Board (IAB). The IAB is the coordinating committee for the design, engineering, and management of the Internet. The IAB has two main committees: the Internet Engineering Task Force (IETF) and the Internet Research Task Force (IRTF). The IETF specifies protocols and recommends Internet standards. The IRTF researches technologies and refers them to the IETF.

Internet Protocol (IP). IP is part of the TCP/IP suite. It is a network-layer protocol that governs packet forwarding from network to network.

Internet. The Internet is a collection of packet-switched networks all linked using the TCP/IP protocol.

Internetwork Packet Exchange (IPX). IPX is the part of Novell's NetWare stack that governs packet forwarding. This network protocol is based on the Xerox Network System (XNS).

internetwork. An internetwork is a collection of several networks that are connected by bridges, switches, or routers, so all users and devices can communicate, regardless of the network segment to which they are attached.

interoperability. Interoperability is the ability of one manufacturer's computer equipment to operate alongside, communicate with, and exchange information with another vendor's dissimilar computer equipment or software.

inverted backbone. An inverted backbone is a network architecture in which the wiring hub and routers become the center of the network, and all subnetworks connect to this hub. In a backbone network, the cable is the main venue of the network, to which many bridges and routers attach.

ISO/IEC 10038 Standard for interconnecting LANs using MAC bridges. Based on IEEE 802.1D and incorporating 802.1i and 802.1m, it was approved in 1993.

ISO/IEC 11802-4 Technical report, not a standard, based on IEEE 802.5j. It covers Token Ring access methods using fiber optic stations. The report was issued in June 1994.

ISO/IEC 15802-2 Common specifications for LAN and MAN management. Based on IEEE 802.1B and 802.1k, it was approved in 1995.

ISO/IEC 15802-4 Standard for LAN and MAN system load protocols. Based on IEEE 802.1E and incorporating 802.1m, it was approved in 1994.

ISO/IEC 8802-2 Standard for logical link control in LAN and MAN connectivity, mainly using bridges. This is based on IEEE 802.2 (1994 edition) and incorporates 802.2a, 802.2b, 802.2d, 802.2e, and 802.5p. This standard replaced the 1989 versions of both standards and was approved in 1994.

ISO/IEC 8802-3 Standard for LAN CSMA/CD access methods and Physical layers. It is based on IEEE 802.3 and incorporates 802.3b, 802.3c, 802.3d, 802.3e, 802.3h, 802.3i, 802.3j, 802.3k, 802.3l,

802.3m, 802.3n, 802.3p, 802.3q, 802.3s, 802.3t, and 802.3v. Approved in 1996, it replaced the 1993 version of the standard.

ISO/IEC 8802-5 Standard for Token Ring access methods and Physical-layer specifications—that is, common Token Ring architecture. It is based on IEEE 802.5, incorporating 802.5b, and was approved in 1995, replacing a 1992 version.

ISO/IEC 8802-9 Standard for LAN interfaces at the MAC and Physical layers. Based on IEEE 802.9, it was approved in 1996.

ISO/IEC ISO is the International Organization for Standardization (www.iso.ch). IEC is the International Electrotechnical Commission (www.iec.ch). The organizations have a joint committee, called JTC1, which has created international networking standards largely based on IEEE's approved standards.

isochronous transmission. An isochronous service transmits asynchronous data over a synchronous data link. An isochronous service must be able to deliver bandwidth at specific, regular intervals. It is required when time-dependent data, such as video or voice, is to be transmitted. For example, Asynchronous Transfer Mode can provide isochronous service.

jitter. Jitter is a kind of distortion of digital signals that takes the form of phase shifts over a transmission medium. Jitter can be thought of as the standard deviation of latency. If latency is constant, then no additional buffering will be necessary for voice or video streams after the initial startup. If jitter is present, buffers must be sized to accommodate the greatest delay, and the initial startup must compensate for the longest latency. See latency.

LAN Manager. LAN Manager was Microsoft's early model of a network operating system based on OS/2. It used NetBEUI or TCP/IP network protocols. LAN Manager supports DOS, Windows, OS/2, and Macintosh clients. Through LAN Manager for Unix, it offered connections to various Unix hosts.

LAN Server. LAN Server is IBM's network operating system that is based on the OS/2 operating system and the NetBIOS network protocol. LAN Server supports DOS, Windows, OS/2, and Macintosh clients.

LANtastic. LANtastic is Artisoft's peer-to-peer, NetBIOS-based network operating system. It supports DOS, Windows, OS/2, Macintosh, and Unix clients.

latency. Signal delay, especially that resulting from transmission and processing.

leased line. A leased line is a transmission line reserved by a communications carrier for the private use of a customer. Examples of leased-line services are 56Kbit/sec or T1 lines.

line of sight. Laser, microwave, and some infrared transmission systems require that no visual obstructions exist in the path between the transmitter and receiver. This direct path is called the line of sight.

local area network (LAN). A LAN is a group of computers, each equipped with the appropriate network adapter and software and connected by cable (or wireless links), that share applications, data, and peripherals. It typically spans a single building or campus.

Local Area Transport (LAT). LAT is Digital Equipment's protocol suite for connecting terminals to an Ethernet network. Because LAT lacks a Network layer, it must be bridged in an enterprise network, not routed.

LocalTalk. LocalTalk is one of Apple's Physical-layer standards. It transmits data at 230Kbits/sec using Carrier Sense Multiple Action with Collision Avoidance (CSMA/CA) over unshielded twisted-pair wire.

Logical Link Control (LLC). OSI Layer 2, the Data-link layer, is divided into the Logical Link Control and the Media Access Control sublayers. LLC, which is the upper portion, handles error control, flow control and framing of the transmission between two stations. The most widely implemented LLC protocol is the IEEE 802.2 standard.

Logical Unit (LU). IBM's LU suite of protocols govern session communication in an SNA network. LU1, LU2, and LU3 provide control of host sessions. LU4 supports host-to-device and peer-to-peer communication between peripheral nodes. LU6.2 is the peer-to-peer protocol of APPC. LU7 is similar to LU2.

LU A Logical Unit is IBM's term for a communications session between physical devices such as computers, printers, or terminals. Some physical devices can simultaneously run multiple LUs. The various LUx session types define the communications protocols used to attach various logical units to an SNA network. LU1 defines communication with text printers. LU2 communicates with 3270-family terminals. LU3 communicates with 3270-family printers. LU4 and LU5 are no longer in use. LU6 supports communications between application programs. LU6.1 assumes that at least one physical device is a mainframe running IBM's CICS or IMS (Information Management System) software, and LU6.2 defines peer-to-peer communication between application software. LU7 defines communication with 5250-family terminals. See IBM 3270, IBM 5250, LU6.2.

LU6.2 LU6.2 is a type of logical unit that supports communication between application programs in an SNA-based network. LU6.2-compliant devices operate as peers within the network and can perform multiple simultaneous transactions over the network. LU6.2 devices can also detect and correct errors. The LU6.2 definition provides a common API for communicating with and controlling compliant devices. APPC is often used to refer to the LU6.2 architecture or to specific LU6.2 features. See APPC, LU.

mail-enabled applications. Mail-enabled applications are a class of software that incorporates e-mail's functionality, but provides additional services, such as workflow automation, intelligent mail handling, or contact management software.

main distribution facility. In a structured wiring system, the main distribution facility is the portion of the wiring that's located in the computer room. From the main distribution facility extends the campus wiring subsystem, which runs to each building.

management information base (MIB). A MIB is a repository of the characteristics and parameters that are managed in a device. Simple Network Management Protocol (SNMP) and Common Management Information Protocol (CMIP) use MIBs to identify the attributes of their managed systems.

Manufacturing Automation Protocol (MAP). MAP is an ISO protocol for communicating among different pieces of manufacturing equipment.

Media Access Control (MAC). The MAC is the lower sublayer of the Data-link layer (Logical Link Control is the upper sublayer), and it governs access to the transmission media.

mesh topology. In a mesh network topology, any site can communicate directly with any other site.

Message Handling Service (MHS). MHS is another name for ISO's X.400 protocols for store-and-forward messaging.

Message Handling System (MHS). MHS is Novell's protocol for electronic mail and other message management, storage, and exchange.

Message Transfer Agent (MTA). In ISO's X.400 electronic messaging protocols, the MTA is responsible for storing messages, then forwarding them to their destinations. The MTA is commonly implemented as the mail server.

Messaging API (MAPI). Using Microsoft's MAPI, application developers can add messaging to any Windows application, and the program can gain access to the message storage, transport, and directory services of any MAPI server.

metropolitan area network (MAN). A MAN covers a limited geographic region, such as a city. The IEEE specifies a MAN standard, 802.6, which uses the Dual Queue, Dual Bus access method and transmits data at high speeds over distances up to 80 kilometers.

MIB In general, a Management Information Base is a schema, or structure, for a repository of characteristics and parameters managed in a network device such as a NIC, hub, switch, or router. Each managed device knows how to respond to standard queries issued by network management protocols. To be compatible with CMIP, SNMP, SNMP-2, RMON, or RMON-2, devices gather statistics and respond to queries in the manner specified by those specific standards. Many managed devices also have "private" MIB extensions. These extensions make it possible to report additional information to a particular vendor's proprietary management software or to other management software that's aware of the extensions. See CMIP, MIB-2, RMON, RMON-2, SNMP.

MIB-2 The expression MIB refers to the original SNMP MIB definition in IETF RFC 1157. The broader MIB-2 (RFC 1213) adds to the number of monitoring objects supported and is included in SNMP-2's MIB. However, SNMP-2's MIB (RFC 1907) is a superset of MIB-2.

mission-critical application. A mission-critical application is one that is crucial to a company's continued operation. As corporations downsize from mainframes, many mission-critical applications are moved to networks.

MMX According to Intel, the acronym "MMX" has no particular meaning, but it's generally inferred to mean "Multimedia Extensions." Specifically, MMX is implemented as a set of new microprocessor instructions for Intel's MMX-enhanced Pentium CPUs and its new Pentium II CPUs. MMX is not specific just to Intel, however, because Advanced Micro Devices, a competing chip manufacturer based in Sunnyvale, CA, has MMX-enabled its new K6 processor family. Many new Windows-based multimedia products for computer telephony or video conferencing are written to take advantage of the new MMX instructions.

multicast. Multicast packets are single packets that are copied to a specific subset of network addresses. In contrast, broadcast packets are sent to all stations in a network.

multimedia. Multimedia is the incorporation of graphics, text, and sound into a single application.

multimode fiber. Multimode fiber-optic cable uses light-emitting diodes (LEDs) to generate the light to transmit signals. Multimode fiber is prevalent in data transmission.

multiplexing. Multiplexing is putting multiple signals on a single channel.

Multipurpose Internet Mail Extensions (MIME). MIME is an Internet specification for sending multiple part and multimedia messages. With a MIME-enabled e-mail application, users can send PostScript images, binary files, audio messages, and digital video over the Internet.

multistation access unit (MAU). A MAU is a multiport wiring hub for Token Ring networks. IBM calls MAUs that can be managed remotely Controlled Access Units, or CAUs.

Narrowband ISDN. Narrowband ISDN is another name for ISDN. Narrowband ISDN offers a smaller bandwidth than the Broadband ISDN services, such as Asynchronous Transfer Mode (ATM) and Switched Multimegabit Data Services (SMDS).

NetBEUI (NetBIOS Extended User Interface). Microsoft's extended version of NetBIOS is called NetBEUI. It is a protocol that governs data exchange and network access. Because NetBEUI does not provide a Network layer, it cannot be routed in a network, which makes building large internetworks of NetBEUI-based networks problematic.

NetBIOS (Network Basic Input Output System). NetBIOS is a networking API developed by IBM that allows programs to access the network and exchange data. Because NetBIOS does not provide Network-layer services, it cannot be routed in a network, which makes building large internetworks of NetBIOS-based networks problematic. Examples of NetBIOS-based NOSs include IBM LAN Server and Artisoft LANtastic.

NetWare Loadable Module (NLM). An NLM is an application that runs on the NetWare server and coexists with the core NetWare operating system. NLMs provide better performance than applications that run outside the core.

NetWare. NetWare is Novell's network operating system. NetWare uses IPX/SPX, NetBIOS, or TCP/IP network protocols. It supports DOS, Windows, OS/2, Macintosh, and Unix clients. NetWare versions 4.x and 3.x are 32-bit operating systems; NetWare 2.2 is a 16-bit operating system.

Network Driver Interface Specification (NDIS). NDIS is a specification, developed by Microsoft and 3Com, for generic device drivers for adapter cards used by LAN Manager and subsequent Microsoft network operating systems.

Network File System (NFS). NFS is Sun Microsystems' file-sharing protocol that works over TCP/IP.

network interface card (NIC). A network interface card is the adapter card that plugs into computers and includes the electronics and software so the station can communicate over the network.

Network layer. The third layer of the OSI model is the Network layer, and it governs data routing. Examples of Network-layer protocols are IP and IPX.

network operating system (NOS). A network operating system is the software that runs on a file server that governs access to the files and resources of the network by multiple users. Examples of NOSs include Banyan's VINES, Novell's NetWare, and IBM's LAN Server.

network. A network is a system of computers, hardware, and software that is connected, and over which data, files, and messages can be transmitted and end users communicate. Networks may be local or wide area.

network-aware application. A network-aware application knows that it is running on a network and has file- and record-locking features.

network-ignorant application. A network-ignorant application has no knowledge that it is running on a network. It lacks file and record locking, and cannot guarantee data integrity in a multiuser environment.

network-intrinsic application. A network-intrinsic application knows it is running on a network and takes advantage of a network's distributed intelligence. For example, a client-server database is a LAN-intrinsic application.

noise. Noise is sporadic, irregular, or multifrequency electrical signals that are superimposed on a desired signal.

Object In the context of network management, an object is a numeric value that represents some aspect of a managed device. An object identifier is a sequence of numbers separated by periods, which uniquely defines the object within a MIB. See MIB.

Object Linking and Embedding (OLE). OLE is Microsoft's specification for application-to-application exchange and communication. It is more powerful and easier to use than Microsoft's older Dynamic Data Exchange (DDE) API.

octet. An eight-bit byte. Internet RFCs refer to octets rather than bytes, presumably because some mainframes and minicomputers have 16-bit or 32-bit bytes.

Open Data-Link Interface (ODI). ODI is Novell's specification for generic network interface card device drivers. ODI enables you to simultaneously load multiple protocol stacks, such as IPX and IP.

Open Shortest Path First (OSPF). The OSPF routing protocol for TCP/IP routers takes into account network loading and bandwidth when moving packets from their sources to their destinations. OSPF improves on the Routing Information Protocol (RIP), but it is not as widely implemented.

Open Systems Interconnection (OSI). The OSI model is the seven-layer, modular protocol stack defined by ISO for data communications between computers. Its layers are: Physical, Data-link, Network, Transport, Session, Presentation, and Application.

open systems. In open systems, no single manufacturer controls the specifications for the architecture. The specifications are in the public domain, and developers can legally write to them. Open systems are crucial for interoperability.

optical drives. Optical drives use lasers to read and write information from their surface. Because of their slow access times, optical drives are used for archiving and other activities that are not particularly time-sensitive. Several types of optical drives are available. CD-ROMs, or compact disk read-only memory, can be remastered. Information can be written to WORM, or write once, read many, disks only once; they cannot be erased. Data can be written to and removed from erasable optical disks.

OS/2. OS/2 is IBM's 32-bit multithreaded, multitasking, single-user operating system that can run applications created for it, DOS, and Windows.

outsourcing. Outsourcing is the process of subcontracting network operations and support to an organization outside your company.

packet switching. In packet switching, data is segmented into packets and sent across a circuit shared by multiple subscribers. As the packet travels over the network, switches read the address and route the packet to its proper destination. X.25 and frame relay are examples of packet-switching services.

packet. A packet is a collection of bits that includes data and control information, which is sent from one node to another.

peer-to-peer. In a peer-to-peer architecture, two or more nodes can directly initiate communication with each other; they do not need an intermediary. A device can be both the client and the server.

personal communications services (PCS). PCS is a marketing category of applications that includes wireless local and personal area communications for portable and desktop computers, wireless notepad and messaging devices, and wireless office and home telephone systems. The FCC is in the process of allotting both licensed and unlicensed frequency ranges for PCS-based devices.

Physical layer. The lowest layer of the OSI model is the Physical layer, and it defines the signaling and interface used for transmission media.

Point-to-Point Protocol (PPP). PPP provides router-to-router and host-to-network connections over asynchronous and synchronous connections. It is considered a second-generation Serial Line Internet Protocol (SLIP).

point-to-point. A point-to-point link is a direct connection between two locations.

Presentation layer. The sixth, or Presentation layer, of the OSI model is responsible for data encoding and conversion.

Primary Rate Interface (PRI). PRI ISDN is a T1 service that supports 23 64Kbit/sec B channels plus one 64Kbit/sec D channel.

propagation delay. Propagation delay is the time it takes for a bit to travel across the network from its transmission point to its destination.

protocol. A protocol is a standardized set of rules that specify how a conversation is to take place, including the format, timing, sequencing and/or error checking.

proxy agent. A proxy agent is software that translates between an agent and a device that uses a different management information protocol. The proxy agent communicates the data to the network manager.

public data network (PDN). A PDN is a network operated by a government or service provider that offers wide area services for a fee. Examples are networks from British Telecom and Infonet.

QOS Quality of Service is a catchphrase for a network that can transport data without losing cells, with predictable end-to-end delay, and with real-time delivery of data once the connection is completed. High-quality multimedia over a network, whether in real-time or merely playing audio or video files from a server, requires a network that can deliver QOS. Protocols such as ATM (Asynchronous Transfer Mode) are designed to deliver multiple levels of QOS. Attempting to deliver QOS using IP requires additional services such as RSVP (Resource Reservation Protocol), which allows bandwidth to be reserved and to be supported on intermediate devices such as routers.

query language. A query language enables users to retrieve information. Structured Query Language (SQL) is a standardized, vendor-independent query language, though most database vendors have proprietary extensions to SQL.

Redundant Array of Inexpensive Disks (RAID). RAID 1 is disk mirroring, in which all data is written to two drives. In RAID 2, bit-interleaved data is written across multiple disks; additional disks perform error detection. A RAID 3 disk drive has one parity drive plus an even number of data drives. Data is transferred one byte at a time, and reads and writes are performed in parallel. Like RAID 3, RAID 4 has a dedicated parity drive, but the data is written to the disks one sector at a time. Also reads and writes occur independently. In RAID 5, the controllers write data a segment at a time and interleave parity among them. (A segment is a selectable number of blocks.) RAID 5 does not use a dedicated parity desk. It offers good read performance, but suffers a write penalty. RAID 1, 3, and 5 are appropriate for networks.

Remote Monitor (RMON). The RMON MIB defines the standard network monitoring functions for communication between SNMP-based management consoles and remote monitors, which are often called probes. RMON extends SNMP by looking at traffic between devices instead of at individual devices. It also facilitates local capture of statistics, history, and even traffic, so that polling activity by the management console can be minimized.

Remote Procedure Call (RPC). An RPC is part of an application that activates a process on another node on the network and retrieves the results.

repeater. A repeater is a Physical layer device that regenerates, retimes, and sometimes amplifies electrical or optical signals.

Request For Comment (RFC). An RFC is the Internet's notation for draft, experimental, and final standards.

request for information (RFI). An end-user company issues an RFI document to ask systems integrators and manufacturers to propose and design a system that will fulfill the corporation's business requirements.

request for proposal (RFP). An end-user company issues an RFP document that asks systems integrators and manufacturers to bid on their network designs and specifications.

requirements analysis. A requirements analysis is the process through which you define and evaluate the business needs of your network system.

return on investment (ROI). Calculating the ROI enables MIS shops to gauge the network's success from a business profit-and-loss standpoint. The savings or benefits of networking projects ought to represent a return on invested capital as good or better than that of the business as a whole.

ring topology. In a ring topology, packets travel in a closed loop. Packets pass sequentially between active stations, and each station examines them and copies any that are intended for it. The packets finally return to the originating station, which removes them from the network.

risk analysis. A risk analysis is the process by which a company analyzes the business and technology risks of installing a new system.

RJ-11. An RJ-11 connector is a four-wire modular connector that is used by the telephone system.

RJ-45. An RJ-45 is an eight-wire modular connector that is used by telephone systems. The eight-pin modular connectors used for 10BaseT UTP cable resemble RJ-45 connectors, but they have substantially different electrical properties.

RMON Groups The original IETF proposed standard for the RMON MIB, RFC 1271 defines nine Ethernet groups: Ethernet Statistics, Ethernet History, Alarms, Hosts, Host Top N ("N" indicates that it collects information on a number of devices), Traffic Matrix, Filters, Packet Capture, and Events. RFC 1513 extends this standard to support Token Ring. See MIB, RMON, RMON Token Ring.

RMON Probe Sometimes called an RMON agent, an RMON probe is either firmware built into a specific network device like a router or switch, or a specific device built for network monitoring and inserted into a network segment. An RMON probe tracks and analyzes traffic and gathers statistics, which are then sent back to the monitoring software. Historically, an RMON probe was a separate piece of hardware, but now RMON firmware is embedded in high-end switches and routers. See RMON.

RMON Remote Monitoring, or RMON, is a set of SNMP-based MIBs that define the instrumenting, monitoring, and diagnosing of local area networks at the OSI Data-link layer. In IETF RFC 1271, the original RMON, which is sometimes referred to as RMON-1, defines nine groups of Ethernet diagnostics. A tenth group, for Token Ring, was added later in RFC 1513. RMON uses SNMP to transport data. To be RMON-compliant, a vendor need implement only one of the nine RMON groups. See MIB, RMON-2, RMON Probe, RMON Token Ring, SNMP.

RMON Token Ring IETF proposed standard RFC 1513 is an extension to the original RMON MIB (RFC 1271), with support for Token Ring. Some sources refer to this standard as RMON TR, but it's generally considered a replacement for the older standard. In RFC 1513, the RFC 1271 Statistics and History monitoring groups have additional specifications for Token Ring, and a tenth group is added to monitor ring configuration and source routing. In 1994, the proposed standard became a draft standard under the designation RFC 1757; many vendors use the RFC 1513 and 1757 numbers interchangeably. See MIB, RMON.

RMON-2 The second Remote Monitoring MIB standard, called RMON-2, defines network monitoring above the Data-link layer. It provides information and gathers statistics at the OSI Network layer and Application layer. Unlike the original RMON, RMON-2 can see across segments and through routers, and it maps network addresses (such as IP) onto MAC addresses. RMON-2 is currently a proposed standard under IETF RFC 2021. To be compliant with RMON-2, a vendor must implement all the monitoring functions for at least one protocol. Note that RMON-2 does not include MAC-level monitoring, and thus it is not a replacement for the original RMON. See RMON.

roll back. A database application's ability to abort a transaction before it has been committed is called a roll back.

roll forward. A database's ability to recover from disasters is called a roll forward. The database reads the transaction log and re-executes all of the readable and complete transactions.

router. A router is a Network-layer device that connects networks that use the same Network-layer protocol, for example TCP/IP or IPX. A router uses a standardized protocol, such as RIP, to move packets efficiently to their destination over an internetwork. A router provides greater control over paths and greater security than a bridge; however, it is more difficult to set up and maintain.

Routing Information Protocol (RIP). RIP is the routing protocol used by most TCP/IP routers. It is a distance-vector routing protocol, and it calculates the shortest distance between the source and destination addresses based on the lowest "hop" count.

RTP The Real-Time Transport Protocol is the IETF's standard for transporting real-time data, such as voice or video, over a packet-based network that doesn't guarantee Quality of Service (QOS). A related standard is RTCP, or the Real-Time Transport Control Protocol, which provides feedback between two units (point-to-point) or a larger group (known as multicast or multipoint). The ITU-T's non-QOS multimedia standards such as H.323 and H.324 are based on RTP/RTCP. See H.323, H.324, QOS.

sag. A sag is a short-term drop (up to 30 seconds) in power-line voltage that typically is in the region of 70 percent to 90 percent of the nominal line voltage.

SDLC In SNA networks, communication links in the Data-Link Control layer use the Synchronous Data-Link Control protocol, which provides half-duplex or full-duplex operation, error detection and recovery, multipoint addressing, and flow control. SDLC is a subset of the ISO-standard HDLC, or High-Level Data-Link Control, protocol. See SNA.

Sequential Packet Exchange (SPX). SPX is Novell's transport protocol, which supports end-to-end connections for IPX networks.

Serial Line Internet Protocol (SLIP). SLIP is used to run IP over serial lines, such as telephone lines.

server. A server is a computer that provides shared resources to network users. A server typically has greater CPU power, number of CPUs, memory, cache, disk storage, and power supplies than a computer that is used as a single-user workstation.

Session layer. The fifth OSI layer, the Session layer, defines the protocols governing online communication between applications.

session. A session is an end-to-end, online communications connection between two nodes.

shielded twisted pair (STP). STP is a pair of foil-encased copper wires that are twisted around each other and wrapped in a flexible metallic sheath to improves the cable's resistance to electromagnetic interference.

Simple Mail Transfer Protocol (SMTP). SMTP is TCP/IP's protocol for exchanging electronic mail.

Simple Network Management Protocol (SNMP). SNMP is a request-response type protocol that gathers management information from network devices. SNMP is a de facto standard protocol for network management. It provides a means to monitor traffic and to set configuration parameters.

single-attachment station (SAS). In FDDI, a single-attachment station is one that is connected to only one of the dual counter-rotating rings. Workstations and other noncritical devices are normally connected using SAS, which is less expensive than dual-attached stations.

single-mode fiber. Single-mode fiber uses lasers, not light-emitting diodes, to transmit signals over the cable. Because single-mode fiber can transmit signals over great distances, it is primarily used in the telephone network, and not for LANs.

SNA mainframe gateways. An SNA mainframe gateway is hardware and software that connects a LAN to an SNA mainframe. It translates between the different systems, making the PC look like a 3270 terminal to the SNA host, so the PC user can access mainframe applications, files, and printers.

SNA Node Types The four most common SNA node types are Type 2.0, Type 2.1, Type 4, and Type 5. Type 2.0 nodes provide end-user services but do not provide intermediate routing services. Type 2.1 nodes can provide both end-user access or intermediate routing; these are also called LEN, or

Low-Entry Networking, nodes. Type 4 nodes provide remote routing and Data-Link Control functions for a Type 5 node. Type 5 nodes provide SSCP (System Service Control Point) management functions and can control Type 2.0, Type 2.1, and Type 4 nodes. See SNA.

SNA The Systems Network Architecture was developed and promoted by IBM. SNA consists of seven layers; from the lowest level they are the Physical, Data-Link Control, Path Control, Transmission Control, Data Flow Control, Presentation, and Transaction layers. These layers are analogous to, but different from, the seven layers in the OSI reference model. SNA is used by IBM's System 390 mainframe and AS/400 and System 3x midrange computers.

SNMP The Simple Network Management Protocol is the most common method by which network management applications can query a management agent using a supported MIB. The first SNMP standard (sometimes referred to as SNMP-1) is spelled out in IETF RFC 1157 and operates at the OSI Application layer. The IP-based SNMP is the basis of most network management software, to the extent that today the phrase "managed device" implies SNMP compliance. RMON and RMON-2 use SNMP as their method of accessing device MIB information. See CMIP, MIB, RMON, RMON-2, SNMP-2.

SNMP-2 A major revision of the original SNMP, SNMP-2 is currently a proposed standard covered by RFC 1902 through RFC 1908. The SNMP-2 MIB—a superset of MIB-2—addresses many performance, security, and manager-to-manager communication concerns about SNMP. For example, SNMP-2 supports encryption of management passwords. See MIB, MIB-2, SNMP.

source routing. Source routing is normally used with Token Ring LANs. In source routing, the sending and receiving devices help determine the route the packet should traverse through the internetwork. The route is discovered via broadcast packets sent between these two points.

source-explicit forwarding. Source-explicit forwarding is a feature of MAC-layer bridges that enables them to forward packets from only those source addresses specified by the administrator.

source-routing transparent (SRT). Source-routing transparent addresses the coexistence of Ethernet, Token Ring, and FDDI. An SRT bridge passes both source-routing and transparently bridged data. The bridge uses source-routing to pass packets with the appropriate embedded routing information, and transparently bridge those packets that lack this information.

spanning-tree algorithm. The spanning-tree algorithm is an IEEE 802.1D technique for configuring parallel MAC-layer Ethernet bridges to provide redundancy. The spanning-tree algorithm manages illegal loops created by multiple parallel bridges.

standby power supply (SPS). A standby power supply is a backup power device that is designed to provide battery power to a computer during a power failure. An SPS experiences small interrupts during switch-over to battery operation.

star topology. In a star topology network, the nodes are connected in a hub and spoke configuration to a central device or location. The "hub" is a central point of failure.

Station Management (SMT). SMT is part of the FDDI specification, and it defines how to manage nodes on FDDI networks.

StreetTalk. StreetTalk is Banyan's distributed global naming and directory service for its network operating system, VINES.

Structured Query Language (SQL). SQL is an ANSI standard query language for extracting information from relational databases. It was originally developed by IBM.

structured wiring. Structured wiring is a planned cabling system which systematically lays out the wiring necessary for enterprise communications, including voice and data. IBM's Cabling System and AT&T Premises Distribution System are two such structured wiring designs. A structured wiring system is made up of horizontal, vertical, and campus subsystems. A horizontal subsystem is the system between the wiring closets and the users' systems. A vertical subsystem or backbone includes the wiring and equipment from the wiring closets to the central equipment room. The campus subsystem interconnects the buildings to a central distribution facility, local exchange carrier, or other point of demarcation.

superserver. A superserver is a computer that is designed specifically to serve as a network server. It typically has multiple CPUs, error-correcting memory, large amounts of cache, large amounts of redundant disk storage, and redundant power supplies. It is designed to provide high speed, high capacities, and fault tolerance. A marketing term.

surge. A surge is a short term (up to 30 seconds) rise in power-line voltage level.

Switched 56. A Switched 56 service is a dial-up connection that uses throughput in 56Kbit/sec increments.

Switched Multi-Megabit Data Service (SMDS). SMDS is a high-speed metropolitan area network service for use over T1 and T3 lines. SMDS' deployment is being stalled by the enthusiasm for Asynchronous Transfer Mode, although SMDS can run in conjunction with ATM.

Synchronous Data Link Control (SDLC). SDLC is IBM's bit-synchronous link-layer protocol. It is similar to HDLC.

Synchronous Optical Network (SONET). SONET will establish a digital hierarchical network throughout the world that will enable you to send data anywhere and be guaranteed that the message will be carried over a consistent transport scheme. The existing telephone infrastructure is digital but is designed for copper lines; SONET is digital and has been designed to take advantage of fiber. SONET offers speeds of 2.5Gbits/sec and higher.

synchronous transmission. A transmission where events occur based on precise clocking, rather than on delimiters whose timing may vary.

Systems Application Architecture (SAA). SAA is IBM's set of rules for computer communications and application development. SAA was designed to help create programs that will run on a wide variety of IBM computing equipment, but it is no longer held out as a panacea for developers who desire universal interoperability.

systems integrator. A systems integrator is a company that is paid to combine disparate pieces of technology into a unified, working system for an end-user company.

Systems Network Architecture (SNA). IBM's protocols for governing communications between terminals, intermediate devices, and mainframes. It was IBM's architecture prior to SAA.

T.120 This ITU-T standard covers the data-sharing component of real-time multimedia over the network and is the basis for such applications as whiteboard sharing and file collaboration during a point-to-point or multipoint session. T.120 is included in video conferencing standards as H.320, H.323, and H.324. See H.320, H.323, H.324.

T1. The ITU-T specifies a four-level, time-division multiplexing hierarchy for the telephone system in North America. T1 provides 24 channels of 64Kbit/sec bandwidth, for a total bandwidth of 1.544Mbits/sec. A T1 circuit can transport voice, video, data, and fax. T1 service sold in 64Kbit/sec increments is called fractional T1.

T2. T2 is the equivalent of four T1s, and it offers 6.3Mbits/sec of bandwidth. Each T2 link can carry at least 96 64Kbit/sec circuits. T2 is not a commercially available service, but it is used within the telephone company's hierarchy.

T3. A T3 circuit carries in one multiplexed signal stream the equivalent of 28 T1 circuits. It provides 44.736Mbits/sec of bandwidth. T3 is not widely used for LANs.

Technical Office Protocol (TOP). TOP is the OSI protocol stack for office automation; it is not widely implemented.

Telnet. Telnet is the TCP/IP protocol for terminal emulation.

terminal emulation. A terminal emulator converts a perfectly capable computer into an enslaved screen and keyboard combination, capable only of raw input and output.

throughput. Throughput is a measure, in bits per second or bytes per second, of the traffic carrying capacity of a channel. LAN and telephone throughput is generally expressed in bits/sec, while computer-based throughput, such as bus capacity and drive I/O rates, are generally expressed in bytes/sec. One byte/sec is usually equal to 8 bits/sec.Given a particular signaling scheme, a channel's throughput is proportional to its bandwidth (expressed in Hz). Despite common usage, maintaining the distinction between bandwidth and throughput is essential to a full understanding of how networks function. (See bandwidth.)

time domain reflectometer (TDR). A TDR is a troubleshooting device that is capable of sending radar-like signals through a cable to check continuity, length, and other attributes.

token passing. Token passing is a network access method that requires nodes to possess an electronic token before transmitting frames onto the shared network medium. Token Ring, Token Bus, and FDDI use token-passing schemes.

Token Ring. Token Ring is the IEEE 802.5 specification for a 4Mbit/sec or 16Mbit/sec network that uses a logical ring topology, a physical star topology, and a token-passing access method. It works with UTP, STP, and fiber-optic cable. Each ring can have up to 256 stations.

token. A token is a pattern of bytes that mediates access on a Token Ring or Token Bus network.

transceiver. A transceiver is a device for transmitting and receiving packets between the computer and the wire. The transceiver is usually integrated directly onto the network adapter card.

Transmission Control Protocol/Internet Protocol (TCP/IP). TCP/IP is the protocol suite developed by the Advanced Research Projects Agency (ARPA), and is almost exclusively used on the Internet. It is also widely used in corporate internetworks because of its superior design for WANs. TCP governs how packets are sequenced for transmission on the network. IP provides a connectionless datagram service. The term "TCP/IP" is often used to generically refer to the entire suite of related protocols.

transparent bridging. Transparent bridging connects similar LANs and is usually used with Ethernet. In transparent bridging, when the station transmits a frame, that frame does not know what path it will take. Instead, the bridges determine the best path at the time the frame is sent. In contrast, in source routing, the path is determined at the start of the transmission, rather than frame by frame.

Transport layer. The Transport layer is the fourth layer of the OSI model, and it provides reliable end-to-end data transport, including error detection between two end user devices. Examples of transport protocols are the Transmission Control Protocol (TCP), Sequenced Packet Exchange (SPX), and Transport Protocol Class 0 (TP0).

Transport Protocol Class 0, Class 4 (TP0, TP4). These protocols are OSI transport protocols. Transport Protocol Class 0 is a connectionless transport protocol for use over reliable networks. Transport Protocol Class 4 is a connection-based transport.

Trivial File Transfer Protocol (TFTP). TFTP is a simplified version of FTP, the TCP/IP file transfer protocol.

tunneling. The process of encasing one protocol in another's format is called tunneling. For example, AppleTalk packets are often enveloped in TCP/IP packet formats for transmission on an enterprise network. Tunneling is also called encapsulation.

twisted pair. Twisted pair is a type of copper wiring in which two wires are twisted around one another to reduce noise absorbtion and signal loss. The Electronics Industry Association/Telecommunications Industry Association (EIA/TIA) specifies a five-level standard for commercial building telecommunications wiring. Category 1 wiring is old-style unshielded twisted-pair telephone cable and is not suitable for data transmission. Category 2 is for use up to 4Mbits/sec; it resembles IBM Cabling System Type 3 cable. Category 3 UTP is specified for speeds up to 10Mbits/sec, and it is the minimum cable required for 10BaseT Ethernet. Category 4 is the lowest grade UTP acceptable for 16Mbit/sec Token Ring. Category 5 is certified for speeds up to 100Mbits/sec, but it can handle speeds of up to 155Mbits/sec. Category 5 cable is suitable for FDDI and other high-speed networks.

two-phase commit. In a distributed database, a two-phase commit ensures data integrity by confirming the successful completion of every step in a transaction before committing any of the steps.

Type 1. The IBM Cabling System specifies different types of wire. Type 1 is a dual-pair, 22 American Wire Gauge (AWG) cable with solid conductors and a braided shield. It is a type of shielded twisted pair.

Type 2. Type 2 is the IBM Cabling System's specification for a six-pair, shielded, 22 AWG wire used for voice transmission. It is the same wire as Type 1, but has an additional four-pair wire.

Type 3. Type 3 is the IBM Cabling System's specification for a single-pair, 22 or 24 AWG, unshielded twisted-pair wire. It is common telephone wire.

Type 5. Type 5 is 100/140 micron fiber; IBM now recommends 125 micron fiber.

Type 6. Type 6 wire is two-pair, stranded 26 AWG wire used for patch cables.

Type 8. Type 8 wire is a two-pair, 26 AWG, shielded cable without any twists; it is commonly used under carpet.

undervoltage. In an undervoltage condition, a lower-than-usual power-line voltage lasts from several seconds to several hours.

uninterruptible power supply (UPS). A UPS is a power conditioning and supply system that affords protection against short-term power outages. A UPS rectifies the incoming AC line voltage to DC, which is then applied to batteries. An inverter, driven by DC power, supplies AC voltage for equipment that requires conditioned power. During outages, the converter is driven by battery power.

Unix. Unix is a 32-bit multitasking, multiuser operating system. Versions of Unix are available for nearly every type of computer platform. Unix was initially popular in universities and research labs, but it is now the basis of many corporate applications.

unshielded twisted pair (UTP). UTP is a pair of insulated copper wires, twisted around each other. UTP is classified into several levels of wire quality suitable for different transmission speeds (see "Category").

user agent (UA). In X.400 mail systems, the user agent is the client component that provides the X.400 envelope, headers, and addressing. The user agent sends the messages to the X.400 mail server, or Message Transfer Agent, which then routes the messages to their destinations.

User Datagram Protocol (UDP). UDP is the connectionless transport protocol within the TCP/IP suite. Because it does not add overhead, as the connection-oriented TCP does, UDP is typically used with network-management applications and SNMP.

V.21. V.21 is the modem standard for the trunk interface between a network access device and a packet network. It defines signalling data rates greater than 19.2Kbits/sec.

V.22, V.22 bis. V.22 is a 1,200-bps duplex modem for use in the public-switched telephone network and on leased circuits. V.22bis is a 2,400-bit modem that uses frequency division multiplexing for use on the public telephone network and on point-to-point leased lines. (The CCITT uses "bis" to denote the second in a series of related standards and "ter" to denote the third in a family.)

V.32, V.32 bis. V.32 are two-wire duplex modems operating at rates up to 9,600bps (with fallback to 4,800bps) for use in the public telephone network and on leased lines. V.32 bis offers speeds in increments of 4800bps, 7200bps, 9,600bps, 12,000bps, and 14,400bps.

V.34. Modems that comply with the V.34 standard can operate at rates as high as 28,800bps.

V.35. Prior to 1988, V.35 was a modem specification that provided data transmission speeds up to 48Kbits/sec. V.35 was then deleted from the V-Series Recommendations.

V.42 error correction, V.42 bis data compression. The V.42 error-correction standard for modems specifies the use of both MNP4 and LAP-M protocols. V.22, V.22 bis, V.26 ter, and V.32 bis may be used with V.42. With V.42 bis compression, data is compressed at a ratio of about 3.5 to 1, which can yield file-transfer speeds of up to 9,600bps on a 2,400-bps modem. Manufacturers can provide an option that will allow a V.42 bis modem to monitor its compression performance and adjust the ratio accordingly.

value-added reseller (VAR). Also called an integrator, a VAR is a company that resells manufacturers' products and adds value by installing or customizing the system.

VAX. Digital Equipment's brand name for its line of minicomputer and workstation hardware is VAX.

vertical wiring subsystem. The vertical wiring subsystem is the part of the structured wiring system that connects the campus wiring system to the departmental wiring system. It runs in a building's risers.

VINES. Banyan's NOS based on a Unix core and TCP/IP protocols. VINES supports DOS, Windows, Mac, and OS/2 clients and is especially popular in large enterprise networks. Its crowning feature is StreetTalk, its distributed directory service.

virtual circuit. A virtual circuit is a shared communications link that appears to the customer as a dedicated circuit. A virtual circuit passes packets sequentially between devices.

Virtual Terminal (VT). VT is the OSI terminal-emulation protocol.

virus. A virus has the ability to reproduce by modifying other programs to include a copy of itself. Several types of viruses exist. Bacteria or rabbits do not explicitly damage files but do reproduce and eat up disk space or RAM. A logic bomb lies dormant in a piece of code or program until a predefined condition is met, at which time some undesirable effect occurs. A password catcher mimics the actions of a normal log-on but catches user IDs and passwords for later use. A Trojan horse is a program that appears to function but also includes an unadvertised and malicious feature. A worm scans a system for available disk space in which to run, thereby tying up all available space.

VMS. VMS is Digital Equipment's proprietary operating system for the VAX.

VTAM The Virtual Telecommunications Access Method is the IBM software running on the host system. It controls an SNA or APPN network and allows communication between the host operating system and user applications. VTAM maintains the directory of network devices and services, and selects the routes between various Network Control Program devices.

vulnerability analysis. A vulnerability analysis is a type of risk analysis in which you calculate the effects of a project's success or failure on your overall business.

wide area network (WAN). A WAN consists of multiple LANs that are tied together via telephone services and/or fiber-optic cabling. WANs may span a city, state, a country, or even the world.

Windows NT. Microsoft's "New Technology" is the company's 32-bit, pre-emptive multitasking operating system that includes support for peer-to-peer file sharing. Windows NT Server provides high-end networking services. Windows 2000 is the name given to NT 5.0.

Windows. Version 3.1 and prior versions of Microsoft's popular 16-bit GUI explicitly ran on top of DOS. Windows 95, previously known as Windows 4.0 and code-named Chicago, is a (primarily) 32-bit OS that integrates DOS and Windows. Windows for Workgroups was Microsoft's peer-to-peer network that used a Windows interface and NetBIOS communications.

wireless LANs. A wireless LAN does not use cable to transmit signals, but rather uses radio or infrared to transmit packets. Radio frequency (RF) and infrared are the most commonly used types of wireless transmission. Spread spectrum is used in the Industrial, Scientific, and Medical (ISM) bands. Most wireless LANs use spread spectrum transmission. It offers limited bandwidth, and users share the bandwidth with other devices in the spectrum; however, users can operate a spread spectrum device without licensing from the Federal Communications Commission (FCC). Some high-frequency RF systems offer greater throughput, but they are used less often because they require an FCC license for the right to transmit.

wiring closet. A wiring closet is a room or closet that is centrally located and contains operating data-communications and voice equipment, such as network hubs, routers, cross connects, and PBXs.

workflow software. Workflow software is a class of applications that helps information workers manage and route their work. It is a special class of groupware or workgroup software.

X Window System (X). X Window System, developed by MIT, is a graphical user system most often implemented on Unix systems.

X.25. X.25 is the CCITT and OSI standard for packet-switching networks that provide channels up to 64Kbits/sec. Public and private X.25 networks can be built. In the United States, common X.25 networks are British Telecom, AT&T, CompuServe, and Infonet.

X.400. X.400 is the OSI and CCITT standard for store-and-forward electronic messaging. It is used for large enterprise networks or for interconnecting heterogeneous e-mail systems. X.400 divides an electronic mail system into a client, called a User Agent, and a server, called a Message Transfer Agent. Message Stores provide a place to store messages, submit them, and retrieve them. Access Units provide communication with other device types, such as telex and fax. Distribution Lists are routing lists.

X.500. X.500 is the OSI and CCITT specification for directory services. For computer users, a directory service provides a function similar to the function the telephone company's white pages provides telephone users. Using a directory service, computer users can easily look up the location of resources and other users.

Xerox Network System (XNS). XNS is Xerox's data-communication protocol; it is the basis for the IPX/SPX network protocols used in NetWare.

Index

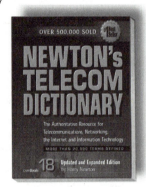